EMERGENCY RESPONDER TRAINING MANUAL FOR THE HAZARDOUS MATERIALS TECHNICIAN

EMERGENCY RESPONDER TRAINING MANUAL FOR THE HAZARDOUS MATERIALS TECHNICIAN

Second Edition

Edited by
KENNETH W. OLDFIELD, M.S.P.H., C.I.H.

Contributing Authors

D. ALAN VEASEY, M.A.ED., M.P.H.
KENNETH W. OLDFIELD, M.S.P.H., C.I.H.
LISA CRAFT McCORMICK, M.P.H.
THEODORE H. KRAYER, M.P.H., C.H.M.M.
BROOKE N. MARTIN, M.P.H.
BATTALION CHIEF SAM HANSEN, E.F.O., E.M.T.-P.
ERVIN ROY STOVER, B.S.
BARBARA M. HILYER, M.S., M.S.P.H. (deceased)

WILEY-INTERSCIENCE

A JOHN WILEY & SONS, INC., PUBLICATION

All drawings by Robin Miller and the authors. All photographs by the authors with assistance from Arizona Winborn, Jr. unless otherwise indicated in figure captions.

Published by John Wiley & Sons, Inc., Hoboken, New Jersey., Hoboken, New Jersey.
Published simultaneously in Canada.

For general information on our other products and services please contact our Customer Care Department within the U.S. at 877-762-2974, outside the U.S. at 317-572-3993 or fax 317-572-4002.

Wiley also publishes its books in a variety of electronic formats. Some content that appears in print, however, may not be available in electronic format.

Library of Congress Cataloging-in-Publication Data

Emergency responder training manual for the hazardous materials technician / edited by kenneth W. Oldfield ; contributing authors, D. Alan Veasey ... [et al.].—2nd ed.
    p. cm.
  Includes index.
  ISBN 0-471-21387-X (cloth)
  1. Hazardous substances–Accidents–Handbooks, manuals, etc. I. Oldfield, Kenneth W. II. Veasey, D. Alan (Dwight Alan)
   T55.3.H3E446 2004
   628.9′2—dc22

                   2004009420

Printed in the United States of America.

10 9 8 7 6 5 4 3 2 1

To the memory of Barbara M. Hilyer who faithfully directed the Workplace Safety Training Program from 1993 until 2002.

CONTENTS

PREFACE

Welcome to the second edition of *Emergency Responder Training Manual for the Hazardous Materials Technician*. This book is designed to address the training needs of personnel who respond to emergencies involving hazardous materials. All the authors are current or former members of the instructor staff in the Workplace Safety Training (WST) program operated by the Center for Labor Education and Research (CLEAR) at the University of Alabama at Birmingham (UAB). The WST program has conducted hazardous materials training since 1988 with grant funding from the National Institute of Environmental Health Sciences (NIEHS). We have used the NIEHS-funded training as a proving ground for developing and constantly improving teaching materials and techniques for hazardous materials training. This textbook was written to share with others what we have learned in order to promote safe and effective response to hazardous materials emergencies.

In addition to 16 years of training over the life of the WST program, this text represents the academic and practical experience of the authors. Four of the eight authors are current or previous emergency responders, with experience in both public sector and private sector response to hazardous materials incidents. Six hold advanced degrees in occupational safety and health-related topics. We also hold numerous relevant professional certifications; such as Certified Industrial Hygienist, Certified Hazardous Materials Manager, and Certified Environmental Trainer certifications, plus numerous training certifications. Most importantly, we bring over 80 years of combined training experience to this book. We have drawn on our varied experience and education to produce a book that covers both the technical and practical ends of the hazardous materials response spectrum.

Emergency Responder Training Manual for the Hazardous Materials Technician is intended to address the training needs of hazardous materials emergency responders in both the public and private sector setting. It is applicable for training personnel through the Hazardous Materials Technician response role, and all prerequisite response roles, as defined in the Occupational Safety and Health Administration's *Hazardous Waste Operations and Emergency Response* standard (29 CFR 1910.120) and the National Fire Protection Association's *Professional Competence of Responders to Hazardous Materials Incidents* standard (NFPA 472). The applicability and use of the textbook are further discussed in Chapter 1 after the various response roles are explained.

With this edition of the book, we are able to bring significant improvements to the information we provided in the first edition. This is due in part to the 12 years of additional research, training, and professional experience we have accrued since the first edition was published. We are also able to provide information related to new technologies and topics that have emerged since the first edition was published. For example, new technologies in areas such as personal protective equipment and hazardous materials control are addressed. One of our most significant new concerns is for emergency response to terrorism incidents, particularly weapon of mass destruction (WMD) events. We addressed this concern by adding a chapter to provide an introduction to terrorism and WMDs, and by adding sections specific to WMD response to chapters devoted to topics such as site control, personal protective equipment, and decontamination.

The authors acknowledge the inherently dangerous nature of hazardous materials emergency response. Because of this, we maintain a primary focus on safety throughout this book, with the greatest emphasis placed on the safety of emergency responders. Only through safe response operations can we accomplish our mission of protecting people, property, and the environment.

D. ALAN VEASEY, M.A. Ed., M.P.H.

Director, Workplace Safety Training
Center for Labor Education and Research
University of Alabama at Birmingham

ACKNOWLEDGMENTS

The authors received valuable assistance in completing this textbook from a number of people, including Judith L. King, Arizona Winborn, Jr., William P. "Bill" Gresham, Jim St. John, Donald Koss, Cedric Harville, Judy McBride, and Robert Heath. This project was partially supported by cooperative agreement #U45ES06155 from the National Institute of Environmental Health Sciences (NIEHS); however, its contents are solely the responsibility of the authors and do not necessarily represent the official views of NIEHS.

1

INTRODUCTION TO HAZARDOUS MATERIALS

Since the dawn of civilization, people have used chemical substances to improve the human condition. With the Industrial Revolution, chemical use accelerated as industry demanded new chemicals for use in the ever-expanding manufacturing processes—a demand that continues to this day, although the industries and processes have changed. The rapid economic expansion following World War II was also accompanied by the rise of the modern petrochemical industry. Today, hazardous materials are commonplace; large quantities are manufactured, shipped, stored, and used throughout the United States and the rest of the world.

Early on, it became quite obvious that these substances, however useful, could also be quite dangerous. Industrialization and the proliferation of chemicals brought with them industrial and chemical accidents, with significant harm to people, property, and the environment. The governments of the United States and other nations responded with laws to protect workers, citizens, and the environment, mandating that employers, manufacturers, and communities take steps to prevent and respond to emergencies involving hazardous materials. This text is intended for those personnel who may be involved in these hazardous materials emergencies.

This chapter introduces the reader to hazardous materials by answering these questions:

- What are hazardous materials?
- Why are we concerned about them?
- How are they harmful?

Emergency Responder Training Manual for the Hazardous Materials Technician, Second Edition, edited by Kenneth W. Oldfield
ISBN 0-471-21387-X Copyright © 2005 John Wiley & Sons, Inc.

The chapter also provides an overview of hazardous materials response and response roles and explains how this book can be used in training responders to fill those roles.

WHAT ARE HAZARDOUS MATERIALS?

Different people and organizations use different definitions to answer this question. In the United States, various governmental agencies have developed different specific terms for hazardous commodities. Some of these definitions are discussed in the following sections along with other terms of interest that may be applied to hazardous materials. The regulatory agencies, terms, and issues mentioned here are discussed further in Chapter 2.

DOT's Terminology: Hazardous Materials

The Department of Transportation (DOT) defines hazardous materials as those substances or materials capable of posing an unreasonable risk to safety, health, or property when transported in commerce. DOT's definition also incorporates materials that other agencies define as hazardous, such as the Environmental Protection Agency's (EPA's) hazardous substances and hazardous wastes. Additional information on DOT terminology is available in Chapter 5 of this text and in the DOT Hazardous Materials Regulations found in Title 49 of the Code of Federal Regulations, parts 171 through 180 (49 CFR 171–180).

EPA's Terminology: Hazardous Substances, Extremely Hazardous Substances, and Hazardous Waste

Hazardous substance is a term applied by the EPA to substances that are designated under the Clean Water Act or the Comprehensive Environmental Response, Compensation, and Liability Act (CERCLA) as posing a threat to waterways or the environment if released. If a hazardous substance is spilled or otherwise released into air, ground, or water in excess of EPA's listed Reportable Quantity (RQ) for that substance, the release must be reported to EPA.

Hazardous waste can be broadly defined as hazardous substances that have no commercial value. Hazardous wastes are regulated by the EPA under the Resource Conservation and Recovery Act of 1976 (RCRA). RCRA authorized the EPA to institute a program of hazardous waste management initiating "cradle-to-grave" tracking of hazardous wastes. Categories of hazardous waste are described in Chapter 5.

Extremely hazardous substances (EHSs) are substances listed in Title III of the Superfund Amendment and Reauthorization Act (SARA) that EPA has determined represent an extreme hazard to communities in which they are released because of their toxic, chemical, or physical properties. Facilities that use, store, or transport EHSs above threshold quantities are required to meet special planning and preparedness requirements in the event of a release. These requirements are discussed further in Chapter 3.

OSHA's Terminology: Hazardous Chemicals

Hazardous chemicals, as defined by the Occupational Safety and Health Administration (OSHA) in its Hazard Communication standard (29 CFR 1910.1200), are chemicals that are hazardous to people in the workplace or to the community if released. A Material Safety Data Sheet (MSDS) for each hazardous chemical used in a workplace must be prepared by the employer or requested from the manufacturer and kept on file in a location accessible to workers and members of the community upon their request. The employer has the responsibility to instruct workers about the chemical hazards with which they work.

Benner's Definition

Ludwig Benner, Jr., a former member of the U.S. National Transportation Safety Board, defined a hazardous material as a substance that "jumps out of its container at you when something goes wrong, and hurts or harms the things it touches." Benner's definition is not recognized by any governmental agencies; however, many emergency responders feel that it is the most functional definition of hazardous materials (Fig. 1.1).

Hazardous Materials Terminology in Common Usage: Hazmats

In everyday language, the term "hazardous materials" is often used in a general sense to describe any substance that may injure responders or other people,

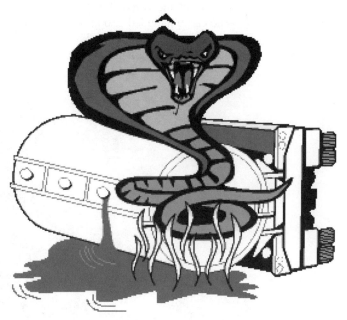

Figure 1.1 Ludwig Benner, Jr. defined a hazardous material as a substance that "jumps out of its container at you when something goes wrong, and hurts or harms the things it touches."

damage property, or harm the environment. This text will follow that convention. Unless stated otherwise, the term "hazardous material" will be used in this text to refer to hazardous chemicals, hazardous substances, hazardous wastes, and any other materials that may cause harm. In some cases, we may shorten the term to "hazmats" for convenience.

WHY ARE WE CONCERNED ABOUT HAZARDOUS MATERIALS?

History has made it clear that hazardous materials incidents are different from other types of emergencies with which responders must deal, such as structure fires, emergency medical calls, and motor vehicle accidents. We all realize that mishandling any emergency service operation can result in needless injury or death to both victims and responders. Hazardous materials response operations are different because responders' mistakes, or failure to take appropriate actions, can result in very large numbers of civilian casualties, massive property damage, and/or environmental damage that can haunt future generations (Fig. 1.2).

To emphasize this point, let's look at a few well-documented hazardous materials incidents. These are only a small sample from a long list of hazardous materials incidents that have resulted in disastrous outcomes.

Texas City, Texas

In 1947, a fire occurred in the hold of the cargo ship *Grandcamp* while it was docked at Texas City, Texas. The cargo included 2300 tons of ammonium nitrate fertilizer. The Texas City Fire Department responded and began trying to suppress the fire. No site control was established, and a large number of civilian curiosity

Figure 1.2 Hazardous materials incidents have the capability to cause tremendous harm if responders make mistakes or fail to take appropriate actions.

seekers gathered at the scene. As the response continued, an attempt was made to smother the fire by battening down the hatches of the ship. This action only increased the heat and pressure within the hold. As a result, an explosion occurred that killed almost 600 people (including the entire Texas City Fire Department), injured approximately 5000, and caused widespread property damage and numerous secondary storage tank and shipboard fires. The mishandling of the response to the *Grandcamp* incident resulted in casualties and property damage at Texas City equivalent to that caused by wartime bombing of a major city.

Waverly, Tennessee

In 1978, a train derailment occurred at the small town of Waverly, Tennessee. A rail tank car containing 27,871 gallons of liquefied petroleum gas (LPG) was critically damaged in the accident. At the time of the derailment, the weather was extremely cold, reducing the pressure of the LPG and, in turn, preventing failure of the tank car. Forty hours after the derailment, responders and rail salvage personnel were attempting to unload the tank car. No precautions, such as evacuating nonessential personnel or employing specialized techniques for unloading the tank car, were utilized. By that time, the weather had warmed considerably, which produced an increase in the pressure within the tank car. As a result, the tank car failed catastrophically during the unloading attempt, releasing the entire cargo of LPG, which immediately ignited. The resulting explosion killed 16 people, severely burned scores of others, and did significant property damage to the town of Waverly.

Bhopal, India and Institute, West Virginia

In December of 1984, 45 tons of methyl isocyanate gas was accidentally released from a chemical plant operated by Union Carbide in Bhopal, India. The released gas formed a deadly cloud that covered a large residential area. As a result, approximately 3500 people lost their lives and 300,000 were injured, with many of the injuries resulting in permanent disabilities. Nine months after the Bhopal disaster, a "near miss" of a similar incident occurred at a Union Carbide facility in the United States. As a result of this "near miss," 135 residents of Institute, West Virginia suffered effects of chemical exposure. The incidents at Bhopal and Institute are usually credited with galvanizing the U.S. government to action to prevent such chemical disasters, resulting in the passage of the Superfund Amendment and Reauthorization Act (SARA) of 1986, as discussed in Chapters 2 and 3.

Kansas City, Missouri

In November of 1988, two fire service engine companies responded to a reported fire in a trailer on a construction site in Kansas City, Missouri. They had no way of knowing that the trailer contained explosives. Shortly after their arrival on scene, the explosives detonated, causing a massive explosion felt for miles. All six firefighters on scene were killed instantly. Two other firefighters approaching the scene were injured when the windshield was blown out of their vehicle at a distance of a quarter of a mile from the scene. Explosives placards were reportedly removed from the trailer when it was brought on site because of concerns about theft.

Birmingham, Alabama

In October of 1997, a fire occurred in a large warehouse in Birmingham, Alabama. Information available at the time of the fire did not indicate the presence of chemicals, and the response was handled as a typical structural fire fighting operation. The warehouse actually contained thousands of gallons of Dursban, a potent pesticide that is especially toxic to aquatic life. A number of firefighters and local residents were exposed to toxic combustion by-products in the smoke. Contaminated runoff from the millions of gallons of water used in the fire suppression operation flowed directly into Village Creek, a major tributary of the Locust Fork of the Black Warrior River. As a result, a major fish kill occurred along a thirty-mile stretch of the waterway.

Your Town, USA

One fact worth noting is that there is always additional room at the end of the list for the names of other cities or locations that may be the scene of future hazmat disasters. One noble ambition is to keep the name of your community off the list.

Another noble ambition is to keep your hazmat response from becoming famous. Hazmat response operations that are safely and efficiently completed typically receive very little media attention. At the same time, response operations that end in disaster are quite likely to be featured on national network news. Wiping out ourselves and our communities is not the way we want to receive national attention!

IN WHAT WAYS ARE HAZARDOUS MATERIALS HARMFUL?

As an emergency responder you are concerned about, and would consider hazardous, any substance that might injure or sicken you and other people; burn or explode; damage property where it has been released; or cause harm to living things in air, water, or soil.

If we subscribe to Benner's definition of a hazardous material as a substance that "jumps out of its container at you when something goes wrong, and hurts or harms the things it touches," then it is important that we understand the types of harm that can result from hazmat emergencies. This will be important for hazard and risk assessment and for formulating strategies for responding to hazmat incidents. We will provide an overview of this topic here and elaborate on it in several subsequent chapters.

Hazardous Materials and Human Health

Hazardous materials can do damage to the bodies of people, including responders, in a variety of ways. This can occur after exposure to hazardous materials in various forms through:

- Inhalation of hazmats into the respiratory system
- Ingestion of hazmats into the digestive system
- Direct contact of hazmats with the skin or eyes
- Absorption into the body through the skin following contact with hazmats

After exposure, hazardous materials may harm us through means such as cell poisoning, irritation and destruction of tissues, reproductive or fetal damage, or cancer. We cover the health hazards of hazmats in detail in Chapter 7.

Chemical Reaction, Fire, and Explosion

As the case histories above illustrate, many of the most infamous hazmat disasters involved explosions and fires. Many hazardous materials will burn or explode readily with a tremendous release of energy in an accident. Some hazmats can undergo uncontrolled chemical reactions to release energy and/or toxic by-products. People exposed to exploding, burning, or reacting hazardous materials may be injured or killed, and property may be severely damaged. We cover the chemical hazards in detail in Chapter 6.

Chemicals in the Environment

One of the major consequences of hazardous materials accidents is contamination of soil, air, waterways, and groundwater. This can result in harm to fish, birds, wildlife, plants, and people, including those who live far downstream or downgrade from the initial location of contamination. The harm resulting from environmental contamination can last for years.

Every attempt should be made to keep hazardous materials out of air, water, and soil. Once chemicals have contaminated the environment, it is often very expensive and difficult—if not impossible—to recover them. In several chapters of this book, we cover procedures that can minimize or prevent environmental contamination.

New Threats: Terrorism Involving Weapons of Mass Destruction

Hazardous materials response has traditionally been geared for dealing with accidental releases of hazardous materials. In recent years, emergency responders have been forced to prepare for incidents in which hazardous materials are used deliberately by terrorists to injure or kill civilians and responders. Examples of such events include:

- The bombing of the Alfred P. Murrah Federal Building in Oklahoma City in 1995
- The Tokyo Subway sarin attack in 1995
- The September 11th, 2001 attack on the World Trade Towers and the Pentagon

Terrorists have used hazardous materials as weapons of mass destruction for generations, although these recent events have focused our attention on this threat as never before. We can think of these events as hazardous materials incidents that are intentionally perpetrated to create a mass casualty incident in order to achieve political or social objectives. Although traditional agents used in chemical, biological, or nuclear warfare may be used in these attacks, industrial chemicals that are routinely stored, used, and transported within our communities may be intentionally released in future attacks. To address these concerns, Chapter 10 is devoted to terrorism and weapons of mass destruction.

HAZMAT RESPONSE 101

In responding to incidents involving hazardous materials, it is important to act in an organized and methodical fashion. If responders rush in and take inappropriate actions, minor incidents may become major disasters and responders themselves may be injured or killed. This is the reason that veteran hazmat responders have long said, "Be part of the solution. Don't become part of the problem."

Responders who act rashly may be exposed to chemicals or physically injured. Other responders must then focus their efforts on dealing with the injured responders rather than responding to the incident. This can have the effect of undermining the response to the hazmat incident and allowing it to get out of control. Because of the disaster potential of hazmat incidents and the high level of threat they represent to responders, special precautions are required when responding to them.

Response Guidelines and Models

The use of response guidelines and models is strongly recommended for hazmat response. These promote methodical decision-making and help to ensure that significant information or actions are not lost or overlooked. They help us to avoid succumbing to a "tunnel vision" approach in which only the most obvious part of a situation is recognized and evaluated. Several established guidelines and models useful in hazmat response are described below.

Benner's DECIDE Process
The DECIDE process was developed by Ludwig Benner, Jr. It consists of six simple steps, and the first letters of the first word of each step spell "DECIDE." The process is useful in several ways in training for hazmat response and is used in several chapters of this text. The steps in the DECIDE process are:

- Determine the presence of hazardous materials
- Estimate likely harm without intervention
- Choose response objectives
- Identify options for achieving objectives
- Do the best option
- Evaluate progress

The Eight-Step Process

One commonly-used model for hazmat response is the Eight-Step Incident Management Process, which was developed by Gregory R. Noll, Michael S. Hildebrand, and James G. Yvorra (Noll, Hildebrand, and Yvorra, 1995). This model for hazmat incident management is based on eight steps:

(1) Site management and control
(2) Identify the problem
(3) Hazard and risk evaluation
(4) Select personal protective clothing and equipment
(5) Information management and resource coordination
(6) Implement response objectives
(7) Decontamination
(8) Terminate the incident

The "Street Smart" Approach to Hazmat Response

Mike Callan developed a practical approach for hazard classification in which potentially hazardous atmospheres are classified as safe, unsafe, or dangerous (Callan, 2002). Safe atmospheres are those that present no significant hazard to responders or the public. Unsafe atmospheres are those in which a hazard is present but the atmosphere is not immediately dangerous and injury or illness is not likely without prolonged or repeated exposure. Dangerous atmospheres pose an immediate danger to the life and health of responders and the public. The use of the Street Smart Approach is explained in Chapter 6.

Response Roles and Procedures

In promulgating the Hazardous Waste Operations and Emergency Response (HAZWOPER) standard (29CFR1910.120), OSHA devoted paragraph "q" of HAZWOPER to hazardous materials emergency response operations. This paragraph established specific procedures, personnel roles, and training requirements for hazardous materials emergency response operations. The personnel roles established in paragraph "q" of HAZWOPER are mirrored in other emergency response standards such as NFPA 472. These roles are summarized below and discussed in greater detail, along with other requirements of the HAZWOPER standard, in Chapter 2.

First Responder Awareness Level

First Responders at the Awareness Level are personnel who are likely to discover or witness a hazardous substance emergency. The most important duties of these personnel are to avoid the hazardous area, isolate the area, and make proper notification to begin the emergency response sequence. The response role of these personnel should involve no potential for exposure to hazards related to an incident.

First Responder Operations Level

First Responders at the Operations Level are personnel who are involved in an initial response for the purpose of protecting people, property, and the environment from hazardous substances. These personnel are trained to respond defensively, instead of actually trying to stop the release at the source. This type of response involves working at some distance from the actual point of release to control the released material, keep it from spreading, and prevent exposures.

Hazardous Materials Technician

Hazardous Materials Technicians are personnel who respond to an emergency to stop or prevent a release of hazardous substances. These personnel assume an offensive emergency response role, as they approach the point of release and plug, patch, or otherwise stop the release at the source (Fig. 1.3).

Hazardous Materials Specialist

Hazardous Materials Specialists are personnel who respond with, and provide support for, hazardous materials technicians. Their duties parallel those of the hazardous materials technicians but require a more specific knowledge of the hazardous substances involved. The hazardous materials specialist also acts as site liaison with federal, state, and/or local government authorities.

On-Scene Incident Commander

On-Scene Incident Commanders are personnel who will assume control of the incident scene beyond the first responder awareness level. The senior emergency

Figure 1.3 Hazardous materials technicians must be adequately trained to safely approach the point of release and stop the release at the source.

response official responding to an emergency will become the individual in charge of the site-specific Incident Command System (ICS). This individual will control and coordinate emergency response activities and communications. Command will be passed up a preestablished line of authority as personnel or officials having greater emergency response seniority arrive.

Other Response Roles
Other response roles described in the HAZWOPER standard include Designated Safety Official, Skilled Support Personnel, and Specialist Employees. These roles are characterized in Chapter 2.

ORGANIZATION AND APPLICABILITY OF THIS TEXTBOOK

The information in this textbook is applicable for training personnel to function in the First Responder Awareness Level, First Responder Operations Level, and Hazardous Materials Technician roles. The primary goal of the text as a whole is to provide Hazardous Materials Technician training, but much of the information included also applies to the Awareness and Operations roles. The response roles form a hierarchy with cumulative training requirements, so that to achieve competency at a given level requires that the training objectives for all lower-level response roles first be achieved.

The information in Chapters 1 through 8, 10, 11, and 16 is appropriate for training First Responder Awareness Level personnel. The information in all chapters except for Chapter 15 is appropriate for training First Responder Operations Level personnel. All chapters of the textbook are appropriate for training Hazardous Materials Technicians.

This textbook is not intended as a sole reference source for Hazardous Materials Technician training. The technician response role is very demanding and requires detailed knowledge in a number of areas to be performed safely and effectively. Specific training requirements may vary depending on the specific type of response operations to be performed. For example, an industrial hazmat response team member may need different hazardous materials containment skills than a municipal firefighter required to respond to transportation incidents. Specialist-level training may be required in some instances.

This text is intended to provide information on a wide variety of topics, with detailed information provided on some topics. The authors realize that it is beyond the scope of this textbook, or any other reference source, to provide all the information needed by all hazardous materials technicians. We intend for the information provided in this text to be supplemented with additional information provided through training and through other sources of information. We reference other information sources throughout the text.

SUMMARY

In this chapter, we have defined hazardous materials, identified some of the ways in which they can cause harm, looked at some basic considerations for responding to hazmat emergencies, and discussed how this textbook can be used in training for hazmat response. We also described a number of case histories in which hazmat responses ended in disasters. All of those incidents resulted in changes in regulations or practices intended to prevent similar disasters from happening again. Unfortunately, those changes came too late for people who were killed, injured, or had their lives destroyed by those events. Regulatory changes to make things safer tend to be disaster driven. The challenge to us as emergency responders is to take a proactive approach by using planning, preparedness, and safe response practices to avoid bad outcomes from hazmat incidents. We hope the information presented in this text helps emergency responders meet that challenge.

2

RESPONSE LAWS, REGULATIONS, STANDARDS, AND OTHER POLICIES

INTRODUCTION

On the morning of April 16, 1947, in Texas City, Texas, longshoremen began loading the French liberty ship *Grandcamp* with ammonium nitrate fertilizer. Within a few hours, workers were alerted to smoke coming from the hold of the ship containing 2300 tons of the fertilizer. A couple of attempts to extinguish the fire were made with water and fire extinguishers, but the fire was too far down in the hold for these efforts to be effective. The fire grew rapidly, and the Texas City Volunteer Fire Department was notified. When the fire department arrived, the firefighters began standard fire fighting procedures.

The captain of the ship was reluctant to use water from the hose lines to extinguish the fire, because ammonium nitrate will ruin when exposed to water. Knowing this, the captain ordered the hatches to the hold battened down. He also opened up the steam lines in this part of the ship in an attempt to smother the fire. During this time, curious town folk began to gather at the dock and the surrounding area. Neither the firefighters nor the ship's captain and crew recognized the imminent disaster.

As recognized by today's standards and described in *Chemistry of Hazardous Materials* (Meye, 1977), ammonium nitrate is a potentially hazardous chemical classified as an "oxidizer." Oxidizers have the ability to generate their own oxygen under certain conditions such as increasing temperature. At that time, ammonium nitrate pellets were coated in paraffin wax and shipped in asphalt-soaked bags to prevent moisture from ruining the product. With the combination of the

Emergency Responder Training Manual for the Hazardous Materials Technician, Second Edition,
edited by Kenneth W. Oldfield
ISBN 0-471-21387-X   Copyright © 2005 John Wiley & Sons, Inc.

oxygen generated by the chemical, the fuel of the paraffin and asphalt, and the increasing temperatures and pressures, the captain had inadvertently built a bomb.

Not long after the captain ordered the hatches sealed, the ship exploded with massive force, killing all 27 members of the fire department, numerous bystanders, longshoremen, and crew. The explosion was so strong that it propelled the ship's anchor a half-mile away. Bystanders near the shore were killed either by the blast or the ensuing tidal wave of seawater generated by the explosion. Hot metal shards and flaming cargo from the ship destroyed a petroleum plant across the bay and started fires all across the city. Pieces of the ship also penetrated the hull of the nearby cargo ship *Highflier*, igniting her cargo: more ammonium nitrate. After the fire on the *Highflier* was discovered, an attempt to tow her out of the harbor was made, but she exploded as well.

In the aftermath of this disaster, it was determined that close to 600 people were dead or missing with civilian injuries estimated as high as 5000. Approximately one-third of the city's homes were destroyed, along with product and property damage estimates totaling $500 million in 1947 U.S. dollars. The contributing factors of this disaster were the lack of planning, lack of respect for the chemicals, transportation practices, and response management. This incident was a sentinel event introducing the age of "technological" disasters in America.

Up to the time of the incident at Texas City and for sometime thereafter, the United States lacked formalized policies at all levels of government for response to chemical and other "technological" disasters. Policy is commonly defined as a definite course of action adopted for the sake of expediency. Most responders recognize major policy "buzzwords" but may not realize the complexity and fragmentation of the U.S. emergency response policy framework (Fig. 2.1).

The primary mechanisms of hazmat response policy at the federal level are laws, regulations, standards, or a combination of these and other policy mechanisms. The major driving policies to date in federal chemical emergency response are laws that are passed by congressional acts such as the Superfund Amendments and Reauthorization Act. Although the U.S. Congress passes these laws, it is the responsibility of the Executive Branch, federal departments and agencies, to execute and enforce them. The respective department or agency enforces the new or amended law with rules, regulations, or regulatory standards to administer the law, thus creating an additional layer of policy. Rules, regulations, and regulatory standards are not actually laws by definition, but they carry the weight and force of law in the field. Throughout this text, rules, regulations, or regulatory standards are commonly cited and are referenced according to their location within the Code of Federal Regulations (CFR). For example, regulations limiting employee exposure to noise are contained in part 1910.95 of Title 29 of the Federal Code. To further complicate the discussion, nonregulatory standards from government agencies such as the National Institute of Occupational Safety and Health (NIOSH) add another layer of policy. There are also numerous industrial and professional associations, such as the American National Standards Institute, that have developed consensus standards. These standards are generally considered recommended practices that are accepted by the industry or profession as the

Figure 2.1 The "buzz words" or abbreviations of terms used by responders in reference to U.S. policies are the equivalent of alphabet soup.

"norm." In some situations departments and agencies of the federal government will reference proven standards or include entire standards verbatim in the respective departments' or agencies' regulations, thus making them regulatory policy.

Federal congressional laws, rules, supporting regulations, and incorporated standards are not the only means of policy making in the United States. Chemical emergency response policies are also established and influenced by the following:

- Executive Office of the President with presidential directives
- State legislation
- Federal, state, and local courts with decisions and precedents
- State and local agencies with their own regulations
- Business interests
- Labor organizations
- Private environmental advocacy groups
- Common laws, initiatives, and referenda

MAJOR FEDERAL POLICY IMPACTING HAZMAT RESPONSE

As sadly demonstrated in Texas City, uniform national response policies were desperately needed. The explosion of the *Grandcamp* and the *Highflier* is an example of a hazmat incident involving fire that started with a response by a volunteer fire department and quickly escalated to the federal level. The increasing frequency of chemical disasters caused the federal government to respond over time with policies that ultimately filter down and impact the local responder. Federal hazmat response policy targets environmental, worker health and safety, transportation, and now homeland security concerns.

National Oil and Hazardous Substances Pollution Contingency Plan

In 1968 the initial National Oil and Hazardous Substances Pollution Contingency Plan (NCP) was developed by the federal government to protect bodies of U.S. waters from oil and hazardous chemicals. The National Contingency Plan was covered by the Clean Water Act (CWA) and eventually required under the Comprehensive Environmental Response Compensation and Liability Act (CERCLA) also known as the "Superfund Act." The National Contingency Plan establishes the roles and responsibilities of federal agencies in response to oil and hazardous substances released to the environment. The National Response System (NRS) is the mechanism for coordinating federal, state, and local governmental responses in accordance with the National Contingency Plan. The National Response System structure primarily consists of the National Response Team (NRT), Federal On-Scene Coordinator, and 13 Regional Response Teams.

The National Response Team consists of 16 federal agencies (Fig. 2.2) with responsibilities and expertise in various stages of emergency response outlined in the National Contingency Plan. The NRT is chaired by the EPA and cochaired by the United States Coast Guard (USCG). The NRT coordinates the actions of the National Response System governmental response by assigning preestablished Federal On-Scene Coordinators (FOSCs) from across the United States. During major environmental emergencies, the FOSC has the authority to command all response activities if warranted, or may serve in a supervisory status either on scene or from a remote location (Fig. 2.3). The FOSC is either a designated agent of the EPA or a commissioned officer of the USCG. To assist the FOSC and expedite activation of the National Response System, Regional Response Teams were established and strategically located in 13 regions of the United States (Fig. 2.4). In addition to response duties, the Regional Response Teams are required to develop and maintain a Regional Contingency Plan that should closely mirror the National Contingency Plan and complement state emergency response plans, area contingency plans, and local emergency response plans (see Chapter 3).

Activation of the National Response System and implementation of the National Contingency Plan are triggered by notifications that must be made by a facility owner, operator, or "person in charge" in the event of an oil spill to water or a release of hazardous substances listed by CWA and CERCLA regulations. In the event of such a release, this notification is made by calling the National Response Center staffed 24 hours by the USCG (800/424-8802) (Fig. 2.5).

The primary goal of the National Contingency Plan and the National Response System is to provide federal support to state and local government response

Figure 2.2   The National Response Team is made up of 16 federal agencies that have specialties and expertise in hazmat response.

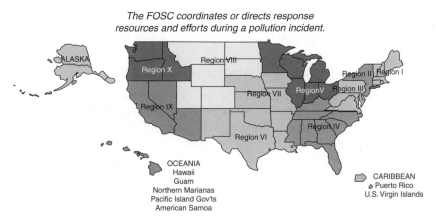

Figure 2.3   The 13 Regional Response Teams (RRTs) of the National Response System.

efforts by utilizing the Incident Command System (also known as the Incident Management System) through a unified command structure. Assistance from the National Response Team's agencies to state and local governments varies from preincident preparedness and planning advice to on-scene command or remote guidance of response activities. Activation of the National Response System is

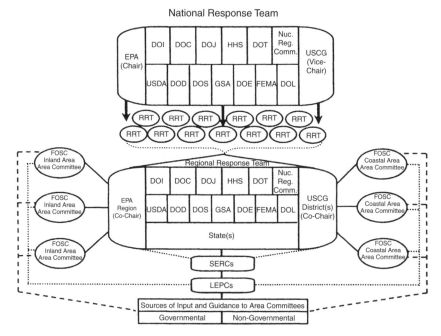

Figure 2.4 The National Response System—planning concept of federal, regional, area, local, and site-specific plans.

intended to provide assistance and coordination to accidental and intentional (terrorist) releases of dangerous chemical, physical, and biological agents.

The Clean Water Act and the Oil Pollution Act of 1990

The Clean Water Act, as amended by the Oil Pollution Act of 1990, is the primary legislation for protecting the waters of the United States from releases of hazardous substances and oil. The CWA authorized the EPA to implement water pollution prevention programs, set water quality standards, and enforce regulations. The Act also required spill response standards to be established. With the growing national concern after the *Exxon Valdez* oil spill in Alaska's Prince William Sound in 1989, the Oil Pollution Act of 1990 amended portions of the CWA. These amendments were designed to improve the nation's ability to prevent and respond to oil releases affecting navigable waters.

Comprehensive Environmental Response Compensation and Liability Act

The U.S. Congress enacted CERCLA on December 11, 1980. This law imposed a tax on the chemical and petroleum industries and provided broad federal authority to respond directly to potential and actual releases of listed hazardous substances that

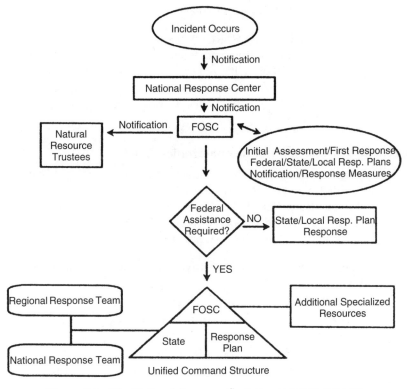

Figure 2.5   The National Response System—response concept.

could endanger the public health, welfare, and environment of the United States. The money collected from the tax went to a trust fund (commonly referred to as the "Superfund") for cleaning up abandoned and uncontrolled hazardous waste sites. The CERCLA legislation provided EPA with authority to regulate the following:

- Prohibitions and requirements concerning the cleanup of closed or abandoned hazardous waste sites
- Liability of responsible parties for releases of hazardous waste at waste sites
- Management of trust fund money to provide for cleanup when no responsible party can be identified

The CERCLA legislation authorized EPA to take two types of response actions. The first action is short-term removal of hazardous waste from sites where prompt response actions are required. The second action authorized is the long-term removal and remediation of hazardous waste to reduce permanently and significantly the threats posed by releases of hazardous contaminants. This second type of action can only be carried out on sites that are listed on EPA's National

Priority List (40 CFR Subpart L 300 Appendix A) as outlined by the National Contingency Plan.

To revise and reauthorize the CERCLA legislation, the U.S. Congress established the Superfund Amendments and Reauthorization Act (SARA) of 1986. Both CERCLA and Title III of SARA required EPA to establish lists of hazardous substances and extremely hazardous substances, respectively, with quantity limits triggering response and planning requirements. See CERCLA regulations in this chapter and in Chapter 3.

Superfund Amendments and Reauthorization Act

The U.S. Congress passed SARA legislation on October 17, 1986, to amend and reauthorize CERCLA and to establish new laws relating directly to environmental research, informing the public of chemical threats, as well as regulating chemical emergency preparedness and response. To accomplish the goals of SARA, the Act was divided into the following five primary titles:

- Title I—Provisions Relating Primarily to Response and Liability (Sections 118–126)
- Title II—Miscellaneous Provisions (Sections 205–213)
- Title III—Emergency Planning and Community Right-To-Know (Sections 301–330)
- Title IV—Radon Gas and Indoor Air Quality Research (Sections 401–405)
- Title V—Amendments of the Internal Revenue Code of 1986 (Sections 501–531.)

The titles of major concern to emergency responders are Title I and Title III.

SARA Title I

SARA Title I had major implications in the field of hazardous waste cleanup, hazardous waste management, and hazardous materials emergency response. This law established the requirement for the Secretary of Labor to develop and establish regulations and standards to protect workers engaged in the previously mentioned activities. OSHA, under the direction of the Secretary of Labor, had one year from the passage of SARA to promulgate such a standard.

Before this legislation and its regulations, the only federally published guidance was the nonmandatory *Occupational Safety and Health Guidance Manual for Hazardous Waste Site Activities* prepared by the National Institute of Occupational Safety and Health (NIOSH), OSHA, USCG, and EPA. To comply with the mandate of this legislation, OSHA developed the Hazardous Waste Operations and Emergency Response or HAZWOPER standard (29 CFR 1910.120).

Other important provisions of SARA Title I included the establishment of grant-funded training programs that were to be administered by the National Institute Singular of Environmental Health Sciences (NIEHS). These grants were to be awarded to nonprofit organizations with the capabilities to provide health and safety training as well as target worker populations engaged in the specified waste and response activities. The University of Alabama at Birmingham's Center for Labor Education and Research was and at the time of publication still is one of the original NIEHS grantees.

SARA Title III

SARA Title III became known as the Emergency Planning and Community Right-To-Know Act (EPCRA). SARA Title III or EPCRA provides the infrastructure at state and local governmental levels needed to plan for chemical emergencies. It also established inventory and release reporting requirements for facilities that use, store, or release chemicals listed as Extremely Hazardous Substances (EHSs) (40 CFR 355 Appendix A). This legislation provided avenues for public access to this information to inform citizens of the potentially dangerous chemicals in their community. See Chapter 3 for a more detailed discussion of SARA Title III.

In summary, the intent of SARA Titles I and III was to close the gap of emergency preparedness and planning at the state and local governmental levels left open by the National Contingency Plan, CERCLA, and other federal policies in existence before 1986.

Resource Conservation and Recovery Act

The Resource Conservation and Recovery Act (RCRA) of 1976 amended the Solid Waste Disposal Act of 1965. Because of increasing concerns over the lack of comprehensive hazardous waste management regulations, RCRA authorized EPA to promulgate standards enforcing the following areas:

- "Cradle-to-grave" management of hazardous waste (from point of generation to final disposal)
- Identification and characterization of hazardous waste
- Tracking of hazardous wastes utilizing the Uniform Hazardous Waste Manifest
- Management of treatment, storage, and disposal facilities (TSDFs)

Additionally, the RCRA legislation also impacts hazmat employers, employees, and emergency responders primarily through EPA regulations covering hazardous waste facility contingency plans and emergency procedures. These regulations are discussed below in this chapter.

Clean Air Act

The Clean Air Act (CAA) as amended in 1990 is the national framework for protecting the air quality of the United States. The amended Act requires EPA to promulgate regulations addressing the following:

- Primary and secondary National Ambient Air Quality Standards
- Hazardous air pollutant emission limits

- Pollution attainment goals with State Implementation Plans and Tribal Implementation Plans
- Regulation and enforcement of motor vehicle emissions and fuels

One important section of the CAA affecting emergency responders is the law requiring facilities storing or utilizing specified hazardous materials and extremely hazardous substances to determine likely "worst-case" scenarios and appropriate emergency response plans. Under CAA Section 112 (r), EPA was granted authority to promulgate and enforce these regulations. CAA Section 112 (r) requires the Administrator of the EPA to establish a list of chemicals that pose extreme risks to human health or the environment if accidentally released. This list can be updated by the Administrator periodically and must be reviewed every five years.

According to the Act, this list was to include specified hazardous materials such as anhydrous ammonia, chlorine, and ethylene oxide, to mention a few. In addition to the chemicals listed in the Act, applicable SARA Title III EHSs and additional chemicals were to be included by the Administrator according to specified decision criteria. The Administrator was also required to establish threshold quantities for each of the listed hazardous chemicals. If facilities store or use amounts of these chemicals in excess of the limits, the owners or operators of the facilities are required to estimate the "worst-case" incident scenario(s) and develop plans specifically to mitigate such incidents. Facility chemical and planning information pertaining to Section 112 (r) is available through EPA to public emergency response agencies on request.

Occupational Safety and Health Act

Before the passage of the Occupational Safety and Health Act (OSHAct) of 1970, no uniform federal safety and health regulations existed. State regulations varied widely, and enforcement proceedings against violators of existing regulations were almost nonexistent. As a result of unsafe and unhealthy workplace conditions, unacceptably large numbers of American workers were experiencing illness, injury, or death. The OSHAct was passed by the U.S. Congress as a means of addressing this problem.

The OSHAct was intended to ensure safe and healthful conditions in the American workplace. The Act requires that employers take steps to protect employees from recognized workplace hazards that are likely to cause illness or injury. If practical, recognized hazards should be completely eliminated from the workplace, such as through the use of engineering controls. If elimination of a hazard is not practical, the employer must provide other measures, such as personal protective equipment, to protect employees. The Act also requires that employees comply with all applicable rules, regulations, and standards pertaining to occupational safety and health.

A number of organizations came into existence as a result of the OSHAct. One of these organizations, the Occupational Safety and Health Administration, is of particular importance and is referred to frequently within this text.

Hazardous Materials Transportation Act

The Hazardous Materials Transportation Act (HMTA) of 1975 passed by the U.S. Congress strengthened the authority of the United States Department of Transportation (DOT). The overarching goal was to provide the U.S. Secretary of Transportation with the ability to establish many new regulations and enhance DOT's enforcement authority. The authorizations established by this law provided for the following regulatory improvements in all modes of transportation:

- Placarding, labeling, marking, handling, and routing regulations
- Regulations for container manufacturers
- Established a national shipper registration program
- Authorized DOT to conduct inspections and levy penalties
- Required the National Transportation Safety Board to report directly to Congress, separating it from DOT
- Defined federal, state, and local regulatory responsibilities
- Training of all employees affecting the transportation of hazardous materials

Hazardous Materials Transportation Uniform Safety Act

In 1990 the Hazardous Materials Transportation Act was modified by the Hazardous Materials Transportation Uniform Safety Act (HMTUSA). As the name implies, this legislation required DOT to promulgate new regulations to enhance uniformity of hazmat transportation safety in the following areas:

- Intrastate commerce
- Minimum required shipping document information
- Required training elements
- Safety permits and procedures
- Improvement of the existing hazmat identification systems
- Shipper and carrier registration fees
- Evaluation of a proposed emergency response telephone system

In addition to the previous legislative elements, HMTUSA required DOT to make grants available to states, territories, and Native American tribes to conduct training of public sector employees who respond to emergencies. HMTUSA Section 117a authorized DOT to develop a training curriculum to accompany the training grant program. DOT complied with the development of the Hazardous Materials Emergency Grant Program (HMEP). These guidelines can be used as a self-assessment tool. The HMEP also stipulates that courses funded by the grant comply with training curriculum goals. The development of these training courses ensured that public sector employees (aka "public responders") could safely and efficiently respond to hazardous materials emergencies. Additional information on the HMEP is available on-line through the U.S. Fire Administration web site (http://www.usfa.fema.gov).

Homeland Security Act

The U.S. Congress passed the Homeland Security Act (HSA) on January 23, 2002. The passage of the HSA was in direct response to the horrific acts of terrorism committed against the United States on September 11, 2001 and the following biological attacks that taxed America's responders.

The HSA created the Department of Homeland Defense (DHS) to improve the nation's capabilities to prepare, prevent, and respond to acts of terrorism in the United States. This new legislation requires the consolidation of numerous existing federal agencies and describes the organizational structure of DHS. In addition to these agencies, the HSA outlines the coordination of DHS with other federal departments and their agencies to protect against all forms of homeland terrorism. The HSA includes legislation to assist state and local governments and responders with improved federal coordination, grant funding, and training programs.

Homeland Security Presidential Directive 5

The President of the United States has the authority to issue directives that are used to establish new policy, decree the commencement or cessation of some action, or ordain the notice to be given to some declaration. Directives are issued in various formats that can include executive orders, proclamations, oral presidential directives, administrative orders, and presidential announcements, just to name a few.

The series of directives involving homeland security issues have been named Homeland Security Presidential Directives (HSPDs) and are issued by the President. HSPD-5 was issued on February 28, 2003 to "enhance the ability of the U.S. to manage domestic incidents by establishing a single, comprehensive national incident management system."

Several policies are established by HSPD-5, but the primary goals of this directive with these policies are twofold:

- This directive ensures that all levels of government in the United States are capable of working efficiently and effectively in mitigating all domestic incidents and emergencies (including domestic terrorism) that require federal response utilizing a national management approach. HSPD-5 outlined the authorities and coordination efforts for all levels of government and federal agencies with the revision of the National Response Plan (NRP).
- Under the auspices of the Department of Homeland Security and pursuant to the Homeland Security Act of 2002, the NRP shall be administered by the utilization of the National Incident Management System (NIMS), which is discussed in Chapter 4. Ultimate authority in the NIMS lies with the Secretary of Homeland Security.

As of fiscal year 2005, HSPD-5 policy requires that all federal departments providing NRP preparedness assistance shall make the adoption of the NIMS mandatory by all state and local agencies seeking available NRP grants and contracts. The Secretary of Homeland Security is responsible for developing the standards and guidelines to determine whether a state or local agency has adopted the NIMS.

REGULATORY DEPARTMENTS AND AGENCIES

As part of the public policy framework within the United States, presidential and congressional departments and agencies are required to promulgate regulations that promote and enforce requirements of the laws. Additionally, these federal departments and agencies may delegate similar authorities to the states. The overall regulatory structure in the United States is beyond the scope of this book, but it is necessary to address the various federal departments and agencies specifically governing emergency response.

The Environmental Protection Agency

EPA was established by the Executive Office of the President and the U.S. Congress in July of 1970. The creation of EPA was in response to public demand for cleaner water, air, and land. EPA's mission is "to protect human health and safeguard the natural environment" of the U.S. by researching and setting national standards for a variety of environmental programs. In addition, EPA offers financial assistance in the form of research grants, environmental research projects, and special programs.

Scope and Authority
The laws established by congressional acts to protect the air, water, and land of the United States grant the EPA the authority to promulgate rules and regulatory standards and manage associated programs and offices. Where national standards are not met, EPA can issue sanctions and take other steps to assist states and tribes in reaching desired levels of environmental quality. The EPA recognizes the rights of the states and the Native American tribes to self-govern their citizens. The EPA shares these responsibilities with the states and the tribes by delegating the authority to ensure compliance with applicable congressional laws. The states and tribes can adopt the standards established by EPA or develop new regulations specific to needs of the state or tribe. If a state or tribe chooses to promulgate its own regulatory standards, EPA stipulates that the regulations of the state or tribe must be as stringent as or more stringent than those of EPA. EPA reserves the right to evaluate the states' and tribes' environmental departments and programs to ensure compliance with the federal laws.

TABLE 2.1   The 12 Subparts of the National Oil and Hazardous Substances Pollution Contingency Plan (National Contingency Plan or NCP)

Subpart	Subpart Title	CFR Location
A	Introduction	40 CFR 300.1–.7
B	Responsibility and Organization for Response	40 CFR 300.100–.185
C	Planning and Preparedness	40 CFR 300.200–.220
D	Operational Response Phases	40 CFR 300.300–.335
E	Hazardous Substance Response	40 CFR 300.400–.440
F	State Involvement in Hazardous Substance Response	40 CFR 300.500–.525
G	Trustees for Natural Resources	40 CFR 300.600–.615
H	Participation by Other Persons	40 CFR 300.700
I	Administrative Record for Selection of Response Action	40 CFR 300.800–.825
J	Use of Dispersants and Other Chemicals	40 CFR 300.900–.920
K	Federal Facilities	Reserved
L	National Oil and Hazardous Substances Pollution Contingency Plan; Involuntary Acquisition of Property by the Government	40 CFR 300.1105 Appendices A–E

Relevant Regulations

Regulations of the National Oil and Hazardous Substances Pollution Contingency Plan. The regulations of the NCP are found in Part 300 of Subchapter J—Superfund, Emergency Planning, and Community Right-To-Know Programs in the Code of Federal Regulations (40 CFR 300). The NCP regulations are further divided into 12 subparts detailing the regulations that emergency responders should be familiar with when responding to oil spills on water and uncontrolled releases of hazardous substances (Table 2.1).

Regulations of the Clean Water Act and the Oil Pollution Act. The regulations promulgated by the CWA and later the OPA of 1990 are designed to protect the nation's navigable waters from sources of pollution. These protective regulations are based on the following:

- Control/elimination of industrial pollution releases with permitting requirements
- Establishing reporting requirements for normal or accidental releases
- Responses to toxic chemical releases and oil spills

The focus of the legislative acts was to require facilities that meet specified criteria regarding petroleum products to develop and maintain a Spill Prevention, Control, and Countermeasure (SPCC) plan [40 CFR 112.7(d) and 112.20–112.21]. EPA has allowed facilities that are required to develop a SPCC plan the option of integrating the plan into an Integrated Contingency Plan (ICP) (see Chapter 3). In addition to the SPCC plan, the response regulations of the CWA and the OPA are to follow the NCP. Ultimately these acts established the regulations of the NRS.

The regulations created by the CWA and the OPA under the authority of EPA can be found in Subchapters D, H, N, and O of Title 40 of the Code of Federal Regulations.

Regulations of the Comprehensive Environmental Response Compensation and Liability Act. The regulations authorized by CERCLA are primarily intended to regulate the cleanup of abandoned or uncontrolled hazardous waste sites. These regulations are outlined in the following manner:

- Designation of hazardous substances, determination of reportable quantities, and notification requirements
- Citizen awards for information on criminal violations under the Superfund
- Procedures for filing claims and reimbursement for cleanup of approved sites
- Worker protection rules for all state and local government responders not covered by a state occupational safety and health plan

These regulations are found in 40 CFR 302–311.

Emergency Planning and Community Right-To-Know Regulations. The regulations promulgated by the EPA under the authority of the EPCRA or SARA Title III are the primary policies detailing hazmat emergency planning and procedures for informing the public of chemical hazards in their community. These regulations can be found in 40 CFR 350–374. The individual Parts cover the following regulatory programs:

- Part 350—Rights of emergency planners and health professionals to request hazardous chemical information normally protected as "trade secrets"
- Part 355—Emergency planning and notification procedures for reportable quantity releases of specified hazardous substances
- Part 370—Procedures requiring facilities to inform the public about the chemicals stored and used in their community
- Part 372—Procedures requiring facilities to inform the public of permitted toxic releases into their community
- Part 373—Procedures requiring the federal government to report hazardous substance activity before the sale or transfer of land by the United States
- Part 374—Authorizations allowing citizens the right to sue anyone in violation of the Act in a civil action

It is very important for hazmat emergency responders to have a strong working knowledge of these regulations to assist them in developing plans and preparing for future hazmat emergencies. The applicability of SARA Title III to responders and its regulations are discussed in Chapter 3 of this text.

Regulations of the Resource Conservation and Recovery Act. The regulatory programs established by EPA to satisfy the RCRA legislation primarily control all management of hazardous waste being generated in the present or in the future. These regulations impact responders and workers through the proper management of hazardous waste to prevent and plan for releases. Hazardous waste facility owners/operators are required to train hazardous waste employees in proper management techniques to ensure the prevention of releases of waste to the environment. They are also required to develop standard operating procedures. In addition to management procedures for normal operations, facilities are required to develop Facility Contingency Plans to help facility owners/operators and employees, along with public emergency responders, to effectively and efficiently respond to hazardous waste spills.

The regulations of RCRA are found in Chapter I of 40 CFR (Parts 260–279). Of these Parts, emergency planners and responders may find it necessary to be familiar with the following:

- Part 261—Subparts C and D (characteristics of hazardous wastes and lists of hazardous wastes)
- Part 262—Subparts B, C, and D (facility manifesting, storage, and record-keeping requirements)
- Part 263—Subparts A, B, and C (regulations describing agreements between EPA and DOT, manifesting procedures, and procedures for releases of hazardous waste in transport)
- Part 265—Subparts B, C, and D (regulations for personnel training, preparedness and prevention, and contingency plans and emergency procedures)

If emergency planners or responders anticipate hazardous waste releases in their jurisdictions, it is recommended that before and during emergency operations they consult with professionals specializing in this field, such as Certified Hazardous Materials Managers (CHMMs).

EPA has allowed facilities required to develop a Facility Contingency Plan the option of integrating this plan into an Integrated Contingency Plan (ICP). Planning requirements of the RCRA regulations and the National Response Teams ICP format are discussed in Chapter 3 of this text.

Clean Air Act—Risk Management Program Rule. As part of the Clean Air Act, EPA was required to promulgate regulations and guidance for chemical accident prevention at facilities using extremely hazardous substances. The Risk Management Program (RMP) Rule (40 CFR 68) was written to comply with Section 112 (r) of the CAA. The RMP Rule requirements complement other industrial standards designed to prevent accidental release of hazardous chemicals to the air and are modeled after OSHA's Process Safety Management standard

(29 CFR 1910.119) (Cox, 2000). The Rule requires companies that use certain flammable and toxic chemicals (40 CFR 68.130) to develop a Risk Management Program that includes the following:

- A hazard assessment, a 5-year history of accidental releases, and evaluations for worst-case scenarios and alternative accident scenarios
- A prevention program detailing safety precautions, maintenance, monitoring, and employee training measures
- An emergency response program that details emergency health care, employee training, and procedures for informing the public and emergency responders should an accident occur

Companies meeting the criteria of the rule and regulations are required to submit to EPA a summary of the program in the form of a Risk Management (RM) Plan, and these plans must be revised and resubmitted every five years. RMPs and the RM Plans are designed to reduce chemical risks in the community by requiring facilities to improve safety of systems, provide public emergency responders with information, and inform the public of chemical hazards in their community.

EPA has allowed facilities required to develop an RM Plan the option of integrating the plan into an ICP (see Chapter 3).

The Occupational Safety and Health Administration

The Occupational Safety and Health Administration (OSHA) was created within the Department of Labor to act as the primary guardian of worker safety and health. As such, OSHA was given the authority to develop and implement workplace safety and health standards.

OSHA is also responsible for enforcing compliance with standards and has the authority to conduct inspections, issue citations, and levy fines. Along with OSHA's enforceable standards, OSHA offers information about safety topics in nonmandatory guidelines. These guidelines are often in addition to established standards and, if followed, greatly enhance established safety and health programs.

Scope and Authority

OSHA's authority to enforce workplace safety and health standards as originally established by the OSHAct covers all private sector workplaces. In 1980, protection under the OSHA standards was extended to cover employees of the federal government. However, OSHA currently has no authority to regulate the safety and health of employees of state and local governments.

Individual states may elect to assume authority from OSHA for regulating occupational safety and health within the workplace. States that have taken advantage of this option are shown in Figure 2.6. State occupational safety and health regulatory programs must be documented in an OSHA-approved written state

States having OSHA - approved plans

Figure 2.6 The shaded states and territory have OSHA-approved occupational safety and health plans as of February 2004.

plan. The protective provisions incorporated into state plans must be at least as stringent as equivalent provisions of the federal OSHA.

To receive OSHA approval, the "state plan" states are required to extend protection under the state plan to all state and local government employees, in addition to private sector employees, within the jurisdiction of the state. Thus, public sector employees, such as law enforcement and fire service personnel, are required to be provided occupational safety and health protection at least equivalent to that required by the federal OSHA within these states.

OSHA Safety and Health Standards

OSHA Safety and Health Standards are legally enforceable sets of industry-specific regulations intended to address concerns for the safety and health of workers. These standards are developed and revised on a constant basis. The opportunity to comment on proposed new standards or revisions is extended to employers, employees, and all other interested parties. OSHA standards are referenced throughout this text.

Numerous safety standards have been developed by OSHA to protect the safety and health of workers involved in construction activities and general labor. Examples of specific safety topics and applicable standards are shown in Table 2.2. From the standpoint of emergency response to releases of hazardous substances, the most significant OSHA standard is 29 CFR 1910.120. This standard is described at length at the end of this chapter.

Compliance with OSHA standards can be difficult for small businesses. Because of this, OSHA offers compliance assistance in the form of resources and education. Additionally, OSHA provides funding to assist all states in the maintenance of consultation programs that help small businesses comply with OSHA regulatory standards.

The Department of Transportation

The U.S. Congress created the Department of Transportation (DOT) on October 15, 1966 with the passage of Public Law 89–670. DOT was given the mission of

TABLE 2.2   Examples of Specific Safety Topics and Applicable OSHA Standards

Safety Topics	Applicable Standards			
	General Industry (29 CFR 1910)		Construction (29 CFR 1926)	
Ventilation		⎧ 1910.94		⎧ 1926.57
Noise	Subpart G	⎨ 1910.95	Subpart D	⎨ 1926.52
Nonionizing radiation		⎩ 1910.97		1926.54
Ionizing radiation	Subpart Z	1910.1096		⎩ 1926.53
Hazardous materials	Subpart H			
Personal protective equipment				
General		⎧ 1910.132		
Eye/face		1910.133		⎧ 1926.102
Hearing	Subpart I	⎨ 1910.95	Subpart E	⎨ 1926.101
Respiratory		1910.134		1926.103
Head		1910.135		⎩ 1926.100
Foot		⎩ 1910.136		
Fire protection	Subpart L		Subpart F	
Materials handling and storage	Subpart N		Subpart H	
Electrical	Subpart S		Subpart K	
Toxic/hazardous substances	Subpart Z			
Trenching and excavation			Subpart P	

ensuring a fast, efficient, accessible, convenient, and safe transportation system that would serve the nation's transportation requirements.

Scope and Authority

DOT oversees the development of national transport regulatory and enforcement policies that promote all modes of transportation. With the exception of RCRA transportation regulations, EPA and OSHA do not have the authority to regulate these hazardous chemicals in transportation. Among many duties, DOT is ultimately responsible for the regulations governing the safe transportation of hazmat for commerce in the United States and to some extent internationally. The DOT serves as the U.S. representative agency when negotiating and implementing international hazmat transportation regulatory agreements with international agencies such as the International Civil Aviation Organization. These international agreements and policies help to ensure uniformity of hazmat transportation regulations among the nations of the world.

Enforcement Agencies and Related Regulations

The Hazardous Materials Transportation Act of 1975 was the legislation granting the Secretary of Transportation the authority to define a hazardous material as any material (either based on quantity or form) that poses an unreasonable risk to health and safety of the public or property when transported for commerce in the United States. As required by this legislation, the Secretary of Transportation promulgates DOT rules and regulations for all modes of transportation.

Later, the HMTA regulations were revised and amended by the requirements of the HMTUSA of 1990. Rules and regulations resulting from both of these Acts are known as the Hazardous Materials Regulations (HMR) (49 CFR 171–180) and are regulated by DOT. As discussed in Chapter 5, the HMR protect responders primarily through the prevention of hazmat releases during transport and facilitation of the rapid recognition and identification of hazmats with shipping documents, labels, placards, markings, hazard classes, and United Nations Identification Numbers. The HMR govern the following topics:

- Classification of hazmats
- Hazmat shipper and carrier operations
- Packaging and container specification for all modes of transportation
- Training and incident reporting procedures
- DOT inspection and enforcement policies focusing on classification, description, marking, labeling, and packaging requirements

HMR enforcement authority is shared among different administrations of the DOT, such as the Research and Special Programs Administration. The U.S. Coast Guard during peacetime operations was once one of these administrations, but as a result of the Homeland Security Act of 2002 the U.S. Coast Guard and its regulations during peacetime are now part of the Department of Homeland Security (Table 2.3).

Over the course of time, the Secretary of Transportation will issue new DOT rules and regulations in the form of dockets. These dockets are announced in the Federal Register, and the regulations will be added to or revise existing regulations of the HMR. Some of these dockets and their regulatory nature are listed in the following:

- HM-181—New and amended hazmat regulations as a result of the HMTUSA legislation
- HM-206—Improvements to the hazardous materials identifications systems
- HM-215—Harmonization with international recommendations
- HM-232—Regulations covering hazardous materials security training of hazmat employees and the development of facility-specific hazmat security plans

The Department of Homeland Security

The Department of Homeland Security (DHS) was created by the Homeland Security Act passed by the 107th U.S. Congress and signed into law by President

TABLE 2.3   The Department of Homeland Security is Divided Among Five Directorates Comprised of Different Agencies

I. The Border and Transportation Security Directorate includes:
   - The U.S. Customs Service (Treasury)
   - The Immigration and Naturalization Service (part) (Justice)
   - The Federal Protective Service (GSA)
   - The Transportation Security Administration (Transportation)
   - Federal Law Enforcement Training Center (Treasury)
   - Animal and Plant Health Inspection Service (part) (Agriculture)
   - Office for Domestic Preparedness (Justice)

II. The Emergency Preparedness and Response Directorate includes:
   - The Federal Emergency Management Agency (FEMA)
   - Strategic National Stockpile and National Disaster Medical System (HHS)
   - Nuclear Incident Response Team (Energy)
   - Domestic Emergency Support Teams (Justice)
   - National Domestic Preparedness Office (FBI)

III. The Science and Technology Directorate Includes:
   - CBRN Countermeasures Programs (Energy)
   - Environmental Measurements Laboratory (Energy)
   - National BW Defense Analysis Center (Defense)
   - Plum Island Animal Disease Center (Agriculture)

IV. The Information Analysis and Infrastructure Protection Directorate includes:
   - Critical Infrastructure Assurance Office (Commerce)
   - Federal Computer Incident Response Center (GSA)
   - National Communications System (Defense)
   - National Infrastructure Protection (FBI)
   - Energy Security and Assurance Program (Energy)

V. The Management Directorate

Note: The Secret Service and the Coast Guard are also located in the DHS, remaining intact and reporting directly to the Secretary of Homeland Security.

George W. Bush in 2002. DHS is the lead federal cabinet agency responsible for protecting the United States and its territories from acts of domestic terrorism in any form.

Scope and Authority

DHS has five major divisions referred to as directorates, each having a different focus. As required by the HSA, the numerous federal agencies consolidated under DHS were assigned to directorates based on the agencies' missions (Table 2.3). Other organizations such as the U.S. Secret Service and the U.S. Coast Guard peacetime operations were included in the DHS and report directly to the Secretary of Homeland Security.

DHS adopted the existing policies of the different consolidated agencies. In addition to these adopted policies, the Secretary of Homeland Security is responsible for promulgating new polices as required to protect the nation from future acts of domestic terrorism. Policies developed under DHS are found in Title 6 of the Code of Federal Regulations.

State and Local Governmental Regulatory Agencies

The emergency response policies and regulatory agencies or departments discussed so far in this chapter are those created at the level of the federal government. However, these are not the only laws, agencies, rules, regulations, and standards applicable to hazmat emergency response in the United States. Because of the fragmentation of government inherent to a democracy, state and local governments have the right to establish their own policies and regulatory agencies that influence emergency response and impact emergency responders.

In most cases when a state or local government has the resources and capability to manage a regulatory program that is required by a federal law, it may establish a regulatory agency or department to do so. As in the case with OSHA, for example, some states have developed a state occupational safety and health program regulating and enforcing worker protection standards (Fig. 2.6). In these cases, the federal agency or department (OSHA) has granted the authority to the state to promulgate, regulate, and enforce the requirements of the applicable law (OSHAct). If a state exercises this option, the primary stipulation for a nonfederal agency (state OSHA) to comply with is that the state's regulatory program be as stringent as or more stringent than the federal rules and regulations. Similar to the relationship between the federal government and state governments, local governments can establish regulatory agencies and promulgate their own policies to comply with applicable federal laws. In all cases, ultimate authority will remain at the federal level, and federal regulators reserve the right to oversee state and local programs. The primary goal in situations where state and local governments choose to set up their own programs is to ensure compliance while tailoring the state and local policies to their respective communities.

In situations in which the state or local government does not have the necessary resources and capabilities to enforce federal policy designed to protect the environment and public welfare, the authority remains with the federal agency or department. It is the responsibility of emergency responders at all levels of government and industry to know the applicable emergency response policies of the United States and their state and local government.

Nonregulatory Standards and Guidance by Governmental Agencies

National Institute of Occupational Safety and Health

The National Institute of Occupational Safety and Health (NIOSH) was created by the Occupational Safety and Health Act of 1970 to conduct research on and make recommendations for the prevention of work-related injuries and illnesses. NIOSH is a federal agency that provides research, information, education, and training in the field of occupational safety and health. The recommendations developed by NIOSH are not legally enforceable; however, OSHA often incorporates various NIOSH recommendations into its regulations. For example, OSHA requires respirators used in the working environment to be NIOSH approved. As discussed in Chapter 9, NIOSH is also responsible for publishing chemical exposure information, such as the Immediately Dangerous to Life and Health (IDLH) values.

Agency for Toxic Substances and Disease Registry

The Agency for Toxic Substances and Disease Registry (ATSDR) was created in 1980 by Congress under the Superfund Act to assess the presence and nature of health hazards at specific Superfund sites. In 1984, ATSDR was given authority by amendments to the Resource Conservation and Recovery Act to conduct public health assessments at hazardous waste storage or destruction facilities. The final piece of legislation that has increased the duties of ATSDR is the passage of the Superfund Amendments and Reauthorization Act of 1986, which expanded ATSDR responsibilities to include environmental public health issues. The current responsibilities of ATSDR have grown to include public health assessments of waste sites, health consultations concerning specific hazardous substances, health surveillance and registries, response to emergency releases of hazardous substances, applied research in support of public health assessments, information development and dissemination, and education and training concerning hazardous substances.

Consensus Standards of Professional Agencies

Consensus standards are documents that have been developed to establish a set of principles that can be used as guidelines for practice for a variety of topics. They are developed by various organizations with participation by interested parties who have a stake in the standards' development and/or use. Consensus standards are available for purchase and are typically bought by an employer for use in a work environment. Consensus standards are voluntary; however, many government regulators often include the use of certain consensus standards by citing them in laws, regulations, and codes. Three organizations that develop consensus standards are the American National Standards Institute, ASTM International, and the National Fire Protection Association.

American National Standards Institute

The American National Standards Institute (ANSI) is a private, nonprofit organization that administers and coordinates the U.S. voluntary standardization and conformity assessment system. ANSI was founded in 1918 by five engineering societies and three government agencies and is supported by both private and public organizations. ANSI standards are developed for a wide variety of topics and include standards for hazardous materials and emergency response.

ASTM International

ASTM International, formerly known as the American Society for Testing and Materials, was founded in 1898. ASTM International is a nonprofit organization that develops voluntary consensus standards for materials, products, systems, and services. The ASTM International standards are used in research and development, product testing, quality systems, and commercial transactions. Various standards relating to hazardous materials and emergency response are developed by ASTM International. For example, F1127-01, the *Standard Guide for Containment by Emergency Response Personnel of Hazardous Material Spills*, can be used to obtain information about a specific topic of hazardous materials response.

National Fire Protection Association

The National Fire Protection Association (NFPA) is a not-for-profit membership organization that uses a consensus process to develop model fire prevention codes and fire fighting training standards, along with health and safety standards. Because of the inherent risks associated with hazardous materials encountered by the fire service, the NFPA has written excellent guidance documents and recommendations to assist the fire service (Table 2.4). These consensus standards incorporate standards of other respected standard-setting agencies such ANSI. Many equipment manufactures use NFPA design and performance criteria to improve function and safety features of their equipment such as PPE that can benefit all responders.

NFPA consensus standards cover many fire service hazmat topics such as the following:

- Hazmat storage and segregation requirements
- Equipment and material compatibility requirements
- Proper handling procedures
- Responder training and operational procedures
- Hazmat identification system
- Health and safety standards
- PPE criteria, training, and use requirements

Most fire departments across the United States and some departments in other parts of the world have designed their departmental policies around selected parts of or entire NFPA standards and guidance documents. Many industrial facilities also rely on these standards for planning, preparedness, and response guidance when governmental regulations are vague or when facilities are trying to improve worker safety. Several key NFPA consensus standards relevant to hazmat response are summarized in the following paragraphs.

Three important hazmat operational and training standards are NFPA 471, 472, and 473. NFPA 471 is the *Recommended Practice for Responding to Hazardous Materials Incidents* standard (NFPA 471, 2002). NFPA 471 outlines the minimum requirements that all response organizations and incident commanders should consider when responding to hazmat incidents. It provides guidance by describing procedures, policies, and application of procedures for specific topics such as planning, PPE, decontamination, safety, and communications. See Chapter 3 for a discussion of planning elements listed in NFPA 471.

TABLE 2.4 NFPA Standards that Apply to Hazmat Preparedness and Response

NFPA Number and Title	Date of Most Recent Edition	Description of Content
NFPA 30—Flammable and Combustible Liquids Code	2003	Applies to all flammable and combustible liquids except those that are solid at 100°F or above. Covers tank storage, piping, valves and fittings, container storage, industrial plants, bulk plants, service stations, and processing plants.
NFPA 53—Recommended Practice on Materials, Equipment and Systems Used in Oxygen-Enriched Atmospheres	2004	Covers the fire and explosion hazards that may exist in oxygen-enriched atmospheres.
NFPA 326—Standard for the Safeguarding of Tanks and Containers for Entry, Cleaning, or Repair	1999	Applies to the entry of underground storage tanks, operating at nominal atmospheric pressure, that have contained flammable or combustible liquids and that might contain flammable or combustible vapors or residues.
NFPA 329—Recommended Practice for Handling Releases of Flammable and Combustible Liquids and Gases	1999	Covers the procedures to be followed in the handling of underground leakage and flammable and combustible liquids. Included are procedures when life and property may be in danger; locating the sources of leakage; testing for underground leaks; tracing liquids underground; and the removal and disposal of such liquids.
NFPA 385—Standard for Tank Vehicles for Flammable and Combustible Liquids	2000	Applies to tank vehicles to be used for the transportation of asphalt or normally stable flammable and combustible liquids with a flash point below 200°F (93.4°C); covers the design and construction of cargo tanks and their appurtenances and sets forth certain matters pertaining to tank vehicles.
NFPA 471—Recommended Practice for Responding to Hazardous Materials Incidents	2002	Outlines the minimum requirements that should be considered when dealing with responses to hazardous materials incidents and specifies operating guidelines for responding to hazardous materials incidents.
NFPA 472—Standard for Professional Competence of Responders to Hazardous Materials Incidents	2002	Covers the requirements for first responder, hazardous materials technician, and hazardous materials specialist.

(continued overleaf)

TABLE 2.4   (*continued*)

NFPA Number and Title	Date of Most Recent Edition	Description of Content
NFPA 473—Standard for Competencies for EMS Personnel Responding to Hazardous Materials Incidents	2002	Identifies the levels of competence required of EMS personnel who respond to hazardous materials incidents. It specifically covers the requirements for basic life support and advanced life support personnel in the prehospital setting.
NFPA 704—Standard System for the Identification of the Hazards of Materials for Emergency Response	2001	Applies to facilities for the manufacturing, storage, or use of hazardous materials. It is concerned with the health, fire, reactivity, and other related hazards created by short-term exposure as might be encountered under fire or related emergency conditions.
NFPA 801—Standard for Fire Protection for Facilities Handling Radioactive Materials	2003	Deals with practices aimed at reducing the risks of fires and explosions at facilities handling radioactive materials and provides requirements for personnel responsible for the design or operation of facilities that involve the storage, handling, or use of radioactive materials.
NFPA 1404—Standard for Fire Service Respiratory Protection Training	2002	Contains minimum requirements for a fire service training program in respiratory protection. These requirements are applicable to organizations providing fire suppression, fire training, rescue and respiratory protection equipment training, and other emergency services including public, military, and private fire departments and fire brigades.
NFPA 1500—Standard on Fire Department Occupational Safety and Health Program	2002	Covers minimum requirements for fire service-related occupational safety and health.
NFPA 1521—Standard for Fire Department Safety Officer	2002	Contains minimum requirements for the assignment, duties, and responsibilities of a safety officer for a fire department or other fire service organization.
NFPA 1561—Standard on Emergency Services Incident Management System	2002	Covers minimum requirements for an incident management system to be used by fire departments to manage all emergency incidents.

TABLE 2.4   (*continued*)

NFPA Number and Title	Date of Most Recent Edition	Description of Content
NFPA 1852—Standard on Selection, Care, and Maintenance of Open-Circuit SCBA	2002	Specifies minimum requirements for the selection, care, and maintenance of open-circuit self-contained breathing apparatus (SCBA) that are used in fire fighting, rescue, and other hazardous duties and that are compliant with NFPA 1981, Standard on Open-Circuit SCBAs for Fire and Emergency Services.
NFPA 1951—Standard on Protective Ensemble for USAR Operations	2001	Applies to the design, manufacturing, and certification of new protective ensembles or new individual elements of the protective ensemble.
NFPA 1981—Standard on Open-Circuit Self-Contained Breathing Apparatus for Fire and Emergency Services	2002	Covers minimum documentation, design criteria, performance criteria, test methods, and certification for open-circuit self-contained breathing apparatus (SCBA) used in fire fighting, rescue, and other hazardous duties.
NFPA 1989—Standard on Breathing Air Quality for Fire and Emergency Services Respiratory Protection	2003	Specifies the minimum requirements for breathing air quality for fire and emergency services organizations that use atmosphere-supplying respirators.
NFPA 1991—Standard on Vapor-Protective Ensembles for Hazardous Materials Emergencies	2000	Covers design criteria, performance criteria, and test methods for vapor-protective suits designed to protect emergency response personnel against exposure to specified chemicals in vapor and liquid splash environments during hazardous chemical emergencies.
NFPA 1992—Standard on Liquid Splash-Protective Ensembles and Clothing for Hazardous Materials Emergencies	2000	Covers design criteria, performance criteria, and test methods for liquid splash-protective suits designed to protect emergency response personnel against exposure to specified chemicals in liquid splash environments during hazardous chemical emergencies.
NFPA 1994—Standard on Protective Ensembles for Chemical/Biological Terrorism Incidents	2001	Specifies minimum design, performance and documentation requirements, and test methods for protective ensembles for personnel responding to incidents involving the release of dual-use industrial chemicals, chemical warfare agents, or biological warfare agents.

NFPA 472 is entitled the *Standard for Professional Competence of Responders to Hazardous Materials Incidents* (NFPA 472, 2002). This standard identifies the required hazmat training levels and their respective required competencies. This standard is intended to reduce and ultimately prevent hazmat exposure, injury, and death of responders in all phases of hazmat response. NFPA also developed NFPA 473, entitled the *Standard for Competencies for EMS Personnel Responding to Hazardous Materials Incidents* (NFPA 473, 2002). NFPA 473 provides training and procedural criteria for EMS responders (at different levels of emergency medical qualifications) when responding to hazmat incidents. See Chapter 3 for a discussion of planning elements listed in NFPA 472 and 473.

Other important NFPA "all hazard" documents applicable to hazmat incidents are the following NFPA standards:

- *NFPA 1500, 2002 Edition—Standard on Fire Department Occupational Safety and Health*
- *NFPA 1201, 2000 Edition—Standard for Developing Fire Protection Services for the Public*
- *NFPA 1561, 2002 Edition—Standard for Fire Department Emergency Management Systems*
- *NFPA 1521, 2002 Edition—Standard for Fire Department Safety Officer*

NFPA 1500 provides further detail supporting the generalized health and safety requirements of the emergency response paragraph of OSHA's HAZWOPER standard [29 CFR 1910.120(q)]. Among other considerations in NFPA 1201, Chapter 10 provides guidance for the implementation of the Incident Management System (IMS) as well as on-scene command considerations. NFPA 1561 and NFPA 1521 can assist in complying with specific requirements of HAZWOPER mandating the use of the Incident Command System (ICS) also known as the IMS and the designation of the Safety Officer [29 CFR 1910.120(q)(3)(i) and (vii)].

Throughout this text, specific NFPA consensus standards are referenced as resources for understanding of the applicable topic and serve as a reference for further information.

THE OSHA HAZWOPER STANDARD

With passage of the Superfund Amendment and Reauthorization Act (SARA) of 1986, the federal government addressed a variety of concerns related to hazardous waste operations and preparedness for emergency response to accidental hazardous substance releases. One major concern addressed was for the safety and health of personnel involved in these highly hazardous types of operations. As discussed above, through SARA Title I, OSHA was directed to develop and put into force a standard designed specifically to protect the safety and health of workers engaged in these types of operations. Under the authority of SARA Title I, OSHA developed and implemented 29 CFR 1910.120, which is entitled "Hazardous

Waste Operations and Emergency Response" and is sometimes referred to as "HAZWOPER."

Purpose and Applicability

The OSHA regulations contained in 29 CFR 1910.120 are intended to protect personnel engaging in various activities involved in hazardous waste and emergency response operations. The general types of operations covered by the standard include the following:

- Cleanup of uncontrolled hazardous waste sites
- Work at RCRA TSD facilities (i.e., facilities involved in treating, storing, and disposing of hazardous wastes in accordance with the requirements of the Resource Conservation and Recovery Act)
- Hazmat emergency response operations

Applicability to Emergency Response Operations

Given the scope of this text, our coverage of 29 CFR 1910.120 will focus solely on the requirements and provisions of paragraph q. This is the part of the standard that applies to emergency response operations at locations such as industrial facilities, chemical plants, and transportation corridors. The provisions of the standard included here apply to emergency response activities intended to stop or prevent an accidental release of a hazardous substance or "material." These requirements apply in all situations involving a response by employees from outside the immediate release area (or other designated responders) to an occurrence that results, or is likely to result, in an uncontrolled release of a hazardous substance. If a release is small enough to be handled with personnel and equipment routinely located in the immediate area of the release, the event is treated as an incidental release, rather than an emergency, and the standard does not apply.

It should be noted that employers are exempt from the requirements of 29 CFR 1910.120 (q) if an emergency action plan (EAP) in accordance with 29 CFR 1910.38 has been developed and implemented by the employer to ensure the safety of employees should an incident occur (see Chapter 3).

In states without an OSHA-approved state occupational safety and health plan (Fig. 2.5), OSHA protection does not extend to employees of state and local governmental agencies. For this reason, employees of agencies such as law enforcement and fire service organizations within these states are not directly covered under 29 CFR 1910.120. To close this loophole, EPA was directed under SARA Title I to issue regulations covering these types of employees. Toward that end, EPA developed a standard (40 CFR 311) that incorporates 29 CFR 1910.120 in its entirety by reference. The EPA standard also covers unpaid state and local employees, such as members of volunteer fire service organizations. Thus the provisions of 29 CFR 1910.120 (q), as described in the following section of this chapter, are applicable to personnel engaging in emergency response to hazardous substance releases in both the private and public employment sectors.

Provisions of 29 CFR 1910.120 Applicable to Emergency Response Operations

Emergency Response Plan

The Emergency Response Plan (ERP) is the preemergency planning requirement of the HAZWOPER standard. The ERP is intended to address response concerns and activities for all likely hazmat releases that emergency responders may encounter. The OSHA has allowed facilities required to develop an ERP the option of integrating the plan into an ICP. The OSHA's ERP requirements and the National Response Teams ICP format are discussed in Chapter 3 of this text.

Emergency Response Chain of Command and Personnel Roles

The Incident Command System. Under 29 CFR 1910.120(q), the Incident Command System (ICS) is required to be used in managing all hazardous material incident responses (see Chapter 4). The senior emergency response official responding to an emergency will become the individual in charge of the site-specific ICS. This individual will control and coordinate emergency response activities and communications. Command will be passed up a preestablished line of authority as personnel or officials having greater emergency response seniority arrive.

The individual in charge of the ICS must identify hazards present and address the following considerations:

- Site analysis
- Use of engineering controls
- Maximum exposure limits
- Hazardous substance handling procedures
- Use of any new technologies

The individual in charge of the ICS must implement appropriate emergency operations based on incident-specific conditions. This person must ensure that personal protective equipment worn is appropriate for the hazards present.

Designated Safety Official. The person in charge of the ICS must designate a safety official who is knowledgeable in the operations involved. The safety official must identify and evaluate hazards and provide direction with respect to the safety of the operations involved.

The safety official has the authority to alter, suspend, or terminate any activities that, according to his or her judgment, involve an IDLH condition and/or an imminent danger condition. The safety official shall immediately inform the IC of any actions needed to lessen the hazards involved.

Skilled Support Personnel. Skilled support personnel are skilled in operating equipment such as earth movers, excavators, and cranes, which may be needed temporarily for immediate emergency support work that cannot be reasonably performed by the employer's regular response personnel. Skilled support personnel may be exposed to hazards during emergency response operations but are not required to be trained as the employer's regular response personnel are. However, skilled support personnel must be briefed before becoming involved in an emergency response. The initial briefing must include the following:

- Use of PPE
- Chemical hazards involved
- Duties to be performed

Skilled support personnel are entitled to all safety and health precautions provided to the employer's regular responders except training.

Specialist Employees. Specialist employees work with specific hazardous substances in the course of their regular job duties and are knowledgeable in the hazards of those substances. These employees may be called on to provide assistance to the individual in charge of response to a hazardous substance release. For example, if two chemicals are accidentally mixed in an incident, a chemist may be called on to predict the reactions that will occur. No specific training requirements for specialist employees are listed in the standard. However, specialist employees are required to receive training appropriate for their area of specialization, or demonstrate competency in their area of specialization, annually.

Specific Personnel Roles. The standard divides emergency response roles into several categories (Fig. 2.7). Responders must be trained for the specific roles that they are expected to perform in the event of an emergency. Specific emergency response roles are described in the section on training.

Emergency Response Procedures

Operations in hazardous areas must be performed in groups of two or more, using the buddy system. The individual in charge of the ICS must not allow anyone except personnel actively performing emergency operations to occupy any area of potential exposure during an emergency.

During emergency operations, back-up personnel must remain on standby with appropriate equipment in case the primary responders require assistance or rescue. Advance first aid support personnel must stand by with medical equipment and transportation available.

After the termination of emergency operations, the person in charge of the ICS must implement appropriate decontamination procedures. All personnel and equipment must be fully decontaminated before leaving the operations area.

Training

Training provided to each responder must be specific to that responder's intended role within the emergency response organization (Fig. 2.7). This training must be completed before the employee is allowed to participate in a response operation.

Figure 2.7   Emergency response roles/certification levels under 29 CFR 1910.120.

Training for emergency response will be provided according to the categories discussed below.

First Responder Awareness Level.   First responders at the awareness level are personnel who are likely to discover or witness a hazardous substance emergency. The most important duty of these personnel is to make proper notification to begin the emergency response sequence. The response role of these personnel should involve no potential for exposure to hazards related to an incident. First responders at the awareness level must be sufficiently trained and/or experienced to demonstrate competency in the following areas:

- An understanding of what hazardous materials are and the risks associated with them
- An understanding of the potential outcomes associated with an emergency involving hazardous materials
- The ability to recognize hazardous materials present in an emergency
- The ability to identify hazardous materials involved in an emergency, if possible
- An understanding of the role of first responder awareness-level personnel in the employer's emergency response plan, including site security and control
- The ability to use the *Emergency Response Guidebook*
- The ability to realize that additional resources are needed for the response operation and to make appropriate notifications to the emergency response communication center

First Responder Operations Level. First responders at the operations level are personnel who are involved in an initial response for the purpose of protecting people, property, and the environment from hazardous substances. These personnel are trained to respond defensively, instead of actually trying to stop the release at the source. This type of response involves working at some distance from the actual point of release to control the released material, keep it from spreading, and prevent exposures. First responders at the operations level must have awareness-level competency and receive at least 8 hours of training, or have sufficient experience to demonstrate competency, in the following areas:

- Knowledge of basic hazard and risk assessment techniques
- Ability to select and use PPE provided to the first responder operational level
- Understanding of basic hazardous materials terms
- Ability to perform basic control, containment, and/or confinement operations within the capabilities of available resources and PPE
- Ability to implement basic decontamination procedures
- Understanding of relevant standard operating procedures and termination procedures

Hazardous Materials Technician. Hazardous materials technicians are personnel who respond to an emergency to stop or prevent a release of hazardous substances. These personnel assume an offensive emergency response role as they approach the point of release and plug, patch, or otherwise stop the release at the source. Hazardous materials technicians must receive at least 24 hours of training incorporating all objectives required for first responder operations-level training. In addition, hazardous materials technicians must be able to demonstrate competency in the following areas:

- Knowledge required to implement the employer's ERP
- Knowledge in the use of field survey instruments and equipment to identify, classify, and verify known and unknown materials
- Ability to function within an assigned role in the Incident Command System
- Knowledge required to select and use proper specialized chemical personal protective equipment provided for emergency response
- Understanding of hazard and risk assessment techniques
- Ability to perform advance control, containment, and/or confinement operations within the capabilities of available resources and protective equipment
- Ability to understand and implement decontamination procedures
- Understanding of termination procedures
- Understanding of basic chemical and toxicological terminology and behavior

Hazardous Materials Specialist. Hazardous materials specialists are personnel who respond with and provide support for hazardous materials technicians. Their duties parallel those of the hazardous materials technicians but require a more specific knowledge of the hazardous substances involved. The hazardous materials specialist also acts as site liaison with federal, state, and/or local government authorities. Hazardous materials specialists must receive at least 24 hours of training incorporating all objectives required for hazardous materials technician training. In addition, the hazardous materials specialist must be able to demonstrate competency in the following areas:

- Knowledge required to implement the local emergency response plan
- Understanding of the use of advanced survey instruments and equipment in the classification, identification, and verification of known and unknown materials
- Knowledge of the state emergency response plan
- Ability to select and use proper specialized chemical PPE provided for emergency response
- Understanding of in-depth hazard and risk assessment techniques
- Ability to perform specialized control, containment, and/or confinement operations within the capabilities of the available resources and protective equipment
- Ability to determine and implement decontamination procedures
- Understanding of chemical, radiological, and toxicological terminology and behavior

On-Scene Incident Commander. On-scene incident commanders are personnel who will assume control of the incident scene beyond the first responder awareness level. Incident commanders will receive a minimum of 24 hours of training incorporating all objectives required for first responder operations-level training. In addition, the on-scene incident commander must be able to demonstrate competency in the following areas:

- Knowledge and ability required to implement the employer's incident command system
- Knowledge required to implement the employer's emergency response plan
- Knowledge and understanding of hazards and risks associated with working in CPC
- Knowledge required to implement the local ERP
- Knowledge of the state ERP and the Federal Regional Response Team
- Knowledge and understanding of the importance of decontamination procedures

Other Considerations for Training. Personnel who receive training as required by the standard must be provided with annual refresher training, or else demonstrate competency in the areas covered by training on an annual basis.

Under 29 CFR 1910.120 it is the employer's responsibility to document that the employee has received training as required for the employee's emergency response role. For first responders at the awareness and operations levels, training requirements can be considered satisfied by an employee's previous experience, provided the employee is able objectively to demonstrate competency in the areas required to be covered by training. If a statement of competency is made, the employer must keep a record of the methodology used to demonstrate competency.

Medical Surveillance and Consultation

OSHA 1910.120 requires that all hazardous materials technicians who are members of organized and designated hazmat teams, and all hazardous materials specialists, receive a baseline physical and be placed under a program of medical surveillance. The standard does not require that first responders at the awareness and operations levels be provided with coverage under a formal medical monitoring program. However, these employees must be provided with medical examinations if they are injured as a result of exposure during an emergency incident or experience symptoms that may be related to exposure. Minimum requirements for medical surveillance programs are presented here.

The standard includes specific requirements for medical examinations performed under the medical surveillance program. These medical examinations must be conducted as follows:

- Before assignment to a hazmat team
- At least annually during hazmat team membership, unless the attending physician believes that some longer interval, not to exceed two years, is appropriate
- After overexposure or the appearance of potentially exposure-related symptoms
- Whenever deemed necessary by the physician
- At the time of reassignment to an area or job that does not require medical surveillance
- At the time of termination

No termination or reassignment examination is required if an employee has received a complete examination within six months before the time of termination or job reassignment and has had no significant exposures or potentially exposure-related symptoms since the exam.

Medical examinations must be performed at no cost or loss of pay to the employee, at a reasonable time and place, and by, or under the direct supervision of, a licensed physician.

The specific content or focus of the medical examination must be determined by the examining physician based on conditions expected to be encountered in the course of response operations. Exams must include a complete or updated medical and work history and must focus on any symptoms that may be related to chemical exposure. Fitness for duty under site conditions (such as use of required PPE under expected temperature extremes) should be emphasized.

Under 29 CFR 1910.120, an employee covered by the medical surveillance program is entitled to receive a written physician's opinion. The results of specific exams and tests will be included if requested by the employee. The opinion must state any medical conditions that require treatment or place the employee at greater risk due to expected hazards and duties. Any recommended work assignment limitations will also be included in the written physician's opinion.

The standard mandates confidentiality of medical examination results. Therefore, specific findings unrelated to occupational exposure cannot be revealed by the examining physician to the employer.

Personnel Protective Equipment
The standard requires that personal protective equipment (PPE) be used as necessary to protect responders from hazards encountered during emergency operations. In addition to conventional safety hazards, employees must be protected from exposure to hazardous substances in excess of applicable exposure limits (see Chapter 12).

Selection and Use of PPE. Selection of PPE must be based on incident-specific conditions and updated as those conditions change or additional information is generated about the incident. A written PPE program is required, and it must incorporate, at a minimum, the following topics:

- Selection of PPE
- Use and limitations of PPE
- Work mission duration
- Maintenance protocols
- Storage procedures
- Decontamination and disposal procedures
- Training and proper fitting
- Donning and doffing
- Inspection procedures
- Limitations during temperature extremes
- Program evaluation

Specific Requirements for PPE. Emergency response personnel exposed to hazardous substances representing a potential inhalation hazard shall wear positive-pressure self-contained breathing apparatus (SCBA) until the individual in charge of the ICS determines through air monitoring that a decreased level of respiratory protection is appropriate. When deemed necessary, approved SCBAs may be used with approved cylinders from other manufacturer's SCBAs provided the cylinders are of the same capacity and pressure rating. All SCBA cylinders must meet applicable DOT and NIOSH requirements.

Personnel must be provided with one of the following methods of respiratory protection for work in Immediately Dangerous to Life and Health (IDLH) atmospheres:

- Positive-pressure SCBA fitted with a full facepiece
- Positive-pressure air-line respirator fitted with a full facepiece and an escape air supply

For operations involving hazards to the skin that may result in an IDLH situation, totally-encapsulating chemical-protective (TECP) suits must be used. These suits must be capable of maintaining a positive internal pressure and resistant to inward gas leakage. Methods to be used in testing TECP suits are described in Appendix A of the standard.

Postemergency Response Operations

After the completion of emergency response operations, it may be necessary to remove hazardous substances, health hazards, and/or contaminated materials (such as contaminated soil) from the site. If so, employers engaging in such activities must comply with all requirements of the standard applicable to hazardous waste site cleanup operations. However, if cleanup is performed on plant property by plant employees, the standard does not apply. In this case, the employees involved should receive training as required in the following:

- 29 CFR 1910.38—Emergency Action Plans
- 29 CFR 1910.134—Respiratory Protection
- 29 CFR 1910.1200—Hazard Communication

Also, any additional safety and health training required by the tasks the employees are required to perform, such as use of PPE and decontamination, shall be provided.

All equipment used in postemergency cleanup must be in serviceable condition. This equipment must be thoroughly inspected before being used.

SUMMARY

Responding to a hazardous material incident is an inherently dangerous job. However, the hazards involved can be minimized if the personnel required to respond are provided with appropriate training and equipment and appropriate procedures are followed during response operations. In addressing the hazards inherent in emergency response operations, the federal government has created policies requiring the employer to provide for the protection of the environment, public health and welfare, and safety and health of response personnel. Consensus standards and recommendations from nonregulatory agencies also provide guidelines and procedures that can be followed to help ensure the safety and health of hazmat emergency responders. It is therefore important for everyone involved in emergency response to be familiar with these policies in order to ensure that they are not violated.

3

PLANNING FOR HAZARDOUS MATERIALS RESPONSE

INTRODUCTION

Planning for hazardous chemical releases has long been a concern of many different public agencies, citizen groups, and industries. Before the mid-1980s, the United States had no official policy on state and local preemergency planning requirements for hazardous materials (hazmat) releases. Up to this point, the only policy in effect was the National Oil and Hazardous Substances Pollution Contingency Plan (NCP) under the Clean Water Act (CWA) and the Comprehensive Environmental Response, Compensation, and Liability Act (CERCLA) aka "Superfund Act." This policy outlined the way in which the federal government would assist state and local governments in response to chemical emergencies, but it provided no specific regulations for the state and local governments to operate with in such an event. As the name implies, CERCLA provides for response, financing, and accountability of toxic waste cleanup efforts by the federal government at hazardous materials disposal and release sites. Other environmental, transportation, and occupational health and safety laws addressed specific issues but stayed away from emergency planning at state and local government levels. This changed with several key industrial events.

Three Mile Island, Pennsylvania 1979—The near catastrophic nuclear core reactor failure of the power plant threatened the entire east coast of the United States.

Bhopal, India 1984—A lack of maintenance, missing safety equipment, and a leaking valve in Union Carbide's Bhopal, India, pesticide plant resulted in the

Emergency Responder Training Manual for the Hazardous Materials Technician, Second Edition, edited by Kenneth W. Oldfield
ISBN 0-471-21387-X   Copyright © 2005 John Wiley & Sons, Inc.

release of 45 tons of methyl isocyanate. The number of people killed has been estimated at 3500, and the number of people injured is estimated at 300,000. If information on chemical inventories had been made to emergency responders, the casualty toll could have been greatly reduced, or better yet, the incident might have been avoided completely.

Institute, West Virginia 1985—A mishap at Union Carbide's pesticide plant in Institute, West Virginia, resulted in the release of aldicarb oxime pesticide. With little planning information known of the chemical's potential to spread and cause injuries, the company underestimated the migration of the vapor plume. The release injured 135 people in the surrounding area.

Something had to be done!

EMERGENCY PLANNING AND COMMUNITY RIGHT-TO-KNOW

By 1986 the recent chemical releases in India and West Virginia had brought the concern over emergency planning to a focal point. In 1986, Congress reauthorized and made additions to CERCLA in new legislation that is now known as the Superfund Amendments and Reauthorization Act (SARA).

History

In anticipation of SARA legislation, the United States Environmental Protection Agency (EPA) established the Chemical Emergency Preparedness Program (CEPP) as a voluntary program to assist local planning groups. The focus of this program was the recognition, identification, and preparation/planning for releases of acutely hazardous materials. This document compiled a listing of acutely hazardous materials, methods for gathering data, and assessing the use of materials in a local area. The CEPP also prompted industries to make information on chemical inventories available to the public that could be useful in planning for accidental releases.

Chemical companies anticipated the federal government's passing of a formal law similar to the CEPP. Some companies used their knowledge of chemical safety to initiate preparedness and response programs. One of the larger programs was the Chemical Manufacturers Association (CMA) Chemical Awareness and Emergency Response Program (CAER), which allowed for the release of information to local authorities to assist in planning. Because programs like CAER were strictly voluntary, chemical safety information was not always complete or fully accurate. This would require the passage of formal federal law.

Superfund Amendments and Reauthorization Act Title III (SARA Title III)

In 1986 Congress enacted the Superfund Amendments and Reauthorization Act Title III (SARA Title III), also known as Emergency Planning and Community Right-to-Know Act (EPCRA). SARA Title III adopted information and suggestions from the CEPP and mandated the establishment of a chemical emergency planning system at the state and local government level. Along with CEPP policies and guidance, EPA now had enforcement authority. The federal government's intent for SARA Title III was to require chemical manufacturers and industry to provide information for emergency planners and responders by raising the awareness of chemical hazards and chemical safety in their local community.

SARA Title III or EPCRA has two major objectives:

(1) Emergency planning for responding to chemical emergencies
(2) Provisions for emergency release notification/hazard communication to the public

The sections in Subtitles A, B, and C of SARA Title III outline the structure of state and local planning authorities as well as how chemical information is to be collected, compiled, and made available to the public. The regulations of SARA Title III legislation may be found in their entirety in 40 CFR Parts 350–372 (Table 3.1).

Framework of SARA Title III

SARA Title III requires that each state must establish a State Emergency Response Commission (SERC). This commission is required to establish local planning districts and direct and maintain a network of Local Emergency Planning Committees (LEPC).

The SERC consists of the state governor and a commissioned group of state governmental departments that would have an active part in planning or responding to a chemical release. Each SERC structure is different, but typically it consists of the governor and state environmental protection, transportation, health, and/or safety departments. Each department may handle a specific section of

TABLE 3.1   The Legislative Contents of SARA Title III

Subtitle A "Community Planning"	Sections 301–305 address state and local planning requirements; substances and facilities covered, notification; response plans; release notification; training; emergency systems
Subtitle B "Community Hazard Communication" or "Community Right-to-Know"	Sections 311–313 address MSDS; chemical inventory requirements; Toxic Release Inventory [TRI]
Subtitle C "Legal Authorizations/ Enforcement"	Sections 321–330 address other laws; trade secrets; provisions for health care providers; public availability of notices and information; enforcement; civil actions; transportation; regulations; definitions; authorizations

SARA Title III, or one department may handle all aspects of the regulations. If a SERC is not appointed, then the governor will serve as the SERC until one is developed. The relationship that EPA has with Native American governments concerning SARA Title III structuring is very similar to that of state governments. Native American governments are required to establish a Tribal Emergency Response Commission (TERC) on federally recognized tribal reservations. Tribal executive officers are required to implement SARA Title III regulations on all tribal lands. The deadline for establishing SERCs/TERCs was set for April 17, 1987. The goals for SERCs are to collect chemical inventories of SARA Title III Extremely Hazardous Substances (EHSs) and CERCLA Hazardous Substances (HSs) and other hazardous chemicals that pose unacceptable risks when stored or transported above planning quantities. Reporting requirements for releases of these chemicals and making this information available to the public for planning uses are also outlined. These listings of chemicals are discussed in detail later in this chapter.

Once the SERCs were developed, they were required to establish planning districts that suited their own local needs. Different states used different methods for distinguishing district boundaries. Some states used county divisional lines, whereas some used geographic separation. The management of these planning districts is not the responsibility of the SERCs but the responsibility of the LEPCs. LEPC members are from various sources of local government, private industry, and other agencies. At a minimum, agencies must be from the following areas:

- Elected state and local officials
- Law enforcement
- Fire and rescue departments
- Civil defense
- Health care and public health professionals
- Environmental groups
- Media
- Hospitals
- Transportation
- Industrial facilities that use hazardous materials and are required to comply with SARA Title III

Some of the organizations' purposes at first glance may seem irrelevant to planning, but groups like media and transportation may be helpful with planning. They could answer a planning question such as: "How do you notify 5000 people to evacuate and appropriate vehicle traffic routes for escape?"

The primary objective of the LEPC is to establish and maintain a local emergency plan specific to the needs of the community. The LEPC will address issues of evaluating resources required to plan and respond to a chemical release. This information should include the following:

- Identification of facilities and chemicals
- Identification of facility owner/operator and/or coordinator
- Extremely hazardous substances transportation routes
- Response procedures specific to locations
- Emergency notification procedures
- Methods for determining releases and probable outcomes
- Contingency plans for evacuation and resource acquisition
- Training for response personnel
- Scheduling simulated release exercises to evaluate the plan

Notification of Chemical Use and Storage

Some of the main duties of a LEPC are to anticipate, recognize, identify, and plan for hazardous material releases. Because the emergency response plan needs to be current, the information used to formulate the plan should be current as well. This is accomplished by facilities completing forms known as Tier 1 or Tier 2 forms and filing the reports with the LEPC. SARA Title III requires facilities to evaluate their inventories and compare them to a list found in SARA Title III (40 CFR 355 Appendix A): "The List of Extremely Hazardous Substances (EHS)." If a facility has extremely hazardous substances stored over certain limits known as Threshold Planning Quantities (TPQ) it is required to provide the LEPC with the amount, storage locations, and appropriate Material Safety Data Sheet (MSDS). The TPQs are determined by EPA based on toxicity, particle size (solids), and airborne mobility of the material in the event of a release. The EHS list should not be considered a complete list of hazardous substances but rather a list of hazardous substances that if released in small amounts can pose an extremely dangerous threat to human health and environment. Even if a chemical is not on the EHS list, but poses a great enough threat and is stored in large enough amounts, it must be considered during emergency planning. The governor or SERC, after public comment, can write to the facility and require them to report this information to the LEPC. With the information provided by facilities, the LEPC can move to formulate or revise an emergency response plan. This information is critical in determining resources, strategy, and tactics that will be required for response to a release.

Emergency Notification

A major goal of SARA Title III, CERCLA, and other environmental laws related to the NCP is emergency release notification. According to CERCLA (40 CFR 302.6) a hazardous substance release over the reportable quantity (RQ) listed in

40 CFR Table 302.4 requires immediate notification by the "person in charge" to the National Response Center at 800/424-8802. In addition to this notification requirement, SARA Title III (40 CFR 355.4) requires immediate notification to be made by the owner or operator of a facility or vehicle for releases of an EHS listed in 40 CFR 355 Appendix A and releases of a CERCLA HS over their respective RQs. This additional notification requirement is to be communicated to the LEPC and/or SERC for fixed facility releases or to the 911 emergency operator or the telephone operator for transportation releases. In either case, it is recommended that an owner, operator, or "person in charge" go ahead and initiate this process as soon as possible rather than waiting. This is especially important when the total volume released is not known but potentially over the RQ of a HS or EHS. This will initiate the process of immediate notification. The initial report should include the following:

- Chemical Name and CAS Registration Number
- Whether it is listed as an Extremely Hazardous Substance
- Estimate of the quantity released
- Time and duration of the release
- Whether the release was to land, air, water, or any combination thereof
- Known or anticipated acute or chronic health effects
- Advice for medical attention for exposed individuals
- Proper precautions (evacuation or shelter in place)
- Name and number of contact person

As discussed in the *Hazardous Materials Management Desk Reference* (Cox, 2000) and listed in 40 CFR 355.40(b)(3), in addition to the information given in the initial notifications required, written follow-up reports should be made to the LEPC and SERC. These additional reports should be made "as soon as practicable." These written reports must contain the following information:

- Updated information that was in the initial report
- Information on response actions taken
- Known or anticipated acute or chronic health risks
- If appropriate, advice regarding medical attention for exposed individuals

Toxic Chemical Release Reporting (TRI Reporting)

EPA felt that the best way to control or minimize permitted everyday releases of hazardous substance from industrial processes is to inform the public of what is being released in their community. Therefore, in addition to emergency planning and reporting, SARA Title III required facilities to report on routine releases with the use of a standardized document known as "Form R." Chemicals found in CERCLA's and EPCRA's hazardous substances lists that are released in "low" permitted levels are to be reported to the EPA or state environmental department and made available to the public for planning and, more importantly, community right-to-know information. The public can gain access to the National Toxic Chemical Release Inventory (TRI) information by written requests sent to their state's environmental protection agency or EPA. Another source of TRI data is the Scorecard Database at www.scorecard.org.

What SARA Legislation Means to Responders

Access to Information

Most large-scale fixed facilities working with hazardous substances have in-house expertise or specially trained first responders familiar with their chemical inventories and where they are stored. However, for public emergency responders to have this type of understanding of surrounding facilities and transportation information, they must be informed of the location, type, and quantities of potential chemical hazards. Under SARA Title III regulations EPA made this information accessible to the responder. Along with the reports filled by facilities to the SERC/LEPC, responders have the additional right to request specific information on chemicals, facility design, facility operation procedures, and even site-assessment inspections. This information can be requested through the SERC/LEPC. Having this type of knowledge of chemical inventories helps responders to plan for staffing, equipment, training, and practice needs.

CAMEO® Computer Program System

With all the information available, public emergency response agencies found it necessary to compile the data learned about specific chemicals in a usable format. There are many sources and computer systems with this type of information, but one of the more popular means was a direct result of SARA Title III. The Computer-Aided Management of Emergency Operations system or CAMEO® was a project initially worked on by the National Oceanic and Atmospheric Administration (NOAA) with assistance from EPA to list reported chemical inventories for a given community.

CAMEO® is a multifunctional, multipurpose computer program system with twelve informational modules and three software applications. The twelve informational modules that CAMEO® integrates and utilizes are shown in Figure 3.1. The three integrated computer software applications that comprise the CAMEO® system or suite are CAMEO®, MARPLOT®, and ALOHA® (Fig. 3.2).

CAMEO® Application and Database. The CAMEO® application is a computer software application containing a database of hazardous chemicals and planning data that is integrated with a search engine to assist planners and on-scene

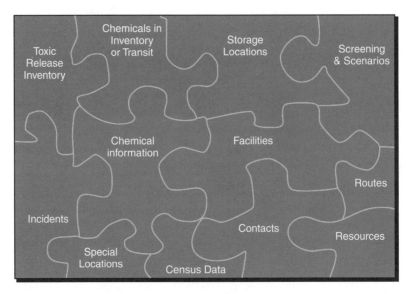

Figure 3.1 The 12 modules that the CAMEO® system integrates and utilizes for providing planners and responders with information.

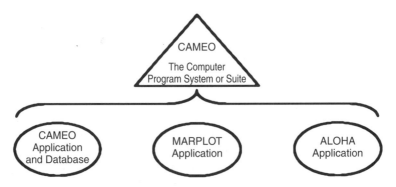

Figure 3.2 The CAMEO® computer program system or suite consists of CAMEO®, MARPLOT®, and ALOHA® applications.

responders. Each chemical entry in the database includes chemical identification information, regulatory information, and Response Information Data Sheet (RIDS) (See Chapter 6). The RIDSs are searchable by the following topics:

- General description
- Properties
- Health hazards
- First aid

- Fire hazards
- Fire fighting
- Protective clothing
- Nonfire response

MARPLOT®. Mapping Applications for Response, Planning, and Local Operational Tasks or MARPLOT® is a mapping application within the CAMEO® system that displays selected hazardous material information and vulnerabilities over base maps. Maps and population data are based on information provided by the U.S. Bureau of the Census.

ALOHA®. The Areal Locations of Hazardous Atmospheres or ALOHA® application is intended to calculate atmospheric dispersion models of chemical vapor clouds based on the following:

- Toxicological or physical characteristics of the hazardous material
- Atmospheric conditions
- Other incident-specific circumstances

Information provided is graphically plotted with the use of MARPLOT® so planners and on-scene responders can estimate the likely plume migration and vulnerable areas.

The integrated chemical data and all three applications make the CAMEO® system an extremely powerful tool for planners and on-scene responders. The CAMEO® system is available on request through the EPA website free of charge.

OSHA Worker Protection
As discussed in Chapter 2 of this text, when legislation for SARA was passed, the U.S. Congress directed OSHA with SARA Title I legislation to develop worker and responder safety standards. In response OSHA promulgated the Hazardous Waste Operations and Emergency Response standard (29 CFR 1910.120) to ensure the protection of emergency workers who respond to hazardous material releases, as well as hazardous waste site and hazardous waste facility workers.

DEVELOPING THE PLAN

As discussed in the National Response Team's *Hazardous Materials Emergency Planning Guide* (NRT, 2001), multiple planning requirements are established by EPA, OSHA, and other federal legislation. In addition to these, there are additional planning regulations and consensus standards. The remaining portion of this chapter discusses defining the appropriate hazmat response plan, considerations for plan development, and exercising the plan.

Defining the Plan

The first step in planning is the recognition by the organization of the need for a plan and the identification of relevant planning regulations. Selection of the appropriate planning requirement(s) is based on which regulations an organization is legally required to follow or elects to adopt, that is, consensus standards. In general, all hazmat response plans have common themes: the safety and health of employees and the public, as well as environmental protection and property conservation.

The major planning regulations, recommendations, and consensus standards that employees and responders will commonly utilize are as follows:

- The Emergency Response Plan as part of 29 CFR 1910.120
- The Emergency Action Plan as described by 29 CFR 1910.38
- The Facility Contingency Plan as described by 40 CFR 265 Subpart D
- The Integrated Contingency Plan defined by the U.S. National Response Team
- National Fire Protection Association Consensus Standards 471, 472, and 473

The Emergency Response Plan

As required by 29 CFR 1910.120, an Emergency Response Plan (ERP) must be developed and implemented by the employer before the commencement of emergency response operations. This plan must be adequate for any emergencies that could reasonably be expected to occur on-site. The plan must be in writing and available to employees, employee representatives, and OSHA personnel.

The ERP must address, as a minimum, the following topics:

- Preemergency planning and coordination with outside parties
- Personnel roles, lines of authority, training, and communication
- Emergency recognition and prevention
- Safe distances and places of refuge
- Site security and control
- Evacuation routes and procedures
- Decontamination
- Emergency medical treatment and first aid
- Emergency alerting and response procedures
- Critique of response and follow-up
- PPE and emergency equipment

Emergency response organizations may use the local and/or state ERPs as part of their ERP, if applicable, to avoid duplication. Emergency response topics that are properly addressed by existing plans required under SARA Title III need not be duplicated for compliance with 29 CFR 1910.120.

The Emergency Action Plan

Most people are familiar with practicing fire or other emergency escape drills at industrial facilities. In these cases, employees and occupants of facilities are simply required to evacuate the building or area of hazard and report to a pre-designated safe area according to a set plan. This type of plan is known as an Emergency Action Plan (EAP).

An EAP is required by the OSHA under 29 CFR 1910.38 for facilities that intend to stop operations and evacuate all personnel in case of a hazardous materials release, fire, or any other type of life safety emergency. This plan describes actions to be taken by employers and employees to guarantee safety without responding to the emergency. All response operations are to be contracted out or handed off to professional response agencies such as a fire department, hazmat team, or other specialists.

EAPs must be written and available to all employees. They should also be offered to responding agencies for review and familiarization before an emergency occurs.

An EAP must address, as a minimum, the following topics:

- Written escape procedures and routes of egress for employees
- Maps and diagrams showing escape routes and areas of hazards (ignition sources, fuel stores, toxic chemical stores) to be avoided
- Written emergency process shutdown procedures for employees who cannot leave equipment and systems until they are safely stopped
- Accountability procedures for personnel to ensure that all employees have left the hazard area
- Procedures to determine whether personnel require emergency medical assistance
- Assignment of rescue and/or medical duties required of certain employees
- Reporting procedures for notifying appropriate emergency response agencies
- Names or job titles to contact for additional information on the EAP or facility operation
- Alarms
- A description of the alarm system as required by OSHA 29 CFR 1910.165
- A description of specific alarm signals used for specific threats, for example, long, continuous alarm for chemical releases or 3 short alarms for fires
- Specific evacuation procedures for all anticipated emergencies
- Training
    - Before the plan is implemented, a sufficient number of employees should be trained by the employer to assist in evacuation and accounting of employees.

- Employees should be trained when they start work initially or change job descriptions or locations in the facility.
- Retraining should occur anytime the plan is changed.

Any facility that uses hazardous materials and expects employees to evacuate the hazard area must have a written EAP. This plan is to be kept on-site and available for employee review at all times. Emergency responders should be informed of this plan. When a facility is developing an EAP, it is advisable to seek input from outside response agencies to avoid problems and confusion before an emergency occurs.

For companies guided primarily by OSHA regulations for emergency planning, a common error is to inadvertently include response actions in an EAP that go beyond evacuation. OSHA intends the EAP to address evacuation and limited shutdown procedures only. Any actions to stop a release or control a spill, particularly at the point of release, must be taken under a full ERP. Employers who assign employees response actions in an EAP may be putting them at risk and certainly are violating OSHA's regulations.

The Facility Contingency Plan

As part of the Resource Conservation and Recovery Act (RCRA) or the "Hazardous Waste Regulations," EPA regulations require owners and operators of hazardous waste interim-status or permitted treatment, storage, and disposal facilities to have a Facility Contingency Plan (RCRA Contingency Plan). The requirements of this plan are found in 40 CFR 265 Subpart D—Contingency Plan and Emergency Response. The RCRA Contingency Plan is intended to set out procedures and actions to be followed by facility personnel and outside assisting agencies. The RCRA Contingency Plan's main goal is to minimize hazards to human health and the environment from fires, releases, and/or explosions from hazardous wastes stored on-site. This plan must be carried out whenever life safety, public health, or environmental health is endangered by any such occurrences.

According to 40 CFR 265.52, a RCRA Contingency Plan must contain at least the following elements:

- Actions to be taken by facility personnel
- Arrangements made with state and local response agencies
- Names, addresses, and phone numbers of qualified emergency coordinators
- Listing, physical description, and storage locations of all emergency response equipment (i.e., fire extinguishers, decontamination equipment, etc.)
- Listing of primary and secondary escape routes for facility personnel

If the facility already has a Spill Prevention, Control, and Countermeasure Plan (SPCC) as required by 40 CFR 112 or Part 1510 of Chapter V or other acceptable spill contingency plan, the owner or operator may satisfy RCRA regulations with the incorporation of hazardous waste management procedures.

The regulations also require that copies of the RCRA Contingency Plan must be made available on-site and provided to state and local emergency response agencies. The RCRA Contingency Plan must be amended whenever regulations or contents of the plan change. Subparts 265.55 and 265.56 describe the generalized responsibilities of the emergency coordinator, such as making notifications, assessment, and reporting of an incident.

The Integrated Contingency Plan

Under the joint leadership of EPA and the U.S. Coast Guard (USCG), the National Response Team (NRT) has established a format and guidance for the development of "One Plan" in response to hazmat releases. This plan is known as the Integrated Contingency Plan (ICP). With a basic outline, matrix of applicable regulations, and appendices, facilities are now able to reduce the confusion of "Which plan applies?" during a hazmat release. Table 3.2 shows the listing of other regulations requiring the development of a chemical hazard response plan that the ICP is intended to satisfy.

ICP Overview. The ICP is intended for all types of facilities to comply with multiple federal planning regulations for response to oil, hazmat, and radioactive releases. "Facilities" are defined as fixed industrial locations (on- or off-shore), transportation vehicles, and pipeline systems. This ICP recommendation by the NRT has allowed for the consolidation of all these plans. With proper coordination with state government, local government, and local response agencies, the ICP can also satisfy their specific planning requirements. The NRT encourages state and local government planners to design contingency requirements that will mesh with the NRT-ICP format.

TABLE 3.2   The National Response Team Agencies' Planning Regulations that are Suitable for Inclusion in the Integrated Contingency Plan Format

EPA— Oil Pollution Prevention Regulation (SPCC and Facility Response Plan Requirements)	40 CFR Part 112.7(d) and 112.20-.21
EPA—Risk Management Programs Regulation	40 CFR Part 68
EPA—Resource Conservation and Recovery Act Contingency Planning Requirements	40 CFR Part 264, Subpart D, and 40 CFR Part 265 Subpart D, and 40 CFR 279.52
RSPA—Pipeline Response Plan Regulation	49 CFR Part 194
USCG—Facility Response Plan Regulation	33 CFR Part 154, Subpart F
MMS—Facility Response Plan Regulation	30 CFR Part 254
OSHA—Emergency Action Plan Regulation	29 CFR 1910.38(a)
OSHA—Process Safety Standard	29 CFR 1910.119
OSHA—HAZWOPER Regulation	29 CFR 1910.120

It is important to note that if facilities choose to do so they can still maintain multiple plans to satisfy each specific regulation, but all NRT agencies recommend the consolidation of these various plans into an ICP. When facilities elect to use an ICP format (NRT format or other acceptable format), they are not relieved of any regulatory burden. Facility planners should read fully and be intimately knowledgeable of all NRT agencies' planning requirements.

Using the NRT-ICP Format. The published notice *The National Response Team's Integrated Contingency Plan Guidance* (61 FR 28642, 1996) details how to develop an ICP utilizing a matrix or decision chart that will assist facility planners. Additional information on the ICP guidance and a copy of the publication are available by calling the RCRA, Superfund, and EPCRA Hotline at 800/424-9346.

The principal steps are as follows:

- First, the plan developers should determine which regulations from Table 3.2 apply to their facility.
- Second, it is necessary to look at each applicable regulation under the regulation heading at the top of the selection matrix (Fig. 3.3) and cross-reference it with the appropriate ICP heading.
- Third, now that the specific regulation is cited in the matrix, the plan developer will need to look up that specific citation in the appropriate regulation and incorporate the facility's information into the plan format. See Figure 3.3 for an example.

Using this means, plan developers can accurately address specific planning requirements and quickly show regulatory compliance by incorporating the decision matrix into the plan annexes.

NRT-ICP Structure. The NRT-ICP structuring philosophy is based on the use of the National Interagency Incident Management System (NIIMS) Incident Command System (ICS). This management system is also commonly referred to as the Incident Management System (IMS) developed by the National Fire Service Incident Management System Consortium (see Chapter 4). This type of management system allows for a common management practice that is the accepted standard throughout the United States. The primary goal of IMS is to ensure the safety of responders as well as the protection of the public, property, and environment. This philosophy is incorporated throughout the NRT-ICP guidance format. The NRT suggests that the ICP format is to be divided into the following three main sections:

(1) Plan Introduction Elements
(2) Core Plan Elements
(3) Annexes

ICP Elements	RCRA (40 CFR part 264, Subpart D, 40 CFR 265, Subpart D, and 40 CFR 279.52)	EPA's Oil Pollution Prevention Regulation (40 CFR part 112)	USCG-FRP (33 CFR part 154)	DOT/RSPA-FRP (49 CFR part 194)	OSHA Emergency Action Plans (29 CFR 1910.38(a)) and Process Safety (29 CFR 1910.119)	OSHA HAZWOPER (29 CFR 1910.120)	CAA RMP (40 CFR part 68)
Section I – Plan Introduction Elements							
1. Purpose and scope of plan coverage	264.51 265.51 279.52(b)(1) 264.52(a) 265.52(a) 279.52(b)(2)(i)				38(a)(1)[1] 119(n) 272(d)	(1)[2] (p)(8) (q)(1)	
2. Table of contents		112.20(h) Appendix F	1035(a)(4)[3] 1030(b)	Appendix A			
3. Current revision date		F1.2	1035(a)(6)				
4. General facility identification information		F1.2 F1.9		194.107(d)(1)(i) 194.113 194.113(b)(1)			
a. Facility name		F1.2	1035(a)(1)				
b. Owner/operator/agent		112.20(h)(2) F1.2 F2.0	1035(a)(3)	194.113(a)(1) A-1			
c. Physical address and directions		112.20(h)(2) F1.2 F2.0	1035(a)(1) 1035(a)(2) 1035(e)	194.113(a)(2) 194.113(b)(3),(4) A-1			
d. Mailing address		112.20(h)(2)	1035(a)(1)	194.113(a)(1)			
e. Other identifying information							

[1]All citations refer to part 1910 unless otherwise noted.
[2]All citations refer to 29 CFR 1910.120 unless otherwise noted.
[3]All citations refer to part 154 unless otherwise noted.
Adapted from The National Response Team's Integrated Contingency Plan Guidance (61 FR 28642, 1996)

Figure 3.3 ICP Development Matrix—an example of planning regulations and the corresponding ICP structure elements.

```
ICP Outline

I. Plan Introduction Elements
        Purpose and Scope of Plan Coverage
        Table of Contents
        Current Revision Date
        General Facility ID Information
                Facility Name
                Owner / Operator (address and phone number)
                Physical Address (city, county, latitude, longitude, minutes...)
                Mailing Address (contact person)
                Other ID Information (EPA ID Number, SIC Code...)
                Contacts for Plan Development / Maintenance
                Phone Numbers for Key Contact(s)
                Phone Number of Facility
                Facility Fax Number
```

Figure 3.4   Outline of the ICP Introduction Elements.

THE PLAN INTRODUCTION ELEMENTS section of the ICP is reserved for basic facility information, such as purpose and scope of the plan, table of contents, current revision dates, and general facility identification information. This information should be presented in a brief and factual manner and will allow for rapid identification of important facility administrative information (Fig. 3.4).

THE CORE PLAN ELEMENTS section should contain the essential response guidance and procedures that would be needed by first responders at the beginning of a response. The procedures would further include the necessary steps to initiate, conduct, and terminate incidents. This section should be kept intentionally small and concise and should be illustrated with flowcharts and checklists so that it could be used as a field operating guide or pocket guide. Information regarding hazard recognition, notification, and initial response (assessment, mobilization, and implementation) would be found in the core plan. Larger facilities with multiple hazards may find it necessary to have references to specific annexes of the plan for further information because the response action would be to a specific hazard or threat (Fig. 3.5).

The NRT suggests listing tiered response levels in the core plan to inform responders of hazard intensity, urgency, vulnerable resources, and required resources (including personnel) automatically in the initial stages of a response. Categorizing possible emergencies in this manner will help responders to rapidly assess the incident without having detailed information. This would be an appropriate subsection for the use of flowcharts or checklists. For example, a company may state that all Level 3 responses call for immediate notification of the fire department. Because this type of incident notification system should be used and properly coordinated with outside or mutual aid agencies, it is critical to use common terminology.

```
┌─────────────────────────────────────────────────────────────────┐
│ ICP Outline                                                       │
│                                                                   │
│ II.  Core Plan Elements                                           │
│          Discovery                                                │
│          Initial Response                                         │
│                 Procedures for Internal and External Notifications│
│                 Establishment of a Response Management System     │
│                 Procedures for Preliminary Assessments            │
│                        ID of Incident Type                        │
│                        Hazards Involved                           │
│                        Magnitude of Problem                       │
│                        Resources Threatened                       │
│                 Procedures for Establishing Strategies and Priorities│
│                        Tactical Priorities                        │
│                            Life Safety                            │
│                            Incident Stabilization                 │
│                            Property Protection                    │
│                        Control Actions (confinement, containment, etc.)│
│                        ID of Response Resources Needed            │
│                        Procedures for Implementation of Tactical Plan│
│                        Mobilization of Needed Resources           │
│                        Sustained Actions                          │
│                        Termination and Follow-Up Actions          │
│                                                                   │
└─────────────────────────────────────────────────────────────────┘
```

Figure 3.5   Outline of the ICP Core Elements.

THE ANNEX section of the ICP should include site-specific information that is usually not considered time sensitive. This planning data should be very detailed in nature and are useful for incident control and demonstrating regulatory compliance. The individual annexes should relate to the basic headings listed in the core plan and will detail the facilities incident command structure. The use of the IMS is the preferred management system, and the functional areas should be listed in the plan with all supporting information. The individual roles of the IMS should have names or titles of individual responders along with potential alternates and contact information. If a facility chooses not to use the IMS structure, it must detail its response management system and list the specific differences.

As seen in Figure 3.6, the annexes will also include the following information:

- Accident investigation procedures and required reports
- Training and exercise logs/schedules
- Plan critique processes
- Plan modification processes
- Spill prevention information
- Regulatory compliance

Other plans and documents may be linked by reference in the development of annexes but should be included in full when information is considered to be necessary for initial response efforts. Linkages by reference may be planning topics such as maps, chemical inventories, and external notification procedures.

ICP Outline

III. Annexes
      Annex 1 – Facility and Locality Information
          Facility Maps
          Facility Drawings
          Facility Description / Layout
               Facility Hazards
               Vulnerable Resources
               Vulnerable Populations (on / off site)
      Annex 2 – Notification
          Internal Notifications to be made
          Community Notifications to be made
          Federal Notifications to be made
      Annex 3 – Response Management System
          General
          Command Structure
          Operational Structure
          Planning Structure
          Logistic Structure
          Finance Structure
      Annex 4 – Incident Documentation
          Post Accident Investigation
          Incident History
      Annex 5 – Training and Exercise Drills
      Annex 6 – Response Critique and Plan Review / Modification Process
      Annex 7 – Accident Prevention
      Annex 8 – Regulatory Compliance and Cross Reference Matrices

Figure 3.6   Outline of the ICP Annexes.

Advantages of the NRT-ICP Format.   At first glance, using an ICP format for documenting a facility's requirements may seem overwhelming because of the wide range of regulations, but this is the most efficient plan when facilities are subject to multiple planning regulations. The time saved in developing and maintaining only one plan when compared to developing and maintaining multiple plans should be easily recognized. Another advantage to the ICP is consistency among the varying planned responses because responders would consult a single plan for multiple types of events. The ICP will help to reduce the confusion as to which plan responders should follow by charting response actions based on releases or events. The use of a single plan will help facilities demonstrate regulatory compliance and helps regulatory agencies by giving them a uniform plan format that is easily comparable to their applicable regulations. With this type of plan, facilities can address issues ranging from inclement weather events impacting operations to hazardous waste releases and everything in between.

NFPA *Hazmat Emergency Response Planning Recommendations*
As discussed in Chapter 2, the National Fire Protection Association has many consensus standards that serve as fire protection codes. These standards are also sound recommendations that municipalities and departments may incorporate as

additional emergency regulations and minimum operating standards. The three NFPA consensus standards that contain hazmat planning recommendations are NFPA 471, 472, and 473. These recommendations are extremely beneficial to the fire service as well as industrial hazmat planning efforts.

NFPA 471. NFPA 471 is the *Recommended Practice for Responding to Hazardous Materials Incidents* standard (NFPA 471, 2002). This standard establishes minimum requirements for hazmat emergency response planning as well as outlining a levels approach to hazmat response. In Chapter 4 of the standard, recommendations are given for establishing fixed facility and likely threat plans as required by SARA Title III. The standard requires the use of planning teams that review existing state and local plans to assist in writing an emergency response plan. At least annually the plan should be evaluated and necessary changes should be made. To assist in evaluations and increase responder competency in utilizing the plan, training exercises for the plan are required annually at a minimum.

Chapter 5 of the standard provides for implementation of a planning guide with several levels. Varying degrees of incident factors are cross-referenced with one of three incident or threat levels. "Incident Level One" is the least hazardous type of event and "Incident Level Three" the most hazardous type of event. Incident factors to be considered in assigning a level are listed below:

- Product identification
- Container size
- Fire/explosion potential
- Leak severity
- Life safety
- Potential environmental impact
- Container integrity

This tiered approach can be useful throughout the development of the plan when addressing issues like minimum training requirements, necessary equipment, and personal protective equipment (PPE).

NFPA 472. NFPA 472 is entitled *Standard for Professional Competence of Responders to Hazardous Materials Incidents* (NFPA 472, 2002). This standard covers the minimum competencies for multiple levels of hazmat training requirements, including response planning and plan implementation when necessary. For example, the Awareness-Level responder does not have specific planning requirements listed but does have responsibilities in plan implementation. The standard covers the competencies for multiple positions in the Incident Management System (IMS) with hazmat responsibilities. The following are brief descriptions of the competencies required by NFPA 472 for the core training levels (Awareness, Operations, and Technician) and the Incident Commander (IC) position.

THE AWARENESS-LEVEL

First Responder's plan implementation requirements include detection and identification of hazmat releases through the use of clues, aides, and tools such as the *Emergency Response Guidebook*. In addition he or she is required to know initial protective and notification actions in accordance with the overarching emergency response plan.

THE OPERATIONS-LEVEL

First Responder's competency requirements include describing the response objectives while identifying defensive and remote options to achieve those objectives according to the plan. At the operations level of training, personnel must demonstrate the ability to select and use the appropriate PPE and identify emergency decontamination procedures in the plan. The competencies for this level of responder also include the ability to establish and enforce site control, initiate the IMS, and perform defensive control techniques (confinement) as identified in the plan.

THE TECHNICIAN or Hazardous Materials Technician level planning competencies include the ability to identify the response objectives while identifying offensive, defensive, and nonintervention options according to the plan. The technician-level responder must demonstrate the ability to identify potential actions or tactical options, select and use PPE, select and perform decontamination procedures, and perform advanced control techniques (containment) as identified in the plan.

THE INCIDENT COMMANDER'S planning competencies include the ability to identify response objectives and options according to the plan—similar to the technician's competencies. Along with the ability to identify potential actions, the Incident Commander (IC) must be able to identify the purposes for different basic and advanced control techniques used to support the plan. The IC must be capable of approving levels of PPE and developing a plan of action given standard operating procedures and local emergency planning requirements. Obviously, the IC must be proficient in implementing the IMS, directing all necessary resources, as well as providing incident information to media and elected officials according to established plans.

Other IMS positions with planning and plan implementation competencies include private sector specialist employees, Hazmat Branch Officer, Hazmat Branch Safety Officer, and Hazmat Technicians with specific specialties. The IC and operational staff mentioned are discussed further in Chapter 4 of this text.

NFPA 473. NFPA 473 is entitled *Standard for Competencies for EMS Personnel Responding to Hazardous Materials Incidents* (NFPA 473, 2002). With regard to planning, this standard addresses levels of competencies for Basic Life Support (BLS) and Advanced Life Support (ALS) emergency medical system (EMS) personnel in a hazmat-prehospital setting. EMS personnel must demonstrate abilities in response planning and plan implementation. The following is a partial list of their required planning-related competencies.

- Preparation to receive hazmat victims
- Triage, treatment, and transport of hazmat patients
- Assessment and care of victims exposed to hazmat agents
- Collection and communication of patient information to receiving hospitals
- Identification of advantages and disadvantages of patient decontamination
- Establishment and management of the EMS Branch activities of the IMS

These descriptions of NFPA 471, 472, and 473 are in no way full descriptions of the standards, but rather generalized overviews of pertinent incident planning competencies and considerations. These standards are very descriptive and useful guidelines in defining requirements and capabilities in an incident emergency response plan. These standards are available online at www.nfpa.org. As mentioned in Chapter 2 of this text, all of the NFPA's codes and standards are available on-line or by mail as the entire series or as individual standards.

Information Research and Evaluation

In the first step of development, the organization has determined the need for a hazardous materials response plan and regulatory requirements for the plan. It is then time to research and continue the development of the plan (Fig. 3.7).

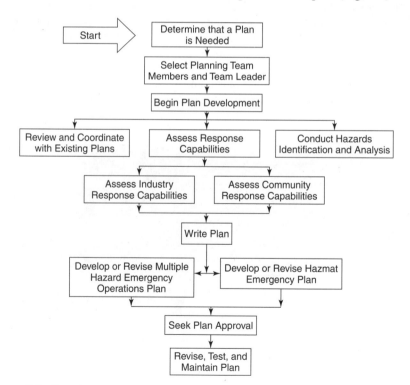

Figure 3.7 Developmental sequence of a hazmat response plan as suggested by the National Response Team's *Hazardous Materials Emergency Planning Guide* (NRT-1).

Even though there are multiple regulatory requirements for the development of hazardous materials response and contingency plans (Fig. 3.3), the intent of this text is to discuss response considerations related specifically to 29 CFR 1910.120(q). For this reason the following discussion will focus primarily on the Emergency Response Plan (ERP) requirements of this OSHA regulation utilizing the National Response Team's recommendations and Integrated Contingency Plan outline (Figs. 3.4–3.6).

Organizing a Planning Team

Per 29 CFR 1910.120 (q) and 40 CFR 311, industrial teams and public emergency response agencies are required to develop and write ERPs for response to uncontrolled hazmat releases. Industrial facilities have some distinct advantages over public response agencies when planning for hazmat releases. Company employees and responders know the identification, location, and facility-specific working procedures of the chemicals present. Company responders are also familiar with the layout, control equipment, and expertise found on-site. Public response agencies have one distinct advantage when planning for fixed facility or transportation hazmat releases; they practice emergency response on an everyday basis. Public response agencies will find it necessary to develop individual hazmat response plans for individual facility locations and transportation routes within their jurisdiction. For these and other reasons these two major groups must share information between themselves and with other affected groups to respond effectively and efficiently to all emergencies, especially hazmat releases (Fig. 3.8).

The second major step in plan development that is sometimes overlooked is the sharing of information in the initial stages of plan development. This sharing

Figure 3.8 Public emergency responders and facility responders must share information for planning and response.

among groups is easier to facilitate if a planning team is developed and led by a knowledgeable plan coordinator. Having a team of people with individual specialties, perspectives, and concerns can greatly assist the person responsible for creating an ERP. Suitable planning team members may include some of the following fields and specialties:

- Health and safety specialists (i.e., industrial hygienists)
- Facility employees and managers
- Public health officials
- Environmental compliance specialists
- Public emergency response officials (fire, EMS, hazmat, law enforcement)
- Public works employees
- Transportation officials

This listing is not all-inclusive, but these groups are commonly the primary groups called on during the initial stages of a hazmat incident.

Although a planning team has substantial advantages, the development of a too-large planning team can lead to complications. The person in charge of the team should give careful consideration to selecting members to avoid slowing the plan research and development with issues such as scheduling conflicts and confusion of responsibilities (Fig. 3.9).

Hazard Analysis

After planning requirements are determined and a planning team has been selected, the third major step will be to perform a hazard analysis to help establish the scope

Figure 3.9 The creation of a planning team promotes a more thorough plan.

or full extent of the plan. For an ERP to be effective, many considerations must be addressed. These considerations include the following:

- Recognition and identification of hazard(s)
- Vulnerability(ies)
- Risk(s)
- Required resources

Planners are ultimately trying to predict the most likely incidents that responders will encounter and anticipate useful information and resources necessary to safely, effectively, and efficiently mitigate the incident. While performing a hazard analysis, planners may utilize the DECIDE process, which is used by responders to make on-scene decisions as discussed in Chapter 1.

Hazard Recognition and Identification. The first part of any hazard analysis is to recognize and identify all potential response hazards, especially those associated with hazmats. This is very similar to "D" or "Determine the presence of hazardous materials" in the DECIDE process. Hazmat emergency response planners must recognize and identify the hazardous materials and the other hazards present by using multiple sources of information.

If available, one of the most important sources of information is a complete chemical inventory. If an inventory is not available, one should be developed. This inventory should include at a minimum the following information:

- Chemical ID information (technical name, common name, CAS No., UN ID No.)
- On-site and remote storage (specify product or waste) with location ID (e.g., Room 401)
- Hazard classification (criteria defined by OSHA, DOT, EPA, NFPA)
- Potential effects of exposure
- Container information (type, design, capacity, construction, number)
- Total quantity in storage or use

Other pieces of information that could prove useful later in the hazard analysis to note on an inventory may include brief descriptions of the following:

- Storage systems
- Processes using hazardous chemicals
- Present release or fire detection systems
- Present spill prevention and protection systems
- Current transportation routes and destinations

The information needed to compile a complete inventory may come from a number of resources. Facilities may have a basic inventory for other purposes such as shipping and billing that can serve as the backbone for the ERP inventory. Other sources of documentation that provide information on the chemicals hazards present include reports required by SARA Title III, environmental discharge permits, waste disposal records, shipping documents, and material safety data sheets (MSDS).

If planners are not familiar with sites or locations, they should make site visits and walk-throughs to have a better understanding of processes and to identify previously unrecognized hazmats and associated hazards. Planners should talk with employees and managers to gain specific information in addition to researching documentation. During these visits, planners should include pictures, maps, sketches, and/or blueprints in their collection of information. These will be useful later when addressing complex response-related issues and, if appropriate, can be included in the core or annexes of the plan.

Other pieces of information to document during the hazard identification investigation that can help identify potential releases and hazards may include the following:

- Documentation of past accidental releases and resulting injuries
- Inherently risky procedures such as hazmat transfer and filling processes
- Maintenance schedules and procedures
- Shutdown or start-up schedules and procedures

Vulnerability Analysis. After planners have identified all foreseeable hazardous materials and other associated hazards, plan creators must perform the second part of the hazard analysis: the vulnerability analysis. Planners will be looking for answers to release- and exposure-related questions such as "What do we have to lose?" and "Who or what will be affected?".

One of the primary concerns for planners to identify is the size of the vulnerable area. Many criteria are involved in making this decision. Planners will need to consider physical, chemical, and environmental parameters such as vapor pressure, temperatures, topography, spill size, and prevailing winds, to mention a few. These and other criteria must be identified or characterized to estimate the area of release. These topics are covered in detail in Chapters 5 and 6.

After identifying the possible size of a vulnerable area, potentially exposed and affected locations and populations must be identified. Typical infrastructures, locations, and associated populations of particular concern may include the following:

- Transportation routes
- Medical care facilities
- Schools
- Water and food sources
- Recreational areas
- Power and communication facilities

- Waste water treatment facilities
- Adjacent industrial processes or facilities
- Residential areas

Obvious locations or populations of major concern are those in close proximity, downhill, and downwind. Facility diagrams or maps showing potential release migration routes are useful tools in a vulnerability analysis and can be included in the plan for future reference. The use of MARPLOT® and ALOHA® with input of potential release conditions can help increase accuracy in estimating potential vulnerabilities and future response objectives.

Risk Analysis. The third part of a hazard analysis for ERP planners to perform is a risk analysis. This step is similar to the "E" or "Estimate likely harm without intervention" from the DECIDE process. For example, say there is a fifty-gallon release of a flammable hazmat such as gasoline on an empty parking lot. If the parking lot naturally confines the spill to a small area by means of topography and construction and there are no possible vulnerabilities present we would estimate a low probability of immediate harm to life and the environment. On the other hand, if this relatively small spill were able to drain to a storm water sewer adjacent to an occupied elementary school, our estimate of likely harm to life and environment without immediate intervention would be an unacceptable level. In both scenarios the hazard is the same, but the vulnerabilities are different, thus changing the outcome of our risk analysis dramatically.

Risk analyses are performed with multiple criteria such as hazard information, potential vulnerabilities, and potential incident factors. Planners must look at all available information to estimate the probability and consequences of a release. The probability of actions or processes likely to result in accidental (or intentional) releases of hazmats should be anticipated or ranked. In addition to the probability of a release, the severity or consequence of such a release should also be ranked. These rankings can be incorporated into a planning matrix or guide similar to the NFPA Planning Guide in NFPA 471. A combination such as low probability–high consequence (or any other variation) will help planners to determine planning emphasis and resource requirements.

Resource Evaluation. Resource evaluation is the fourth and final part of a hazard analysis. Evaluators must consider resources discovered and identified during investigations and recognize any likely deficiencies. If resource deficiencies are noted, they must be corrected as soon as possible and necessary measures to compensate for them must be listed in the plan (e.g., mutual aid agreements).

PERSONNEL are among the most important resources utilized and managed in a hazmat response. Evaluations must be made of each incident type and response strategy to define the minimum level of training required by 29 CFR 1910.120(q) and NFPA 472. Planners may determine that some potential releases will require an offensive strategy, thus requiring responders be trained to the hazmat technician level. Other foreseen responses may be initially defensive in nature and require operations-level training at a minimum. In any case, the minimum level of training must be determined and stated in the plan. For information on training requirements see Chapter 2.

In addition to training requirements, the sheer number of required personnel must be considered when establishing the strategy and scope of the plan. For example, planners should avoid plans relying on advanced control techniques (see Chapter 15) when an insufficient number of trained hazmat technicians are available. Sometimes companies and response agencies limit their planned response to a defensive or operations-level response strategy despite having some technician-level trained responders. In these instances, the plan may rely on mutual aid agreements to provide responders for implementing offensive strategy and tactics.

The personnel consideration that is commonly overlooked is experience. The fact that responders have recently completed technician-level training does not mean that they have the necessary experience or practice to carry out an aggressive plan. A less aggressive plan relying more on outside assistance may be initially developed and adjusted with time as responders' experience levels increase.

PERSONAL PROTECTIVE EQUIPMENT (PPE) are tangible resources that must be evaluated for planning purposes. PPE encompasses respiratory protection, chemical protective clothing (CPC), and other forms of protective equipment, as discussed in Chapter 12. Planners must address PPE needs for all anticipated emergencies based on the chemical hazards present and the situations or environments that a release may occur. This information may be included in the ERP by incorporation or reference to the written Personal Protective Equipment Program required by 29 CFR 1910.120(q)(10), 29 CFR 1910.120(g)(3–5), and 29 CFR 1910 Subpart I. In addition to these OSHA regulatory requirements, references from other PPE plans and programs required by relevant NFPA and ANSI consensus standards should be included.

It may be appropriate to include PPE selection guides in the ERP. These guides, along with consultation with health and safety professionals, will prove beneficial in establishing required PPE inventories. Once the appropriate type and quantity of PPE are established, the minimum inventory quantity and storage location(s) should be outlined or referenced in the plan.

One specific PPE resource that is often underestimated is breathing air. Almost all hazmat responses will require the use of a positive-pressure self-contained breathing apparatus. For this reason planners must outline in detail how responders may acquire additional breathing air for extended responses. Most public response agencies accomplish this by relying on portable cascade filling systems, compressors, and/or extra cylinders. If this is not feasible, planners can include or reference mutual aid agreements through which additional breathing air can be provided to the scene.

EMERGENCY MEDICAL SERVICES (EMS) resource evaluations must be performed in anticipation of injury that may result during all likely hazmat releases and related consequences. If the potential for injury to responders and/or the public exceeds the available in-house resources identified, planners must outline how to provide sufficient EMS resources and capabilities. At a minimum, EMS resource evaluations should identify the requirements for the following concerns:

- Health and safety of EMS personnel
- Injured victims (anticipated number, population, type of injury, etc.)
- Medical supplies and drugs
- Number of trained personnel required
- Level of EMS training required
- Triage, transport, and treatment resource considerations

After it is determined that there are resource deficiencies, planners can outline how to provide necessary resources by referencing mutual aid agreements, contracts, purchasing supplies, and/or additional training. For additional information regarding EMS operations and considerations, see Chapters 4, 7, 8, and 13.

SPILL CONTROL RESOURCES are another primary concern during the process of resource evaluations. These evaluations should be based on foreseeable strategies, tactics, and tasks that will be deployed by responders to mitigate a hazmat incident. Some response plans that anticipate only defensive or confinement techniques will only require spill control equipment such as booms and sorbents in sufficient quantity. An offensive spill response plan for the same release will most likely require sufficient quantities of control equipment for containment in addition to defensive control equipment.

Once control equipment resources have been evaluated, additional required resources should be identified and purchased. If planners foresee the need for specialized spill control or heavy equipment that is not available on site, the availability and quantity must be identified and acquisitions procedures established. Spill control equipment storage locations, minimum inventory, equipment replacement procedures, and acquisition procedures should be outlined in the plan. See Chapters 14 and 15 for more information regarding spill control techniques and equipment.

COMMUNICATION requirements are a crucial response resource to be evaluated. In some instances planners may determine that voice command and hand signals may be sufficient, whereas other situations may require more complex and flexible communication equipment. In either case, communication resources should be assessed and deficiencies corrected. Resulting inventories of equipment and communication systems should be described in the plan. Common communication system considerations include the following:

- System interoperability with other responding agencies' equipment
- Need for intrinsic safety features
- Distance system is required to communicate across

- Public address needs
- Hands-free operation or voice activation
- Number of channels/frequencies and privacy
- Interferences (ambient noise, shielding, etc.)
- System or unit redundancy
- Durability and simplicity
- Compatibility with required PPE
- Charging and power source requirements
- Other forms of communications (computers, faxes, remote imaging)

Other resources, in addition to the resources previously discussed, must be considered as well. Some of the common resources and considerations requiring evaluation are the following:

- Hazmat and fire detection equipment or systems
- Hazmat monitoring equipment
- Decontamination equipment and resources
- Security and control resources
- Procedures for recovery of all equipment used

It bears repeating that those who are developing an ERP should consider the equipment inventories and response capabilities of outside facilities and public response agencies. These outside groups may provide necessary resources that otherwise cannot be acquired. These evaluations should include inventories, mutual aid agreements (existing and potential), and mutual aid capabilities.

Analysis Techniques
Throughout the hazard analysis process involved in creating an ERP, planning team members will find it beneficial to use common analysis techniques or approaches. As part of OSHA's Process Safety Management of Highly Hazardous Chemicals standard (29 CFR 1910.119), industries utilizing and storing listed hazardous chemicals are required to perform hazard analysis and document their rationale. In this standard OSHA outlines useful hazard analysis techniques suitable for providing planning rationale. The following is a brief overview of OSHA's suggested process analysis methods.

"What If" Method. Planning team members will probably use this method automatically. The idea is for the members to look at specific systems utilizing hazmats and ask themselves what the consequence(s) would be if a foreseeable action were to occur. The use of checklists outlining possible actions and cross-referencing with potential hazards is beneficial. This is a good technique as long as planners are careful to keep the list of actions within the realm of possibility.

Hazard and Operability Study (HAZOP) Method. This method is intended to be a multidiscipline team effort utilizing leadership to direct brainstorming efforts to identify causes and consequences of specified abnormal operations and potential human errors.

Failure Mode and Effects Analysis (FEMA) Method. Planning teams may use this method to evaluate consequences and hazards due to different equipment failure modes or sequences within a system or process. After potential consequences are identified, they are then ranked according to severity of outcome. This ranking method or a similar variation can be integrated into a planning matrix or guide to establish planning priorities.

Fault Tree Analysis Method. This method starts with a likely outcome or incident and works backward to determine possible combinations of system and equipment failures. This method is used to establish a probability of occurrence utilizing system-specific historical data such as maintenance schedules, failure rates, and operator experience.

Even though the methods mentioned in this standard are related to hazard analyses of equipment and systems, with modifications these and other techniques can be used by planners and incident mangers for evaluation of hazmat response processes and procedures.

Commodity Flow Study Method. A Commodity Flow Study (CFS) is a hazard analysis method intended to assess types and volumes of hazmats moving through communities or jurisdictions. For some areas transportation incidents are the only source of hazmat risks, whereas other regions must consider transportation risks in addition to those of fixed facilities. Primary hazards analysis information collection efforts should focus on the following:

- Identification of designated or restricted hazmat corridors
- Characterization of substances, frequencies, container types, and capacities
- Specification of location, length, and priority of traffic routes and modes
- Proximities of hazmat traffic routes to vulnerable areas, facilities, and populations
- Identification and characterization of transportation terminals such as truck stops and airports
- Traffic accident information and trends for a specific region and route (with and without hazmat involvement)

As discussed in EPA's *Hazards Analysis on the Move* (EPA Document Number 550-F-93-004, 1993) there are several suggested survey methods and processes, such as placard surveys, that planners can use to collect data. After performing a vulnerability analysis, planners can use the hazard data collected to estimate the likelihood and severity of a transportation release.

Research Existing Plans

The fourth major step in the ERP development process is researching existing plans. The planning team should research and evaluate all relevant plans for two primary reasons:

(1) Pertinent information included in these existing plans can be utilized, incorporated, or referenced in the ERP to be created. Why reinvent the wheel when you can steal it? Most response agencies at all levels encourage and promote the shared use of existing information.

(2) The ERP being created should complement and provide interoperability with existing emergency plans of governmental agencies, local public response agencies, and other potentially affected industrial facilities. Hazmat response planning teams should strive to maintain the highest level of uniformity with relevant existing plans. These efforts will increase life safety, expedite incident stabilization, and reduce environmental damage.

Research of existing emergency response plans should include the following sources of information:

- U.S. Department of Homeland Security National Response Plan
- State Emergency Response Plan
  - Associated plans of SERC representative agencies
  - SERC field operating guides (FOGs)
  - State Homeland Security Department plans
- Local Emergency Response Plan
  - Associated plans of LEPC representative agencies
  - LEPC and local emergency management agency FOGs
- Old existing plans within the same facility or agency
- Industrial Facility Plans
  - Other plant locations with different ERPs within the same parent company
  - ERPs of surrounding or potentially affected industrial facilities
- Public Response Agencies
  - ERPs of surrounding jurisdictions
  - Mutual aid agreements between agencies
  - Existing standard operating procedures
- Professional Networks
  - Professional emergency response, environmental, or health and safety conferences
  - Professional associations

Writing the Plan

The fifth major step in the development of the ERP is writing the plan. To help a planning team along with this process, the use of an accepted and proven outline or format is critical. The outline recommended by the National Response Team for the development and creation of an ICP (Figs. 3.4–3.6) serves as an excellent outline for writing an ERP. There are numerous steps in writing the plan according to the NRT recommendations that will not be individually addressed in this section; however, major writing topics are discussed.

Structuring the Written Plan

As discussed earlier, if the plan is to utilize the NRT suggested format, it must be written with the three major sections of plan introduction, core plan, and plan annexes. The information collected by the team should be evaluated on the basis of its relevance to these three sections. In situations in which the planning team is revising an existing plan or including information taken from other plans, planners must carefully consider the appropriate section in which the information is to be included.

The introductory section of this format is intended for brief and factual information on facility or agency administration concerns. The core plan section is reserved for important response information that is critical for quick reference by response managers. The annex section is primarily intended for less time-sensitive supporting and regulatory compliance data.

If the information collected during the research and evaluation phase is not included in the ERP with consistency and proper structuring, critical information could be misinterpreted or overlooked during a response. The use of a uniform plan structure such as the NRT format during the writing phase will prevent informational problems and increase the plan's usefulness.

Regulatory Compliance. While the ERP is being written, writers must ensure that all regulatory planning requirements are satisfied by the information collected and described in the plan. As stated in the discussion of the ICP overview, this major concern can be accomplished with the aid of the ICP Development Matrix (Fig. 3.3). This matrix cross-references the plan sections and outline headings with the appropriate regulation. By using this matrix, writers will include the correct information in the correct place, thus increasing regulatory planning compliance.

If other regulations' planning requirements are to be included later, the writers can simply look at the matrix to find the regulatory requirement and see whether those required elements have already been addressed. Any deficiencies can be quickly recognized and corrected with the inclusion of necessary information.

Public response agencies intending to respond to hazmat emergencies at locations can also benefit from the NRT-ICP design. If departments must plan for multiple facilities they can simply develop multiple core plan sections specific to the facilities of concern and adjust the annex section accordingly. This will ultimately help to ensure a uniform approach to all hazmat responses and increase regulatory compliance.

Plan Details. After the planning team has fully analyzed all the relevant response information collected, and deficiencies are corrected, the detailed information should be included in an appropriate fashion. Writers should consider the use of flowcharts, outlines, diagrams, and checklists rather than pages of detailed text when possible (Fig. 3.10). When responders are in "emergency mode," the use of visual aides and lists that convey the proper information increases the efficiency of decision-making processes. This quick-reference information should be included in the core plan section, with supporting information shown in the annexes.

All mutual aid agreements and contracts should also be outlined or referenced in the plan. It may be necessary to give a brief description of the agreement and activation procedures in the core section and provide the actual written agreement or justification for referencing the agreement in the appropriate annex.

Response Management Information. Response management is an example of a large and potentially complex topic for which a simple checklist and a flowchart in the core plan can aid in providing sufficient information to initiate and deploy the IMS. The annex section is better suited to giving a full description of IMS to be utilized during a response for compliance and reference purposes.

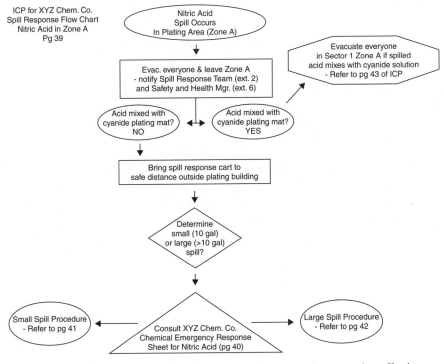

Figure 3.10 Flowcharts and diagrams of planned responses increase plan effectiveness during "response mode."

PLAN APPROVAL

After the ERP is developed and written, it is necessary to distribute it among response personnel, other planning groups, and all referenced mutual aid groups for review and comment. This is necessary if the new or revised ERP will affect or is dependent on other emergency plans such as another facility's EPA contingency plan or a LEPC's community ERP. The ERP's coordinator or team should specify a period of time for review and a suitable deadline for return of the ERP with comments. During this time members of the planning team should be available to answer any question that may arise during the review process. Once reviews are complete and comments are made, the planning team should make any necessary changes.

After the ERP is finished and reviewed, it is critical to get endorsements and approval of the plan at the highest level of the organization's management. The plan should be adopted as policy. Without acceptance by upper management, the best written plan is virtually useless. All parties involved in a hazmat response must know and should expect that the plan will be enforced by the organization with no exceptions.

When the plan has been reviewed and approved it must be distributed to all assisting and responding groups affected by the plan. The plan must be on file and available for review by all employees, responders, and regulatory agencies' inspectors.

PLAN REVIEW AND MODIFICATION PROCEDURES

Now that the ERP has been approved and is in place, the plan must not remain dormant. As outlined in the ICP Annexes (Fig. 3.6) and required by OSHA, the ERP should outline procedures for response critique, plan review, and plan modification. Specific mechanisms that require plan evaluation and possible modification include the following:

- Programmed review on a regular basis—sufficient to keep plan up to date.
- Changes in facility processes
- Changes in inventories described in the plan
- Changes in agreements described in the plan
- Changes in response technologies
- Instances where the plan fails

EXERCISING THE PLAN

A successful response to a hazardous materials incident will require aspects of most or all of the topics found in this text at one time. How can the emergency response team confirm that it will be able to function as a whole to successfully conduct response operations? This section of the planning chapter will focus on the concept of exercising the emergency response plan by practicing response skills in a training course environment.

Benefits of Conducting Exercises

The development of a facility's ERP and the training of emergency responders are both processes. In both, the participants will benefit from the opportunity to apply what is learned or developed. The potential benefits of exercises include the following:

- The opportunity to assess the capabilities of an individual or group
- Hands-on application of the ERP and/or hazardous materials response training to improve responder confidence
- The opportunity to detect deficiencies in the plan or training before an actual emergency
- The opportunity for departments, organizations, and agencies to work together before the tension of an actual emergency

It is hoped that an industrial facility will not have many actual hazardous materials emergencies. Of those that do occur, no two will be exactly the same. There will be enough unknowns and uncertainties during an actual incident without having to wonder whether members of a team can work together or properly use equipment. An actor can learn his part in a play without the other members of the cast, but he will not perform well if he does not rehearse with the other performers before opening night. Likewise, response team members should rehearse the ERP before the occurrence of an actual emergency.

Levels of Exercise

Consider a football team. When they hit the field on game day, it is actually the culmination of a continuous process of learning and practicing. The players all learn the functions of their positions and the plays the team will run. They get together and listen as the coaches map out a strategy for the game. Then the players get out on the field and practice the individual components of the game such as throwing passes, blocking, and rushing the passer. Finally, they have scrimmages and practice games. By the day of the game, they don't know exactly what will happen, but they know they're as ready as they can be. When the game is over, they start the process over again the next day, beginning with an evaluation of the previous game.

In the same way, a hazardous materials team must rehearse the ERP in stages. Four levels of exercise are described by the National Response Team (NRT-2, 1990) and by Kelly (1989) in discussions of emergency preparedness. These are:

- Orientation seminar
- Tabletop exercise

- Functional exercise
- Full-scale exercise

These exercises follow a pattern that is similar to the example of the football team. They start with the basic, individual components of the plan and build on these to produce a whole, team effort. Table 3.3 compares these levels.

A first step in the process of exercising the components of an ERP is determining the need for such exercises. Figure 3.11 is a sample checklist of the status of the ERP components. The checklist should be reviewed regularly and updated as conditions or personnel change, or as exercises are conducted.

In the following sections, examples of the exercises will refer to the hypothetical metal fabrication plant shown in Figure 3.12. The major chemical or physical concerns in each area are indicated, although it is recognized that smaller quantities of other hazardous materials would be present. Methylene chloride is stored in a 2500-gallon above-ground tank, xylene in drums in the storage area, and oxygen and acetylene in 150-pound steel cylinders. The plant is a subsidiary of a corporation that has similar facilities in other states.

Orientation Seminar

The orientation seminar is typically used as an opportunity to introduce a new or recently modified ERP to individuals who may have a part in a response. This type of exercise is the most basic and typically involves an informal discussion of the ERP and the general actions to be taken during an emergency response. Each of the various groups that may be involved in a response to a hazmat incident has the opportunity to see how its role fits into the overall plan and becomes familiar with the roles and capabilities of the other groups.

In a broad sense, the lectures and discussions during training courses serve as the orientation seminar (Fig. 3.13). The courses are intended to provide the emergency responder with an opportunity to discuss and learn how to safely conduct emergency responses to hazmat incidents. A specific orientation discussion may be used to introduce higher-level exercises, such as when specific instructions in how to patch drums are given.

Tabletop Exercise

Tabletop exercises are used to practice group problem-solving skills, and they may take a variety of forms (Fig. 3.14). During training, this type of exercise focuses on particular problems or decision-making processes. For instance, the participants may be given a chemical name and asked to find specific information about hazards from various reference materials. They could also be given a scenario and asked to write a safety plan that would require them to plan for safety procedures and personal protective equipment. In any case, the exercise may take the form of an ongoing scenario in which the trainees respond to new information as the scenario unfolds.

Emergency planners can use tabletop exercises to bring together various groups that may work together during a response. The purpose is to work out specific

TABLE 3.3 A Comparison of Exercise Levels

	Orientation	Tabletop	Functional	Full Scale
Uses	Introduce new or revised plan: acquaint new personnel with plan	Practice group problem solving; study specific scenario; observe information sharing	Evaluate any function; reinforce established procedures; test seldom used resources	Demonstrate cooperation; test procedure and equipment capability; media attention; evaluate overall response capabilities
Hazards	High profile	Any priority	To highlight function	Highest priority
Numbers of on-going activities	Single functions	One or two functions	Few to several disparate functions	All disparate functions for a particular scenario
Types of activity	Walk through; identify roles and responsibility	Problem solving; brain storming; resource allocation task coordination	Hands-on decision making; coordination; communication	Actual field operations; field command coordination
Degree of realism	None	Scene setting with scenario narrative and low-key messages	Limited; some simulated messages	Intense, full transmission of simulated messages

Adapted from Birmingham Jefferson County EMA (1991)

Plan Component	New	Updated	Exercised	Used in Emergency	Dormant
Plan Introductory Elements					
Core Plan Elements					
Plan Annexes					
Standard Operating Procedures					
Resource List					
Site Maps					
Reporting Requirements					
Notification / Warning Procedures					
Mutual Aid Agreements					
Fire Brigade					
Security					
Off-site Coordination					
Air Surveillance Procedures					
Emergency Shut-down					
Damage Assessment Techniques					

Figure 3.11   The need for exercises in particular areas can be determined with a checklist.

roles and responses to various potential incidents. These exercises should explore the potential involvement of different departments of the facility or plant without the stress of an actual response. During this discussion, the needs or deficiencies of the ERP can be revealed in a nonthreatening environment.

In the metal fabrication plant, the emergency planning committee may conduct a tabletop exercise to discuss the specific procedures to be followed in the event of a spill and ignition of xylene in the paint storage room. The exercise could involve personnel from such groups as are described in the plan development section of this chapter. The tabletop exercise would be an opportunity for these groups to plan a consolidated response to such a spill or fire before one actually occurred.

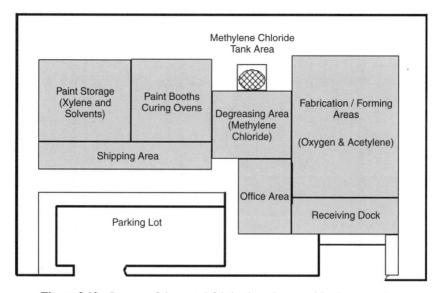

Figure 3.12   Layout of the metal fabrication plant used in the examples.

Figure 3.13   Orientation exercises communicate information, often in a discussion or lecture setting.

Figure 3.14   Tabletop exercises practice the problem-solving skills of groups.

In designing tabletop exercises, it is important to select certain objectives to be met and to design into the exercise information, clues, or messages that will trigger the desired response. This does not mean that the controllers or instructors necessarily lead the discussion or that written exercises are simple fill-in-the-blanks. Rather, the exercise should be introduced by clearly stating the focus to the participants and then left to take its own course toward the objectives. Occasional direction from instructors may be necessary when the discussion gets off course or when an important response is missed. Computer software such as CAMEO can provide an excellent format for tabletop exercises.

Functional Exercise

Some emergency response procedures or activities cannot be adequately demonstrated in a tabletop or discussion exercise. To practice these, one must do them. For instance, donning respirators and other personal protective equipment, plugging or patching leaking containers, or operating decontamination lines are all procedures that should not be performed for the first time during an actual emergency. In training courses, these types of functional exercises are often referred to as "hands-on" activities.

In practicing the ERP at the fabrication plant, the exercise may focus on specific actions such as initiating the warning system, establishing on-scene incident command, or evacuating areas of the plant. The selection of functions to be tested will be based on the response capabilities and the potential hazards at the facility. For instance, if the plant has no fire brigade and plant personnel are not expected to take offensive actions to fight fires, exercises in procedures for entering the hazard area would not be appropriate. An exercise in the replacement of broken valves on the methylene chloride tank system while wearing appropriate personal protective equipment would be a possible functional exercise.

The functional exercise may include more than one focal activity, such as having participants repair containers while wearing protective equipment (see Fig. 3.15). However, care must be taken to avoid diminishing the effectiveness of the exercise by including too many activities. Functional exercise should not become full-scale scenarios. Having several exercises with well-chosen objectives is preferable to having one in which participants are overwhelmed or rushed through important points.

Examples of hands-on activities that may be used in emergency response training include:

- Inspecting and donning a self-contained breathing apparatus and performing ordinary activities that involve some physical exertion, such as playing Frisbee
- Simulating chemical spills with water and practicing confinement procedures
- Practicing various containment procedures and going through decontamination in full Level A protection
- Calibration of and measurement with direct-reading instruments

Full-Scale Exercise

A full-scale exercise offers the opportunity to practice the full implementation of the ERP for specific, likely scenarios. Training courses can use the simulated incident response to pull together and apply all of the concepts presented in the course. The exercise necessitates the cooperation and integration of the various response groups and adds the stress of time. The degree of realism involved

Figure 3.15 Functional exercises practice individual activities that may be part of an emergency response.

depends on the exercise organizers and such factors as time, budget, and personnel availability. A full-scale exercise may be as large or as small as desired. The important feature is that it involves all personnel and procedures required to respond to an emergency situation.

To the extent that budget and time constraints allow, the actual equipment, materials, and procedures involved in the response should be incorporated into the exercise to enhance realism. However, some activities may be simulated if they (a) are part of the normal daily operations of the personnel (e.g., traffic control by police officers or normal structural fire fighting techniques for firefighters), (b) involve the use and disposal of expensive consumable items, or (c) would normally take a longer time to complete than can be incorporated into the time allotted for the exercise.

An important way to enhance the realism of the exercise is to use visual or audio props where possible. These props, such as dry ice or smoke bombs, can provide visible clues of a hazard that is being simulated (Fig. 3.16). It is important that the props be tied to a specific hazard that is designed into the scenario, and not simply thrown in for effect. Intentionally sending participants on a "wild goose chase" with misleading clues or information does not serve the purpose of the exercise.

Figure 3.16 The use of props, such as smoke generators, helps make the exercise realistic.

Some specific areas that should be tested by the exercise include the following:

- Implementation of the Incident Management System
- Site control and zoning procedures
- Coordination of emergency services from off-site
- Hazard recognition and assessment, including use of reference documents and monitoring equipment that will be available to the responder
- Selection and use of proper personal protective equipment
- Communication, both primary and back-up
- Logistical and administrative functions
- Decontamination, including control of any fluids generated for proper disposal
- Containment and confinement procedures that will likely be performed by the responders
- Termination procedures

A scenario that may serve as the basis of a full-scale exercise at the metal fabrication plant is that a drum of xylene is spilled in the paint storage area. A spark causes ignition of the vapors, and the building becomes engulfed in flames. The personnel involved in the exercise could practice the response to this emergency, from the initial warning and notification of the fire department through the termination of the incident and the reporting requirements for regulatory agencies. The actions could be triggered by planned messages carried over the emergency communications system, the known hazards in the area and surrounding area, and the normal actions called for in the standard operating procedures (SOPs) of the ERP as the incident unfolds. The concerns for potential hazards from the degreasing and fabrication areas would have to be addressed, as would such concerns as toxic vapors produced during the fire. The entire response as simulated in the exercise could then be evaluated at its conclusion.

The Exercise Process

Even an apparently simple written or tabletop exercise must be developed, conducted, evaluated, and followed up to be effective as a testing or learning tool. Each part of this process takes time and effort; how much time will depend on the complexity of the exercise.

Development

The first step in the process of developing an exercise is to establish the objective(s) to be met. The objectives of training exercises may be to provide a practical application of a subject that has been discussed or presented in a lecture. It may be to test the trainee's ability to perform a particular task or function, or it may ultimately be to give the trainee an opportunity to practice all areas covered in a course during a simulated incident, thereby fulfilling several objectives.

A facility's emergency planning team may desire to evaluate the level of understanding of the emergency response plan by the various groups that may be called on to participate in an emergency response. These groups may include the following:

- Health and safety department
- Production supervisors
- Maintenance
- Spill control team
- Security
- Equipment operators

Groups from outside the plant that may be involved include the following:

- Local fire department
- Emergency medical services
- Corporate health and safety personnel
- Hospital emergency room personnel
- Law enforcement agencies
- Local emergency planning committee (LEPC)
- State or federal environmental protection agencies
- State or local politicians

Representatives from these groups should participate in planning the exercises that will involve them. This will ensure that the exercise activities provide a realistic test of their functions during the response. Practicing the emergency response plan also offers the chance to look for flaws or gaps in its implementation.

An exercise may fail because the participants are not given the information and direction necessary to respond effectively. By clearly stating the objectives first, the exercise can be designed to accomplish these. For instance, if the objective of a written training exercise is to select the proper direct-reading instrument for a scenario, the description should not emphasize information important to spill confinement techniques.

Another danger is that an exercise may be too ambitious, expecting too much from the participants. For example, participants may be asked to write a safety plan based on a scenario that describes too many or unrealistic hazards. The likely end is that the participants will not take the exercise seriously and little or nothing will be accomplished. For an exercise to be effective, the participant must feel that it simulates a task or activity he or she may actually participate in or accomplish during a response.

It is important not to present a complex exercise without preceding it with simpler exercises that deal with its components. For instance, if a training course does not conduct functional exercises before the day when a full-scale incident simulation is conducted, this will result in a confused exercise in which participants are unable to efficiently perform individual tasks. Instead of demonstrating an imperfect, but successful, response to an incident, the exercise frustrates and discourages the participants.

Once the objectives are established, the specific content of the exercise can be created. For functional or full-scale exercises requiring equipment and supplies or integration of several groups, the planning should start early. The whole exercise must be well thought out. Possible response actions other than those specifically intended by the planner should be predicted, so that exercise controllers are not caught off guard during the exercise itself.

Conducting the Exercise

The way in which the exercise is presented and managed will directly affect its success. Three roles of people are involved in conducting an exercise: controllers, simulators, and evaluators (Kelly, 1989). In tabletop exercises conducted during training courses, a single instructor may be able to fill all these roles. Obviously, larger exercises will require more people. All people involved in conducting the exercise must be familiar with their roles and with the exercise as a whole.

The controller keeps the exercise on course by monitoring activities and intervening when necessary to maintain professionalism or prevent distractions. The simulator is an "actor" who provides realism by portraying groups that may be involved during a response but are not included in the exercise. Simulators will be interacting with the participants throughout the exercise to provide information and may play such roles as a chemical manufacturer's representative, an injured person, or a media representative. The evaluator's role is to observe the

Figure 3.17 Evaluators should observe exercise activities without intervening.

exercise and provide an objective assessment of what areas went well and what areas need improvement (Fig. 3.17). The evaluator usually will not intervene in the exercise while it is under way, but he or she observes and then reports after the exercise has been terminated.

If the exercise is well prepared, the participants can work through the activities and/or respond to information as it is presented with little assistance from controllers or simulators. During tabletop exercises, discussion, whether in large or small groups, is the primary activity. The controller should allow participants to discuss the activity or scenario openly with minimal interference. However, controllers should be very attentive to the progress of the incident in order to intervene quickly to prevent distraction, provide additional information, or simply answer questions.

Functional and full-scale exercises typically require several people. The activity may be spread out over a wide area or in several locations. Functional activities may involve one task, such as inspecting and donning respiratory equipment, but may have many people participating. Full-scale incidents may involve many people performing many activities. In both instances, sufficient personnel should be available to act as controllers, simulators, and evaluators.

Full-scale and functional exercises should be realistic simulations of activities to be conducted during an emergency response. Therefore, the exercises should be conducted at locations that may be real incident sites, under appropriate time constraints, and with equipment that will be used. Response activities may be triggered by the mere presence of labeled hazards (e.g., a drum labeled "acetone" to initiate concern for flammability hazards), the simulation of events or hazards (e.g., water leaking from a container labeled as a hazardous chemical), or by messages provided at specific times by exercise controllers (e.g., "An explosion has been reported in the supply warehouse"). The controllers and simulators should ensure that the pace of the response is realistic and that participants maintain a professional attitude. However, the actual direction and involvement of the controllers should be kept to a minimum to allow the participants to find their own solutions.

The exercise should be closely monitored by the controllers to ensure that at no time are the health and safety of the participants, controllers, or general public endangered. Signals or code words that are different from the communications used in the exercise should be established that will stop the exercise if unsafe conditions arise or if a real emergency occurs. Dangerous response activities that might be considered during a real incident should be simulated during the exercise. Exercise controllers should be instructed to observe participants for signs of trouble, such as symptoms of heat stress.

Evaluation
If the exercise activities are important to practicing a response, then the evaluation is equally important to learning from the exercise. The evaluation of exercise activities can take place at two levels. First, the participants may conduct a postincident analysis or critique as a part of the exercise, if time permits. This would be an extension of the incident simulation, in which the response action as a whole is reviewed to learn from mistakes and to reinforce successful actions. The exercise controller should monitor the critique to ensure that it does not become a "blame" session or that personality conflicts do not deter constructive criticism. This level of evaluation causes participants to practice looking critically at the performance of the response team.

Second, the exercise should be evaluated from the perspective of the exercise staff (controllers and evaluators). This level of evaluation should assess how well the objectives set forth for the exercise were met, as well as the performance of individuals or groups involved in the response. The exercise staff should be able to compare the decisions and actions observed during the response to those expected according to the emergency response plan. For exercises conducted during training courses, the exercise staff can use this opportunity to evaluate the effectiveness of the training course in preparing trainees for emergency responses.

In addition to evaluating the participants and the emergency response plan in the exercise, the exercise itself should be evaluated. Lessons learned during one exercise can be used to improve future exercises. Exercises should not be one-time events but should be conducted regularly at all levels (orientation through full scale). The frequency should be determined by the emergency response planning committee based on an assessment of previous exercises. Therefore, an exercise evaluation should be conducted that incorporates the comments of participants and exercise staff alike, as well as any nonparticipating observers who may have been invited. A written report should be prepared for distribution to all parties involved that describes the exercise, a summary of needs (planning, training, equipment, and/or personnel) disclosed during the exercise, and recommendations for follow-up activities.

Follow-up

It is unlikely that a simulated incident will produce a perfect hazmat response. The value of an exercise is lost if follow-up activities are not conducted to correct problems or deficiencies uncovered during the exercise. The process begins with the compilation of the exercise final report. For exercises intended to test a facility's emergency response plan, the report should summarize the following:

- The objectives of the exercise
- The pertinent points raised during the postincident analysis
- The lessons learned
- Deficiencies in planning, training, or equipment
- Specific recommendations for improving the response operations or the ERP

Once the report and recommendations have been distributed, the emergency response planning committee should monitor the progress of implementation of the recommendations. Also, any changes in the emergency response plan that result from the exercise must be distributed to the appropriate personnel. Recommendations for improvements in the exercise process itself should be incorporated into planning for future exercises.

Exercises conducted during a training course, with participants from various companies, will also involve follow-up activities. The follow-up will be largely the responsibility of the individual trainee, although an observant exercise staff may detect areas in which trainees need additional work. The trainee should be encouraged to reflect on the exercise and honestly note areas where he or she did not understand what was going on. These areas should be studied further. Instructors and exercise staff should be able to provide additional assistance or references to assist the trainee, even after the course has been completed.

SUMMARY

In history there are many well-documented case studies of hazmat responses that suffered from a lack of proper preparedness and planning. The SARA legislature and its promulgated standards and regulations are the attempts of the U.S. Government to prevent the lack of emergency planning by facilities and state and local governments. The primary focus of these planning regulations is to ensure life safety, incident stabilization, and environmental protection. Development of a successful hazmat emergency response plan is a multistep process. Once this process is complete, the plan must be documented, understood by all involved parties, and utilized during hazmat emergency responses to be effective.

In addition to a "good" plan, exercises of the plan are a critical part of the overall concept of planning. A well-planned and executed exercise need not result in a flawless response to a simulated hazardous materials incident to be a success. Exercises can be conducted at four levels: orientation seminar, tabletop, functional, or full-scale exercises. The exercise will involve development, execution, evaluation, and follow-up activities. The success of an exercise is determined by the performance of the participants and the degree to which deficiencies and problems are resolved after the exercise. However, failure to conduct exercises of an emergency response plan can result in unnecessary confusion during an actual emergency.

4

INCIDENT MANAGEMENT SYSTEM

INTRODUCTION

One of the most challenging aspects of responding to a hazardous materials emergency is managing the response effectively. This is especially true of major incidents because bringing the incident under control may require the activities of large numbers of personnel to be carefully coordinated. Different groups of responders may be required to carry out different actions simultaneously at different locations. Operations may need to be sustained over a long period of time, and resource requirements may be huge. Effective incident management is also important for minor incidents because it promotes safer response. In some cases, it can prevent minor incidents from becoming major incidents.

As described in Chapter 1, we have a documented history of hazmat incidents that had disastrous outcomes because of inappropriate actions by responders. In many of those incidents, bad outcomes were a direct result of a lack of effective incident management. The history of hazmat response also contains a number of successful case histories in which bad outcomes were minimized or avoided. Effective incident management is a common denominator of all successful emergency response operations. To emphasize these points, let's take a look at several case histories in which incident management aspects were critical to the outcomes of response operations either for better or worse.

Crescent City, Illinois, 1970—A boiling liquid expanding vapor explosion (BLEVE) occurred as a result of the derailment of a railcar containing liquefied

Emergency Responder Training Manual for the Hazardous Materials Technician, Second Edition,
edited by Kenneth W. Oldfield
ISBN 0-471-21387-X   Copyright © 2005 John Wiley & Sons, Inc.

petroleum gas (LPG). In the aftermath of the destruction it was determined that the massive property destruction totaled over 3 million dollars. Sixty-four people were injured, and over 1500 people were evacuated; fortunately, there were no fatalities.

Crescent City was staffed by two fire stations, but by the end of the incident 34 fire departments totaling 250 responders and 58 pieces of fire fighting apparatus were needed to extinguish the fires and treat the injured. Incident command was initially established; unfortunately, one of the first people inured in the response was the Incident Commander. This led to problems with additional arriving units developing individual strategies and freelancing tactics.

Waverly, Tennessee, 1978—A derailed tank car containing LPG ruptured, causing a massive explosion. In the aftermath of the catastrophic fireball 16 people were killed, including the police chief, fire chief, and a Tennessee Emergency Management Agency employee. Over 200 people were injured.

In after-incident analysis, mutual aid responders were reported to say that failure to announce and identify the command post as well as an overall lack of organization were major contributing factors that interfered with response efforts.

Oklahoma City, Oklahoma, 1995—An act of domestic terrorism beyond comprehension occurred at the Murrah Federal Building in downtown Oklahoma City. Over 170 people, including 19 children, were killed as a result of the blast. Over 600 people were injured. Twenty-six fire departments along with 10,000 volunteers assisted in the response and recovery efforts.

The Incident Management System was implemented and successfully used throughout this protracted incident. In addition to the use of the Incident Management System, a multiagency command center was established, containing representatives from every agency involved in the effort.

New York City, New York, 2001—The deadliest attack on American soil killed over 3000 civilians, more than 100 law enforcement officers, and 343 firefighters and emergency medical responders. The magnitude and complexity along with the loss of the Incident Command, Command Staff, and communication early in this act of terrorism overwhelmed one of the most response-capable fire departments in America.

Vestavia Hills, Alabama, 2003—Having considered a number of famous hazmat incidents, let's examine one you may never have heard of, in which effective incident management played a key role. Vestavia Hills is a suburb south of Birmingham, Alabama, with a medium-sized career fire department. The firefighters and paramedics heard something they had never heard before, "Airplane Crash," come over their radios. A ten-passenger aircraft had crashed into a creek adjacent to a large apartment complex, instantly killing the pilot and the sole passenger. A large amount of aviation fuel was released into the creek. In a matter of minutes the city's fire and police departments were fully committed to the response.

Vestavia Hills Fire Department mandates the routine use of the Incident Management System and requires the use of unified command (as described below) when warranted. The Incident Management System was immediately established, and the Incident Commander evaluated incident priorities and established an appropriate strategy for the response. After personnel safety considerations and exposures were addressed, the strategy was quickly identified by the Incident Commander as property conservation because of the fuel spill into the creek and the fact that there were no survivors to be rescued and no civilians were in danger. The creek flows through several jurisdictions, and the fuel spill was quickly migrating downstream and threatening a recreational lake in a residential subdivision in an adjacent city. Because of the nature of the incident and the multijurisdictional challenges, 13 different agencies and 8 fire and police departments were represented in the unified command structure. The successful mitigation and scene management of the incident were directly related to the implementation of the unified Incident Management System.

HISTORY OF THE INCIDENT COMMAND SYSTEM

In the early 1970s, a series of major wildland fires in southern California prompted municipal, county, state, and federal fire authorities to form an organization known as Firefighting Resources of California Organized for Potential Emergencies (FIRESCOPE). Organizational difficulties involving multiagency responses were identified by FIRESCOPE. Other difficulties included ineffective communications, lack of accountability, and lack of a well-defined command structure. Their efforts to address these difficulties resulted in development of the original Incident Command System (ICS) for effective incident management. Although originally developed for wildland settings, the system ultimately evolved into an all-risk system, appropriate for all types of fire and nonfire emergencies.

Since the development of the ICS, the fire service has experienced several challenges in understanding its application. As a result, inconsistencies in the system began to develop; other hybrid systems came into existence, further distancing a common approach to incident command. A single incident management system is critical for effective command and control of major incidents. At these incidents, a single department may interface with other agencies on the local, state, and federal level. To reduce the inherent confusion that may be associated with larger-scale incidents, a common system is a must.

Recognizing the challenges the fire service faced in applying a common approach to incident command, the National Fire Service Incident Management System Consortium was created. Developed in 1990, its purpose is to evaluate an approach to developing a single Command system. The Consortium consists of many individual fire service leaders, representatives of most major fire service organizations, and representatives of federal agencies, including FIRESCOPE. One of the significant outcomes of the work done by the Consortium was the identification of the need to develop operational protocols within ICS, so that response personnel could apply ICS as one common system. In 1993, as a result of this, the IMS Consortium completed its first document: *Model Procedures Guide for Structural Firefighting*. This text led to the use of the term "Incident Management System" (IMS), which is essentially synonymous with the ICS with subtle changes in philosophy. FIRESCOPE adopted this in principle as an application to the Model FIRESCOPE ICS. The basic premise is that the organizational structure found in the FIRESCOPE ICS is now enhanced with operational protocols that allow the nation's response personnel to apply the IMS effectively regardless of where in the country they are assigned.

It is important to note that the IMS *Model Procedures Guide for Structural Firefighting* has had other applications or modules similar to the structural fire fighting applications that have been in place for some time. These create a framework for other activities to operate in and further enhance the use of IMS. As an example, there is the *Model Procedures Guide for Hazardous Materials Incidents* (NFSINC, 2000), as well as other guides.

Since the 1970s, the value of the Incident Command System/Incident Management System as an emergency management tool has been proven repeatedly by emergency response agencies that have implemented it in hazmat emergencies. For this reason the Occupation Safety and Health Administration (OSHA), the Environmental Protection Agency (EPA), and the Executive Office of the President have developed regulatory directives that require the use of the IMS at hazardous materials incidents, as well as in response to a weapon of mass destruction event or national-scale disaster of any origin (see Chapter 2).

BASICS OF THE INCIDENT COMMAND SYSTEM

The IMS is applicable to users throughout the country. With its basic common elements it is simple enough for a small organization at a small incident, or the it can be expanded to coordinate the activities of several organizations at a major incident. It can be used in incidents that require (1) single jurisdiction/single agency involvement, (2) single jurisdiction with multiagency involvement, or (3) multijurisdiction/multiagency involvement.

Four characteristics that are critical to the proper functioning of the Incident Management System can be remembered by the acronym USDA. They are:

Unity of Command—To minimize confusion and conflict, the IMS is designed so that an individual responder working within the system receives orders from only one person and reports to only one person.

Span of Control—In the IMS model system the ratio is considered 1 supervisor:5 subordinates, as discussed below.

Division of Labor—Tasks are assigned to individual responders based on their training and ability as indicated by their job descriptions.

Accountability—The location and status of all personnel, equipment, and supplies are monitored at all times through the system.

The system is intended to control all actions taken in response to an incident. The IMS does not allow for "freelancing', which refers to an individual or a group doing what he/it wants to without the permission and knowledge of the incident commander or a supervisor. Discipline is a very important aspect of the IMS that helps to promote efficiency, teamwork, and safety.

Components of the IMS

The IMS consists of eight essential components. Interaction of these components is vital to the proper performance of the IMS. The IMS components are defined and characterized in the following sections of this chapter.

Comprehensive Resource Management

To bring a hazardous materials incident under control, it is important that all available resources be managed and utilized in an effective and efficient manner. These resources may include personnel, equipment, and supplies required to perform a response operation. The IMS serves as a mechanism for centralized control and coordination of these resources, to provide for an efficient response.

Modular Organization

The organization structure of the IMS is modular and is intended to expand or unfold from the top down based on the managerial needs of an incident. This provides the IMS with the flexibility required to be applicable to both major and minor incidents.

Common Terminology

It is important that terminology be utilized consistently by all personnel required to operate within the IMS. For this reason, standardized terminology is used for personnel roles, equipment, and facilities used in emergency response operations. Terms such as "operations section," "hot zone," "decontamination," and "command post" should therefore have the same meaning for all personnel involved in a response operation. This is especially important given the potential for multiple agencies to be involved in an incident.

Unified Command Structure

For major incidents requiring a multiagency response, all involved agencies must have input into the processes of establishing response objectives, selecting strategies, and planning for tactical operations. This is the purpose of a unified command structure.

Manageable Span of Control

As a general rule, a person in a managerial role during a response operation can efficiently direct and supervise the activities of three to seven subordinates. Because of this, a span of control of five responders is used as a general rule of thumb in establishing the IMS. If a reasonable span of control is exceeded, the managerial skills of response personnel can be overwhelmed, leading to a breakdown of control within the system. Thus, as the IMS unfolds or "builds down" and greater numbers of response personnel become involved in an operation, managerial responsibilities must be delegated to various response personnel according to the emergency response chain-of-command. This will ensure that no individual's span of control is exceeded and that centralized control and coordination of all activities under the IMS can be maintained.

Consolidated Action Plan
Efficient incident response requires a well-defined plan of action. Action plans must cover tactical and supporting activities required for performing the hazmat operation. It is important that all personnel and agencies involved in an incident response operate under action plans that are consolidated so that duplication of activities and contradictory actions are avoided.

Integrated Communications
Clear, concise, and expedient communications are vital for effective response operations. We can think of the communication system as the central nervous system of IMS. The IMS must provide a means for controlling and coordinating all incident communications to form an integrated communications system. Communications are more fully discussed below.

Designated Incident Facilities
Various types of facilities may be established at the scene of an incident to facilitate the response. The specific type and location of these facilities will vary according to the requirements of an incident. Two very commonly used incident facilities are the command post and staging areas (see Chapter 11).

Major Functional Areas of the IMS

The IMS is divided into five major functions that must be addressed regardless of the numbers of responders and/or complexity of the response. The five major functional areas of IMS are command, operations, logistics, planning, and finance (Fig. 4.1). Activities within each of the major functional areas are performed by a separate section of the IMS. It is important for all response personnel to understand the types of duties performed under each of the functional areas and the way in which the functional sections interact within the system as a whole.

Command
Command is ultimately responsible for the safety of responders, victims, and other civilians; stabilizing the incident; and property conservation. This involves

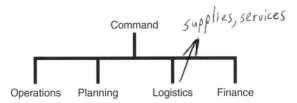

Figure 4.1 The IMS allows control and coordination of resources in 5 major functional areas.

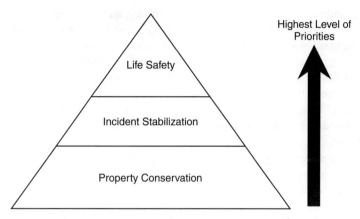

Figure 4.2   The Pyramid of Incident Priorities.

sizing up or assessing the situation, establishing strategic goals, and ordering and allocating personnel, equipment, and other resources as required to achieve those goals.

The "Pyramid of Incident Priorities" (Fig. 4.2) should be used in establishing strategic goals so that the highest priority is given to life safety issues, with incident stabilization given secondary priority, and property conservation given the lowest priority. For example, the pyramid of priorities would require that rescuing viable victims of an incident be given the highest priority before resources are diverted to stabilizing the incident or protecting the environment.

In addressing life safety issues, the "Pyramid of Survivability" (Fig. 4.3) should be used to establish priorities so that one's own safety is given the highest priority, with secondary priority given to the safety of other responders, and the lowest priority given to the safety of victims involved in the incident. When the pyramid

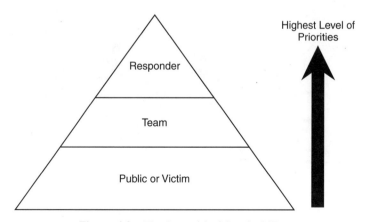

Figure 4.3   The Pyramid of Survivability.

of survivability is ignored responders may undertake foolhardy rescue attempts that result in their being injured or exposed. As a result, attention must be diverted to care for them, which in turn undermines the effectiveness of the overall response with adverse affects on the safety of other responders and victims.

The Incident Commander (IC) is the official in charge of the command section and is thereby ultimately responsible for management of the entire response operation. If required by the size or complexity of the incident response, the IC may be assisted by command staff personnel (as described below) in managing the incident.

The IC is directly responsible for all responsibilities that are not formally delegated to a subordinate in charge of a command staff role or functional area. The IC must realize that failure to address any one of the functional areas will most likely lead to an inefficient and/or ineffective mitigation effort. An easy way for an IC to remember this is to arrange the functional areas in the following order:

- Finance
- Logistics
- Operations
- Planning

The Command will **FLOP** if these are not addressed!

It has been said that ICs must remember the "Turtle on the Post Theory"—When you see a box turtle sitting atop a fence post (Fig. 4.4), one thing is for certain: He had help getting there. ICs must remember this analogy and utilize all the help or resources available to them when mitigating any incident. It is impossible to operate any system of management acting alone.

Figure 4.4   Incident Commanders should remember the "turtle on the post theory" and utilize all available help or resources. It is impossible to conduct a response operation acting alone.

Finance

Finance is involved with the financial aspects of an incident response. Financial concerns include supplies expended during a response operation, the purchasing of materials required, timekeeping for response personnel, and other monetary matters.

Logistics

The logistics section provides equipment, supplies, and services needed to support the response operation. This may require that various items, ranging from earth moving equipment to food for responders, be procured and transported to the incident scene.

Operations

Operations direct all tactical operations undertaken to bring a hazardous materials emergency under control. This involves direct response at the hands-on level, such as performing procedures intended to control, confine, and contain hazardous material releases.

Planning

Planning is responsible for collecting and evaluating information related to an incident and developing action plans and alternative plans. Information and advice provided by the planning section is utilized by command in decision-making.

Interaction of the Functional Areas of the IMS

Command must be clearly established at the beginning of all emergency response operations. The command post location must also be clearly established. The other major functional areas of the IMS may or may not be formally delineated depending on the specifics of the incident. For example, assume that the first-arriving team of responders has reached the scene of a minor hazmat incident. In accordance with the requirements of 29 CFR 1910.120 (see Chapter 2), the senior official of the team assumes the role of on-scene commander and the team is immediately able to stabilize the incident. Because of the simple nature of this incident, the acting on-scene incident commander was able to manage all functional areas of the IMS by carrying out all duties related to finance, logistics, operations, and planning required to bring the incident under control.

In contrast to the previous example, response to a major incident may require that the actions of many teams of responders be controlled and coordinated over a long period of time. The modular organization of the IMS allows the managerial system to shrink or expand as needed so that all personnel involved in managing the system are able to maintain a reasonable span of control.

For example, assume that the first-arriving industrial response team has reached the scene of a major hazmat incident. The senior official of the team assumes the role of on-scene IC, begins sizing up the scene, and immediately calls for additional help. As personnel having greater seniority arrive at the scene, command will be relinquished by the acting IC and passed up a preestablished line of authority, as specified in the emergency response chain of command. As progressively larger numbers of personnel become involved in the response operation, it will be necessary to formally delineate the functional sections subordinate to command and to assign someone to manage each of the sections. Thus, for a major incident, management of the major functional areas of the IMS may require the involvement of a planning section chief, an operations section chief, a logistics section chief, and a finance section chief in addition to the incident commander (Fig. 4.5). Furthermore, the IC may require direct assistance from command staff personnel, as described below, in carrying out his or her duties.

As additional personnel become involved in a response operation, the IMS must expand or "build down" as required to maintain an appropriate span of control. Thus an operations chief may deploy group supervisors. Likewise, unit leaders can be established to perform as directed by a group supervisor (Fig. 4.5). As the system expands, it is vital that the response roles of all key personnel within the IMS be clearly delineated. An effective way to do this is through the use of vests that identify personnel roles (see Fig. 4.10 below). As response personnel begin to bring the incident under control and fewer personnel are needed for the operation, the IMS can be shrunk progressively by reversing the process of expansion.

IMS Communications

Communications equipment that enhances interoperability is required for efficient functioning of the IMS. Examples may include radio, telephone (both on- and

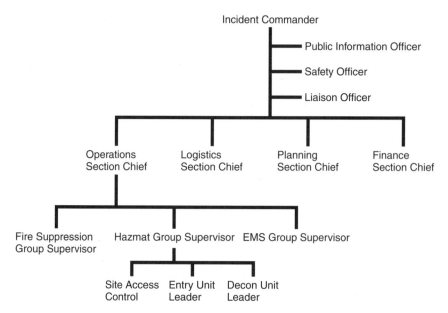

Figure 4.5   IMS managerial roles with hazmat group expanded.

off-site), and public address equipment. Other items such as cellular telephones, E-mail, and fax machines may also be used. All channels of communication must be controlled and coordinated through the IMS to form an integrated communications system. Such a communication system may involve fixed centers, landlines, mobile and hand-held units, and personnel trained to operate the equipment. During emergency response activities, a specific radio channel should be dedicated solely for communication between the incident commander and staff officers, sector chiefs, and other key personnel who are positioned at any location other than the command post.

Communication procedures should be established during preparation of the emergency response plan and standard operating procedures. Failure of communication equipment or the communication system as a whole must be addressed. For example, runners may be deployed with messages as a last resort to maintain communication. Backup communication procedures should be developed for critical communications in the event of primary communication failure. An example would be using horn blasts or sirens to warn personnel to evacuate the site.

The most common root cause of problems within the IMS is failure to communicate effectively. It is important for communications to be well integrated so that all personnel can clearly understand them. This is especially important for multiagency responses, because the use of nonstandardized codes can produce a breakdown of communication. One simple approach to this problem is to use plain English rather than "10 codes" for all communications during multiagency response operations.

Security of communication channels and the communication system as a whole must also be considered. Responders have long been cautioned to bear in mind that two-way radio transmissions are not secure and can be monitored by anyone with a scanner. Confidential or controversial information can be inadvertently released to the public in this way. Given recent heightened concerns for terrorism, this concern becomes even more critical, because terrorists may monitor communications to identify how to best disrupt the response operation. For perpetrators seeking to do so, the communication system itself may be a prime target.

COMMAND OF THE INCIDENT RESPONSE

Single and Unified Command Structures

Command may be single or unified depending on the incident. If the incident is located in one jurisdiction and only one agency is involved in the response operation, the single command system will be used. In the single command system one designated official will act as commander. For example, assume that a release occurs at a fixed facility. The facility hazmat team responds under the command of the designated on-scene IC and brings the incident under control, preventing any hazardous chemicals from exiting the facility property. This is an example of response by a single organization (the in-house hazmat team) utilizing the single command system.

As another example, assume that a release occurs at a fixed facility that has taken advantage of an exclusion from emergency response planning requirements provided by 29 CFR 1910.120 (q) (1). The local fire department is notified and, in keeping with the conditions of the exclusion, employees are evacuated from the workplace and are not allowed to assist in the response operation. The local fire department responds with a designated member of the department, usually the fire chief or ranking fire officer, acting as IC and rapidly brings the incident under control without harm to the environment or threat to the off-site population. This is another example of a single agency (the local fire department) utilizing the single command system.

The unified command system is used when an incident involves more than one jurisdiction and/or more than one agency is involved in the response. For example, assume that a major release occurs at a fixed facility located near the boundary line separating two municipalities. The release represents a potential threat to off-site population in both municipalities and to the environment. A long-term response operation is required involving the facility hazmat team, the local fire departments from both municipalities, the county emergency management agency, and the state environmental regulatory agency. In the unified command system the individuals designated by their respective jurisdictions/agencies must make joint decisions concerning objectives, strategy, and priorities.

In most public-sector hazardous materials incident response operations, the local fire department will assume command and the ranking fire officer will be the IC. As the incident unfolds, it is not uncommon to change from a single command system to a unified command system as various agencies are notified and begin to arrive at the incident site. As representatives of environmental agencies and other regulatory agencies with legal authority arrive on the scene, a unified command system can develop, or the arriving agencies may agree to continue under a single command system with the agencies acting as staff advisors to the IC. The designated IC may be from the fire service, law enforcement, emergency management, or some other organization involved in the response operation.

A clearly understood, preestablished chain-of-command is vital for all personnel and agencies that become involved in a response operation. For example, assume that a major incident is underway at a fixed industrial facility. The highest-ranking member of the facility emergency response organization has assumed the role of on-scene IC and initiated the response operation in accordance with the site emergency response plan. The local fire department has been summoned and is en route to the scene. Because of the nature of the incident, the fire department's involvement will be vital in bringing the incident under control. In this type of response, there is no time for confusion or conflict regarding who will be in command after the arrival of the public sector responders from off-site. Some states have laws intended to prevent this type of conflict by mandating that the ranking fire official at the scene of any emergency will be in command. These types of potential conflicts must be resolved during preemergency planning before incidents occur. Toward this end, the private sector response organization must work through agencies such as the local emergency planning committee to interface its emergency response chain-of-command with those of applicable emergency response agencies from the public sector. The IMS allows for development in the single, unified, or area command system. However, it is vital that all agencies and organizations agree on the evolution of the system and work together for successful termination of the incident.

Area Command Structure

An area command structure is often implemented to manage multiple incidents that are already being mitigated by separate Incident Management Systems or the management of a large incident that has several IMS teams assigned to it. If the incidents involve multiple jurisdictions, a unified area command must be established. For example, area command needs to be established to manage independent multiple incidents drawing from the same pool of resources (Fig. 4.6).

The IMS allows for development in either the single, unified, or area command system. However, it is vital that all agencies and organizations agree on the evolution of the system and work together for successful termination of the incident.

National Incident Management System–HSPD 5

In 2003, the President of the United States issued Homeland Security Presidential Directive-5 (HSPD-5). The purpose of HSPD-5 is to increase the ability of the

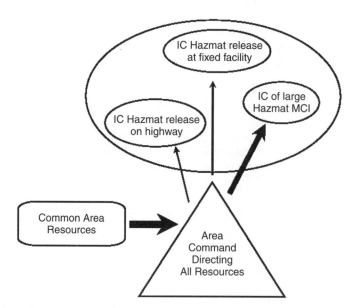

Figure 4.6   An area command structure may be implemented to manage multiple incidents that are already being mitigated by separate Incident Management Systems or to manage a large incident that has several IMS teams assigned to it.

United States to manage large-scale domestic incidents. As a part of the National Response Plan (NRP) (see Chapter 2) HSPD-5 calls for the creation of a National Incident Management System (NIMS), which provides a standardized system for implementing the NRP.

The NIMS is tailored after the proven National Inter-agency Incident Management Systems (NIIMS). NIMS consists of five major components, which are listed below:

- Command and incident management
- System preparedness
- Reserve management
- Communications, information, and intelligence management
- Science and technology management

The use of the NIMS by federal agencies responding to terrorist or other major domestic incidents helps to provide uniform incident management across different agencies. The use of NIMS is also crucial when private, local, and state responders are already working and need federal assistance, or when an on-site federal agency is requesting assistance from other federal agencies.

COMMAND STAFF RESPONSIBILITIES

An important part of the IMS for major incidents is the command staff. This is probably the most overlooked part of the IMS and in many ways can be the most important part. The command staff consists of the information officer, the safety officer, and the liaison officer (Fig. 4.5).

Safety Officer

Responder safety first... responder safety always!

The safety officer is a member of the command staff. His role is to ensure that safety procedures are established and carried out throughout the incident response. The safety officer's specific functions are as follows:

- Obtain briefing from the IC
- Identify hazardous situations related to the incident
- Participate in planning meetings
- Review incident action plans
- Identify potentially unsafe situations
- Exercise emergency authority to stop and prevent unsafe acts
- Investigate accidents that occur within the incident area
- Review and approve the medical plan
- Maintain a unit log

The safety officer's importance cannot be overstated, as this person is responsible for the safety of all personnel involved with the incident response and the safety of the incident itself. The safety officer is the only person that can countermand an order given by the IC or the operations officer, and he should not hesitate to do so if, in his opinion, an operation is unsafe. However, any time an order is countermanded by the safety officer, the IC must be notified immediately.

In larger incidents, the safety officer may be aided by a safety staff, such as a medical officer, sector safety officers, and a decon safety officer. A hazardous materials sector safety officer is commonly designated for hazmat response operations. This staffing allows the safety officer to remain at the command post (CP) with the IC during planning sessions to offer safety suggestions. When a safety staff is developed, the safety staff members have the same authority to terminate unsafe operations as the safety officer but also carry the same responsibility to notify the IC immediately upon such action.

It has long been a goal of emergency services to be part of the solution and not part of the problem. The safety officer and his staff will play an important role in achieving this goal when they are assigned in a timely manner and perform properly.

Public Information Officer

The public has the right to know what is going on during an emergency. In many cases, accurate information about an incident can allay needless fear and anxiety on the part of the public. In some cases, the media may be able to assist responders by relaying information to the public on actions they should take because of the incident. In addition to being accurate, information must also be timely. Incident information can be compared to bread and contrasted with wine (Fig. 4.7). Like bread, information must be served when fresh to be good. Unlike wine, information does not improve with age.

The public information officer's (PIO) function is to control the location of the press and release accurate information concerning the incident as it is cleared by the IC. Toward this end, the PIO should establish a press area, preferably in a location where the press can safely see what is happening without jeopardizing the safety or effectiveness of the operation. From this location the PIO can meet with the press on a regular basis and give information and updates as needed.

Under no circumstances should the press be allowed to roam at will during the emergency phase of the incident. The PIO must be aware that some members of the press will try to do this. Members of the media can be most easily controlled by being provided accurate, timely information and good opportunities to secure photographs, film footage, and interviews with responders, if feasible.

The PIO should maintain a unit log of activities during the course of the incident.

Liaison Officer

Another member of the command staff is the liaison officer, who can be thought of as the "diplomat of the incident embassy." His responsibility is to coordinate the

Figure 4.7 Incident information is like bread, not wine. Like bread, information must be served when fresh to be good. Unlike wine, information does not improve with age.

involvement of agencies such as fire, law enforcement, Red Cross, public works, utilities, environmental regulators and cleanup contractors, who may become involved in an incident response. The specific functions of the liaison officer are:

- Obtain a briefing from the IC
- Provide a point of contact for assisting/cooperating agency representatives
- Identify agency representatives from each agency including communications link and location
- Respond to requests from incident personnel for interorganizational contacts
- Monitor incident operations to identify current or potential interorganizational problems
- Maintain a unit log

Larger incidents involving several agencies will require a liaison officer because the IC simply cannot keep up with which agencies are involved at various stages of the operation. A competent and capable Liaison Officer will be able to advise the IC of which agency can perform specific tasks and notify those agencies of their assigned tasks. The Liaison Officer is therefore an invaluable asset to the IC during the course of larger-scale response operations.

THE OPERATIONS SECTION

The operations section is responsible for carrying out tactical operations to achieve the strategic goals established by command. In minor incidents the Incident Commander may choose to manage the operations section directly. However, for larger incidents, the Incident Commander will need to appoint an Operations Section Chief to oversee the operations section (Fig. 4.5). Think of the Operations Section Chiefs as the quarterbacks of the IMS team–they deliver the IC's game plan.

Structure of the Operations Section

The size of the response operation will dictate the structure of the operations section. For a small hazmat spill confined to a small area (by the nature of the material, the geographic area of the spill, the amount of material, or a combination of these), operations may be very small, with the hazmat group utilizing an entry unit and a decon unit. For larger spills, the operations section may need to be subdivided into Divisions and Groups. Divisions are geographic subdivisions, and groups are functional subdivisions.

Division or Sectoring at the Incident Scene
Divisions are created to divide an incident into geographic areas of operations. For example, if we have a hazmat spill in a warehouse, we might want to establish a front division of operations and a rear division of operations. A synonym for "division" is the term "sector." Using the above example, there would be a front sector and a rear sector.

It is common practice in emergency operations to use sectoring. However, it is very important for the operations officer to appoint a sector supervisor for each sector created. It should be noted that some emergency response organizations also use the term "sector" as a synonym for the term "group," as defined in the following section.

Operational Groups
Groups are established to divide the incident into functional areas of operation. The most common groups at a hazmat incident are:

- Hazmat
- Fire suppression
- EMS

The hazmat group will be subdivided into site access control, entry, and decontamination units (Fig. 4.5). The fire suppression group will concentrate on extinguishing or controlling fire at the incident scene or preventing fire from occurring. The EMS group is responsible for the health and safety of emergency responders working at the incident and victims of the incident who need emergency medical care. EMS personnel may be directly involved in rescue operations. This will depend on the level of training and personal protective equipment available to the EMS personnel.

Basic Considerations for Operations

Control of Staging

One of the most important functions of the IMS is staging. Proper staging prevents freelancing by not allowing incoming apparatus to arrive on-scene without direction and placement from the IC. There are two levels of staging in the IMS.

Level I Staging. Typically used in fire and mass casualty response efforts, the first apparatus on-scene reports incident conditions and assumes command. Other responding apparatus are directed to stage at a location specified by the IC or by preestablished incident command procedures before arriving on-scene. Depending on the type of incident and mode of command, these vehicles are to stage at the nearest safe location. As needed, the IC then directs apparatus to the scene. This type of staging approach may be useful in some hazmat applications.

Level II Staging. Level II staging is more commonly used in responding to major hazmat incidents because of their inherent complexities. To prevent the incident site from becoming congested, the IC must effectively manage all arriving personnel and equipment. This may be accomplished by the IC directing the Operations Officer (or possibly the Logistics Officer) to appoint a Staging Officer or, as commonly practiced in the fire service, the Staging Officer is the first arriving officer at the staging area. The Staging Officer designates a staging area near the incident to temporarily locate response resources. The Staging Officer also manages and releases resources as requested to support the strategy established by the IC.

Communication Considerations for Operations

There must be clear communication between the Incident Commander and the Operations Section Chief at all times. In most cases the Operations Chief will be located at the command post. This makes communication much easier and allows the Operations Chief to have first-hand knowledge of what is happening in planning and logistics. If the Operations Section Chief is located someplace other than the command post, a dependable means of communication must be maintained at all times between operations and command. It will also be critical for the Operations Section Chief to be able to communicate effectively with subordinate Group Supervisors and for Group Supervisors to be able to communicate with Unit Leaders.

Communications between personnel working in the hot zone and their supervisor are absolutely critical. For this reason, a standard practice is to designate a radio channel solely for communications between entry teams and the team supervisor. For personnel operating in the hot zone, establishing standard operating procedures before an incident, and clearly stating entry objectives in preentry briefings, can reduce the need for communication during an entry. However, if the entry team will be required to work out of sight of the supervisor directing the team, radio check or other procedures must be set up to confirm their status at regular intervals. If communication cannot be confirmed during an entry, it may be appropriate to assume that the team members are in trouble and the backup team may need to enter. Standard Operating Procedures (SOPs) should address procedures to be followed in the event that communication equipment fails during an entry.

The SOPs should cover backup communication procedures to be used in the event of communication equipment failure. Backup procedures may include the use of signboards or hand signals for line-of-sight communications. Emergency alarm signals should be designated for warning entry team members to immediately evacuate the hot zone or take other protective actions. The signals should be brief, limited in number, distinct from any ordinary signals that may be used, and rehearsed regularly to be effective.

Transfer of Information During Shift Changes for Prolonged Operations

If the incident is prolonged, assistants must be appointed to keep operations running while the operations chief sleeps. The most important aspect of the prolonged operation is good record keeping and proper transfer of information as shifts change. Each time someone assumes a job, there is a tremendous chance for information to be lost or at least not passed along. The officers in the Incident Command System must continuously guard against this loss of information.

Operations for Mass Casualty Incidents Related to Hazmat or WMD

What is the common bond in the Crescent City, Oklahoma City, Waverly, NYC 911, and other noted chemical and terrorist incidents? The large numbers of casualties and the requirement for additional resources needed to mitigate such large-scale incidents.

A Mass Casualty Incident (MCI) is not a new concept to the emergency response services. An accepted definition of MCI in the first responder community is any type of incident regardless of size that depletes the on-scene resources and temporarily overwhelms the responding agency. Mass casualty scene management begins with the implementation of IMS and rapidly addressing the incident priorities to ensure a timely triage.

Mass Casualty Triage. The difference in MCI triage and routine incident triage is the need to determine the "most good for the largest number of victims." The Simple Triage and Rapid Treatment system or START system is one of the easiest and most widely accepted mass casualty triage methods. This system was developed in 1983 in Newport Beach, California, and is designed to work within the IMS. The process only requires 30 to 60 seconds per patient and can be performed by basic life support (BLS)-trained emergency medical technicians (EMT) allowing advanced life support (ALS)- trained paramedics to treat critically injured patients.

The START system is based on three functions: respiration, perfusion, and mental status (RPM). The typical triage ticket (Fig. 4.8) is designed for the most

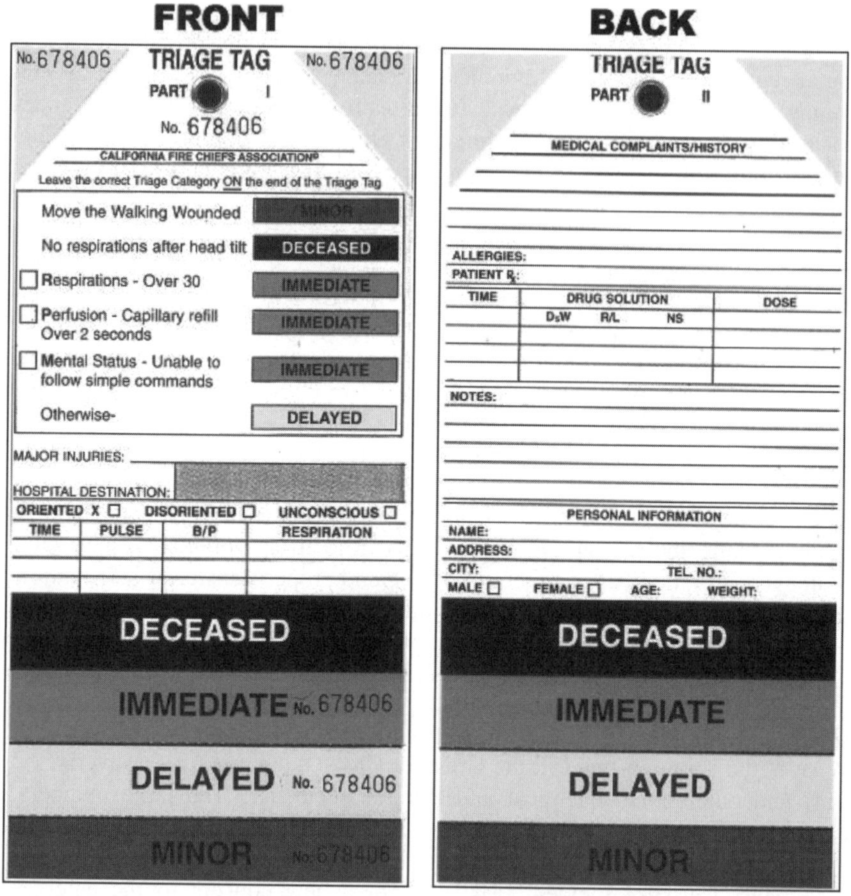

FRONT

BACK

Figure 4.8 The typical triage ticket is designed for the most common form of triage involving trauma and burns. (Courtesy of the California Fire Chiefs Association).

common form of triage involving trauma and burns. However, increased threats of a terrorist event involving weapons of mass destruction (WMD) have created the need for the development of a new "all hazards" triage ticket or tag.

In 2002 Disaster Management Systems, Inc. developed the "All Risk Triage Tag" (Fig. 4.9). This new tag can be used with the existing START system. This triage tag is widely accepted in the emergency response community and endorsed by the California Fire Chiefs Association and California Metropolitan Medical Response System (MMRS) cities. Triage tickets or tags are important pieces of incident documentation and critical to the expedient and appropriate medical care of victims (Fig. 4.10). For information regarding START kits, contact your local medical supply companies, and for more information on the triage tag, contact Disaster Management Systems, Inc. by visiting their website at www.triagetags.com.

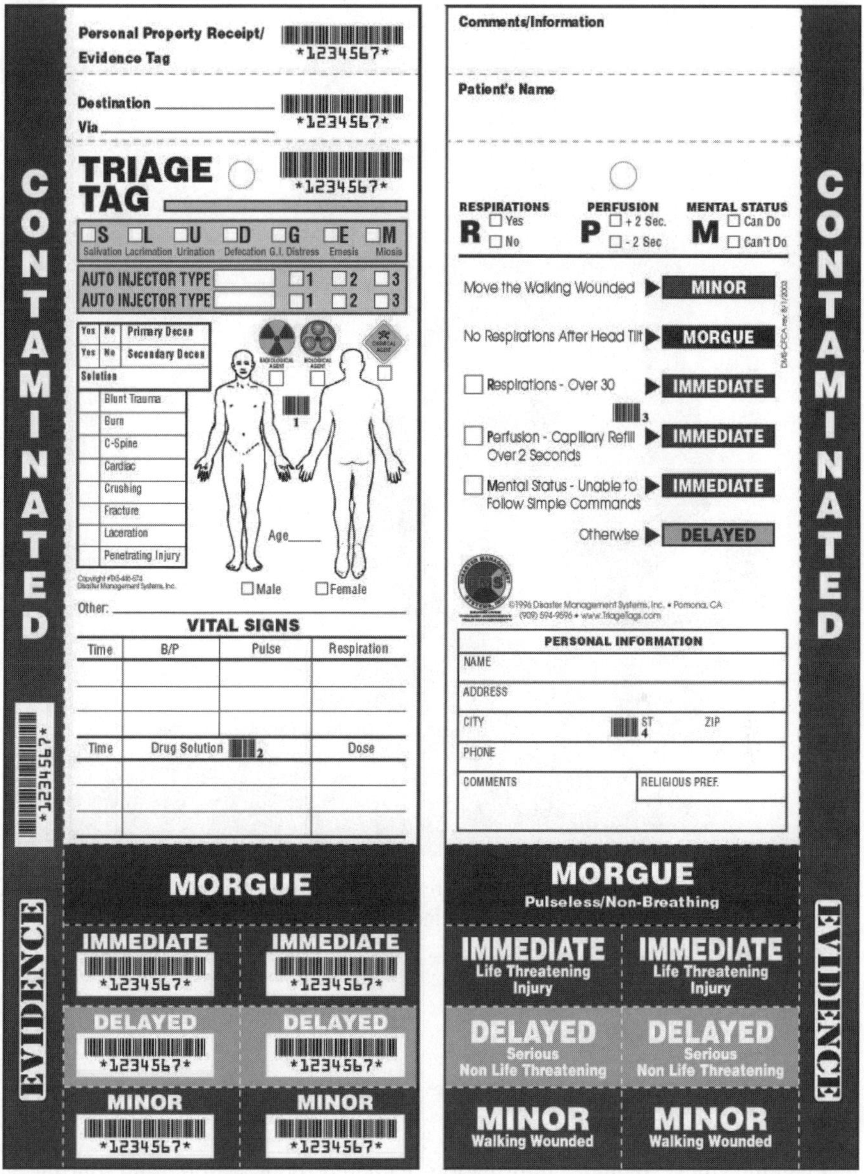

Figure 4.9 The "All Risk Triage Tag" by Disaster Management Systems, Inc. can be used with the START system and is suitable for responding to terrorist events involving weapons of mass destruction (WMD) as well as common trauma and burns. (Courtesy of Disaster Management Systems, www.triagetags.com).

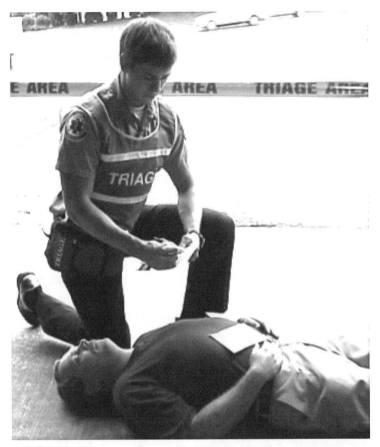

Figure 4.10 Triage tickets or tags are critical to the expedient and appropriate medical care of victims and are also important for incident documentation. Note the use of vests to identify IMS roles.

THE PLANNING SECTION

Failing to plan is planning to fail!

Planning must include information gathered both before and during an incident, because preplanning is a critical aspect of emergency response. The site-specific emergency response plan should include general provisions for safely terminating any hazmat emergency that can reasonably be anticipated. The plan can then be updated or fine-tuned based on information gathered during size-up of the actual incident.

At the scene of any operation, every effort should be made to gather as much useful information as possible before mitigation operations begin. This information-gathering and decision-making process begins as emergency response units approach the incident location and proceeds until the operation is completed. For example, placards, labels, container types, and release locations may be identified as the hazard area is surveyed by the initial emergency responders. Based on this, information can be obtained from MSDS sheets, various reference documents, and agencies such as Chemtrec. Information can be provided by computer databases or through the Internet or by E-mail at the incident scene.

For incidents involving hazmat transportation vehicles, shipping papers should be obtained as soon as possible, sometimes by the initial response team and sometimes by the recon team (if entering into a hazardous area, using PPE is required). As the incident unfolds, information from other agencies, such as the shipper or the manufacturer of the product involved, may lend helpful information. Detailed information on information gathering and decision-making is contained in Chapters 5 and 6.

Assigning a Planning Section Chief in the early stages of a major incident is of the utmost importance. As information is gathered, the Planning Section Chief (and his staff in major incidents) can review the information and help the incident commander and operations officer in decision-making.

Armed with as much information as possible before operations, personnel undertake recon, rescue, confinement, and containment operations. During these operations, especially recon, more information is gathered and additional decisions concerning the incident must be made. The incident commander must depend on his staff to constantly gather and review information and offer advice on decision-making. Thus staff meetings are almost constant during a hazmat incident.

With a constant flow of information and continuous decision-making, the incident commander guides the incident through its course. The ultimate responsibility for each decision rests with the incident commander.

THE LOGISTICS SECTION

As decisions are made and instructions are issued to the operations section, personnel, equipment, supplies, and services must be available so that operations can function as required. Logistics provides the supplies, services, and facilities required to support the entire response operation. Logistical needs may include:

- Operational support, such as supplying breathing air, heavy equipment, fuels, sorbent materials, and other items needed for the response operation
- Facilities, such as the command post, required at the scene to support the operation
- Communications, ensuring that communication equipment required is available and sufficient for the needs of the operation
- Rest and rehabilitation for personnel involved in the operation

These needs require a logistics section headed by a Logistics Section Chief answering to the Incident Commander and coordinating with the Operations

Section Chief. The Logistics Section Chief acts as the supply sergeant of the response and must be able to get anything at any time, as dictated by the situation, the Incident Commander, and the Operations Section Chief.

All reserve personnel, equipment, and supplies should be located at a staging area. As personnel, equipment, and supplies arrive on the scene, they should be directed to report to staging and wait for deployment as the operations section needs them. The logistics officer and his staff should log personnel, equipment, and supplies in and out of the staging area and coordinate this logging activity with the safety officer and his staff. This is especially important in accounting for personnel.

At the command post, constant communication between the Incident Commander, Operations Section Chief, Safety Officer, Planning Section Chief, and Logistics Section Chief must take place. The Logistics Section Chief must know the quantity, quality, and type of personnel, equipment, and supplies that will be needed and should be provided this information well in advance if possible.

THE FINANCE SECTION

Another major functional section of the incident command section is the finance section. Nothing involving hazardous materials incident response is free except advice. Hazmat operations cost money, and someone must pay. Even if the responsible party is on the scene and accepting the responsibility for payment, someone must keep a record of what is being used so that a statement can be prepared. This information may prove vital if claims are filed. For major response operations, the Incident Commander should appoint a Finance Section Chief who acts as the "bean counter" for the response, with the responsibility of putting a price on everything that is expended in the response operation.

PUTTING IT ALL TOGETHER

Managing an emergency operation can be very challenging. It is critical for the Incident Commander, and all other personnel involved in managing the incident response, to make sure that all areas of concern are addressed. One longstanding practice for ensuring that nothing "falls through the cracks" is the use of incident management tactical work sheets and checklists when responding to an incident. Generic work sheets are available from a number of commercial sources. Work sheets can also be readily created, allowing them to be adapted to the specific needs or preferences of the responders who will use them (Fig. 4.11). The use of work sheets helps to promote a comprehensive approach to incident response and to avoid taking inappropriate actions based on a misperception of the situation. Misperceptions can occur when responders fail to keep an open mind and allow

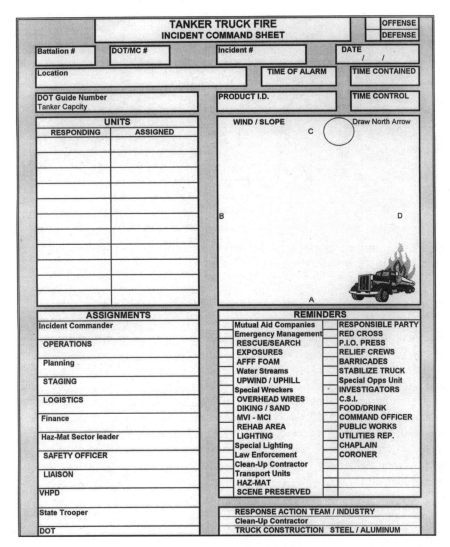

Figure 4.11 The use of tactical work sheets promotes a comprehensive approach to incident response. This work sheet was developed by the Vestavia Hills, AL Fire Department specifically for use in responding to highway cargo tank fires.

preconceived notions to cloud the perception of events and judgment regarding appropriate response options. We can think of response options as tools in the responder's toolbox—it is important to maintain a variety of options to select from. Keep in mind that if the only tool in the Incident Commander's toolbox is a hammer, everything tends to look like a nail (Fig. 4.12).

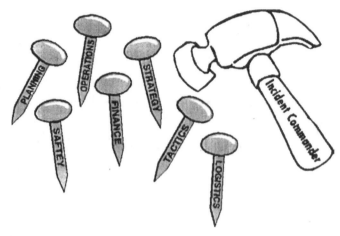

Figure 4.12   If the only tool in the Incident Commander's toolbox is a hammer, everything about the incident tends to look like a nail.

SUMMARY

In conclusion, history has shown that disastrous events can happen suddenly anywhere, anytime. They can range from a simple routine emergency to the most complex of incidents. The person in charge must handle the incident efficiently and effectively. For that reason, there is an overwhelming need for an emergency management system that will work on all types of incidents, a system that will function effectively. IMS meets these needs. It can be used on a small, simple, routine incident or a large, complex emergency incident. The Incident Management System has been used by emergency response agencies of all types and sizes across the United States. It is an effective management system that has been validated through years of application.

5

INCIDENT ASSESSMENT AND INITIAL ACTIONS OF RESPONDERS

The driver of a tractor trailer was approaching the heavily populated residential outskirts of a large metropolitan area when he discovered a fire in his truck. The driver pulled onto the shoulder of the interstate highway and was quickly joined by police and hazmat-trained firefighters. There were no placards on the truck, and, because responders were unable to retrieve the shipping papers from the cab in the fire, the driver was the only source of information about the contents of the trailer. The driver was carrying a load for a landscaping company and stated that the material was urea nitrate. DOT classifies urea nitrate as a flammable solid when it is shipped wetted with at least 20 percent water by mass. It is classified as an explosive if it is shipped dry or wetted with less than 20 percent water by mass. In response, the interstate was closed and a large residential area along both sides of the highway was evacuated while attempts were made to contact the shipper. Extensive manpower and funds were expended, and the public was greatly inconvenienced. Responders finally contacted the shipper, who faxed accurate information about the shipment. The cargo was actually urea—a soluble source of nitrogen used as fertilizer. Urea is not flammable, not a hazardous material, and not regulated during transport.

In this example, bad information resulted in needless expense and inconvenience. In similar instances, responders or members of the public have been injured or killed because of inaccurate information and inappropriate actions. To prevent this, first responders to a hazmat incident must take several important

Emergency Responder Training Manual for the Hazardous Materials Technician, Second Edition, edited by Kenneth W. Oldfield
ISBN 0-471-21387-X   Copyright © 2005 John Wiley & Sons, Inc.

initial actions, all of which require accurate information. As a minimum, first responders must:

- Recognize that a hazardous materials emergency is under way
- Make notification to initiate assistance from more highly trained responders, such as the hazmat response team
- Identify the hazardous materials involved, if it can be done safely
- Establish a safe perimeter around the incident and secure the scene
- Avoid exposure to the hazards of the incident and prevent others from becoming exposed
- Continue to assess the situation and take appropriate initial response actions

This chapter discusses several ways to identify hazardous materials, and how you can use this information to determine the initial actions that you should take.

RECOGNIZING AND IDENTIFYING HAZARDOUS MATERIALS

There are several ways in which you, as an emergency responder, can safely and dependably recognize the presence of hazardous materials and tentatively identify them. Other methods may be undependable or hazardous to you (Fig. 5.1).

Some of our most valuable clues for initial response to hazmat incidents are mandated by federal regulations. Shipping papers and labels, placards, and container markings all provide accurate assessment information for transportation incidents.

Figure 5.1   When assessing hazmat incidents, avoid becoming part of the problem.

In this section, we begin with an overview of relevant regulatory considerations. Then we will consider several ways that hazmats can be detected and identified, including:

- Personal knowledge, emergency response plans, and facility maps
- Shipping papers
- Placards, labels, and other hazard identification markings
- Occupancy and location of the incident
- Container configurations and features
- Detection equipment and sampling of hazardous materials
- Biological indicators
- Human senses

Overview of Regulatory Requirements Related to Incident Assessment

From a global perspective, a number of entities are involved in regulating the shipping of hazardous materials. Within the United States, the Department of Transportation (DOT) has primary regulatory authority for hazardous materials transportation by all modes. The International Civil Aviation Organization (ICAO) has produced technical instructions for air shipment of hazardous materials that apply internationally. The International Maritime Organization (IMO) has produced the International Maritime Dangerous Goods (IMDG) Code, which regulates the international transport of hazardous materials by water. Canadian shipments are regulated by the Canadian Transportation of Dangerous Goods (TDG) Code. The United Nations (UN) has developed recommendations applicable to all modes of international transport. In the past, significant differences existed between DOT hazard classifications used in the United States and hazard classifications used elsewhere in the world. During the 1990s, DOT regulations were revised to bring the DOT hazard classes into closer agreement with international hazard classes by making greater use of the IMO system. In this section, we focus mainly on hazard classifications used by DOT, but we also consider EPA classifications for pesticides and hazardous wastes.

DOT Hazardous Materials Regulations and Hazardous Materials Table
DOT Hazardous Materials Regulations are listed in 49 CFR 171–180. Each hazardous material regulated by DOT has a proper shipping name and a four-digit identification number assigned to it.

The proper shipping names and identification numbers for the materials regulated by DOT are listed in the Hazardous Materials Table, found in 49 CFR 172.101. Most identification numbers are recognized internationally and are preceded by the letters "UN," but some are only recognized in North America and are preceded by the letters "NA." In most cases the number designates a single hazardous material, but in some cases it may identify several materials with very similar hazardous properties.

DOT Hazard Classes, Divisions, Packing Groups

DOT classifies hazardous materials according to the types of hazards they pose during transport. The DOT hazard classification system consists of nine hazard classes, with most of the classes being subdivided into two or more divisions (Table 5.1). The placards and labels, as shown inside the covers of this textbook and used on shipping containers, generally coincide with the DOT hazard classes and divisions.

TABLE 5.1   DOT Hazard Classification System

Class 1—Explosives	
Division 1.1	Explosives with a mass explosion hazard
Division 1.2	Explosives with a projection hazard
Division 1.3	Explosives with predominantly a fire hazard
Division 1.4	Explosives with no significant blast hazard
Division 1.5	Very insensitive explosives; blasting agents
Division 1.6	Extremely insensitive detonating articles
Class 2—Gases	
Division 2.1	Flammable gases
Division 2.2	Nonflammable, nontoxic * compressed gases
Division 2.3	Gases toxic* by inhalation
Division 2.4	Corrosive gases (Canada)
Class 3—Flammable Liquids [and Combustible Liquids in the U.S.]	
Class 4—Flammable Solids; Spontaneously Combustible Materials; and Dangerous When Wet Materials	
Division 4.1	Flammable solids
Division 4.2	Spontaneously combustible materials
Division 4.3	Dangerous when wet materials
Class 5—Oxidizers and Organic Peroxides	
Division 5.1	Oxidizers
Division 5.2	Organic peroxides
Class 6—Toxic Materials and Infectious Substances*	
Division 6.1	Toxic* materials
Division 6.2	Infectious substances
Class 7—Radioactive Materials	
Class 8—Corrosive Materials	
Class 9—Miscellaneous Dangerous Goods	
Division 9.1	Miscellaneous dangerous goods (Canada)
Division 9.2	Environmentally hazardous substances (Canada)
Division 9.3	Dangerous wastes (Canada)

*The words "poison" or "poisonous" are synonymous with the word "toxic."

Source: Emergency Response Guidebook, 2000

TABLE 5.2   DOT Packing Groups

Packing Group	Degree of Danger
PG I	Great danger
PG II	Medium danger
PG III	Minor danger

The Hazardous Materials Table assigns hazmats in some hazard classes to one of three Packing Groups (PG), based on the degree of danger presented by the material (Table 5.2). Materials assigned to PG I represent a great danger. Materials assigned to PG II represent medium danger. Materials assigned to PG III represent a minor danger.

EPA Hazard Classifications

EPA classifies substances based on the type of harm they pose to the environment or to living things. EPA has developed a list of hazardous substances and established Reportable Quantities (RQs) for each. EPA requires that releases in excess of the RQ be reported to the agency. DOT treats EPA's hazardous substances as hazardous materials during transport if an amount exceeding the RQ for the substance is shipped in a single container.

EPA has created a specific classification system for labeling pesticides, as discussed in the section on container labels below. EPA has developed a list of substances that must be disposed of as hazardous wastes if disposal is required. Under EPA regulations, hazardous wastes are classified based on whether they are ignitable, corrosive, reactive, or toxic. Table 5.3 shows EPA hazardous waste categories and how they are defined. In addition to the substances listed by chemical name, any substance meeting EPA's criteria for ignitability, corrosivity, reactivity, or toxicity is required to be treated as hazardous waste for disposal without regard to the specific identity of the substance.

Other Assessment-Related Regulatory Requirements

Other regulations related to topics such as emergency planning and hazard communication include provisions that can be helpful to responders conducting assessment of hazmat incidents. These regulatory requirements are described in Chapters 2 and 3.

Personal Knowledge, Emergency Response Plans, and Facility Maps

In some situations, you may have personal knowledge of the locations and identities of hazardous materials that may be involved in an emergency incident. In other situations, you may be able to rely on the personal knowledge of other people you consider trustworthy. As an example, assume that you work at a facility and know that a specific container is only used for the storage of number 2 fuel oil. If you see a liquid product leaking from that container, you can be relatively sure that the product is number 2 fuel oil. For off-site responders, documentation

TABLE 5.3   EPA Hazardous Waste Categories

40 CFR...	Hazard Code	Code Category	Definition or Description
§261.21	D001	Characteristically ignitable	Wastes meeting any one of the following: <ul><li>Liquid other than an aqueous solution containing <24% alcohol and with a flash point <140°F</li><li>Not a liquid, but under standard temperature and pressure capable of causing fire</li><li>Defined as an ignitable gas in 49 CFR 173.300</li><li>Defined as an oxidizer in 49 CFR 173.151</li></ul>
§261.22	D002	Characteristically corrosive	Wastes meeting any one of the following: <ul><li>pH ⩽ 2 or ⩾ 12.5</li><li>Ability to corrode steel at a rate >0.25 in. per year @ 130°F</li></ul>
§261.23	D003	Characteristically reactive	Wastes meeting anyone of the following: <ul><li>Normally unstable and readily undergoes violent change without detonating</li><li>Reacts violently with water</li><li>Forms potentially explosive mixtures with water</li><li>When mixed with water generates toxic gases, vapors, or fumes sufficient to cause harm</li><li>Cyanides- or sulfides-bearing wastes when exposed to pH conditions between 2 and 12.5 generate toxic gases, vapors, or fumes sufficient to cause harm</li><li>Capable of detonating if subjected to irritating source or heat under confinement</li><li>Readily capable of detonation or explosive decomposition or reactive at standard temperature and pressure</li><li>Classified as a forbidden explosive by 49 CFR 173.51 or a Class A or B explosive by 49 CFR 173.53 and 88, respectively</li></ul>

TABLE 5.3 (*continued*)

40 CFR...	Hazard Code	Code Category	Definition or Description
§261.24	D004-D043	Characteristically toxic	Wastes determined to be toxic by EPA accepted testing methodology containing any one of the listed hazardous components in Table 1 of 40 CFR 261.24
§261.31	"F Codes"	Listed hazardous wastes from nonspecific sources (processes)	Such as, spent halogenated and nonhalogenated solvents, wastewaters, wastewater treatment sludges
§261.32	"K Codes"	Listed hazardous wastes from specific sources (processes)	Such as, distillation bottoms, spent filter media, etc., from common specified process
§261.33	"P Codes"	Listed toxic and/or reactive acutely hazardous wastes	Discarded commercial chemical products, off-specification species, container residues, and spill residues
§261.33	"U Codes"	Listed toxic, reactive, ignitable, and/or corrosive hazardous wastes	Discarded commercial chemical products, off specification species, container residues, and spill residues AND subject to the small quantity generator status in 40 CFR 261.5 (a) and (g)

such as emergency response plans and chemical storage and use maps may be dependable sources of information (see Chapter 3).

When you are dealing with information based on the personal knowledge of others, consider the source. Caution is in order, and a personal judgment may be required regarding the dependability of the information. Remember: The best time to resolve any confusion about the identity of hazardous materials is during the preemergency planning phase.

Shipping Papers

Public sector responders are called on to deal with hazmat incidents involving commodities being transported. Private sector emergency response brigade members at an industrial facility may also be involved in transportation incidents, such as those occurring during loading or unloading of hazardous materials for transport. In such an incident, the hazardous materials involved may be bound for another facility on the transporter's route, so that personnel at the scene of the incident are unfamiliar with them. In transportation incidents, information contained in shipping papers may provide you with a specific identification of the hazardous materials involved.

TABLE 5.4   Shipping Papers

Mode of Transportation	Shipping Paper	Location of Papers	Party Responsible
Air	Air waybill	Cockpit	Pilot
Highway	Bill of lading	Cab of vehicle	Driver
Rail	Consist and waybills	Engine	Conductor
Water	Dangerous cargo manifest	Bridge or pilot house	Captain or master

The Bill of Lading is the standard shipping paper for highway transport (Table 5.4). For rail shipments, two types of shipping papers are used, the Train Consist and Waybills. The Train Consist lists all railcars making up a train, identifies the cars that transport hazardous materials, and may identify the hazmats on board. The Waybill is the shipping document for an individual railcar. The Dangerous Cargo Manifest is the primary shipping paper for water transport. The Air Waybill with Shipper's Certification for Restricted Articles is used for air shipment. For shipments of materials regulated by EPA as hazardous wastes, the Uniform Hazardous Waste Manifest is the required shipping paper.

For highway transportation, the shipping papers should be in the possession of the driver (Table 5.4), and are usually in the driver's side door pocket during operation. They are supposed to be within arm's reach of the driver at all times while the vehicle is under way. For rail transport, the shipping papers should be in the possession of the conductor, who should be in the locomotive while the train is under way. During air shipment of hazardous materials, the pilot in command should have the Air Waybill in the cockpit of the plane. For water transport, the shipping papers should be in the possession of the Captain or Master on the bridge of a vessel or the pilot house of a tugboat. During barge transport, the papers may be in a pipelike container mounted on the barge.

Regardless of the mode of transportation, the shipping papers must show the basic description of the commodity being transported, including:

- Proper shipping name
- Hazard class and division
- Hazardous material identification number
- Packing group

Other useful information, such as the quantity of hazardous material being shipped, must also be included. Hazardous and nonhazardous materials may be included in the same shipment, but hazmats must always be clearly delineated on shipping paper entries.

Some Division 2.3 and 6.1 materials are considered exceptionally toxic through inhalation. For these materials, the words "Poison–Inhalation Hazard" must be entered on shipping papers, and packagings must be marked "Inhalation Hazard" or display an inhalation hazard label or placard.

Additional information for the benefit of emergency responders must be printed on the shipping papers or carried with them in the form of Material Safety Data Sheets or other written documents. Emergency information documents must contain, at a minimum, the following:

- Description of the material (technical name)
- Immediate hazards to health
- Risk of fire or explosion
- Immediate precautions to take in the event of accident
- Immediate methods for handling fires
- Initial methods for handling spills or leaks
- Preliminary first aid for exposure victims

A 24-hour emergency response contact telephone number must be entered on shipping papers. The telephone must be monitored at all times by someone who is knowledgeable of the hazards and characteristics of the material being shipped. The shipper may list the number of an agency such as CHEMTREC if the agency has been given all the required information on the material. CHEMTREC is an information service operated by the Chemical Manufacturers Association that provides emergency information by telephone to assist responders.

For rail transportation, the standard transportation commodity code (STCC or "stick" code) is a code system used to assign numbers to commodities being shipped. STCC numbers are commonly listed on rail shipping papers. STCC numbers beginning with the digits "49" are hazardous materials.

Shipping papers can obviously provide a lot of valuable information for initial assessment. You should place a high priority on accessing them early on, but only if you can do so safely.

Labels, Placards, and Other Hazard Identification Markings

In many instances, your first indication of the presence of hazardous materials will be labels, placards, or markings on hazardous materials containers or storage areas. Labels and placards commonly required by DOT are shown inside the covers of this textbook. In some cases, these indicators may allow a specific identification of the materials involved. In other cases, they only provide the general hazard categories. Always determine the specific identity of the hazardous materials involved as soon as safely possible. In this section we describe the most commonly available systems for hazard recognition and identification.

One handy reference for label and placard familiarization is the DOT Chart 12 currently available through the U.S. Department of Transportation's Research and Special Programs Administration. The chart is revised periodically, with a new number assigned for each revision, so that future versions will be designated DOT Chart 13 and higher.

Labels and Markings on Nonbulk Containers

A nonbulk container is defined by the Department of Transportation as a packaging with a maximum capacity of 450 liters (119 gallons) or less as a receptacle for a liquid; a maximum net mass of 400 kilograms (882 pounds) or less or a maximum capacity of 450 liters (119 gallons) or less as a receptacle for a solid; or a water capacity of 454 kilograms (1000 pounds) or less as a receptacle for a gas. Examples of nonbulk containers include cans, boxes, carboys, cylinders, and drums (see Fig. 5.10 below).

Nonbulk containers hold a variety of hazardous materials, including some from every hazard class. DOT regulations include extensive rules for labeling nonbulk containers. These containers must be designed to DOT specifications to safely contain hazardous materials. Hazard class labels and other labels and markings on these containers can be extremely valuable to you in seeking to identify materials involved in an incident.

Hazard Class Labels. Hazard class labels are standard labels that identify the DOT hazard class of the contained material. In some cases, labels indicate the hazard class division of the product. Labels are required to be at least 100 mm (3.9 in.) on each side. Each label has a unique design and provides information in four different ways. The primary hazard is indicated by a label that shows (1) text identifying the hazard class/division, (2) a hazard symbol, (3) the hazard class number, and (4) a color coding for the hazard (Fig. 5.2). However,

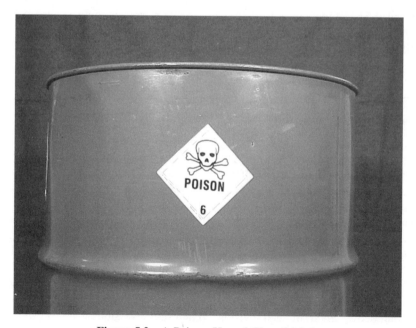

Figure 5.2   A Poison, Hazard Class 6 label.

text indicating hazard class/division may be omitted. Labels commonly used in domestic shipment are shown inside the covers of this textbook.

If a hazardous material meets the definition of more than one hazard class, additional labels must be used on the container to indicate the secondary or subsidiary hazard of the contents. Beginning October 1, 2005, subsidiary hazard class labels will be required in rail and highway transport.

Container Markings. DOT regulations require that all nonbulk containers bear certain markings pertaining to the product within the container. These markings include:

- The proper shipping name
- The four-digit UN or NA identification number
- Markings indicating the UN specifications to which the container is constructed
- Instructions and cautions
- Any additional information required by the DOT hazardous materials regulations for the specific product involved.

Inhalation Hazard Labels and Markings. DOT requires special markings for certain Division 2.3 and 6.1 materials that are considered exceptionally toxic through inhalation. Nonbulk packagings containing these materials must be marked "Inhalation Hazard" or display an inhalation hazard label.

Mixtures in a Container. When two or more chemicals are mixed inside one container, they may form a new chemical that has a name. If not, the proper shipping name is based on the properties of the mixture. For example, the proper shipping name of two flammable liquids that have been combined—but which do not form a new, named chemical—is "Flammable Liquid, N.O.S." where N.O.S. stands for Not Otherwise Specified. The regulations require that the two components in a mixture that contribute most greatly to the hazard must be identified by name on the label.

Pesticide Container Labels. Pesticides are regulated by the Federal Insecticide, Fungicide, and Rodenticide Act (FIFRA) and are labeled under an EPA terminology system that differs from the DOT labeling system for nonbulk containers (Fig. 5.3). Pesticides are generally defined as chemicals that are designed to kill some type of living organism, like an insect, weed, or fungus, which is considered to be a pest. Any chemical that kills other living things will most likely have some bad health effects on humans. Some of them are very toxic in small amounts. A signal word on the label indicates the degree of toxicity of the pesticide:

- DANGER is used for the most toxic pesticides.
- WARNING is used for moderately toxic pesticides.
- CAUTION indicates a relatively minor degree of toxicity.

Figure 5.3   A pesticide container label.

Pesticide labels (Fig. 5.3) show the trade name (which is not useful for looking up hazard information, because it is not the actual chemical name); the Environmental Protection Agency registration number (which can be used to identify the chemical by the actual chemical name), and hazard and precautionary statements. Active ingredients are identified by chemical name, and these are the names that you can use in consulting references about health, fire, and reactivity hazards of the pesticide.

Placards and Markings on Transport Vehicles and Bulk Containers

DOT has established specific requirements for warning the public and emergency responders when hazmats are transported in bulk containers. A bulk container is defined by DOT as a packaging with a maximum capacity of greater than 450 liters (119 gallons) as a receptacle for a liquid; a maximum net mass of greater than 400 kilograms (882 pounds) or a maximum capacity of greater than 450 liters (119 gallons) as a receptacle for a solid; or a water capacity of greater than

454 kilograms (1000 pounds) as a receptacle for a gas. Common examples of bulk containers used in hazmat transportation include cargo tanks, rail tank cars, and intermodal tank containers, which are covered later in this chapter.

Placards. Placards are 10-3/4-inch signs that are affixed to both ends and both sides of vehicles carrying hazardous materials (Fig. 5.4). For most hazard classes and divisions, placards are very similar to labels in terms of lettering, color coding, symbols, and hazard class numbers. However, some labels have no equivalent placard and vice versa. Placards generally convey information in the same four ways that labels do: text indicating hazard class/division, hazard class number, hazard symbol, and color coding. However, text may be omitted from some placards and some placards may display four-digit identification numbers (Fig. 5.5). Examples of placards used in domestic transport are shown inside the covers of this textbook.

For purposes of placarding, hazardous materials are divided into two types: those for which placards are required when they are transported in any quantity (Table 5.5) and those for which no placards are required when the gross weight of hazardous materials in nonbulk packages on the vehicle is less than 1,001 pounds (Table 5.6). In essence, this provides an exemption from placarding requirements for loads with less than 1001 pounds of hazardous materials from hazard classes listed in Table 5.6. In addition, no placard is required for shipments of combustible liquids in nonbulk packages or for hazardous materials classified as Otherwise Regulated Material Category D (ORM-D) materials. This can cause

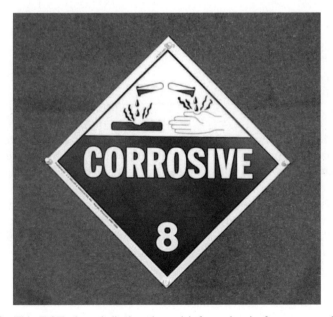

Figure 5.4 This DOT placard displays hazard information in four ways: color, hazard symbol, text, and hazard class number.

Figure 5.5   This bulk container placard shows the 4-digit identification number.

TABLE 5.5   Hazard Classes/Divisions that Require Placards for any Quantity

Hazard Class or Division	Placard Name
1.1	EXPLOSIVES 1.1
1.2	EXPLOSIVES 1.2
1.3	EXPLOSIVES 1.3
2.3	POISON GAS
4.3	DANGEROUS WHEN WET
5.2 (Organic peroxide, Type B, liquid or solid, temperature controlled)	ORGANIC PEROXIDE
6.1 (Inhalation Hazard, Zone A or B)	POISON INHALATION HAZARD
7 (Radioactive Yellow III label only)	RADIOACTIVE

problems for emergency responders who may encounter transport vehicles that contain hazardous materials but are legally unplacarded. Do not presume that no hazmats are present based solely on the fact that a vehicle has no placards!

Inhalation Hazard Placards and Markings. DOT requires special markings for certain Division 2.3 and 6.1 materials that are considered exceptionally toxic through inhalation. Bulk containers transporting these materials must be marked "Inhalation Hazard" or display an inhalation hazard placard.

The Dangerous Placard. If materials from two or more hazard classes listed in Table 5.6 are being transported in placardable quantities in nonbulk containers on the same truck, the "Dangerous" placard may be used instead of the individual placards. However, if 1000 kg (2205 lbs) or more of one of the materials is

TABLE 5.6   Hazard Classes/Divisions that Require Placards for 1001 lbs. or More

Hazard Class or Division	Placard Name
1.4	EXPLOSIVES 1.4
1.5	EXPLOSIVES 1.5
1.6	EXPLOSIVES 1.6
2.1	FLAMMABLE GAS
2.2	NONFLAMMABLE GAS
3	FLAMMABLE
Combustible Liquid	COMBUSTIBLE
4.1	FLAMMABLE SOLID
4.2	SPONTANEOUSLY COMBUSTIBLE
5.1	OXIDIZER
5.2 (Other than organic peroxide, Type B, liquid or solid, temperature controlled)	ORGANIC PEROXIDE
6.1 (Other than inhalation hazard, Zone A or B)	POISON
6.2	(None)
8	CORROSIVE
9	CLASS 9 [§172.504(f)(9)]
ORM-D	(None)

loaded at one facility, the placard for that hazard class must also be on the truck. If you approach a vehicle bearing the Dangerous placard keep in mind that you cannot determine what is inside the truck without further information from the shipper or the shipping papers.

Special Placarding Situations. A placard must remain on a bulk container or bulk transport vehicle even when it is empty. Some volatile, flammable materials leave behind vapors that are quite dangerous, and many toxic chemicals are hazardous through exposure to residues. Only when the tanker has been cleaned and purged, or cleaned and filled with a nonhazardous material, is the placard allowed to be removed.

Hazardous Materials Identification Numbers. DOT requires the four-digit hazardous material identification number to be displayed on certain hazmat shipments. All bulk containers such as highway cargo tanks, rail tank cars, and intermodal tank containers are required to display the four-digit hazardous material identification number. The four-digit number must also be displayed by vehicles that contain 4000 kg (8820 lbs) in nonbulk packages of a single hazardous material having the same proper shipping name and identification number. It is also required for shipments of 1000 kg (2205 lbs) of certain materials that are poisonous by inhalation.

The four-digit number may allow you to identify the material from a significant distance. The four-digit number is displayed across the middle of the placard in place of the hazard class text (Fig. 5.5) or displayed on a separate orange panel adjacent to the placard. The four-digit numbers allow you to identify hazardous materials commonly shipped in your jurisdiction. Use them before an emergency occurs to become familiar with the hazmats in your area and the types of containers in which they are commonly shipped.

Stenciled Commodity Names. For a number of materials shipped in rail tank cars, DOT requires that the commodity name be stenciled directly onto the tank car, as discussed in the section on railcars below. Stenciled commodity names are also required by DOT for other modes of transportation in certain cases. In addition to stencil-required commodities, an owner wishing to dedicate a container solely to transportation of any commodity may voluntarily stencil the name of the commodity onto any bulk transport container. A container displaying a stenciled commodity name cannot legally be used to transport any other commodity.

Company Names on Transport Containers. Some bulk transportation containers may have the names of known manufacturers or distributors of hazardous materials. For example, across the United States, BOC Gases is widely recognized as a supplier of industrial and medical specialty gases and cryogenics, Amerigas is recognized as a major supplier of propane, and Chevron is readily recognized as a motor fuel supplier. In addition to nationally recognized names such as these, you should be aware of the names of other manufacturers or distributors active in your area. In some cases, you may be able to contact the transporter or shipper of a product directly to seek information on the identity of the product.

Intermodal Container Hazard Identification Codes. Hazard identification codes may be found in the top half of orange panels found on some intermodal bulk containers. These codes are referred to as "hazard identification numbers" under European and some South American regulations. The four-digit identification number is found in the bottom half of the orange panel. Intermodal hazard identification codes are explained later in this chapter in the section on intermodal containers.

U.S. Military Warning Symbols
The United States Department of Defense has developed a system of warning symbols for hazardous materials as shown in Figure 5.6. These military markings will primarily be seen on structures and containers at U.S. military facilities.

Hazard Warning Systems Used at Fixed Facilities
Two systems are commonly used at fixed facilities to indicate the presence of hazardous materials. The National Fire Protection Association's (NFPA) 704 System was developed by NFPA to warn firefighters of the dangers of hazardous materials they may encounter during emergencies at fixed facilities. The Hazardous Materials Identification system (HMIS) was developed by adapting the NFPA 704 system for use in warning workers of the hazards of materials with

U.S. Military Markings

Chemical Hazard Symbol

RED = Toxic Agents
(e.g. tabun, sarin,
mustard agent)

YELLOW = Harassing Agents
(e.g. tear gas
smoke-producing agent)

WHITE = Illuminating Devices
(e.g. white phosphorus,
ethyl aluminum)

Apply No Water

**Wear Protective Mask
or Breathing Apparatus**

Mass Detonation

**Explosive with
Fragmentation Hazard**

Mass Fire Hazard

**Moderate
Fire Hazard**

Figure 5.6 These markings may be seen at U.S. military facilities.

which they work. In addition, other warnings may be present in fixed facilities such as commodity names on piping systems and vessels.

NFPA 704 System. You may see NFPA 704 labels on specific hazardous materials containers. In some cases, you may see them on fences or doors to identify general hazards of a storage area. Figure 5.7 shows an NFPA label. It is a diamond divided into four smaller diamonds, each representing a type of hazard. Health hazards are indicated in blue at the far left, flammability is shown at the top in the red diamond, reactivity with other chemicals appears on the right in yellow,

Figure 5.7   The NFPA 704 hazard identification system is used at fixed facilities.

and the lower white diamond provides information on specific hazards. Ratings range from 0 to 4, with 4 indicating the highest hazard level in each category, as shown below.

Health Hazards (Blue)

4 Materials too dangerous to health to expose firefighters in self-contained breathing apparatus (SCBA) and turnout gear, which is not adequate protective clothing.
3 Materials extremely hazardous to health, but areas may be entered with extreme care. SCBA should be worn and no skin should be exposed.
2 Materials hazardous to health, but areas may be entered with full-face SCBA.
1 Materials only slightly hazardous to health. It may be desirable to wear SCBA.
0 Materials that on exposure under fire conditions would offer no hazard beyond that of ordinary combustible material

Flammability (Red)

4 Very flammable gases or very volatile flammable liquids
3 Materials that can be ignited under almost all normal temperature conditions
2 Materials that must be moderately heated before ignition will occur
1 Materials that must be preheated before ignition can occur
0 Materials that will not burn

Reactivity (Stability) (Yellow)

4 Materials that are readily capable of detonation or explosive decomposition or explosive reaction at normal temperatures and pressures

3 Materials that are capable of detonation or explosive decomposition or explosive reaction, which require a strong initiating source or must be heated under confinement before initiation

2 Materials that are normally unstable and readily undergo violent chemical change but do not detonate. Includes materials that undergo chemical change with rapid release of energy at normal temperatures and pressures or that can undergo violent chemical change at elevated temperatures

1 Materials that are normally stable but that may become unstable at elevated temperatures and pressures or that may react with water with some release of energy, but not violently

0 Materials that are normally stable even under fire exposure conditions and that are not reactive with water

Note that reactivity classification under the NFPA 704 system is based on the stability of the material by itself or in contact with air or water. It does not consider the chemical's potential for reacting if it mixes with other chemicals.

Other Information

This diamond provides information on specific hazards. Symbols are used to relate hazards such as water reactivity, radioactivity, and the presence of an oxidizer.

HMIS System. The HMIS system is a variant of the NFPA 704 system, with a few significant differences. The HMIS labels are rectangular instead of diamond-shaped. Under the HMIS system, the fourth quadrant, which displays specific hazard information under the 704 system, may contain symbols indicating the type of protective equipment required for working with the substance.

Occupancy and Location of the Incident

Even locations that seem innocuous may harbor hazmats. For example, local hardware stores commonly store hazmats such as flammable paints and solvents, oxidizing pool chemicals and fertilizers, and toxic pesticides. Dry cleaners use cleaning solvents that can generate highly toxic gases in fires. Water treatment plants typically store large quantities of chlorine. The authors of this textbook work at a university that hosts a large research and medical complex with thousands of laboratories and several hospitals, all of which are likely locations for hazmats.

When specific information on the presence of hazardous materials is lacking, it may be possible to infer the likely presence of hazmats based on the location and type of business being conducted at the scene of the incident. Public sector responders might reasonably suspect that any call to a location such as a highway, rail line, airway facility, waterway facility, tank farm, pipeline location, or industrial facility could involve hazardous materials. Certain types of chemicals are routinely used, and can routinely be inferred to be present, in certain types of industries, as shown in Table 5.7. Structures and containers at fixed facilities may be labeled with company names that are readily associated with certain types of hazardous materials (Fig. 5.8).

The best practice is to identify the location of hazardous materials during pre-emergency planning and avoid unpleasant surprises later. Always remain alert for the presence of hazmats. Significant quantities of hazardous materials have been discovered stored illegally in residential occupancies. This has become much more common in recent years because of the proliferation of illegal drug manufacturing.

Container Recognition

Emergency responders always place a high priority on establishing the specific identity of the hazardous materials involved in an incident. In some cases, we lack this information initially. In such cases, the hazmat containers themselves may be the only clue to the presence of hazardous materials (Fig. 5.9). Container configuration, design, and features may provide initial clues to the general types of hazardous materials involved. A general knowledge of containers may also allow us to realize that the contents have been incorrectly identified.

Figure 5.8 Seemingly innocuous occupancies may harbor hazmats, as evident from this cryogenic storage tank containing liquid oxygen at a hospital.

TABLE 5.7 Examples of Hazardous Materials Used in Various Industries

Industry	Material	Hazard
Paintings and coatings—retail and manufacturing	Resins—e.g., alkyds, acrylics, vinyls	Flammable/combustible liquids
	Pigments—dry metallic powders, e.g., aluminum and zinc dust	Flammable/combustible solids
	Solvents—e.g., benzene, toluene, xylene	Flammable/combustible liquids/carcinogenic
	Nitrocellulose—low-nitrated cotton fibers	Flammable solid
	Aerosols—e.g., propane, isobutene	Flammable gas/propellant
Food processing	Refrigeration equipment—compressed gas ammonia refrigerant Freon	Flammable/explosive Volatile/toxic Toxic
	Lubricating oils, cooking oils	Combustible liquids
	Alcohol, solvents	Flammable liquids
	Dust—e.g., sugar, flour, grains, starches, artificial sweeteners	Combustible solids
Mining	Methane	Explosive gas
	Coal dust	Explosive solid
	Coal, metal sulfide ores	Spontaneously combustible
Pulp and paper	Pulp processing chemicals—e.g., sodium hydroxide, sodium carbonate, sodium sulfide	Water reactive
	Bleaching chemicals—Chlorine Methanol	Promotes combustion/toxic Flammable liquid
	Pulp processing residue—"Black liquor"	Explosive liquid
Bulk grain handling	Grain dust—e.g., corn, alfalfa, barley	Flammable/explosive solid
	Pesticides	Toxic/flammable liquid/products of combustion
Construction sites	Blasting agents, low and high explosives	Explosive
	Fuels	Flammable/combustible liquids
Semiconductor manufacturing	Organic liquids—e.g., butyl, acetone, toluene	Toxic/flammable
	Gases—Halogen, germane Cyanide, carbon monoxide	Toxic (irritant) Toxic (asphyxiant)
	Metals—Cadmium Mercury Cobalt	Toxic (irritant) Toxic (poison) Toxic (carcinogen)
	Acids	Corrosives

Figure 5.9 In some incidents, the containers involved may provide the only initial clues to the presence of hazardous materials.

It is important to realize the limitations of this type of information. Container characteristics may, at best, indicate the general type of material involved. This information may help us to decide on appropriate initial actions, but it is very general and limited in nature. Always determine the specific identity of the hazmats involved as soon as it is possible to do so safely.

In this section of the chapter, we will relate container characteristics to likely contents for nonbulk containers, pipes and pipelines, storage tanks, motor carriers, railcars, intermodal containers, and other containers. Relevant advanced training in container recognition and response is highly recommended for emergency responders. One source for such training for rail, highway, intermodal, and marine containers is the Transportation Technology Center, Inc. (TTCI) operated by the Association of American Railroads (AAR) at Pueblo, Colorado.

Nonbulk Containers
As discussed in the section above on Labels and Markings on Nonbulk Containers, nonbulk containers hold a variety of hazardous materials, including some in every hazard class. The DOT-mandated markings and labels are the best indicators of their contents. Use of markings and labels as clues to contents requires making the assumption that the container holds the original material for which it was designed. During shipment of hazardous materials from the manufacturer or packager to the original user, this assumption is probably valid. During later use of the containers, it may not be a reliable assumption.

Recognition of Contents. Common nonbulk containers include cans, bottles, boxes, carboys, cylinders, and drums (Figs. 5.10 and 5.11). The shape of a

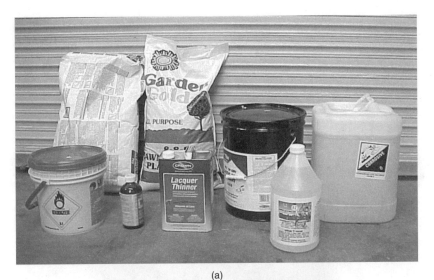

(a)

(b)

Figure 5.10 Examples of common nonbulk containers include: (a) bags, bottles, and pails, which may contain various solid and liquid materials, (b) carboys, which are designed to contain corrosive liquids, and (c) 30-gallon open-head carbon steel drum and 55-gallon closed-head plastic drum. (Fig. 5.10(b) courtesy of Cedric Harville).

(c)

Figure 5.10 (*continued*)

container and the material from which it is constructed may be indicators of the contents the container was designed to hold. Table 5.8 relates container design to likely contents for common nonbulk containers.

In the case of drums, whether the drum is of open-head or closed-head design is significant, because open (or removable)-top drums usually contain solids and closed (or "bung")-top drums usually contain liquids. The material the container is made of may also provide important clues. For example, we typically associate carbon steel drums with noncorrosive products, unless the drum has a plastic liner to protect the steel from corrosion. Likewise, we generally associate plastic containers with corrosive products, although such commodities as food products are also shipped in plastic containers. Drums of exotic materials such as aluminum or stainless steel are significantly more expensive than regular carbon steel drums and may contain unusually hazardous materials. Keep in mind that these are general assumptions that will not always hold true. Exceptions can always be encountered.

Safety Features. Cylinders are the only nonbulk containers typically fitted with pressure relief devices, except that some drums may be equipped with bung caps fitted with pressure relief valves. Cylinders generally have a pressure relief valve,

Figure 5.11 Gas cylinders may contain compressed gases or gases that have been liquefied.

TABLE 5.8 Nonbulk Containers

Containers	Description	Possible Contents
Bags	Multiwall paper, plastic, or paper lined with plastic	Solids (e.g., corrosive, flammable, oxidizers, poison, blasting agents)
Bottles (jars)	Glass or plastic with stopper or lid; hold up to several gallons	Any types of hazardous materials; solids or liquids
Carboys	Glass, plastic; often inside cushioned boxes or protective cage	Liquids; primarily used for corrosives
Cylinders	Metal; may be color coded, but this is not required by law and therefore is not a reliable indicator	Pressurized gases
Pails	Metal, plastic; hold 1–5 gallons	Any types of hazardous materials; solids or liquids
Drums	Metal, fiberboard, plastic; hold 5–85 gallons	Any types of hazardous materials; solids or liquids

set to relieve pressure greater than a predetermined safe amount, or a fusible plug with a soft metal core that will melt at high temperatures to provide a vent. Both these types of devices are designed to prevent cylinders from exploding or becoming unguided missiles because of increased internal pressure such as during flame impingement. Some relief valves reset to the closed position once internal pressure has dropped, and some continue releasing gas until internal pressure equals ambient pressure. Fusible plugs, once melted, remain open and must be replaced.

Pipes and Pipelines

Pipes may contain various materials including electrical wiring, water, sewage, telephone or television cable, and various chemical products in liquid or gas form. Pipes are used to convey hazardous materials for transfer and processing within fixed facilities and for transportation cross-country.

Piping Systems in Fixed Facilities. Piping systems in fixed facilities should be well identified on facility storage and use maps. Pipes within fixed facilities may be marked by color code and/or by label to identify the product they contain. However, no universal color code system exists for pipes. Avoid relying on color as an indication of the contents of piping systems at fixed facilities unless you are sure of the specific system used and that pipes are accurately identified.

Pipelines. Pipelines are indicated on small-scale maps, such as topographic and county road maps. On topo maps, the designation is a dotted line marked "PIPELINE." On county road maps, pipelines are usually indicated by a broken line with "G" or "OIL" between the line segments.

Because pipelines are almost always underground, the emergency responder must rely on aboveground clues to ownership and contents. Pipelines are usually kept cleared of brush and trees and present a wide, visible path. One petroleum pipeline company located in the southeastern United States, for example, mows its lines every two years. Federal regulations require the lines be observed from a plane 26 times a year to detect encroaching construction and digging operations that may damage them.

Pipeline owners mark their lines well, because the major cause of leakage is accidental damage to the lines by outside parties. Figure 5.12 shows a typical pipeline marker. It provides information about the pipeline contents, identifies the pipeline operator, and provides an emergency telephone number.

Pipeline materials, size, and operating pressure vary widely according to the commodity being transferred and other factors. The American Society of Mechanical Engineers has responsibility for guiding pipeline design, construction, and operations. Documents governing pipeline systems are written by this group, working through committees of the American National Standards Institute (ANSI), and are available from ANSI.

PETROLEUM PRODUCT PIPELINES transport petroleum products from their source, often in Texas, to destinations all over the country at a speed of 3–5 miles

Figure 5.12 Pipeline markers can provide valuable information in an emergency.

per hour. Products include 26 grades of gasoline, 6 grades of kerosene, and 10 grades of home heating oil and diesel fuel. Different petroleum products are commonly transported through the same pipeline. This is accomplished by placing a separating device called a "pig" between the different products, or by simply allowing the adjacent products to intermingle at the interface.

Pipelines that transport petroleum products are made of improved steel, with corrosion-resistant waterproof coatings. Heavier-wall pipe is used in sensitive areas such as near waterways and for river crossings. The pipe also has cathodic corrosion protection. Automated inspection tools, which are also called "pigs," are sent through the pipeline at intervals. Corrosion pigs detect corrosion, and caliper pigs measure the thickness of the pipe.

Pipelines for petroleum products are typically 32 to 40 inches in diameter. Typical pressures range from 550 to 600 psi on the main lines and 200 psi on stub lines.

GAS TRANSMISSION PIPELINES transport commodities such as natural gas, propane, and anhydrous ammonia as gases under pressure. Transportation of gases in this manner requires them to be cooled and condensed. In large transmission lines they must be transferred under high pressure.

Natural gas and propane are odorless by nature. To aid in detecting leaks, an odorant is added to these products as they enter the pipeline system. The odorant adds the "skunky" odor that is commonly associated with natural gas and propane. Keep in mind that these substances have no warning properties before being odorized.

There are numerous plants in the United States where natural gas is condensed at a ratio of 615:1 by chilling it to −260 degrees F. It is then stored in insulated, aboveground storage tanks or, in some parts of the country where geologic formations are suitable, in natural underground spaces. When the gas is needed, it is warmed and put into the pipeline.

Pipelines for cross-country natural gas transportation are typically 30 to 40 inches in diameter and may have pressures in excess of 1000 psi. Natural gas distribution lines within communities are typically 2 inches or larger, with smaller lines running to residential gas meters and into homes. Pressures within these systems may range from 600 psi in distribution lines to 0.25 psi at the point of end use.

PIPELINE SAFETY FEATURES are intended to provide rapid detection of leaks. Pipeline products are monitored and controlled by computers. Pipelines are equipped with low-pressure alarms designed to alert pump station operators to leaks, but even immediate shutdown of a line may allow a considerable amount of product loss, especially if the break is at or near the lowest section in the line. Specific safety features vary according to the product transported by the pipeline.

In most geographic areas, there is a central Line Location Center whose members own pipelines and will assist in locating one of their lines or providing emergency information. The main function of these centers is to mark the pipelines on request to prevent damage during activities such as construction, excavation, and drilling. In the event of an emergency, they may be able to provide valuable information to responders.

Storage Tanks

Large storage tanks are used to hold almost every type of product. As with non-bulk containers, the design of the container may provide emergency responders with valuable clues to the contents. A reference devoted specifically to responding to incidents involving storage tanks is Hildebrand and Noll's *Storage Tank Emergencies* (Hildebrand and Noll, 1997).

Recognition of Contents. Fixed facility storage tanks are used to hold all kinds of products including solids, liquids, and gases. Figure 5.13 shows storage tanks designed to store products that commonly exist in the liquid state under ambient temperature and pressure conditions. Figure 5.13 also relates the type of liquid products likely to be contained in each type of storage tank. Some storage tanks are designed to store gases that have been liquefied, either by being compressed into a liquid state under high pressure or by being supercooled to a cryogenic

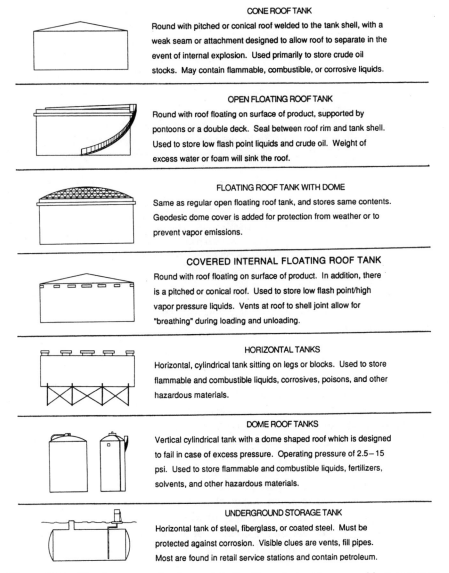

CONE ROOF TANK

Round with pitched or conical roof welded to the tank shell, with a weak seam or attachment designed to allow roof to separate in the event of internal explosion. Used primarily to store crude oil stocks. May contain flammable, combustible, or corrosive liquids.

OPEN FLOATING ROOF TANK

Round with roof floating on surface of product, supported by pontoons or a double deck. Seal between roof rim and tank shell. Used to store low flash point liquids and crude oil. Weight of excess water or foam will sink the roof.

FLOATING ROOF TANK WITH DOME

Same as regular open floating roof tank, and stores same contents. Geodesic dome cover is added for protection from weather or to prevent vapor emissions.

COVERED INTERNAL FLOATING ROOF TANK

Round with roof floating on surface of product. In addition, there is a pitched or conical roof. Used to store low flash point/high vapor pressure liquids. Vents at roof to shell joint allow for "breathing" during loading and unloading.

HORIZONTAL TANKS

Horizontal, cylindrical tank sitting on legs or blocks. Used to store flammable and combustible liquids, corrosives, poisons, and other hazardous materials.

DOME ROOF TANKS

Vertical cylindrical tank with a dome shaped roof which is designed to fail in case of excess pressure. Operating pressure of 2.5–15 psi. Used to store flammable and combustible liquids, fertilizers, solvents, and other hazardous materials.

UNDERGROUND STORAGE TANK

Horizontal tank of steel, fiberglass, or coated steel. Must be protected against corrosion. Visible clues are vents, fill pipes. Most are found in retail service stations and contain petroleum.

Figure 5.13   Fixed storage tanks that contain liquid products at low or ambient pressures.

liquid state. Figure 5.14 shows the shape and likely contents of storage tanks designed to store liquefied compressed gases at pressures up to 500 psi or to maintain cryogenic liquids at temperatures below −130 degrees F.

Safety Features. The low-pressure tanks shown in Figure 5.13 have weak seams or welds at the top of the tank. This allows release of contents upward in the event of increased pressure, to prevent catastrophic failure of the tank. The cone

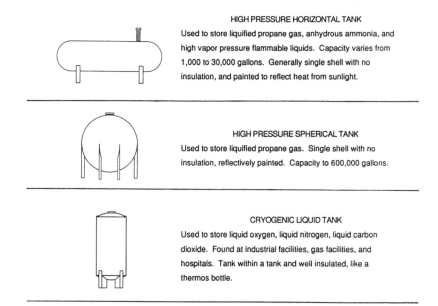

HIGH PRESSURE HORIZONTAL TANK
Used to store liquified propane gas, anhydrous ammonia, and high vapor pressure flammable liquids. Capacity varies from 1,000 to 30,000 gallons. Generally single shell with no insulation, and painted to reflect heat from sunlight.

HIGH PRESSURE SPHERICAL TANK
Used to store liquified propane gas. Single shell with no insulation, reflectively painted. Capacity to 600,000 gallons.

CRYOGENIC LIQUID TANK
Used to store liquid oxygen, liquid nitrogen, liquid carbon dioxide. Found at industrial facilities, gas facilities, and hospitals. Tank within a tank and well insulated, like a thermos bottle.

Figure 5.14   Fixed storage tanks that store liquefied compressed gases under high pressures or refrigerated gases under very low temperatures.

roof tank is constructed with a weak weld at the seam between the sides and roof. Dome roof tanks have a roof that is designed to fail in case of excess pressure. Floating roof tanks allow the release of excess pressure around the roof/tank seal. Covered internal floating roof tanks are designed to vent any pressure or vacuum that may develop between the floating roof and the cover during on-loading or off-loading. High-pressure tanks are equipped with pressure relief devices as described for other high-pressure containers.

Tank Confinement Systems. Although not a safety feature of the tank itself, confinement systems prevent further dispersal of the product from a storage tank in case of leakage from the tank. Federal regulations require a confinement system that will confine 10 percent of the contents in all the tanks in the confinement area, or the entire contents of the largest tank, whichever is greater.

Motor Carriers

With the exception of certain chemicals specified by DOT, virtually any of the thousands of chemicals in existence today may be transported on the highways. Any of the nonbulk containers previously discussed in this chapter may be transported by highway in van freight trailers, flatbed trailers, or other vehicles. In addition, highway cargo tanks may transport hazardous materials in bulk quantities as solids, liquids, or gases. Intermodal containers and specialized containers discussed later in this section may also be used in highway transport. Highway motor carriers fall into a few categories as defined by DOT specifications. The classification logic for bulk containers used in highway transportation

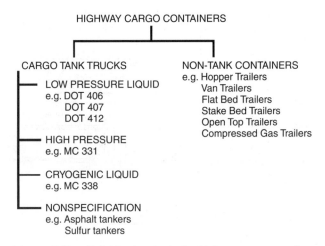

Figure 5.15 Classification logic for highway motor carriers.

is shown in Figure 5.15. Being familiar with the motor carrier designations will make it possible to determine the types of materials most likely to be transported in each. Remember that box semitrailers can carry a wide variety of hazardous materials in nonbulk packages. Also, refrigerated containers, or "refers," have an onboard refrigeration unit with a self-contained fuel supply that may be damaged and released in an accident.

Motor Carrier Recognition and Nomenclature. Common highway cargo tanks (Fig. 5.16) can be roughly classified as bulk liquid cargo tank trucks (DOT 406, DOT 407, DOT 412), liquefied compressed gas tank trucks (MC 331), and cryogenic liquid tank trucks (MC 338). Other carriers, such as compressed gas trailers (see Fig. 5.22 below), may also be used. Additionally, hazardous materials in dry bulk form may be transported in pneumatic or "hopper" trailers. Nonspecification tank trailers are used to carry certain materials such as asphalt and molten sulfur.

Figure 5.16 shows common highway cargo tanks and gives information about the design and likely contents of each. DOT has regulations addressing the design and safety features of the transportation containers that are based on a code of specifications established by the American Society of Mechanical Engineers. All cargo trailers will have a specification plate mounted at the front of the trailer either on the driver or passenger side. The specification plate identifies the type of trailer and provides additional important information such as cargo tank capacity.

Keep in mind that the guidelines provided here on container recognition are general and that exceptions may be encountered. Although most DOT 406 carriers, for example, are easily distinguishable from DOT 407s, not all of them are. You cannot rely completely on visual recognition to estimate likely contents, but you can use this general information as part of the hazard and risk assessment process until more specific information is available.

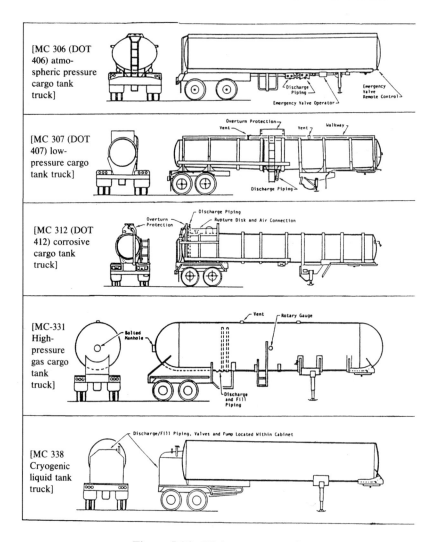

Figure 5.16   Highway cargo tanks.

DOT 406. The DOT 406 specification tanker is used most often to transport petroleum products such as gasoline and fuel oil. Most have an oval cross section (Fig. 5.16), although round tanks are occasionally used. To carry multiple grades of fuel, DOT 406s typically have three to five compartments separated by bulkheads. Each compartment has a separate manway for top access (Fig. 5.17a) and a separate bottom discharge fitting (Fig. 5.17b). Internal baffles within the tanks add strength and prevent liquid surges. Product is usually loaded and unloaded through the bottom valves at the center of the trailer. Capacity of the DOT 406 ranges from 2000 to 9500 gallons, and the design pressure is 3 psi. A reference devoted specifically to responding to emergencies involving the DOT 406 is Noll,

Flammable and combustible liquids such as gasoline and diesel fuel, and class B poisons	Oval or round rear cross section, pressure less than 3 psi. Usually single shell aluminum.
Capacity: 2,000 - 9,500 gallons	
Products with vapor pressures not more than 40 psi at 70°F Flammable liquids and mild corrosives	Circular cross section, pressures up to 25 psi. Double shell construction, usually steel
Capacity: 2,000 - 8,000 gallons	
High density liquids and corrosives	Cylindrical cross section, narrow diameter, external ribs. Steel, stainless steel, or aluminum; often lined to resist degradation or reaction to contents.
Gases which are liquefied by pressure. Compressed gases and some very hazardous liquids	Cylindrical cross section, usually larger diameter than MC 312. Hemispherical heads; smooth surface without external ribs. Pressure 100 - 500 psi.
Cryogenic gases liquefied by refrigeration, such as liquid helium at -425°F	Cylindrical cross section, hemispherical heads, smooth surface. Container within a container, as a thermos bottle. Pressure 23.5 - 500 psi.

Figure 5.16 (*continued*)

Hildebrand, and Donahue's *Gasoline Tank Truck Emergencies* (Noll, Hildebrand, and Donahue, 1996).

The DOT 406 was previously known by the motor carrier designation "MC 306." Several years ago, DOT upgraded specification requirements and changed the designation to DOT 406. Many hazmat responders still refer to this container as the MC 306.

DOT 407. Also known as the "chemical trailer" and "the workhorse of the chemical industry," the DOT 407 is used to carry a very wide variety of liquid products

(a)

(b)

Figure 5.17 The DOT 406 cargo tank typically has several compartments with (a) a separate set of manway fittings for each compartment and continuous rollover protection running the entire length of the tank on top of the tank and (b) a separate discharge for each compartment beneath the tank.

Figure 5.18 The sheet metal jacket covering the insulated DOT 407 has a horseshoe-shaped cross section. Note also the centrally mounted splash box and the guards for rollover protection on top of the tank.

such as flammable liquids, Division 6.1 toxic materials, and mild corrosives. They also commonly transport nonhazardous commodities. DOT 407s have a round cross section (see Fig. 5.16), although insulated 407s will have a horseshoe-shaped cross section (see Fig. 5.18). Most are constructed of stainless steel, although some may be of lined aluminum or mild steel. External circumferential rings or reinforcing ribs are visible on uninsulated DOT 407 tanks, as shown in Figure 5.16. Insulated DOT 407 tank trucks typically have a highly polished stainless steel outer jacket that covers the insulation and conceals the supporting rings, as shown in Figure 5.18. DOT 407s usually have a capacity of 6000 to 7000 gallons and a design pressure of 40 psi.

The DOT 407 was previously known by the motor carrier designation "MC 307." Many hazmat responders continue to refer to it as an MC 307.

DOT 407s are usually single-compartment tanks fitted with a top-mounted manway and other fittings that are usually contained within a protective flashing or "splash box" located on top of the tank (Fig. 5.19a). DOT 407s are usually fitted with a single bottom discharge valve located at either the center or the rear of the trailer (Fig. 5.19b). DOT 407s with more than one compartment are rare, but when encountered they will have a set of fittings for each compartment.

It is easy to mistake an uninsulated DOT 407 for a DOT 412, because visible supporting rings are considered a key feature of the DOT 412 (see Fig. 5.16). The uninsulated DOT 407 can usually be distinguished from the DOT 412 based on the following characteristics:

- The DOT 407 usually has a larger diameter than the DOT 412.
- Top-mounted fittings are usually located at the middle of the DOT 407 and at the rear of the DOT 412.

(a)

(b)

Figure 5.19   DOT 407s usually have a single compartment with (a) a single set of fittings on top of the tank and (b) a single discharge underneath the tank.

- Supporting ribs tend to be more widely spaced (up to 5 feet apart) on the DOT 407 and closer together (no more than 3 feet apart) on the DOT 412.

Keep in mind that these distinctions are general rules, and exceptions will be encountered. The best way to identify the motor carrier specification will be to read the specification plate, if this can be done safely.

DOT 412. The DOT 412 is designed to carry heavy corrosive products including concentrated acids and bases. The trailer resembles the uninsulated DOT 407 except that it is usually smaller in diameter because corrosives are very heavy liquids (Figs. 5.16 and 5.20). The trailer is constructed of mild steel or stainless steel and will be lined unless the material is immune to attack by the product or is thick enough to withstand 10 years of service without corroding to a minimum thickness specified by regulations. The DOT 412 is typically designed for top unloading and usually does not have a bottom discharge valve. The top-mounted manway and other fittings are contained in a protective box that is usually mounted at the rear of the trailer (Fig. 5.20). DOT 412s may have a band of black corrosion-resistant paint around the tank at the dome. DOT 412s typically have a capacity of 5000 to 6000 gallons and a design pressure of 75 psi.

The DOT 412 was previously known by the motor carrier designation "MC 312." Many hazmat responders continue to refer to it as an MC 312.

MC 331. The MC 331 carrier is designed to transport liquefied compressed gasses such as propane or anhydrous ammonia. It has a large-diameter, uninsulated tank constructed of 3/8-in. steel. The tank is round in cross section, with hemispherical heads (Fig. 5.16). Valves and other fittings are located beneath the tank in a protective cage situated either immediately forward of the rear wheels or at the

Figure 5.20   The DOT 412 typically has visible stiffening rings and a rear-mounted set of fittings designed for top unloading.

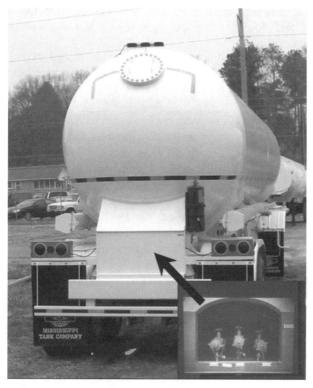

Figure 5.21   MC 331 cargo tanks typically have hemispherical heads with a bolted manway in the rear head and well-protected valves (inset photo). (Courtesy of Don Koss of Innovative Safety and Training Systems).

rear of the trailer between the frame sills (Fig. 5.21), and a small bolted manway is usually located in the rear head. Two-thirds of the tank's upper surface must be painted with white or aluminum paint to minimize solar heating. MC 331s have design pressures ranging from 100 to 500 psi and capacities ranging from 2500 gallons (for "bobtail" delivery trucks) to 11,500 gallons.

MC 338. Gases that have been liquefied by refrigeration to temperatures of −130 degrees F or below are transported in MC 338 carriers, which are commonly called "cryogenic trailers." The trailer is constructed like a thermos bottle with an inner and an outer shell. A vacuum exists between the two shells, and the container is heavily insulated. All valves and fittings are contained in a cabinet located at the rear of the MC338 trailer. The cabinet frequently extends beyond the back end of the tank, resulting in a "camel hump" appearance (Fig. 5.16), but is sometimes recessed into the tank shell. MC 338s transport cryogenic liquids such as liquid carbon dioxide, liquid nitrogen, liquid oxygen, and liquid hydrogen.

Compressed Gas Carriers. Compressed gas carriers are used to transport compressed gases that are not liquefied. The gases must be compressed to a very high

Figure 5.22 The compressed gas trailer consists of a number of cylinders mounted in a trailer frame and manifolded together. (Courtesy of Don Koss of Innovative Safety and Training Systems).

pressure to make transporting them economically feasible. The compressed gas carriers typically consist of a number of cylinders that are mounted to a trailer and manifolded together at the rear of the trailer (Fig. 5.22). For this reason, they are commonly referred to as "tube trailers." A variety of gases, including carbon dioxide, nitrogen, argon, oxygen, and hydrogen, are commonly shipped by tube trailer. Design pressures range from 3000 to 5000 psi, so that tube trailers represent a significant physical hazard in addition to any chemical hazards posed by the gases they transport.

Pneumatic or Dry Bulk Carriers. Dry powder materials are carried in bulk quantity in hopper trailers (Fig. 5.23). They have a distinctive "W" shape when seen in profile, with discharge valves located at the bottom of each compartment.

Figure 5.23 Pneumatic or dry bulk cargo trailers transport dry, granular products.

Product is removed from the trailer by gravity or is "blown out" by air or gas pressure (hence the name pneumatic) through the product valves. The trailers are designed to operate at pressures up to 15 psi. Hazardous materials commonly transported in this type of container may include oxidizers such as ammonium nitrate fertilizer and corrosives such as dry caustic soda.

Motor Carrier Safety Features. Motor carriers have safety features that vary slightly among the different specification tanks but can generally be described as:

- Roll-over protection for manways, valves, and other accessories
- Cages and bumpers to protect the tank, valves, and accessories in the event of a collision
- Pressure and vacuum relief valves
- Internal valves and remote shut-off switches to prevent accidental product loss

The specifications for these and other features are outlined in Sections 173 and 178 of Code of Federal Regulations (CFR) Title 49 and are based on ASME standards. Below is a general discussion of these features and the specifics of each type of carrier.

ROLLOVER PROTECTION is intended to protect valves, gauges, manways, piping, or other items protruding above the tank's upper surface if the container overturns. For this reason, specifications require that these openings and accessories be either recessed into the tank or protected by guards capable of supporting two times the loaded weight of the trailer. DOT 406s typically have continuous parallel guards running the entire length of the tanker to protect the manways to the multiple compartments (Fig. 5.17). DOT 407s and DOT 412s usually have guards located immediately forward of and behind the manway. These are sometimes incorporated into the flashing or "splash box" surrounding the manway fittings. Valves and fittings may be located on the top of the tank outside these guards as long as they do not extend higher than a line drawn from the top of the guard to the end of the tank. Manways for the MC 331s and MC 338s, if present, are not located on the top of the tank, so rollover protection is not typically seen on these tankers.

PRESSURE AND VACUUM RELIEF VALVES are intended to prevent damage to the tank caused by the buildup of excessive pressure or vacuum within the tank. It is especially important to prevent overpressurization, which can result in catastrophic container failure. Each cargo tank type is designed to withstand a maximum pressure, above which a pressure relief device will activate. The design pressure for each carrier type is dependent on the products it will carry.

Vacuum relief valves are designed to let air into the tank to relieve any vacuum that may develop. In the DOT 406, the vacuum relief valve is typically incorporated into the dome lid. In the DOT 407, it may be part of the "Christmas tree" assembly, which is a combination device with both pressure and vacuum relief valves and a compressed air fitting mounted in the top of the tank, or as a stand-alone valve on the tank's upper surface.

Pressure relief valves are spring-operated valves designed to relieve pressure buildup within the tank. Pressure may build as the product heats up, leading to increased vaporization. Product may be warmed by the sun or may be heated during a fire. In any case, the valve is designed to open when the pressure in the vapor space exceeds a specified percentage of the maximum allowable working pressure for the tank and to close when the pressure drops to the specified level.

In addition to reclosable spring-loaded valves, a trailer may have fusible/ frangible disk caps installed in vents for each compartment to provide additional venting capacity. These are designed to rupture in the event of fire or extreme tank pressure conditions and do not reseal after rupture, but must be replaced. DOT 407s carrying flammable liquids must have at least one such cap in addition to the normal spring-loaded safety valve(s).

EMERGENCY PRODUCT FLOW CONTROL is provided by shut-off valves located within the tank that are designed to fail to the closed position. Shut-off valves may be opened by mechanical (cable), hydraulic (fluid), or pneumatic (air) assemblies that are connected to an operating arm.

The emergency shut-off systems have control levers or plugs that the driver or emergency personnel can use to activate the shut-off system (Fig. 5.24). In the event of an uncontrolled release, moving the control to the emergency position

Figure 5.24 Emergency shut-off devices (see lower right inset photo) on cargo trailers allow any open discharge valves to be closed instantaneously.

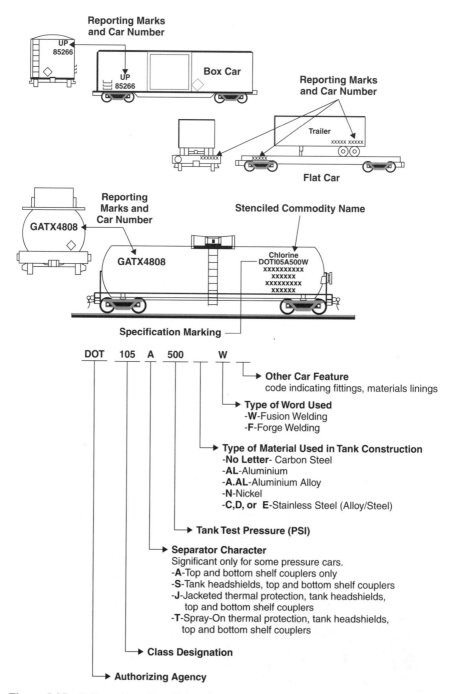

Figure 5.25 Railcars have identifying reporting marks and car numbers, and tank cars have specification numbers and other markings.

will immediately close any open discharge valves on the trailer. All tank trucks are required to have at least one emergency shut-off control located at the front of the trailer just behind the driver's door. Some motor carriers have additional shut-off controls at other locations on the trailer. A fusible link located in the area of the emergency shut-off valve is designed to melt in the event of a fire to shut off product flow.

Discharge valves are recessed into the tank and designed to fail to the closed position. Discharge nozzles or other fittings attached to the valves are designed to shear off in an accident, leaving the valve intact.

Railcars

Railcars are used to transport a wide variety of hazardous materials, typically in trains consisting of many cars, some of which may contain large quantities of hazmats. This can make product identification and incident assessment very challenging for rail incidents (Fig. 5.9).

Railcar Markings and Orientation. A dependable way to identify the contents of railcars is through the use of consists and waybills, as discussed in the section on shipping papers earlier in this chapter. Each railcar has a unique identifier called "reporting marks," which consists of the car initial and number stenciled onto the car as shown in Figure 5.25. The reporting marks are stenciled onto both sides (toward your left as you face either side) and both ends of the railcar. Reporting marks are used to index the information in shipping papers to the specific cars making up a train and are critical for identifying the contents of specific railcars with shipping papers. When assessing a rail incident, make every effort to safely obtain the reporting marks and relative positions for all railcars involved. As in other cases involving hazardous materials transportation, placards and four-digit identification numbers can also be used to identify contents of railcars.

For a number of materials shipped in rail tank cars, DOT requires that the commodity name be stenciled directly onto the tank car, toward the right end as you face each side, as shown in Figure 5.25. These chemicals include:

- Acrolein
- Anhydrous ammonia
- Bromine
- Butadiene
- Chlorine
- Chloroprene (when transported in DOT 115A specification tank car)
- Difluoroethane (may be stenciled "Dispersant Gas" or "Refrigerant Gas" in lieu of name)
- Difluoromonochloromethane (may be stenciled "Dispersant Gas" or "Refrigerant Gas" in lieu of name)
- Dimethylamine, anhydrous
- Dimethyl ether

- Ethylene imine
- Ethylene oxide
- Formic acid
- Fused potassium nitrate and sodium nitrate
- Hydrocyanic acid
- Hydrofluoric acid
- Hydrogen
- Hydrogen chloride
- Hydrogen fluoride
- Hydrogen peroxide
- Hydrogen sulfide
- Liquefied hydrogen
- Liquefied hydrocarbon gas (may also be stenciled "Propane," "Butane," "Propylene," or "Ethylene")
- Liquefied petroleum gas (may also be stenciled "Propane," "Butane," "Propylene," or "Ethylene")
- Methyl acetylene propadiene, stabilized
- Methyl chloride
- Methyl chloride-methylene chloride mixture
- Methyl mercaptan
- Monomethylamine, anhydrous
- Motor fuel anti-knock compound or anti-knock compound
- Nitric acid
- Nitrogen tetroxide
- Nitrogen tetroxide-nitric oxide mixture
- Phosphorus
- Sulfur trioxide
- Trifluorochloroethylene (may be stenciled "Dispersant Gas" or "Refrigerant Gas" in lieu of name)
- Trimethylamine, anhydrous
- Vinyl chloride
- Vinyl fluoride, inhibited
- Vinyl methyl ether, inhibited

The name of any chemical may be stenciled onto tank cars by owners wishing to dedicate the car solely to transportation of the chemical named. As long as a commodity name is stenciled on a tank car, that car cannot legally be used to transport any other substance.

For orientation purposes, the end of a rail car that contains the handbrake is referred to as the "B" end and the opposite end of the car is referred to as the

"A" end. If you stand at the B end of a car facing the A end, the side of the car to your right is designated the right side of the car and the side to your left is designated the left side of the car. Keep this rule of thumb in mind during activities such as noting and reporting the location of damage on railcars.

Given the high degree of damage and confusion that can result from rail accidents, first responders may be unable to identify the contents of railcars directly through the methods previously discussed (Fig. 5.9). Shipping papers may be destroyed or missing, and placards or markings may be torn from cars or obscured by the position of cars in a derailment. In these instances, you may be able to make initial inferences about the types of products railcars may contain based on the configuration and features of the cars involved.

Overview of Railcars. Various types of cars are utilized in rail transportation. They are generally analogous to the highway transport containers discussed above. For example, boxcars on the rail line may transport the same nonbulk hazmat containers as the van trailer on the highway. Refrigerated containers, or "refers," have an onboard refrigeration unit with a self-contained fuel supply that may be damaged and released in an accident. Some containers used in highway transport, such as intermodal portable tanks, ton containers, and "piggy back" highway trailers, are also used in rail transport. The classification logic for most rail cars commonly used to transport hazardous materials is shown in Figure 5.26. An overview of railcars is shown in Figure 5.27 and described in Table 5.9. Each type of railcar has certain distinctive features and is used to transport certain types of materials. Keep in mind that these features can serve at best only as clues to the general types of materials the cars may contain. A thorough coverage of rail containers is available through the *GATX Tank and Freight Car Manual* (GATX, 1994).

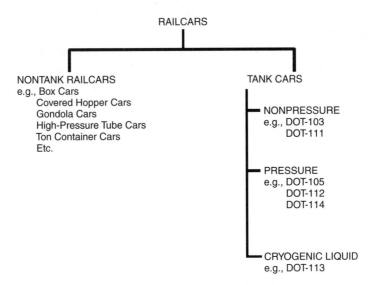

Figure 5.26 Classification logic for railcars.

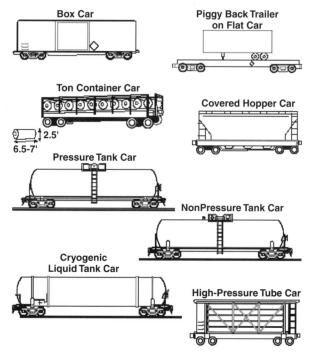

Figure 5.27   Overview of railcars.

Overview of Tank Car Classification. Rail tank cars are the major focus of attention in this section, because the properties and quantities of hazardous materials commonly transported in tank cars lend themselves to major incidents. Tank cars are required to meet specific design and construction standards in accordance with the types of hazardous materials they are intended to contain. Given a general knowledge of these standards, certain inferences can be made about the possible contents of tank cars involved in an incident.

Tank cars can be classified as general service (or nonpressure), pressure, and cryogenic liquid tank cars. Nonpressure tank cars are readily distinguished by exposed manways, valves, and fittings on top of the car. Pressure cars are distinguished by a protective dome on top of the car that conceals all valves and fittings. Cryogenic cars have a completely clean top profile, with no exposed fittings, pressure domes, or other discontinuities on top of the car. As described in Table 5.10, tank cars falling within each of these categories have distinguishing design characteristics and features in keeping with the types of materials that they commonly transport.

Tank cars have specification stencils that indicate the classification of a car. As shown in Figure 5.25, the DOT specification (and various other information relating to tank car construction) is shown in the markings required to be stenciled onto all tank cars. The specification stencil is located on both sides, toward your right as you face either side of the car.

TABLE 5.9   Overview of Rail Containers

Type of Cars	Description	Type of Lading	Typical Hazardous Contents
Boxcars	Large, box-shaped cars with completely enclosed sides and top. Loaded through sliding doors on sides.	Nonbulk packages such as bags, bottles, boxes, carboys, and drums. May house bulk containers (e.g., refrigerated liquid tanks or ton containers) in some instances.	Items representative of most hazard classes can be encountered in boxcars.
Covered hopper cars	Large, completely enclosed cars with flat or rounded sides and flat or angular ends. Top has two or more loading hatches, and bottom has two or more sloping bins with unloading attachments.	Dry bulk solids.	Ammonium nitrate fertilizer, soda ash, adipic acid, caustic soda, sodium chlorate
Gondolas	Short, open top cars, with solid floor and fixed sides and ends.	Typically bulk solids, but may contain ton containers in some instances.	Contaminated dirt, flake sulfur
Tank cars	Large cylindrical tanks, with circular cross section. Ends vary from rounded to almost flat. Design varies for pressure, nonpressure, and cryogenic contents.	Liquids or gases	Various materials representing flammable gas, flammable liquid, corrosive, oxidizer, poison, and other hazard classes.
High pressure (tube cars)	Constructed of 25 to 30 seamless steel cylinders mounted horizontally in a 40-foot frame with open sides. Test pressure from 3000 to 5000 psi.	Compressed gases	Helium, hydrogen
Ton container cars	Specially designed flat cars that carry ton containers. (Cylindrical containers approximately 2.5 feet in diameter and 6.5–7 feet long.) Ton containers hold 180 to 320 gallons of liquefied gas at pressures of 500 to 1000 psi. Ton containers may also be transported in box cars or gondola cars.	Liquefied gases	Chlorine, anhydrous ammonia, sulfur dioxide, phosgene

(*continued overleaf*)

TABLE 5.9   (*continued*)

Type of Cars	Description	Type of Lading	Typical Hazardous Contents
Specialized	Distinctive containers with unusual shapes. Dedicated to the transportation of a single hazardous material and designed specially for the hazards of that material.	e.g., water-reactive	e.g., calcium carbide
	(e.g., calcium carbide: flat car with vertical bins having conical tops)	flammable solid.	
	(e.g., radioactive materials: low cars with extremely heavy packaging)	Radioactive materials.	

General Service (Nonpressure) Tank Cars. General service or nonpressure cars commonly transport a wide variety of hazardous materials that exist in the liquid state under ambient temperature and pressure conditions (see Table 5.10), as well as various nonhazardous commodities. The general service cars most commonly encountered are those of the DOT 103 and 111 class designations, with the DOT 111 being far more common then the DOT 103. Cars of the DOT 103 class are readily distinguished from those of the DOT 111 class by the expansion dome atop the 103 (Fig. 5.28). The expansion dome allows the shell of the DOT 103 car to be completely filled, whereas "outage" must be left within the shell of the DOT 111 car to allow for expansion of the product. Avoid mistaking the expansion dome on a DOT 103 for the pressure dome on a high-pressure railcar. The DOT 103 can be readily distinguished by the presence of exposed valves and fittings on top of the expansion dome.

As noted in Table 5.10, the major distinguishing feature of nonpressure tank cars is the fact that valves and other fittings are typically exposed along the top and bottom of the tank (see Fig. 5.28). However, there are exceptions to this general rule. For instance, nonpressure cars used to transport corrosives typically have no bottom fittings or discontinuities, except for wash out plugs. Also, cars used to transport nitric acid have a protective dome that encloses the manway fittings. Corrosive cars are sometimes also distinguished by a band of corrosion-resistant paint around the manway, or by a characteristic staining of the tank around the manway.

Examples of valves and fittings used on nonpressure cars are shown in Figure 5.28. Nonpressure cars may be either uninsulated or insulated (as described below). Insulated cars may have heating coils, mounted either internal or external to the tank, for steam heating to assist in off-loading viscous products. Some tank cars have a protective lining on the inner surface of the tank for protection from corrosive products.

TABLE 5.10 Rail Tank Car Types and Features

Type of Car	Distinguishing Features	Common DOT Class Designations	Commonly Transported Hazmats	Capacity (gallons)	Tank Test Pressures	Pressure Relief Devices
General service (nonpressure) Tank Cars	Valves and other fittings on top of tank are typically exposed. Exposed fittings beneath cars are common. May be insulated or uninsulated.	103 (with expansion dome) 111 (without expansion dome)	Various flammable and combustible liquids, poison Bs (class 6.1), corrosives, oxidizers, organic peroxides	4000 to 45,000	60 to 100 psi.	Safety relief valve (generally set at 75% of tank test pressure) and/or safety vent with frangible disk (set at 100% of tank test pressure). Safety vent not allowed for flammables or poisons.
Pressure tank cars	Valves and fittings on top of tank enclosed in protective dome. Fittings beneath tank (bottom discontinuities) generally not permitted. May be insulated or uninsulated. May be equipped with thermal protection.	105 (insulated) 112 and 114 (uninsulated)	Chlorine, LPG, anhydrous ammonia, vinyl chloride	6000 to 42,000	100 to 600 psi.	Safety relief valve, set to open at 75% of tank test pressure. (Safety vent with frangible disk not allowed).
Cryogenic liquid tank cars	Constructed of tank within a tank, with heavy insulation and vacuum in space between tanks. No fittings exposed above or beneath car. All valves and fittings located in ground level cabinets on sides or ends of car.	113	Hydrogen, ethylene, argon, nitrogen, oxygen	15,000 to 30,000	60 to 175 psi.	Inner tank fitted with pressure relief valve and safety vent (set at tank test pressure). Outer tank fitted with safety vent set at 16 psi.

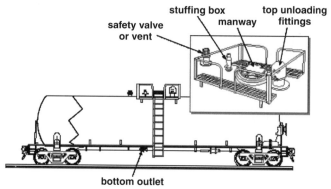

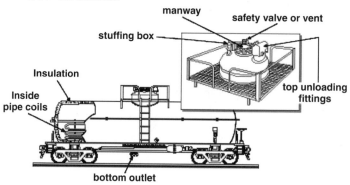

Figure 5.28   Nonpressure tank cars.

It is worth noting that, although classified as " nonpressure," these cars may contain significant pressure during normal operation. If suddenly released, this pressure can constitute a hazard to personnel through physical injury or through chemical splash or spray.

Pressure Tank Cars. Pressure tank cars carry various flammable and nonflammable liquefied compressed gases, including Division 2.3 gases that are toxic by inhalation. Pressure cars always contain hazardous materials. Some of our most infamous hazmat incidents have occurred when flame impingement on pressure cars containing liquefied petroleum gases produced Boiling Liquid Expanding Vapor Explosions, or BLEVEs, resulting in injury or death to responding personnel and massive property damage.

The pressure cars most commonly encountered are those of the DOT 105, 112, and 114 class designations. The DOT 105 class cars are insulated, whereas the DOT 112 and 114 classes are uninsulated (Figs. 5.29–5.31). Cars of all three

DOT - 105

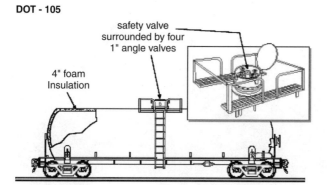

DOT - 112 With Jacketed Thermal Protection

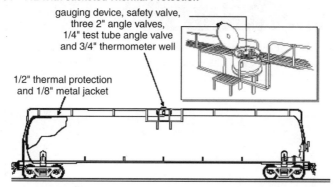

Figure 5.29 High-pressure tank cars.

classes may be equipped with thermal protection against flame impingement. The major distinguishing feature of pressure cars is the fact that all valves and fittings on top of these cars are contained in a protective dome and bottom fittings or discontinuities are typically absent. Examples of valves and fittings used on pressure cars are shown in Figure 5.29.

Responders should be aware that, in addition to the primary hazards of the substances contained in pressure cars, the contents of these cars are also hazardous because of the pressure under which they are contained and their potential to produce very cold temperatures if released. This may result in frostbite to responders or damage to personnel protective equipment.

Cryogenic Liquid Tank Cars. Cryogenic liquid tank cars are used to transport gases that have been converted to the liquid state through supercooling. Cryogenic cars are best represented by cars of the DOT 113 class designation. As described in Table 5.10, cryogenic cars utilize a "thermos bottle" design to maintain refrigerated gases in the liquid state for up to 30 days at relatively low pressures. Cryogenic liquid tank cars are distinguished by a complete lack of fittings or

Figure 5.30   The jacket covering the insulation on this DOT 105 is evident from the flattened appearance of the head, the seams running around and across the head, and the flashing that conceals the point where the frame bolster joins the tank shell (arrow).

Figure 5.31   This DOT 112 is unjacketed, leaving the tank shell visible, as evident from the hemispherical appearance of the heads and the visible point where the frame bolster joins the tank shell (arrow).

discontinuities along the tops and bottoms of the cars (Fig. 5.27). All valves and fittings are contained within cabinets located near ground level toward opposite ends of the car. A few cryogenic tank cars are completely housed within boxcars.

Tank Car Safety Features. A number of safety features may be incorporated into tank cars. These include features intended to prevent the overpressurization of tanks, the puncture of tanks during derailments, and damage to tanks or contents due to heat.

PRESSURE RELIEF DEVICES used on tank cars include safety relief valves and safety vents (Table 5.10). Safety relief valves are spring-loaded so that they open when internal pressure exceeds the valve setting and close automatically once excess pressure has been relieved. In contrast, safety vents are fitted with a frangible disk that shatters to vent excess pressure. Once activated, safety vents remain open until the frangible disk is replaced.

Both types of pressure relief devices are found on nonpressure cars. However, safety vents are not allowed on cars that carry flammables or poisons. Pressure tank cars are fitted with safety relief valves or with combination devices, which have a frangible disk and a spring-loaded valve that closes after venting is completed. Cryogenic liquid tank cars are fitted with both types of devices (see Table 5.10).

PUNCTURE PREVENTION DEVICES. Two types of safety devices are utilized to reduce the probability of the heads of tank cars being punctured by the couplers of adjacent cars during train wrecks. These devices are top and bottom shelf couplers and head shields, which function as shown in Figure 5.32. Note that head shields may be mounted externally to the tank or incorporated into the head or jacket of the car.

Shelf couplers are required for all tank cars that transport hazardous materials. Head shields are required, in addition to shelf couplers, for all cars that transport flammable compressed gases or anhydrous ammonia. The separator character included in specification markings indicates whether or not these features are incorporated into pressure tank cars (Fig. 5.25).

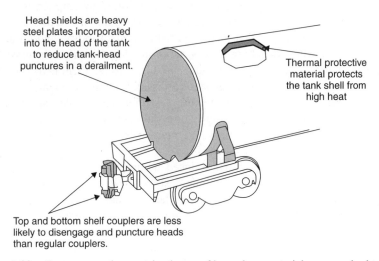

Head shields are heavy steel plates incorporated into the head of the tank to reduce tank-head punctures in a derailment.

Thermal protective material protects the tank shell from high heat

Top and bottom shelf couplers are less likely to disengage and puncture heads than regular couplers.

Figure 5.32   Cars transporting certain classes of hazardous materials are required to have special safety features.

THERMAL PROTECTION AND INSULATION. Thermal protection is required, in addition to shelf couplers and head shields, for all pressure cars that transport flammable compressed gases. Thermal protective materials may be contained within a protective sheet metal jacket, giving the tank car the same external appearance as an insulated car. Thermal protective material may also be sprayed directly onto the shell of an unjacketed tank car. The sprayed-on thermal protective material will give the tank shell a rough outer surface. The presence and type of thermal protection is indicated by the separator character in the specification marking of pressure tank cars (Fig. 5.25). Thermal protection is designed to protect the tank shell from flame impingement for a limited amount of time, assuming the thermal protective layer is intact.

Insulated tank cars are fitted with several inches of insulation held in place by a sheet metal jacket (Fig. 5.30). Insulation of tank cars is designed to moderate the effects of ambient temperature changes on the product inside the car. All DOT 105 pressure cars and some nonpressure cars are insulated. Insulation is not the same as thermal protection and is not intended to protect against flame impingement. Some pressure tank cars are equipped with both insulation and thermal protection. Pressure tank cars without insulation or thermal protection are required to have at least the upper two-thirds of the tank covered with light-reflective paint.

Note that it is not possible to distinguish between an insulated tank car and a car with jacketed thermal protection by visual examination. In both cases, the jacket conceals the identity of the material beneath it. The distinction can only be made by viewing the separator character in the specification marking (Fig. 5.25).

Intermodal Containers

Intermodal containers have become popular as a means of transporting many commodities. As the name implies, they can be transported by ship, truck, or train with equal ease and used as a storage container at the point of end use. Intermodal containers offer advantages that include portability, stackability, and no need for product transfer when switching transportation modes. Given the increasing use of these containers, emergency responders are likely to see them involved in hazmat incidents. This section gives general information on the recognition of these containers and some of the safety features. A reference devoted specifically to responding to emergencies involving intermodal containers is Noll, Hildebrand, and Donahue's *Hazardous Materials Emergencies Involving Intermodal Containers* (Noll, Hildebrand, and Donahue, 1995).

Intermodal containers were first introduced in Europe and are used extensively in international trade, so they are commonly designed to international specifications. IMO regulations contained in Section 13 of the International Maritime Dangerous Goods Code include specific design requirements for intermodal containers used in international transportation. DOT regulations contained in 49 CFR Parts 170 through 179 include design requirements for intermodals used domestically. All intermodal containers used in the United States must be built to the DOT specifications, unless exempted from the requirements by DOT. Standards of a number of other agencies may come into play as well. For example, all portable tank containers that are accepted for transport by the rail industry must meet the requirements of the Association of American Railroad's requirements specified in AAR 600.

Intermodal containers can generally be classified as intermodal freight containers, portable tank containers, and intermediate bulk containers. Each of these categories is discussed below.

Intermodal Container Markings. All intermodal containers used in domestic transportation must meet the DOT requirements for bulk container placards and markings as discussed above in this chapter. Intermodal freight and tank containers are required to have stenciled reporting marks and container numbers that identify the container, plus additional markings for each type of container as described below.

Intermodal tank containers shipped from Europe may display a hazard identification code. The hazard identification code will be found in the top half of orange panels attached to the containers. The four-digit U.N. identification number is found in the bottom half of the orange panel. These codes are called "hazard identification numbers" under European regulations. The intermodal hazard identification codes are explained in Figure 5.33, and examples of various codes with their meanings are shown in Table 5.11.

HAZARD IDENTIFICATION CODES
DISPLAYED ON SOME INTERMODAL CONTAINERS

Hazard identification codes, referred to as "hazard identification numbers" under European and some South American regulations, may be found in the top half of an orange panel on some Intermodal bulk containers. The 4-digit identification number is in the bottom half of the orange panel.

33
1203

The hazard identification code in the top half of the orange panel consists of two or three figures. In general, the figures indicate the following hazards:

2 – EMISSION OF GAS DUE TO PRESSURE OR CHEMICAL REACTION
3 – FLAMMABILITY OF LIQUIDS (VAPORS) AND GASES OR SELF-HEATING LIQUID
4 – FLAMMABILITY OF SOLIDS OR SELF-HEATING SOLID
5 – OXIDIZING (FIRE-INTENSIFYING) EFFECT
6 – TOXICITY OR RISK OF INFECTION
7 – RADIOACTIVITY
8 – CORROSIVITY
9 – RISK OF SPONTANEOUS VIOLENT REACTION

- Doubling of a figure indicates an intensification of that particular hazard (i.e., 33, 66, 88).
- Where the hazard associated with a material can be adequately indicated by a single figure, the figure is followed by a zero (i.e., 30, 40, 50).
- A hazard identification code prefixed by the letter "X" indicates that the material will react dangerously with water (i.e., X88).

Figure 5.33 These hazard identification codes may be found on some intermodal containers of foreign origin.

TABLE 5.11   Examples of Hazard Identification Codes Displayed on Some Intermodal Containers

Hazard ID Code	Hazard Identification Code Meaning
20	Inert gas
22	Refrigerated gas
223	Refrigerated gas, flammable
225	Refrigerated gas, oxidizing (fire intensifying)
23	Flammable gas
236	Flammable gas, toxic
239	Flammable gas that can spontaneously lead to violent reaction
25	Oxidizing (fire intensifying) gas
26	Toxic gas
263	Toxic gas, flammable
265	Toxic gas, oxidizing (fire intensifying)
266	Highly toxic gas
268	Toxic gas, corrosive
30	Flammable liquid
323	Flammable liquid that reacts with water, emitting flammable gas
X323	Flammable liquid that reacts dangerously with water, emitting flammable gas
33	Highly flammable liquid
333	Pyrophoric liquid
X333	Pyrophoric liquid that reacts dangerously with water
336	Highly flammable liquid, toxic
338	Highly flammable liquid, corrosive
X338	Highly flammable liquid, corrosive, that reacts dangerously with water
339	Highly flammable liquid that can spontaneously lead to violent reaction
36	Flammable liquid, toxic, or self-heating liquid, toxic
362	Flammable liquid, toxic that reacts with water, emitting flammable gas
X362	Flammable liquid, toxic, that reacts dangerously with water, emitting flammable gas
368	Flammable liquid, toxic, corrosive
38	Flammable liquid, corrosive
382	Flammable liquid, corrosive, that reacts with water, emitting flammable gas
X382	Flammable liquid, corrosive, that reacts dangerously with water, emitting flammable gas
39	Flammable liquid that can spontaneously lead to violent reaction
40	Flammable solid, or self-reactive material, or self-heating material
423	Solid that reacts with water, emitting flammable gas

Source: Emergency Response Guidebook, 2000

Intermodal Freight Containers. Intermodal freight containers, or "box" containers, are generally analogous to the highway transport containers discussed above. For example, intermodal freight containers may transport the same nonbulk hazmat containers as the van trailer on the highway. They are typically 20 to 40 feet long, 8 feet wide, and 8 or 8.5 feet high.

Intermodal freight containers are designed with corner castings that are used as points of attachment for securing the containers to the transport vehicle and that serve as lifting points for equipment designed to handle intermodals. Intermodal containers are only intended to be lifted and moved by the corner castings.

Some intermodal freight containers are refrigerated with diesel fuel, electricity, or cryogenic liquids, any of which may pose a hazard to responders. Electricity and cryogenics are usually externally supplied.

Freight containers used in domestic shipment must meet all DOT requirements for placards and markings, as previously discussed in this chapter. In addition, they are required to have reporting marks and container numbers stenciled toward the right side, as you face either side, and on both ends (Fig. 5.34). As with rail cars, the intermodal reporting mark indicates ownership and the container number identifies the container. Other required markings indicate the country of registry and the size and type of container (using an ISO size-type code). Additional information is available through *The Official Intermodal Equipment Register*, published quarterly by K-III Information Company, Ltd., 424 West 33rd Street, New York, NY 10001-2604.

Intermodal Tank Containers. Portable tank containers (Fig. 5.35) are usually constructed as a single metal tank inside a rectangular metal frame. All of the tanks are registered with the International Container Bureau in France. They are bulk carriers and, therefore, must have all placards and markings required by DOT

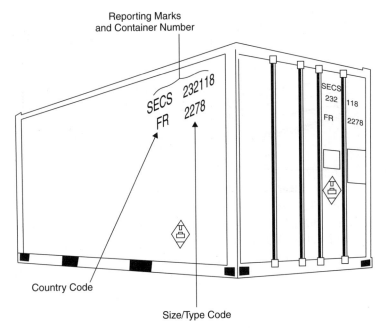

Figure 5.34 Intermodal freight containers are required to have stenciled markings.

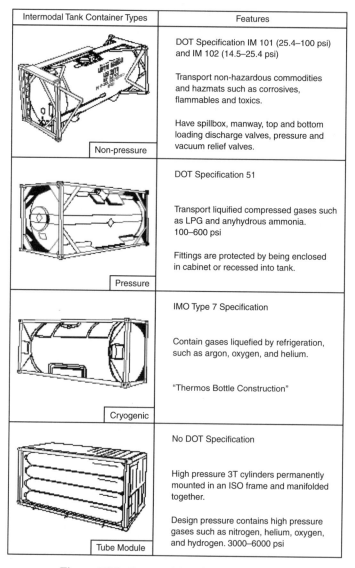

Intermodal Tank Container Types	Features
Non-pressure	DOT Specification IM 101 (25.4–100 psi) and IM 102 (14.5–25.4 psi) Transport non-hazardous commodities and hazmats such as corrosives, flammables and toxics. Have spillbox, manway, top and bottom loading discharge valves, pressure and vacuum relief valves.
Pressure	DOT Specification 51 Transport liquified compressed gases such as LPG and anyhydrous ammonia. 100–600 psi Fittings are protected by being enclosed in cabinet or recessed into tank.
Cryogenic	IMO Type 7 Specification Contain gases liquefied by refrigeration, such as argon, oxygen, and helium. "Thermos Bottle Construction"
Tube Module	No DOT Specification High pressure 3T cylinders permanently mounted in an ISO frame and manifolded together. Design pressure contains high pressure gases such as nitrogen, helium, oxygen, and hydrogen. 3000–6000 psi

Figure 5.35   Intermodal tank container types.

regulations for domestic hazardous materials transport. In addition, tank containers have additional markings (Fig. 5.36), including:

- Reporting marks and container numbers used to indicate ownership and identify the specific container, as previously described for intermodal freight containers
- Specification markings that indicate the standard to which the container is constructed

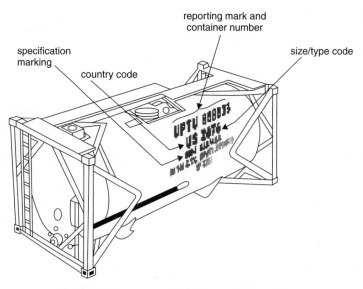

Figure 5.36   Intermodal tank container markings.

- DOT exemption markings identifying any DOT exemptions under which the container is operated in domestic transportation
- AAR-600 markings indicating that the container meets the requirements of the Association of American Railroads for domestic shipment by rail
- Country, size, and type markings that identify the country of registry, size, and pressure range of the container. Additional information is available through *The Official Intermodal Equipment Register* as described in the section on intermodal freight containers
- Tank and valve test dates

In addition, DOT requires the proper shipping name to be stenciled in letters at least 2 inches high on opposing sides of the container for all domestic hazardous materials shipments. Portable tanks may also display additional placards required by IMDG regulations.

Intermodal tanks have two types of frame designs, box and beam. In Figure 5.35, the nonpressure tank shows the beam-type design and the other three tanks show the box-type design. The box-type frame encloses the entire tank in a cagelike framework. The beam-type frame uses frame structures only at the ends of the tank and relies on the shell of the tank to act as a beam to support the weight of the cargo. All tank intermodals are fitted with corner castings to be used for lifting and securing the container. If they must be repositioned in an emergency, they should only be lifted by the corner castings or corner posts and then only once in communication with the owner. For reference, the end of the tank fitted with valves is considered the rear of the tank and the left and right sides are oriented as you face the rear.

Over ninety percent of intermodal tanks are constructed of stainless steel. Most have a single compartment with no internal baffles. Multicompartment units are rare. Tank capacity usually does not exceed 6340 gallons, and the dimensions are usually 20 feet long, 8 feet wide, and 8 feet or 8.5 feet high; however, much larger tank containers may be encountered rarely.

A data plate containing specification, approval, and operational information is attached at the rear of the container. In addition, a tube containing documents such as shipping papers, cleaning information, or an MSDS may also be permanently mounted near the data plate.

Portable tanks may be lined to protect the tank material from the product. Some portable tank containers are insulated and some include refrigeration units, as described above for freight containers. Some have heating units to assist in off-loading viscous products. Heating units are either electrical or use steam coils, which may be installed internally or externally to the tank. Either type may represent a hazard to responders in an emergency. An electrical control box located at the rear end of the tank contains controls for the heating unit.

Intermodals have fittings, valves, and other features that are analogous to those previously described for highway cargo tanks. However, the intermodals commonly have British Standard Pipe Thread, British Standard Whitworth, or Metric thread patterns as opposed to the National Pipe Thread pattern that is standard in the United States.

Intermodal tank containers can be generally classified as nonpressure, pressure, or specialized portable tanks. These are shown in Figure 5.35 and are discussed below.

NONPRESSURE INTERMODAL TANK CONTAINERS (Fig. 5.35) constitute 90 percent of the portable tanks in use. They commonly transport a wide variety of both nonhazardous and hazardous liquid products. The nonpressure containers typically have a capacity of 5000 to 6300 gallons. The most common types of nonpressure tank containers in domestic transport are the IM 101 (designated as IMO-type 1 internationally) and the IM 102 (designated as IMO-type 2 internationally). Although these containers are built to different specifications, it is not possible to distinguish between them visually, except by noting the specification marking on the container markings or data plate.

IM 101 containers have maximum allowable working pressures from 25.4 to 100 psi. They are commonly used to transport liquid hazardous materials, including toxics, corrosives, and flammables with flash points of less than 32 degrees F, as well as various nonhazardous commodities.

IM 102 containers have maximum allowable working pressures from 14.5 psi to 25.4 psi. They most commonly contain nonhazardous products but are also used to transport hazmats, including toxics, corrosives, and flammables with flash points from 32 to 140 degrees F.

Nonpressure tank containers have fittings similar to those found on DOT 407 and 412 cargo tank trucks (Fig. 5.37). These include a manhole and cover contained within a splash box that drains through open pipes. Both top and

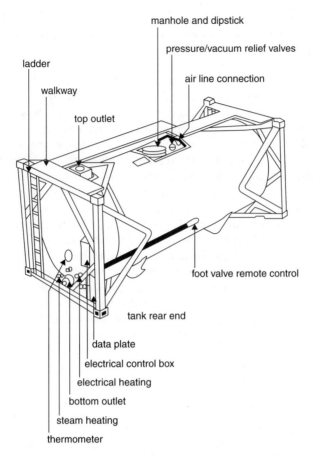

Figure 5.37   Nonpressure intermodal tank container fittings.

bottom valves are used for unloading product from portable tanks. For hazardous materials shipments, bottom discharges must have two valves in series and the inner valve must be internally mounted. The external valve must have a liquid tight closure, such as a blind flange, threaded cap, or cam-lock cap. Airline connections are used for pressure unloading, vapor return, and blanketing the product with inert gas. They may also be fitted with thermometers for measuring the temperature of the product.

Nonpressure tank containers are fitted with safety devices similar to those provided for DOT 406, 407, and 412 highway cargo tanks. Two pressure/vacuum relief valves are mounted on top of the containers. The relief valves may have a rupture disk between the product and the valve to protect the spring in the valve from the product. In that case, a gauge may be present to detect a ruptured disk. If present, the gauge should read zero unless the disk has failed. Nonpressure portable tanks feature emergency remote shut-off devices that close the inner discharge valve and that can be activated from the front end of the container.

Fusible links or nuts attached to the remote shut-offs are designed to melt at temperatures of 250 degrees F or greater to activate the shutoff. Discharge valves have shear sections designed to fail, leaving the valve intact.

PRESSURE INTERMODAL TANK CONTAINERS used in domestic shipment are typically constructed to DOT Specification 51, and are generally equivalent to IMO Type 5 containers used internationally (Fig. 5.35). We can think of the "Spec 51" containers as generally equivalent to MC 331 highway cargo tanks in terms of design and likely contents. They typically have design pressures from 100 to 500 psi. Capacity is usually 4500 to 5000 gallons, although much larger ones may be encountered rarely. These containers are far less common than nonpressure intermodals but may be used for domestic transport of hazmats such as LPGs, anhydrous ammonia, high-vapor-pressure flammable liquids, and pyrophoric liquids.

Pressure intermodals can be readily distinguished from nonpressure containers because the fittings are protected by being recessed or enclosed by a protective cover (Fig. 5.35). The fittings include loading and unloading valves for both liquid and vapor product, and each type of valve must be clearly marked. Other fittings include gauging devices, sample lines, and thermometer wells.

Pressure tank containers have top-mounted pressure relief devices intended to prevent overpressurization of the container. The pressure relief devices may be contained in a compartment designed to protect them from the elements. Eduction pipes for liquid valves on pressure intermodals are fitted with excess flow valves intended to prevent the discharge of liquid product in the event of a broken valve. Vapor valves may also have excess flow valves.

SPECIALIZED INTERMODAL PORTABLE TANKS include those designed to transport cryogenics and nonliquefied gases. These are rare in domestic transportation.

Cryogenic intermodal containers (Fig. 5.35) are built to the IMO Type 7 specification, and are generally analogous to the MC 338 highway cargo tank. They utilize a "thermos bottle" type construction, as previously described for the MC 338 highway cargo tank and are designed to contain gases liquefied through refrigeration, such as helium, argon, and oxygen.

Tube modules consist of high-pressure T3 cylinders permanently mounted in an ISO frame and manifolded together (Fig. 5.35). Test pressures range from 3000 to 5000 psi. Tube modules transport nonliquefied gases such as nitrogen, helium, oxygen, and hydrogen.

Intermediate Bulk Containers. Intermediate bulk containers (IBCs), commonly referred to as "totes," have become very popular as an alternative to nonbulk containers such as 55-gallon drums. DOT issued new regulations in 1994 that formally recognized this type of container. IBCs can be divided into three basic types: all-metal tanks, polyethylene and steel tanks, and flexible containers (Fig. 5.38).

ALL-METAL TANKS are constructed completely of metal and are usually cubic in shape, although cylindrical ones are also used. They can be readily stacked and

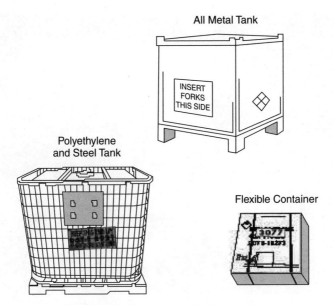

Figure 5.38   Intermediate bulk containers.

moved with forklifts. They are commonly used to ship flammable liquids and solids and have a capacity of 300 to 400 gallons. All-metal tanks have a top fill opening and frequently have a bottom discharge as well. Any pressure relief devices present, such as a fusible plug, are incorporated into the fill lid. Fill lids are a common source of leakage in emergencies.

POLYETHYLENE AND STEEL TANKS consist of a polyethylene tank within a steel frame, making them easily stacked and moved with forklifts. They are commonly used to transport corrosives and will hold 300 to 400 gallons of product. Polyethylene and steel tanks have a top fill opening and usually have a top discharge.

FLEXIBLE CONTAINERS consist of a bag of cloth or plastic that has a tied opening at the top and a bottom discharge. They may have loops or rings for lifting on top and are often mounted on skids for handling by forklift. They are designed to transport dry bulk products and have a capacity of up to 2000 pounds. Some may be enclosed in heavy cardboard boxes and attached to pallets. They are not authorized for liquids unless given an exemption by DOT. They may transport DOT Class 9 hazardous materials.

Other Hazardous Materials Containers
Other containers may contain hazmats. Some of these are described below.

Ton Containers. Ton containers are cylindrical pressure vessels that are approximately 2.5 feet in diameter and 6.5 to 7 feet long (Fig. 5.39). They have a one-ton

Figure 5.39   Chlorine ton container.

water capacity and are designed to contain liquefied compressed gases such as chlorine, anhydrous ammonia, sulfur dioxide, and phosgene. Ton containers are transported by both rail and highway and are used as storage vessels at the point of end use. They are commonly used for chlorine at water treatment plants.

Ton containers are fitted with a liquid valve and a vapor valve in one head of the container. A protective cap covers the valves during transport. Ton containers typically have three fusible plugs in each head designed to melt and vent excess pressure in a fire. Ton containers for phosgene reportedly have no pressure relief devices.

Cryogenic Vessels. Cryogenic vessels (Fig. 5.40) are bulk containers designed to maintain gases that have been liquefied through refrigeration in the liquid state. They are commonly used in the industrial, medical, and research fields and may be encountered in transportation or at fixed facilities. Cryogenic vessels are constructed like a thermos bottle to maintain the contents in a cryogenic state and are fitted with pressure relief valves designed to maintain a set internal pressure by releasing gas. They are commonly mounted on casters to make them easy to move. Cryogenic vessels commonly contain liquid nitrogen, argon, oxygen, and helium.

Calcium Carbide Casks. Special casks or bins are used to transport calcium carbide (Fig. 5.41). They are designed to keep the product dry, because calcium carbide reacts violently with water to produce heat and large quantities of acetylene gas. These containers are commonly attached to rail flat cars or flatbed highway trailers for transport.

Figure 5.40   Cryogenic vessel.

Figure 5.41   Calcium carbide containers.

Shipping Containers for Radioactive Materials. Radioactive materials are commonly shipped, just as other hazardous commodities are. Two different types of containers may be used, type A or type B packages (Fig. 5.42a and b), depending on the level of hazard associated with the commodity being shipped.

TYPE A PACKAGES are used to ship radiological materials in amounts that are not life-threatening (Fig. 5.42a). Steel buckets or drums and fiberboard, wooden, or metal boxes may be used for type A packages. Approved type A containers will display the marking "U.S. D.O.T. 7A-TYPE A." Type A packages are designed to withstand only normal shipping and handling, not vehicle accidents or fires.

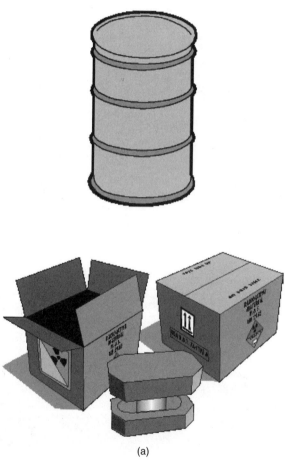

(a)

Figure 5.42 Containers for radioactive materials include type A packages (a), which are used to ship radiological materials in amounts that are not life-threatening and are designed to withstand only normal shipping and handling, and type B packages (b), which are used to ship life-threatening amounts of radioactive materials and are designed to remain intact through transportation accidents, crashes, and fires.

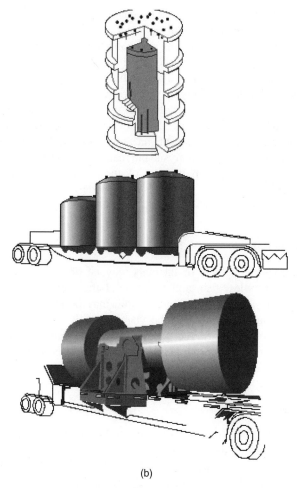

(b)

Figure 5.42   (*continued*)

TYPE B PACKAGES are used to ship life-threatening amounts of radioactive materials (Fig. 5.42b). They are very rugged containers that are designed to remain intact through transportation accidents, crashes, and fires. Some type B packages are designed to be mounted to transport vehicles.

Air Monitoring and Sampling of Hazardous Materials

Responders should employ air monitoring equipment any time a hazardous material release is suspected. Direct-reading instruments (DRIs) are available that can detect the presence of airborne hazards such as combustible gases or vapors, abnormal oxygen atmospheres, and certain specific chemicals. DRIs can warn of the presence of hazards that would require special protection for the responders while further assessment is being conducted. Air monitoring is discussed in detail in Chapter 9.

Sampling and analysis of hazardous materials is of limited usefulness for identifying hazardous materials during emergency response operations. Valid sampling requires specialized training and equipment, and the sample collection process may expose personnel to unidentified hazards. Definitive laboratory analysis of samples can provide a positive identification but is too time-consuming to be very useful for most emergency response operations. Field identification kits are available that use flowcharts and various test materials. These kits, which are sometimes referred to as "Hazcat" kits, can be used to identify, or at least determine the chemical group of, hazardous materials. Results are perhaps less accurate than laboratory analysis but can be obtained in about 30 minutes. Users must be trained in the use of these kits. It is worth noting that opinions on the usefulness of these kits vary widely among people who have used them.

Biological Indicators

In some situations the effects of hazmats may be obvious in dead wildlife—mammals, birds, reptiles, fish, and insects—at the incident scene. Vegetation, such as grass in an area of contaminated soil, may be killed through prolonged exposure. Hazardous materials may be visible as oily sheens or other types of discoloration in lakes, streams, and puddles of water. In all hazmat situations, use your common sense, and incorporate indicators such as these into the recognition and assessment process.

Human Senses

Generally speaking, the human senses are not dependable indicators of the presence of hazardous materials. One exception to this statement is the sense of sight, which can be used in conjunction with the methods described in this chapter to safely recognize and identify hazmats. Relying on the other senses, harking back to the "scratch and sniff" approach to hazmat identification (as shown in Fig. 5.1) can get us into trouble.

This is especially true for the sense of smell. Some hazmats, such as anhydrous ammonia and chlorine, have very low odor thresholds and can be smelled at concentrations well below hazardous levels. Other hazmats, however, have odor thresholds well above hazardous levels, so that when the concentration is high enough for you to detect the chemical by smell you will already be overexposed or subject to flash fire. Many hazardous materials have no odor, taste, sense of irritation, or any other type of warning property at any concentration. Other hazmats, such as hydrogen sulfide, cause olfactory fatigue, so that an odor may be detected initially followed by loss of the ability to smell the contaminant. The sense of smell varies widely from person to person, and some people cannot smell certain odors at all. In a worst-case scenario, the sense of smell may provide the only warning you have that hazmats are present, but it should never be relied on solely.

INITIAL ASSESSMENT AND ACTIONS

Incident assessment should begin well before an emergency incident occurs. This has long been expressed by the fire service adage "size up begins with preplanning." Studying relevant response factors during the preemergency planning stage enables us to act more quickly and efficiently when an emergency occurs. This allows some emergency response research and decision-making to be done when ample time is available and no adrenalin is flowing. A good preplan will provide general response guidance and limit assessment and decision-making to that required to fine-tune the preplan. In some cases, responders may be required to "start from scratch" with little or no advance guidance in conducting initial assessment and determining initial actions to take in response to an incident. In either case, some degree of decision-making is involved.

Using the DOT *Emergency Response Guidebook*

The *Emergency Response Guidebook* (ERG) is published by the U.S. Department of Transportation's Research and Special Programs Administration for use by fire fighters, police, and emergency services personnel during the initial stages of response to hazardous materials incidents. It is distributed free of charge to public sector organizations, and an electronic version can be accessed on the internet at www.hazmat.dot.gov. The ERG is a guide for initial actions in response to a hazmat incident. It is primarily focused on transportation accidents involving truck or rail transport but may have limited applicability to other types of incidents such as those occurring at fixed facilities. The ERG also lists sources of information for decisions on further action and cleanup. Most of the ERG is divided into four major sections: yellow-bordered pages, blue-bordered pages, orange-bordered pages, and green-bordered pages.

General ERG Instructions
The ERG begins with instructions, use considerations, and safety precautions for using the book in responding to an incident. These initial pages are plain, in that they do not have color-coded borders. As an emergency responder, it is critical for you to be familiar with these instructions and able to use the ERG effectively in first response to a hazmat incident. At the same time, you should realize the limitations of the generalized information contained in the ERG. Always try to get more detailed information as soon as possible.

Note that the information provided here is intended as a general description of how the ERG can be used and is not intended as a substitute for specific training in using the ERG. This information is specific to the 2000 edition of the ERG. The ERG is periodically updated, and new editions are published.

The instructions with the ERG instruct us to first identify the material. We can do so by the four-digit UN/NA identification number or by the commodity name listed on shipping papers, container stencils, or other sources.

Using the Yellow and Blue Sections: Listings by UN/NA Number or Material Name

If we have identified the product by UN/NA identification number, we are instructed to look the number up in the yellow section of the ERG. The four-digit numbers are listed in the yellow section in numerical order, along with the name of each material (Table 5.12). The guide numbers in the yellow section refer us to the appropriate hazard and action guide in the orange section.

If we have identified the product by name, we are instructed to look the name up in the blue section of the ERG. The same information just described for the yellow section is included in the blue section, except that the materials are listed alphabetically by name instead of numerically by identification number (Table 5.13).

Entries with the letter "P" following the guide number in the yellow and blue sections indicate that the material may undergo violent polymerization if heated or contaminated. Explosives are not listed individually by name or four-digit identification number in the ERG. They are covered under Hazard Guides 112 and 114, as explained in the instructions to the ERG.

Materials that are highlighted in the yellow and blue sections are Toxic Inhalation Hazards (TIHs) or materials that may react with water to produce toxic gases. The highlighted materials are also listed in the Table of Initial Isolation and Protective Action Distances in the green section. Special instructions apply for highlighted entries. If we encounter highlighted entries in either the yellow or blue sections, we are instructed to immediately turn to the green section and look up the material. If not, we are directed to proceed to the orange section. We will describe use of the green section later.

Using the Orange Section: Hazard and Action Guides

In the orange section of the ERG, materials are grouped under various guide numbers according to their expected behavior during an emergency. Each guide includes generalized information on the hazards we can expect and the actions we should take during initial response to the incident (Fig. 5.43). The guides can be read quickly, and they give important facts regarding fire and explosion hazards, health hazards, public safety, and emergency actions recommended for small or large fires and small or large spills and first aid treatment of exposure victims.

Dealing with Situations in Which Information is Sparse

The ERG also provides guidance for situations in which we are not able to identify the material by UN/NA number or by name. In such a case, if we can see a placard on the damaged container, we are directed to locate the placard in the table of placards in the front of the ERG and determine the appropriate guide to consult in the orange pages. If we have no UN/NA number, name, or placard for the material, but we can see that a common highway or rail container is involved, the ERG directs us to turn to the charts showing container silhouettes in the front section of the ERG and determine the appropriate guide to consult in the orange pages. Keep in mind that this approach is only recommended for a

TABLE 5.12 Example Page from 2000 *Emergency Response Guidebook*, **Yellow Section**

ID No.	Guide No.	Name of Material	ID No.	Guide No.	Name of Material
1570	152	Brucine	1584	151	Cocculus
1571	113	Barium azide, wetted with not less than 50% water	1585	151	Copper acetoarsenite
			1586	151	Copper arsenite
1572	151	Cacodylic acid	1587	151	Copper cyanide
1573	151	Calcium arsenate	1588	157	Cyanides, inorganic, n.o.s.
1574	151	Calcium arsenate and Calcium arsenite mixture, solid	1588	157	Cyanides, inorganic, solid, n.o.s.
1574	151	Calcium arsenite, solid	1589	125	CK
1574	151	Calcium arsenite and calcium arsenate mixture, solid	1589	125	Cyanogen chloride, inhibited
			1590	153	Dichloroanilines
			1590	153	Dichloroanilines, liquid
1575	157	Calcium cyanide	1590	153	Dichloroanilines, solid
1577	153	Chlorodinitrobenzenes	1591	152	o-Dichlorobenzene
1577	153	Dinitrochlorobenzene	1592	152	p-Dichlorobenzene
1578	152	Chloronitrobenzenes	1593	160	Dichloromethane
1578	152	Chloronitrobenzenes, liquid	1593	160	Methylene chloride
1578	152	Chloronitrobenzenes, solid	1594	152	Diethyl sulfate
1578	152	Nitrochlorobenzenes, liquid	1594	152	Diethyl sulphate
1578	152	Nitrochlorobenzenes, solid	1595	156	Dimethyl sulfate
1579	153	4-Chloro-o-toluidine hydrochloride	1595	156	Dimethyl sulphate
			1596	153	Dinitroanilines
1580	154	Chloropicrin	1597	152	Dinitrobenzenes
1581	123	Chloropicrin and methyl bromide mixture	1598	153	Dinitro-o-cresol
			1599	153	Dinitrophenol, solution
1581	123	Methyl bromide and chloropicrin mixtures	1600	152	Dinitrotoluenes, molten
1581	123	Methyl bromide and more than 2% chloropicrin mixture, liquid	1601	151	Disinfectant, solid, poisonous, n.o.s.
			1601	151	Disinfectant, solid, toxic, n.o.s.
1582	119	Chloropicrin and methyl chloride mixture	1601	151	Disinfectants, solid, n.o.s. (poisonous)
1582	119	Methyl chloride and chloropicrin mixtures	1602	151	Dye, liquid, poisonous, n.o.s.
1583	154	Chloropicrin, absorbed	1602	151	Dye, liquid, toxic, n.o.s.
1583	154	Chloropicrin mixture, n.o.s.	1602	151	Dye intermediate, liquid, poisonous, n.o.s
			1602	151	Dye intermediate, liquid, toxic, n.o.s.

Source: 2000 *Emergency Response Guidebook*

Note: If an entry is highlighted in either the yellow-bordered or blue-bordered pages AND THERE IS NO FIRE, go directly to the Table of Initial Isolation and Protective Action Distances (green-bordered pages) and look up the ID number and name of material to obtain initial isolation and protective action distances. IF THERE IS A FIRE, or IF A FIRE IS INVOLVED, go directly to the appropriate guide (orange-bordered pages) and use the evacuation information shown under PUBLIC SAFETY.

TABLE 5.13   Example Page from 2000 *Emergency Response Guidebook*, **Blue Section**

Name of Material	Guide No.	ID No.	Name of Material	Guide No.	ID No.
3-Chloroperoxybenzoic acid	146	2755	Chlorosilanes, n.o.s.	155	2985
Chlorophenates, liquid	154	2904	Chlorosilanes, n.o.s.	155	2986
Chlorophenates, solid	154	2905	Chlorosilanes, n.o.s.	156	2987
Chlorophenolates, liquid	154	2904	Chlorosilanes, n.o.s.	139	2988
Chlorophenolates, solid	154	2905	Chlorosilanes, water-reactive,	139	2988
Chlorophenols, liquid	153	2021	flammable, corrosive, n.o.s.		
Chlorophenols, solid	153	2020	Chlorosulfonic acid	137	1754
Chlorophenyltrichlorosilane	156	1753	Chlorosulfonic acid and sulfur	137	1754
Chloropicrin	154	1580	trioxide mixture		
Chloropicrin, absorbed	154	1853	Chlorosulphonic acid	137	1754
Chloropicrin and methyl bromide	123	1581	Chlorosulphonic acid and sulphur	137	1754
mixture			trioxide mixture		
Chloropicrin and methyl chloride	119	1582	1-Chloro-1,2,2,2-tetrafluoroethane	126	1021
mixture			Chlorotetrafluoroethane	126	1021
Chloropicrin and nonflammable,	123	1955	Chlorotetrafluoroethane and	126	3297
nonliquefied compressed gas			ethylene oxide mixture, with not		
mixture			more than 8.8% ethylene oxide		
Chloropicrin mixture, flammable	131	2929	Chlorotoluenes	130	2238
Chloropicrin mixture, n.o.s.	154	1583	4-Chloro-*o*-toluidine hydrochloride	153	1579
Chloropivaloyl chloride	156	9263	Chlorotoluidines	153	2239
Chloroplatinic acid, solid	154	2507	Chlorotoluidines, liquid	153	2239
Chloroprene, inhibited	131P	1991	Chlorotoluidines, solid	153	2239
1-Chloropropane	129	1278	1-Chloro-2,2,2-trifluoroethane	126	1983
2-Chloropropane	129	2356	Chlorotrifluoroethane	126	1983
3-Chloropropanol-1	153	2849	Chlorotrifluoromethane	126	1022
2-Chloropropene	130P	2456	Chlorotrifluoromethane and	126	2599
2-Chloropropionic acid	153	2511	Trifluoromethane azeotropic		
α-Chloropropionic acid	153	2511	mixture with approximately 60%		
2-Chloropyridine	153	2822	chlorotrifluoromethane		
Chlorosilanes, corrosive, flammable,	155	2986	Chlorpyrifos	152	2783
n.o.s.			Chromic acetate	171	9101
Chlorosilanes, corrosive, n.o.s.	156	2987	Chromic acid, solid	141	1463
Chlorosilanes, flammable, corrosive,	155	2985	Chromic acid, solution	154	1755
n.o.s.					

Source: 2000 *Emergency Response Guidebook*
Note: If an entry is highlighted in either the yellow-bordered or blue-bordered pages AND THERE IS NO FIRE, go directly to the Table of Initial Isolation and Protective Action Distances (green-bordered pages) and look up the ID number and name of material to obtain initial isolation and protective action distances. IF THERE IS A FIRE, or IF A FIRE IS INVOLVED, go directly to the appropriate guide (orange-bordered pages) and use the evacuation information shown under PUBLIC SAFETY.

worst-case situation when no better information is available. If we have no other information to go on, but we suspect that hazardous materials are involved in an incident, or if the Dangerous placard is present, the ERG instructs us to use Guide 111.

Using the Green Section: Situations Involving Toxic Inhalation Hazards

Once we identify that a material has a highlighted entry in either the yellow section or the blue section, the ERG directs us to consult the green section immediately. The green-edged pages of the guidebook contain the Table of Initial

Isolation and Protective Action Distances (Table 5.14) and the Table of Water-Reactive Materials Which Produce Toxic Gases (Table 5.15). These tables list materials that are seriously hazardous by inhalation.

A set of detailed instructions and considerations for using the green section is located just in front of the green-edged pages. It is critical that we fully understand these instructions. The information in the green section is useful only for the first 30 minutes of an accident involving hazardous materials without fire.

GUIDE 154	Substances – Toxic and/or Corrosive (Non-Combustible)	ERG 2000

POTENTIAL HAZARDS

HEALTH

- **TOXIC**; inhalation, ingestion, or skin contact with material may cause severe injury or death.
- Contact with molten substances may cause severe burns to skin and eyes.
- Avoid any skin contact.
- Effects of contact or inhalation may be delayed.
- Fire may produce irritating, corrosive and/or toxic gases.
- Runoff from fire control or dilution water may be corrosive and/or toxic and cause pollution.

FIRE OR EXPLOSION

- Non-combustible, subtance itself does not burn but may decompose upon heating to produce corrosive and/or toxic fumes.
- Some are oxidizers and may ignite combustibles (wood, paper, oil clothing, etc.).
- Contact with metals may evolve flammable hydrogen gas.
- Containers may explode when heated.

PUBLIC SAFETY

CALL Emergency Response Telephone Number on Shipping Paper first. If Shipping Paper not available or no answer, refer to appropriate telephone number listed on the inside back cover.
- Isolate spill or leak area immediately for at least 25 to 50 meters (80 to 160 feet) in all directions.
- Keep unauthorized personnel away.
- Stay upwind.
- Keep out of low areas.
- Ventilate enclosed areas.

PROTECTIVE CLOTHING

- Wear positive pressure self-contained breathing apparatus (SCBA).
- Wear chemical protective clothing which is specifically recommended by the manufacturer. It may provide little or no thermal protection.
- Structural firefighters' protective clothing provides limited protection in fire situations ONLY; It is not effective in spill situations.

EVACUATION

Spill
- See the Table of Initial Isolation and Protective Action Distances for highlighted substances. For non-highlighted substances, increase, in the downwind direction, as necessary, the isolation distance shown under "PUBLIC SAFETY".

Fire
- If tank, rail car or tank truck is involved in a fire, ISOLATE for 800 meters (1/2 mile) in all directions; also, consider initial evacuation for 800 meters (1/2 mile) in all directions.

Source: 2000 Emergency Response Guidebook

Figure 5.43 An example of a Hazard and Action Guide from the orange section of the DOT ERG.

ERG 2000   **Substances – Toxic and/or Corrosive**   **GUIDE**
(Non-Combustible)   **154**

EMERGENCY RESPONSE

FIRE

Small Fires
- Dry chemical, CO_2 or water spray.

Large Fires
- Dry chemical, CO_2, alcohol-resistant foam or water spray.
- Move containers from fire area if you can do it without risk.
- Dike fire control water for later disposal; do not scatter the material.

Fire involving Tank or Car/Trailer Loads
- Fight fire from maximum distance or use unmanned hose holders or monitor nozzles.
- Do not get water inside containers.
- Cool containers with flooding quantities of water untill well after fire is out.
- Withdraw immediately in case of rising sound from venting safety devices or discoloration of tank.
- ALWAYS stay away from tanks engulfed in fire.

SPILL OR LEAK

- ELIMINATE all ignition sources (no smoking, flares, sparks or flames in immediate area).
- Do not touch damaged containers or spilled material unless wearing appropriate protective clothing.
- Stop leak if you can do it without risk.
- Prevent entry into waterways, sewers, basements or confined areas.
- Absorb or cover with dry earth, sand or other non-combustible material and transfer to containers.
- DO NOT GET WATER INSIDE CONTAINERS.

FIRST AID

- Move victim to fresh air.
- Call 911 or emergency medical service.
- Apply artificial respiration if victim is not breathing.
- **Do not use mouth-mouth method if victim ingested or inhaled the substance; induce artificial respiration with the aid of a pocket mask equipped with a one-way valve or other proper respiratory medical device.**
- Administer oxygen if breathing is difficult.
- Remove and isolate contaminated clothing and shoes.
- In case of contact with substance, immediately flush skin or eyes with running water for at least 20 minutes.
- For minor skin contact, avoid spreading material on unaffected skin.
- Keep victim warm and quiet.
- Effects of exposure (inhalation, ingestion or skin contact) to substance may be delayed.
- Ensure that medical personnel are aware of the material(s) involved, and take precautions to protect themselves.

Figure 5.43   *(continued)*

After 30 minutes the distances in the green tables may not be adequate because of ongoing migration of the hazard. For situations involving fire, the physical threat of container explosion and projection of shrapnel may exceed hazards due to chemical toxicity. In cases involving fire, evacuation distances given in the appropriate orange section guide for a container involved in fire should be used.

The Table of Initial Isolation and Protective Action Distances lists materials numerically by the four-digit identification number and also shows the name of the material for each entry. The table includes isolation and protective action distances for both small and large spills (Table 5.14). Small spills would be those

TABLE 5.14 Example Page from 2000 ERG, Table of Initial Isolation and Protective Action Distances

ID No.	NAME OF MATERIAL	SMALL SPILLS (From a small package or small leak from a large package)							LARGE SPILLS (From a large package or from many small packages)							
		First ISOLATE in all directions		Then PROTECT persons downwind during—						First ISOLATE in all directions		Then PROTECT persons downwind during—					
				DAY		NIGHT					DAY		NIGHT				
		Meters	(Feet)	Kilometers	(Miles)	Kilometers	(Miles)		Meters	(Feet)	Kilometers	(Miles)	Kilometers	(Miles)		
1541	Acetone cyanohydrin, stabilized **(when spilled in water)**	30 m	(100 ft)	0.2 km	(0.1 mi)	0.2 km	(0.1 mi)		95 m	(300 ft)	0.8 km	(0.5 mi)	2.1 km	(1.3 mi)		
1556	MD **(when used as a weapon)**	30 m	(100 ft)	0.3 km	(0.2 mi)	0.8 km	(0.5 mi)		125 m	(400 ft)	1.3 km	(0.8 mi)	3.5 km	(2.2 mi)		
1556	Methyldichloroarsine	30 m	(100 ft)	0.2 km	(0.1 mi)	0.3 km	(0.2 mi)		60 m	(200 ft)	0.5 km	(0.3 mi)	1.0 km	(0.6 mi)		
1556	PD **(when used as a weapon)**	30 m	(100 ft)	0.2 km	(0.1 mi)	0.2 km	(0.1 mi)		30 m	(100 ft)	0.2 km	(0.1 mi)	0.3 km	(0.2 mi)		
1560	Arsenic chloride	30 m	(100 ft)	0.2 km	(0.1 mi)	0.3 km	(0.2 mi)		60 m	(200 ft)	0.6 km	(0.4 mi)	1.4 km	(0.9 mi)		
1560	Arsenic trichloride															
1569	Bromoacetone	30 m	(100 ft)	0.2 km	(0.1 mi)	0.3 km	(0.2 mi)		95 m	(300 ft)	0.8 km	(0.5 mi)	1.9 km	(1.2 mi)		
1580	Chloropicrin	60 m	(200 ft)	0.5 km	(0.3 mi)	1.3 km	(0.8 mi)		185 m	(600 ft)	1.8 km	(1.1 mi)	4.0 km	(2.5 mi)		
1581	Chloropicrin and Methyl bromide mixture	30 m	(100 ft)	0.2 km	(0.1 mi)	0.5 km	(0.3 mi)		125 m	(400 ft)	1.3 km	(0.8 mi)	3.1 km	(1.9 mi)		
1581	Methyl bromide and Chloropicrin mixtures															

TABLE 5.14 (*continued*)

ID No.	NAME OF MATERIAL	SMALL SPILLS (From a small package or small leak from a large package)						LARGE SPILLS (From a large package or from many small packages)					
		First ISOLATE in all directions		Then PROTECT persons downwind during—				First ISOLATE in all directions		Then PROTECT persons downwind during—			
				DAY		NIGHT				DAY		NIGHT	
		Meters	(Feet)	Kilometers	(Miles)	Kilometers	(Miles)	Meters	(Feet)	Kilometers	(Miles)	Kilometers	(Miles)
1581	Methyl bromide and more than 2% Chloropicrin mixture, liquid	30 m	(100 ft)	0.3 km	(0.2 mi)	1.1 km	(0.7 mi)	215 m	(700 ft)	2.1 km	(1.3 mi)	5.6 km	(3.5 mi)
1582 1582	Chloropicrin and Methyl chloride mixture Methyl chloride and Chloropicrin mixtures	30 m	(100 ft)	0.2 km	(0.1 mi)	0.8 km	(0.5 mi)	95 m	(300 ft)	1.0 km	(0.6 mi)	3.2 km	(2.0 mi)
1583	Chloropicrin, absorbed	60 m	(200 ft)	0.5 km	(0.3 mi)	1.3 km	(0.8 mi)	185 m	(600 ft)	1.8 km	(1.1 mi)	4.0 km	(2.5 mi)
1583	Chloropicrin mixture, n.o.s.	30 m	(100 ft)	0.3 km	(0.2 mi)	1.1 km	(0.7 mi)	215 m	(700 ft)	2.1 km	(1.3 mi)	5.6 km	(3.5 mi)
1589	CK (**when used as a weapon**)	60 m	(200 ft)	0.6 km	(0.4 mi)	2.4 km	(1.5 mi)	400 m	(1300 ft)	4.0 km	(2.5 mi)	8.0 km	(5.0 mi)

Source: 2000 Emergency Response Guidebook

involving a single small package such as a 55-gallon drum. Large spills involve a spill from a single large container such as a cargo tank or tank car, or multiple spills from many small packages.

The ERG instructs us to first evacuate everyone within the initial isolation zone, which is the area located within the isolation distance in all directions from the spill (Fig. 5.44). We are directed to have them move away from the spill in a crosswind direction to the distance specified.

Once the initial isolation zone is evacuated, the ERG directs us to protect anyone within the downwind protective action zone. The table lists protective action distances for daytime and nighttime conditions for both large and small spills (Table 5.14). Protective action distances tend to be greater for nighttime releases because atmospheric conditions are less conducive to dispersal of the released material to safe levels. The protective action distance indicates the size of the downwind protective action zone (Table 5.14). People within the protective action zone must either be evacuated or be protected in place. Protection in place is discussed in Chapter 11.

Some of the materials listed in the Table of Initial Isolation and Protective Action Distances produce toxic inhalation hazards when spilled in water. Additional information on those materials is available in the Table of Water-Reactive Materials Which Produce Toxic Gases, located in the last part of the green section (Table 5.15).

Other Considerations for Using the ERG

In the 2000 version of the ERG, information was added to assist us in responding to a criminal or terrorist incident involving chemical or biological agents. Chemical warfare agents were added to the Table of Initial Isolation and Protective Action Distances. Specific information on criminal/terrorist use of chemical/biological agents was also appended to the end of the guidebook and added to the glossary. The 2000 ERG can be used in conjunction with the federal Emergency Response to Terrorism Job Aid as described in Chapter 10.

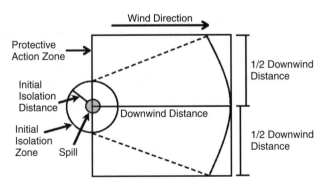

Figure 5.44 This drawing is provided in the instructions for using the Table of Initial Isolation and Protective Action Distances from the green pages of the DOT ERG.

TABLE 5.15   Example Page from 2000 ERG, Table of Water-Reactive Materials Which Produce Toxic Gases

Materials Which Produce Large Amounts of Toxic-by-inhalation (TIH) Gas(es)
When Spilled in Water

ID No.	Guide No.	Name of Material	TIH Gas(es) Produced
1162	151	Dimethyldichlorosilane	HCl
1242	139	Methyldichlorosilane	HCl
1250	155	Methyltrichlorosilane	HCl
1295	139	Trichlorosilane	HCl
1298	155	Trimethylchlorosilane	HCl
1340	139	Phosphorus pentasulfide, free from yellow and white phosphorus	H_2S
1340	139	Phosphorus pentasulphide, free from yellow and white phosphorus	H_2S
1360	139	Calcium phosphide	PH_3
1384	135	Sodium dithionite	H_2S, SO_2
1384	135	Sodium hydrosulfite	H_2S, SO_2
1384	135	Sodium hydrosulphite	H_2S, SO_2
1397	139	Aluminum phosphide	PH_3
1412	139	Lithium amide	NH_3
1419	139	Magnesium aluminum phosphide	PH_3
1432	139	Sodium phosphide	PH_3
1433	139	Stannic phosphides	PH_3
1541	155	Acetone cyanohydrin, stabilized	HCN
1680	157	Potassium cyanide	HCN
1689	157	Sodium cyanide	HCN
1714	139	Zinc phosphide	PH_3
1716	156	Acetyl bromide	HBr
1717	132	Acetyl chloride	HCl
1724	155	Allyl trichlorosilane, stabilized	HCl
1725	137	Aluminum bromide, anhydrous	HBr

Chemical Symbols for TIH Gases:

Br_2	Bromine	HF	Hydrogen fluoride	PH_3	Phosphine
Cl_2	Chlorine	HI	Hydrogen iodide	SO_2	Sulfur dioxide
HBr	Hydrogen bromide	H_2S	Hydrogen sulfide	SO_2	Sulphur dioxide
HCl	Hydrogen chloride	H_2S	Hydrogen sulphide	SO_3	Sulfur trioxide
HCN	Hydrogen cyanide	NH_3	Ammonia	SO_3	Sulphur trioxide

Decision-Making In Emergencies

It is recommended that we use a structured approach to decision-making during hazmat response operations, as discussed in Chapter 1. Ludwig Benner's DECIDE process has been used widely as a model for structuring the steps taken in response to a hazmat emergency. The DECIDE process includes six steps:

(1) Detect the presence of hazardous materials.

(2) Estimate likely harm without intervention.

(3) Choose response objectives.

(4) Identify options.

(5) Do the best option.

(6) Evaluate progress.

Let's take a look at each step in the process to see how it can be used in assessment and response to a hazmat incident.

Step 1: Detect the Presence of Hazardous Materials

If the materials involved or likely to become involved in an emergency incident are hazardous, response to the incident will be subject to applicable OSHA, DOT, and EPA regulations, as described in Chapter 2. If they are not hazardous, the incident can be treated as a normal fire or nonhazardous spill and the regulations for response and reporting do not apply. The term hazardous materials, or hazmats, as used here could include hazardous materials, hazardous chemicals, hazardous substances, and hazardous wastes, as defined in Chapter 1.

At the beginning of this chapter, we covered a variety of ways in which we may be able to recognize and identify hazardous materials involved in an emergency. If emergency response plans are available and hazardous materials containers and storage areas are properly labeled or placarded, the presence of hazmats will be obvious. If plans are not in place, or materials are not labeled or placarded, it cannot be assumed that no hazards are present without further investigation. If there is no clear indication of a hazard, but things just don't seem "right," trust your instincts and proceed with caution until the situation is determined to be safe. Remember that this step of the DECIDE process is the most important one to responder safety, and it initiates the proper flow of all decisions that follow.

Step 2: Estimate Likely Harm Without Intervention

In this step, responders try to determine what bad outcomes will result if the incident is allowed to run its course. Before this can be done, the material must be identified by name, so that its hazardous properties can be determined. Failing this, at least the hazard class must be determined. It is impossible to estimate harm without knowing the name of the chemical or at least the hazard class.

To estimate likely harm, one must think of the possible undesirable outcomes. Ask yourself:

- What has been released?
- How much has been or may be released?
- What are the possible migration routes for the released product?
- What are the hazards to people, property, and the environment?

This step will require the use of reference materials that have been gathered in preparation for an incident. At an absolute minimum, references should include the DOT *Emergency Response Guidebook*, described above in this chapter. It is hoped that additional information sources will also be available. These may include information sources described above in this chapter, such as emergency information from shipping papers, information on chemical hazards as described in Chapter 6, and information on health hazards as described in Chapter 7. Consider physical hazards also, as described in Chapter 8.

Step 3: Choose Response Objectives

Having figured out what harm is likely to result from the incident, we can then develop strategic objectives for the response operation. Response objectives are intended to prevent or minimize potential bad outcomes. Objectives might include stopping a leak, preventing fire if the chemical involved is flammable, preventing exposure to people in a nearby community, or keeping a spilled liquid out of a storm drain.

It will be important to choose either an offensive or a defensive response strategy. In cases in which sufficient emergency response resources are available, an offensive strategy intended to contain the release at the source may be feasible (see Chapter 15). In situations in which an offensive response is not feasible, a defensive strategy may be the only alternative. A defensive approach might include calling for additional help from off-site and then preventing exposure to potentially affected persons, performing remote confinement operations (see Chapter 14) to limit the spread of the contaminant, and gathering additional information about the incident while waiting for help to arrive. An important factor in formulating response objectives will be whether or not victims remain in hazardous areas.

In choosing response objectives, consider what can reasonably be accomplished with available personnel, equipment, training, and personal protective equipment. Don't establish objectives that are impossible to achieve.

Step 4: Identify Options for Achieving the Response Objectives

Once the response objectives have been chosen, we must identify tactical options for achieving those objectives. Options might include shutting down a chemical supply line to stop a release, suppressing vapors to prevent ignition of a flammable liquid, using evacuation or protection in place to prevent exposure to people in a community, or diking a spill to keep the material out of a storm drain.

At this point, be sure to identify all feasible options for achieving the response objectives. At the same time, keep in mind that the resources available will limit the tactical options.

Step 5: Do the Best Option

In this phase of the DECIDE process, the best option of all those that have been identified is chosen. At this point, we should have a firm answer to the question, "What can we reasonably expect to safely achieve, using the resources available to us, to limit harm resulting from the incident?"

Always evaluate all options from the standpoint of responder safety. Discard any options that involve undue risks to personnel, even if it means remaining in a defensive mode and allowing the incident to run its course. Remember the emergency response adage: To be part of the solution, don't become part of the problem.

Step 6: Evaluate Progress

As response actions are taken, monitor and evaluate progress carefully. Is the action plan working to accomplish the chosen objectives? Are operations being carried out safely? Use any new information about the incident gained as the response operation proceeds in the evaluation process.

If progress is lacking or responders are at risk, don't hesitate to halt operations and revisit previous steps in the DECIDE process. It may be necessary to select different tactical options or even choose different strategic objectives for the response operation. This is especially true when hazards are discovered that were unknown at the time the course of action was selected.

Conducting a Perimeter Survey

If you are a first responder on scene, once you have established command and made an initial report, try to determine what, how much, and where hazardous material has been released, but do so safely. Instead of rushing into the situation, establish site control, as described in Chapter 11, and conduct an initial survey from a safe location beyond the perimeter of the hazardous area. Use information about the incident from facility contacts, workers, or other knowledgeable people.

A good pair of binoculars is a must for conducting a perimeter survey. When conducting the survey, try to answer the following questions.

Are any People in Immediate Danger?

If workers or civilians are in unsafe areas, they will need to be ordered to exit immediately. If they have been injured or exposed in the incident, they may not be able to evacuate under their own power. In that case, a determination will need to made quickly as to whether rescue should be attempted. Rescues should only be attempted if victims are viable and adequate protective measures are available for the rescuers. This will be an important strategic decision for the IMS command function, as discussed in Chapter 4.

What Is the Identity of the Material Involved?

This important question may be answered during the initial perimeter survey by various means described in the Recognition and Identification section of this chapter. Remember that to accurately and completely assess the risks inherent in the incident, the name of the product must be known. A major portion of risk assessment involves predicting the behavior of the product based on its chemical and physical properties. These are best obtained by looking up the product by name in the reference sources described in Chapters 6 and 7.

If it is not possible to identify the chemical by name, try to determine the hazard class. If that fails, try to make an inference based on container recognition, occupancy and location, or other methods covered in this chapter.

If specific information from an emergency plan is not available at this point, use the information from the DOT *Emergency Response Guidebook*, as a minimum, and try to obtain more specific reference information as soon as possible. Remember that without some sort of identification of the hazard, it will not be possible to proceed to Step 2 of the DECIDE process and estimate the likely harm resulting from the incident.

Where Is the Point of Release?

Exactly where is the released material coming from or, at least, what is the location of the released material? The exact location of a leaking container may provide a valuable clue regarding its contents if a chemical storage and use map is available. Also, someone who works in that area can be contacted to find out what materials are stored or used there. This information will be important for making decisions regarding site control and evacuation or protection in place.

Where Is the Released Material Headed?

Topography will determine the direction of liquid flow and may affect vapor or gas movement by influencing air movement. For example, valleys may "funnel" wind currents, and ridges may block or divert the wind. If the release is a gas, vapor, or dust, wind speed and direction are important.

Temperature affects the volatility of a liquid and the rate at which the liquid will evaporate. High temperature will also hasten loss of integrity of a damaged container by expanding the chemical inside it. Rain, of course, will enhance runoff, will disperse spilled materials more widely, and may react violently with water-reactive chemicals.

The ongoing migration of the material will affect decisions regarding site control and zoning. The ultimate destination of the spilled chemical will be determined by its chemical and physical properties. Predicting dispersal of materials is addressed more completely in Chapter 6.

What Are the Hazards to Responders and Others?

The final step to be taken before anyone enters the potential hazard zone is an evaluation of the risks the situation presents to responders. On the basis of this evaluation, responders on the entry teams will choose protective gear and develop a plan for a safe entry. This should also include an evaluation of whether personnel are currently staged at safe locations. If not, site control measures should be reassessed.

Making Entry and Conducting an On-Site Survey

If the perimeter survey is inconclusive, or indicates the need for further evaluation of the incident scene, a closer on-site survey may be appropriate. If the hazards have not been completely characterized before entry, the maximum personal protective equipment should be worn and only properly trained personnel should enter.

The on-site survey is used to verify and confirm the findings of the perimeter survey; an entry plan setting out specific goals of the survey should be formulated. Key objectives of the survey are to monitor the air, make visual observations, and observe the released materials.

Monitor the Air

Before the selection of response action options and the choice of appropriate personal protection procedures and gear, intrinsically safe, portable instruments should be used to determine the presence of

- Oxygen-deficient or oxygen-enriched atmosphere
- Flammable gases or vapors
- Toxic vapors or gases
- Radioactive materials

The use of air monitoring equipment is fully described in Chapter 9.

Visual Observations

During the on-site survey, make careful visual observations of conditions in the hazard area and report the resulting information for use in assessment and decision-making. Use all of the methods discussed above in this chapter to identify hazardous materials. Pay special attention to hazardous materials containers or containment systems during this survey.

All sizes and types of containers may be subjected to stress during an emergency. Look for containers that may be stressed by heat, if fire is a real or potential part of the incident. Look also for physical factors such as mechanical damage and for chemical stress such as corrosion or chemical reaction. Stress can strain or deform a container beyond the limit of its ability to adapt to the stress, causing it to breach. It is vital to assess the integrity of a container before approaching it.

During the on-site survey, focus on safety with the primary concern being your safety and that of other entry team members. This is especially true when previously undetected hazards are discovered during the entry. In addition to chemical hazards and health hazards, don't forget about physical threats such as slip, trip, and fall hazards or high noise levels. Be sure to relay the hazard information you gather so that others can remain safe.

Observation of Released Material

Observations of the released material may be helpful, especially if its identity is in doubt. Physical state, color, visible vapor emission, bubbling, and other properties or behaviors may help to confirm the identity of a product.

Bubbling, smoking, and other visible behaviors may indicate that a chemical reaction between the spilled chemical and water, air, or another spilled chemical is occurring. If uncontrolled reactions are evident, responders should be wary and remain at a safe distance. Always be on the lookout for fire or other chemical reactions, as these will be important considerations for hazard assessment.

Sampling and Analysis of Released Materials

Collect samples of unknown materials only if no other avenue exists to identify them. Do this only if you have been trained in safe sampling and rapid analysis can be made. A sample that has to be sent to a laboratory for analysis may take weeks to provide results, and obviously that is too long to wait in an emergency. Considerations for sample collection and analysis are discussed in the section of this chapter on Air Monitoring and Sampling of Hazardous Materials.

SUMMARY

In this chapter, we have considered initial assessment of a hazardous materials incident and the initial actions we should take in response. It is vital for all responders, from first responder awareness level to the most highly trained responders, to be able to perform these functions. In the chapters that follow, we take a detailed look at a number of additional topics that may be important for bringing a hazmat incident under control. As we cover these more advanced topics, keep in mind that they all build on the initial assessment and actions described in this chapter. These are critical factors for establishing a safe and stable basis for ongoing response operations.

6

CHEMICAL HAZARD ASSESSMENT

INTRODUCTION

In Pennington, Alabama, two contract workers were killed and eight others injured while working on a construction project at a pulp and paper mill. They were overcome by hydrogen sulfide (H_2S) gas that was vented from the sewer through a nearby fiberglass manhole cover. The H_2S was generated when a drain from a tank truck unloading area delivered sodium hydrosulfide into an acidic process sewer system. Mixing the sodium hydrosulfide with the acidic material in the sewer generated deadly H_2S gas. Clearly, knowledge of the sewer system in the mill, of the types of chemicals being received and potentially drained into the system, and of the incompatibility of these products with the acidic materials already in the sewer could have prevented this incident from happening.

As discussed earlier in this book, hazardous materials incidents include some of the same dangers encountered in other types of emergencies. However, hazmat emergencies are distinguished from other kinds of emergencies by the presence of chemicals or other substances that, by nature of their quantity, concentration, or properties, can harm humans who encounter them.

Emergency responders must know what to expect from the chemicals involved in a hazardous materials incident before they can evaluate the hazards present in the incident. The likelihood of reaction, fire, explosion, and toxicity, as well as the mobility of the material on land, in air, and in water, are all determined by the chemical makeup of the product. To control an incident safely and efficiently,

Emergency Responder Training Manual for the Hazardous Materials Technician, Second Edition, edited by Kenneth W. Oldfield
ISBN 0-471-21387-X   Copyright © 2005 John Wiley & Sons, Inc.

a responder must understand enough chemistry to be able to predict the behavior of hazardous materials. This chapter contains a very basic discussion of chemical properties and behavior, which will be but just another tool in the toolbox used by first responders in assessing the risk involved in responding to a hazardous materials incident.

ESTIMATE LIKELY HARM WITHOUT INTERVENTION

In this, the second step of the DECIDE process, responders try to determine what will happen if they do not intervene, or just do nothing. Before this can be done, the material must be identified by name so that its hazardous properties can be determined and predictions about its behavior can be made. It is impossible to estimate the likely harm without knowing the name of the chemical or at least to which hazard class it belongs.

As stated earlier in this book, to estimate likely harm, one must think of the possible undesirable outcomes. Ask yourself:

- What has been released?
- How much has been or may be released?
- What are the hazards to people, property, and the environment?
- What are the possible migration routes for the released product?

This step will require the use of reference materials, some of which have already been mentioned, such as the DOT *Emergency Response Guidebook* and shipping papers, and others such as Materials Safety Data Sheets, the NIOSH *Pocket Guide to Chemical Hazards*, and the Response Information Data, or RIDs, found in the CAMEO Database that are discussed below in this chapter.

Atoms and Molecules

The most basic form of all chemicals is the atom. You probably already know that all chemicals are made up of some combination of just over 100 different kinds of atoms called elements. All atoms are made up of protons (positive charge), neutrons (neutral charge), and electrons (negative charge). The number of protons an atom contains indicates which element it is. Each atom is most stable when it has the same number of protons as electrons, which creates a net neutral charge. Atoms then combine with other atoms either of the same element or of a different element to form molecules and compounds. The forces that attract and hold atoms together are what give chemicals their unique characteristics or properties. That is why there are thousands of different types of chemicals, some of which are hazardous.

Physical State

Hazardous materials can be liquids, solids, gases, or even sludges (a solid suspended in a liquid). The way that different kinds of atoms are joined together by forces called bonds to form compounds is how we have different kinds of chemicals. For instance, the two elements hydrogen and oxygen, which are both gases under normal conditions, are bonded together to form water (H_2O), which is a liquid under normal conditions. There are other kinds of forces, not as strong as bonds, that attract molecules to each other. This is how we can have a glass full of water molecules. Otherwise, the universe would be just a huge jumble of every different kind of chemical.

Another important law of chemistry to keep in mind is that every molecule is always in motion, even the molecules that make up a seemingly still, solid rock. The amount of motion is related to the heat energy the molecule contains. So the more a molecule is heated, the more it moves. Therefore, cold molecules (less heat) move less and are able to pack in close together. When they do, they are in what we know as a solid physical state.

As more heat energy is applied, the molecules move more and must spread out more, although the forces still attract like molecules to each other. This less dense form can change shape and flow but still holds together. We call this a liquid physical state. Although it flows and it is harder to control, it is still visible.

When the liquid is heated, this adds even more energy to the molecules and causes them to move even more. At the surface of the liquid, some of these molecules are moving so much that they overcome the forces attracting them to the other similar molecules and they enter the air. They are now individual molecules suspended in air in what we call the vapor or gas state. Now they are so far apart that they cannot be seen, and they move wherever the air moves.

Throughout all of these changes in physical state, the chemical make up of the material never changes. For instance, ice (solid), water (liquid), and steam (gas) are all what we call water (H_2O), only in different states. It is the same for every other kind of chemical. Adding or removing heat can change the chemical's physical state without changing its chemical makeup.

The other factor affecting physical state is atmospheric pressure. If pressure is applied to a gas, it can be compressed into a liquid form of the same chemical. This is discussed below in this chapter in the section on liquefied gases. However, compressing a liquid will not change it into a solid form.

Part of the nature of the hazard in an incident depends on the form of the material; for example, a gas leaking from a container will disperse into the surrounding environment and could displace breathable air or spread downwind. A spilled or leaked liquid will spread out and pool in low-lying areas and possibly vaporize into the air, whereas a spilled solid generally will form a pile more or less in the vicinity of the spill. As you can see, the form of the spilled material can greatly influence the type and magnitude of the response.

Solids

The primary concerns with spilled solids on land are their flammability, radioactivity, and air reactivity. Solids will not disperse very far unless they change state, as a refrigerated solid will do if it reaches its melting point, a subliming solid will do if it reaches its sublimation temperature, or a water-soluble solid will do if it spills into water or encounters rain. Sublimation of a substance is a conversion between the solid state and the gaseous state without an intermediate liquid stage. The odors of naphthalene (once used for moth balls) and paradichlorobenzene (used as moth balls and solid bathroom deodorizers) are caused by sublimation. Dry ice, solid carbon dioxide, is another common product that sublimes. The toxicity and flammability of the vapors must be assessed if subliming solids are spilled. If solids spill into water, characteristics of solubility and reactivity must be considered.

Solids can enter the atmosphere and can be hazardous. It is possible for a finely divided solid to be dispersed as a dust by wind or in the smoke of a fire. However, responders can easily be protected during a hazardous materials incident from these potentially toxic materials by the use of respirators that provide clean breathable air as discussed in Chapter 12. Responders should be concerned about flammability. The increased surface area of very finely ground metals, organic powders, or dusts can produce extreme fire and explosion hazards in air. Examples include metal powders, finely powdered plastics, coal dust, and grain dust.

Flammable solids, such as white or red phosphorus, can change chemically when burned, becoming oxides and/or other compounds that can damage skin or respiratory tract linings.

Specific terms are used to describe these airborne particles or solids: dusts, fibers, fumes, and smoke.

Dusts

Dusts are the small particles formed by grinding, scraping, sanding, filing, or otherwise mechanically breaking up larger solids. They have no particular shape and range in size from 0.001–1 mm. They can come from inorganic sources like metals, coals, and limestone or organic sources like wood, plants, flour, animal skin, or hair.

Fibers

Fibers are particles that are longer than they are wide. Asbestos is the most familiar example of a hazardous fiber, although fibers from natural materials such as cotton and wool are also common hazards.

Fumes

When solids are heated until they evaporate, their vapors rise into the cooler air and condense back into small particles called fumes.

Smokes

Most of us are familiar with smoke particles. These particles are formed when materials burn incompletely and are dispersed in the air by the heat of the combustion reaction.

Liquids

Materials in the liquid state may be released on land or into water; both types of spills may allow volatile liquids to evaporate into air.

Liquids are able to flow away from a leak, thereby extending the hazard area. Almost all the properties that are described below should be considered before dealing with every spilled liquid. Volatile liquids create additional problems because they evaporate and disperse in air, expanding the hazard zone even further.

The viscosity of a liquid may be a factor in incident response, because the more viscous (thick and sludgy) liquids tend to flow more slowly and to stick to clothing and equipment more readily. Sludges that have thickened because of evaporation may be quite concentrated and are possibly more dangerous than the original liquid.

Gases and Vapors

Vapors

Vapor is the gaseous state of a chemical that is a liquid under normal conditions. Vapors are formed when liquids evaporate. Raising the temperature will cause the molecules that make up the liquid to move faster, so that more evaporation occurs. So as ambient temperatures increase, you can expect more vapors to be present on a site where a liquid has been released. Volatile chemicals are liquids that have high evaporation rates and will readily get into the air if their container is opened or they are released.

Compressed Gases

A compressed gas is any material that, when enclosed in a container, has an absolute pressure of more than 40 pounds per square inch (psi) at 70°F, or an absolute pressure exceeding 104 psi at 130°F, or both. Compressed gases can be either pressurized gases or liquefied gases, either in the gas or liquid state, depending on the amount of pressure inside the container.

Liquefied Gases. Gases that are liquefied when compressed exist in a liquid-vapor relationship inside their containers. One property of such a chemical is its critical pressure, the pressure above which the gas molecules will condense into a liquid as they are pushed closer together. Transporting or storing the liquefied gas above its critical pressure allows a container to hold a greater quantity of the material. Examples of liquefied gases are ammonia, butane, chlorine, propane, propylene, and Freon. Another property of a liquefied gas is its expansion ratio. A breached container of liquefied gas may spew liquid that changes to the gas state as the pressure is released and expands considerably. This creates a large hazard zone.

Pressurized Gases. Pressurized gases are those compressed gases that remain in the gas state when compressed. Examples of pressurized gases are air, hydrogen, methane, nitrogen, helium, and oxygen.

Acetylene

Acetylene is a gas that does not readily fit into either category. Commercial acetylene cylinders contain diatomaceous earth saturated with acetylene gas dissolved in liquid acetone. Because acetylene is very reactive chemically and is shock sensitive, dissolving the gas in acetone decreases its sensitivity and reactivity, thus making it possible to handle the gas safely. The diatomaceous earth packed into the cylinder keeps the acetylene from forming pockets of gas inside the cylinder.

Cryogenic Gases

Cryogenic gases have boiling points of less than −150°F and are transported, stored, and used as liquids. A large volume of gas can be stored as a liquid in a much smaller volume at low temperatures. The hazards of the cryogenic gases relate to the nature of the particular gas, the expansion ratio of vapor from liquid, and the extreme cold that results as the released liquid vaporizes. Cryogenic gases include fluorine, oxygen, methane, helium, hydrogen, and nitrogen. Note that several of these gases may also be shipped under high pressure rather than low temperature. Responders must observe containers to recognize the hazards of shipping or storage methods used for gases.

Changes in Physical State

Problems often develop when chemical reactions or changes in temperature cause a released material to change physical state. A leaking liquid may change to the gas form, with a tremendous increase in volume. Chlorine, for example, has an expansion ratio of 460 to 1, meaning that the contents of a cylinder of pressurized, liquefied chlorine will expand upon release to fill a space 460 times the size of the capacity of the cylinder. Containers heated in a fire may leak or experience a BLEVE (boiling liquid expanding vapor explosion) due to an increase in the volume of contained gas or liquid. The expansion ratio of a chemical (the volume of its liquid form compared to the volume of its gas form) is known for all chemicals and is an important consideration in an incident involving the release of gases liquefied by pressure.

PROPERTIES OF CHEMICALS

Because all atoms of a certain element have the same structure, they also have the same properties. The structure and properties make behavior predictable. Atoms combine in predictable ways to form compounds, based on atomic structure and how atoms share or exchange electrons. The combination is always the same if the elements and conditions are the same, so a particular compound always contains the same ratio of elements.

The compound water, for example, is always made up of one oxygen atom and two hydrogen atoms. Carbon tetrachloride ("tetra" meaning four) is always one carbon atom with four chlorines attached (Fig. 6.1).

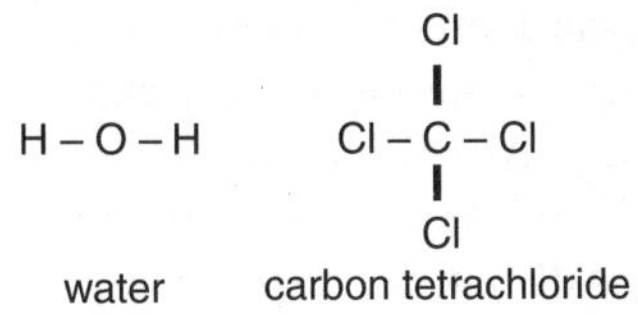

$$H-O-H \qquad\qquad Cl-\underset{\underset{Cl}{|}}{\overset{\overset{Cl}{|}}{C}}-Cl$$

water   carbon tetrachloride

Figure 6.1 Molecules of water and carbon tetrachloride.

TABLE 6.1 Approximate Temperature Conversions

°F	°C	°F	°C
32	0	80	27
36	2	85	29
40	4	90	32
45	7	95	35
50	10	100	38
55	13	125	52
60	16	150	66
65	18	175	79
70	21	200	93
75	24	212	100

Because a compound is always made of the same atoms in the same ratio, that compound always has the same properties whenever we find it. We refer to these properties as "characteristic properties," because no two different compounds have exactly the same properties.

Physical and chemical properties determine the behavior of elements and compounds. A hazmat responder should always find out the properties of a chemical as soon as possible to determine how the chemical will behave with regard to fire, explosion, reactivity, dispersion, and human health hazards. Several references that can be used to get this information are discussed below in this chapter.

When assessing properties, it is important to note the temperature and temperature scale at which the property was determined. Some properties are very temperature dependent, and some indication of the dangers can be gained by considering which will occur at the ambient temperature in the area of the incident. In the United States, the Fahrenheit scale is used in describing weather and ambient temperature, but most scientific references use the centigrade scale. Responders making hazard assessments must be able to convert temperatures from one scale to the other to make good judgments about potential hazards (Table 6.1).

Flammability

Workplace fires have been the cause of many deaths and permanent injuries. Three things are needed for a fire to occur: fuel, oxygen, and energy (ignition source). You may have heard of the fire triangle theory, which assigns each of fuel, oxygen, and energy to a side of a triangle and states that if any one side of the triangle is removed or eliminated a fire will not occur. With this in mind, properties that should be immediately consulted when trying to determine whether a chemical will burn or explode are flash point, lower and upper flammable/explosive limits, and boiling point. Other properties that may be helpful in making a determination of flammability and explosiveness include vapor pressure and vapor density.

Flash Point

A chemical's flash point (Fl.P.) is the minimum liquid temperature at which enough vapors are present above the surface of the liquid to ignite. In other words, the flash point is the lowest temperature at which a liquid product will give off enough vapors to form an ignitable mixture with the air above the surface of the liquid. At temperatures above the flash point, if an ignition source is present, the vapor-air mixture above the surface of the liquid could flash and/or sustain combustion. The lower the flash point, the greater the fire hazard of a material. If the flash point is lower than the air temperature, the material is an immediate fire hazard and all ignition sources must be removed from the area.

Examples: Gasoline Fl.P. = −45°F

Ethylene glycol (antifreeze) Fl.P. = 232°F

Flash point is determined by one of two general methods as reported in hazardous material and chemical references, closed cup or open cup. The closed-cup method prevents vapors from escaping and therefore usually results in a flash point that is a few degrees lower than in an open cup. Because the two methods give different results, one must always list the testing method when listing the flash point.

In classifying liquids as flammable or combustible, both DOT and OSHA use flash point as the determination. The DOT classifies liquids with a flash point of 141°F or lower as flammable and liquids with a flash point greater than 141°F and less than 200°F as combustible. OSHA classifies liquids with flash points of 100°F or lower as flammable and liquids with a flash point greater than 100°F and less than 200°F as combustible.

Lower and Upper Explosive/Flammable Limits

The terms "flammable limits" and "explosive limits" are used interchangeably in most references. The lower flammable limit (LFL) or lower explosive limit (LEL) is the minimum percent vapor in air (% by volume in air) that is required to form an ignitable mixture in air. Below the LEL the vapor-air mixture is too "lean" to burn, (i.e., there is not enough fuel to complete the fire triangle). The lower a substance's LEL, the less it has to vaporize to create a vapor-air mixture that will burn or explode. Therefore, chemicals with a lower LEL pose a greater flammability hazard.

The upper flammable limit (UFL) or upper explosive limit (UEL) is the maximum percentage of flammable vapors or highest concentration that can form an ignitable or explosive mixture in air. Any concentration greater than the UEL in air is too rich to be ignited, that is, there is so much fuel that now there is not enough oxygen present to complete the fire triangle.

The flammable/explosive range is the range of concentrations of the material in the air which will support a fire, or the percentage difference between the UEL and the LEL. For example, the LEL for acetylene gas is 2.5% and the UEL is 100%; therefore, concentrations below 2.5% in air will not burn. But acetylene has an extremely wide flammable/explosive range, as indicated by its UEL of 100% in air, which makes it extremely dangerous.

We can conclude that the lower a chemical's flash point, the lower its LEL and the wider its flammable range, the greater its fire hazard.

Boiling Point

The boiling point (BP) of a liquid is the temperature at which the liquid begins to rapidly change to a vapor at normal pressure (which is 760 mmHg or 1 atm at sea level). If the boiling point is lower than the temperature of the air around the chemical, the chemical will boil and vaporize on release. This does not imply that vapors are produced only at or above the boiling point temperature; they are generated at lower temperatures as indicated by a property called "vapor pressure" that will be discussed shortly.

> Examples: Water BP $= 212°F$
>
> Hydrogen chloride BP $= -121°F$
>
> 1,1,2-Trichloroethane BP $= 237°F$

Looking at the above examples, water and 1,1,2-trichloroethane are liquids at normal temperature and pressure. This is because their boiling point temperatures are much greater than the temperature in a normal room. Hydrogen chloride, on the other hand, is a gas at room temperature and pressure. We know this because its boiling point is much lower than average room temperatures.

Vapor Pressure

Vapor pressure (VP) is the pressure placed on the inside of a closed container exerted by the saturated vapor in the "head" space above the liquid (Fig. 6.2). The units of vapor pressure in most references are millimeters of mercury (mmHg) or atmospheres (atm). When discussing units of pressure, most of us are more comfortable thinking of pressure in pounds per square inch (psi). But, unfortunately, references do not report vapor pressure in psi. For reference think of the following:

$$14.7 \text{ psi} = 760 \text{ mmHg} = 1 \text{ atm}$$

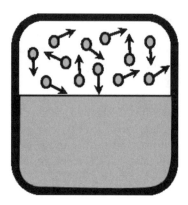

Figure 6.2   Vapor pressure is the pressure that saturated vapor places on the inside walls of a closed container.

Figure 6.3   Evaporation.

If a person is standing at sea level, there is a column of air above her being pulled by gravity and pressing down on her at 14.7 psi or 760 mmHg or 1 atm. If we pour some drinking water in an open container located at sea level, the column of air above the surface of the water is pushing down at 760 mmHg but the water is pushing up only at about 21 mmHg at 75°F. As the temperature of the water increases, so does the vapor pressure and more and more of the water molecules break free from the surface of the water or evaporate (Fig. 6.3). If the temperature is increased to 212°F, water's boiling point, then the vapor pressure of the water is equal to 760 mmHg. The water is pushing up as hard as the air column is pushing downward.

Most references report vapor pressure at 68°F unless otherwise noted. Chemicals with high vapor pressures at ambient temperatures are considered volatile. They will be more likely to cause their containers to explode when heated, because their vapor pressure increases as temperature increases. The closer a liquid's vapor pressure is to 760 mmHg, the more volatile the chemical and the more readily the chemical evaporates.

Examples: Benzene VP $= 75$ mmHg at $68°$F

Acetone VP $= 180$ mmHg at $68°$F

Carbon dioxide VP $= 56.5$ atm at $68°$F

Looking at the examples above it is easy to see that acetone evaporates at a rate that is over twice that of benzene at $68°$F. They are both liquids at $68°$F because their vapor pressure does not exceed 760 mmHg. However, carbon dioxide would immediately vaporize if released into a room that is $68°$F. Remember that 1 atm $=$ 760 mmHg. If we raise the temperature of any of our examples, we would expect a vapor pressure higher than what was given in the reference.

Vapor Density

The vapor density (VD) of a chemical tells us whether a volume of the vapor weighs more than the same volume of air. Many references do not list the vapor density for chemical, but it can be easily estimated by dividing the molecular weight of the chemical by the molecular weight of air, which is approximately 29.

Vapors of chemicals with vapor densities greater than 1.0 (or with molecular weights greater than 29) will sink to the ground or floor and may accumulate and displace breathable air. If a heavy vapor is flammable, it can travel along the ground to an ignition source and ignite and the flames can travel back to the leaking container.

Vapors with density less than 1.0 (or with molecular weights less than 29) will rise and float away and, depending on atmospheric conditions, can dissipate rapidly into the air. Very few chemicals have vapor densities of less than 1. Examples include:

- Hydrogen
- Helium
- Acetylene
- Ammonia
- Methane
- Ethane
- Ethylene
- Carbon monoxide

Nitrogen has a density close to that of air, so it can rise or fall depending on wind and other atmospheric conditions.

In some references, the density of a chemical that is a gas at normal temperature and pressure is reported as a relative gas density or RGasD. An RGasD greater than 1 tells us that the gas will sink to the ground, whereas an RGasD less than 1 tells us that the gas will rise.

Solubility

Solubility indicates the tendency of a chemical to dissolve evenly in a liquid. The liquid may be water or an organic solvent such as benzene or alcohol. Sometimes solubility is described in words, and sometimes a number is given, indicating the percentage of the material that will dissolve.

Hazardous materials references only list solubility of a material in water because this is what first responders are usually concerned with. Solubility in water indicates the amount of a material that will dissolve in water, volume to volume. For example, acrolein has a solubility of 40% in water. This means that if a gallon of acrolein is mixed with a gallon of water, 40% of the acrolein will dissolve in the water and the other 60% will form a separate layer. However, if a gallon of acrolein is poured into a large body of water, it will all be dissolved because the volume of water is much greater. Materials that are completely soluble in water no matter the volume-to-volume ratio or that are soluble in water in all proportions are said to be miscible.

Insoluble or slightly soluble materials will form layers when they enter the water. These layers will either float or sink depending on their specific gravity and the properties of the body of water (e.g., temperature, salinity) (see Table 6.2).

Specific Gravity

Specific gravity is the density of the product divided by the density of water, with water density defined as 1.0 at a certain temperature. An insoluble material with specific gravity less than 1.0 will float; one with specific gravity greater than 1.0 will sink (Table 6.2). Specific gravity is sometimes given as density, even though the terms do not have the same meaning.

Corrosives

The property that gives a measure of the acidity or alkalinity of a solution is pH. The pH of a chemical is a measure of a solution's hydrogen ion concentration. The higher the hydrogen ion concentration of a solution, the lower the pH or the more acidic is the substance. The pH scale is logarithmic and extends from 0 to 14, where 7 is defined as neutral. Solutions with a pH less than 7 are acidic, and those with a pH greater than 7 are alkaline or basic. Because the pH scale

TABLE 6.2 Chemical Behavior Due to Solubility and Specific Gravity

Chemical	Soluble in Water	Specific Gravity	Behavior in Water
Gasoline	No	0.7	Floats
Trichloroethane	No	1.3	Sinks
Sulfuric acid	Yes	1.8	Dissolves

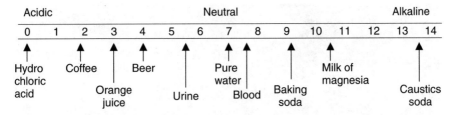

Figure 6.4   The pH scale.

is a logarithmic scale, a solution with a pH of 1 is 10 times more acidic than a solution with a pH of 2 and a solution with a pH of 2 is 10 times more acidic than a solution with a pH of 3. Therefore, a solution with a pH of 1 is 100 times more acidic than a solution with a pH of 3. Figure 6.4 compares the pH of several common substances.

Materials with pH of 2.0 or lower, or 12.5 or higher, are classed as corrosive by the U.S. Department of Transportation and are labeled and placarded as such.

PREDICTING DISPERSAL OF HAZARDOUS MATERIALS

Using reference documents to identify the health, fire, reactivity, corrosivity, and radioactivity hazards of chemicals in an emergency response incident is a basic step in assessing the risks posed to responders and the community. The chemical and physical properties of hazardous materials determine these risks and also determine how the material will move (disperse) in land, water, and air after a release (Fig. 6.5).

Predicting the dispersal of released hazardous materials is an important part of risk assessment and must be done to "estimate likely harm without intervention." Dispersal is highly dependent on the physical state of the material during and immediately after the release.

Gas or Vapor Dispersal in Air

There are some special considerations for emergency response to incidents that involve gases. Today many gases are routinely shipped and stored in a variety of forms and containers, so it is likely that they will be encountered in incidents. Gases and vapors may be released into the air by direct venting from containers or as vapors from volatile solids or liquids. Factors emergency responders must consider when attempting to predict the pathway of gas and vapor releases are:

- Travel distance and direction
- Duration of the discharge
- Mixing of the gas or vapor with air (dilution)

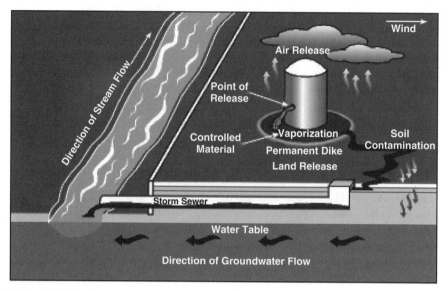

Figure 6.5 Liquids released on land may contaminate air, soil, groundwater, and surface water.

Travel Distance and Direction
Travel distance can be predicted by multiplying current wind speed by time. Direction is assessed by observation of a wind sock. In an emergency, anything that will blow from a pole may be used to estimate wind direction, even if it is not totally accurate.

Duration
Duration refers to the length of time the release continues. The two basic types of discharges are considered to be instantaneous (a puff or cloud) and continuous (a plume).

Dilution
A puff or cloud will move with the wind, at similar direction and speed. As it begins to mix with the air its concentration will begin to drop. The area of the cloud will become larger as this mixing occurs, but the concentration will decrease. As the cloud mixes and grows, even the densest concentration of contaminant toward the center of the plume will at some point drop below the concentration that is considered harmful to the general public. Exposure limits and other guidelines used to protect against harmful exposure to airborne chemicals are discussed in Chapter 9.

Ground-level contaminant concentration in an instantaneous discharge decreases as the cloud moves away from the point of release. Figure 6.6 illustrates the decrease. At the same time, the cloud expands and the hazard zone grows larger, as shown in Figure 6.7.

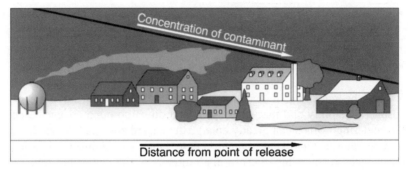

Figure 6.6 Plume concentration decreases as the plume mixes with air.

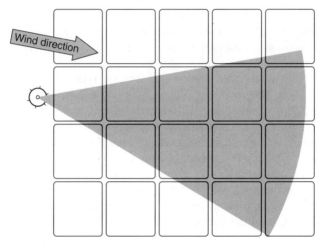

Figure 6.7 Plume hazard area enlarges as the plume moves downwind.

In a continuously released plume, the concentration downwind will be relatively constant for a period of time approximately equal to the duration of the release. A period of time will elapse before the leading edge of the plume reaches a certain location, as is true for an instantaneous emission, and a similar length of time will pass after the release is ended before the trailing edge leaves the same location.

Responders will have to predict the hazard zone, the area in which the concentration of contaminant is too high to allow people to remain there unprotected. Selecting the hazard zone depends on the three factors listed above (wind distance and direction, release duration, and plume dilution), the predicted chemical concentration and the size of the cloud, and the four factors listed below.

- Amount of the discharge. Generally speaking, the larger the release, the longer and wider the zone.

- Prevailing atmospheric conditions. These conditions include temperature, strength of sunlight, and wind speed and direction.
- Gas or vapor density relative to air
- Height of the discharge

It must be remembered that terrain or buildings affect wind direction and atmospheric stability and that wind speed and direction often change. Changing or "meandering" winds can greatly enlarge the hazard zone. A "vulnerable zone" should be identified and kept in mind in case winds should change. In a location that is subject to changing winds from all directions, the vulnerable zone may be a circle around the point of release.

Lighter-than-air plumes will float up if atmospheric conditions do not prevent this, and they soon will dilute or disperse to a nonhazardous concentration. Heavy plumes will hug the ground as they move and may accumulate to higher concentrations, possibly displacing breathing air or posing a potential for ignition. Typical ignition sources are vehicles, spark-generating friction, pilot lights, and cigarettes.

A release from an elevated portion of a tank or from a pressure relief pipe of a stack may affect the ground-level concentration of the contaminant. In such a case, it will make a great deal of difference whether the relative vapor density is higher, lower, or the same as the density of air.

Hazard zone size and direction can be calculated with the Areal Locations of Hazardous Atmospheres (ALOHA) application. ALOHA is intended to calculate atmospheric dispersion models of chemical vapor clouds based on the physical characteristics of the hazardous material, the atmospheric conditions, and other incident-specific circumstances such as tank size, volume of chemical in tank, and location of leak on the container.

Liquids Released into Water

In predicting the dispersion of liquids that are spilled or flow into water, several properties of the chemical must be known. Boiling point, vapor pressure, solubility, and specific gravity will determine where the liquids will go.

The boiling point and vapor pressure of the material will determine whether part of the material will boil off or vaporize from water. A container dropped into water will release its contents if it ruptures. If the water temperature is above the chemical's boiling temperature, the chemical will vaporize rapidly as it enters the water from the breached container and the vapors will bubble up through the water into air. Volatilization may occur well below the boiling point, but at a much slower rate.

The solubility of the material will determine whether or not it will dissolve, and if so, at what rate. Solubility is quantified by number in some references, where the portion that will dissolve in water is given as a decimal or percentage. Other references use qualitative terms such as "insoluble" or "partially soluble." Designation of a chemical as insoluble should not be taken as an absolute, as many so-called insoluble materials will partially dissolve after enough time has elapsed. Both liquids and vapors can dissolve in water. In bodies of water that are turbulent, the mixing action may increase the rate of dissolution. Also, insoluble materials may become physically mixed with water if sufficient turbulence or wave action is present.

The insoluble portion of a liquid will sink or float, in still water, depending on the density or specific gravity of the material. Specific gravity and density do not describe the same measurement but are sometimes reported interchangeably, because the numerical values are the same at certain temperatures. These terms can be defined as follows. Density is the mass of a substance divided by the volume it fills; specific gravity is the density of a substance divided by the density of water.

Water is assigned a specific gravity of 1.0 at normal temperature, so a liquid with specific gravity greater than 1.0 will tend to sink and a liquid with specific gravity of less than 1.0 will tend to float. Materials with specific gravity close to 1.0 may be dispersed throughout the water column. In turbulent water, mixing will occur and decrease or break up the floating layer. The "slick" may be less visible, but the chemical is still present and dispersing.

When a flammable liquid enters a sewer or storm drain, it can pose a fire or explosion hazard. A water-reactive liquid may produce the same hazards by generating a flammable gas. Toxic liquid contaminants pose a threat to all life forms exposed to them on land or in water.

The effects of boiling point, vapor pressure, specific gravity, and solubility on the dispersion of chemicals in water can be seen in Table 6.3, and a further discussion of each of these properties can be found below in this chapter.

Liquids Released on Land

Liquids may percolate through soil and will flow downhill on the surface or down sewers and storm drains. Because they flow, liquids may contaminate groundwater, surface water, and soil. Liquids may also vaporize, thereby contaminating the atmosphere (Fig. 6.5).

Whether a released liquid moves readily through soil depends on the properties of the chemical and the properties of the soil onto which it spills. An emergency responder is not expected to determine the transport of liquids through soil but should instead attempt to prevent chemicals from remaining on soil long enough to penetrate.

Liquids that are stored or shipped at cold temperatures may change to the vapor state when released. If the boiling point of the liquid is below ambient (outside the container) temperature, the liquid will begin to boil on release. Even if the boiling point is above ambient temperature, the rate of evaporation will increase as the liquid warms.

Liquefied gases, which are liquefied because of the pressure in the containers in which they are stored and shipped, will behave differently on release, depending on whether the point of release is below or above the surface of the liquid. Liquids

TABLE 6.3   Predicting Dispersal of Chemicals in Water

Boiling Point	Vapor Pressure	Specific Gravity	Solubility	Expected Behavior in Water
Below ambient	Very high	Any	Insoluble	All liquid will rapidly boil from surface of water.
Below ambient	Very high	Below that of water	Low or partial	Most liquid will rapidly boil off, but some will dissolve. Some of the dissolved liquid will evaporate.
Below ambient	Very high	Any	High	At least 50% will rapidly boil off; the rest will dissolve. Some of the dissolved liquid will evaporate later.
Above ambient	Any	Below that of water	Insoluble	Liquid will float, forming a slick. Those with significant vapor pressure will evaporate over time.
Above ambient	Any	Below that of water	Low or partial	Liquid will float but will dissolve over time. Those with significant vapor pressure may simultaneously evaporate over time.
Above ambient	Any	Below that of water	High	Liquids will rapidly dissolve in water up to the limit (if any) of their solubility. Some evaporation may take place over time if vapor pressure is significant.
Above ambient	Any	Near that of water	Insoluble	Difficult to assess. May float on or beneath surface or disperse through the water column. Some evaporation may occur from surface over time if vapor pressure is significant.
Above ambient	Any	Near that of water	Low or partial	Will behave as above at first and eventually dissolve. Some evaporation may take place over time.
Above ambient	Any	Any	High	Will rapidly dissolve up to the limit (if any) of their solubility. Some evaporation may take place over time.
Above ambient	Any	Above that of water	Insoluble	Will sink to the bottom and stay there. May collect in deep water pockets.
Above ambient	Any	Above that of water	Low or partial	Will sink to the bottom and then dissolve over time.
Above ambient	Any	Above that of water	High	Will rapidly dissolve up to the limit (if any) of their solubility. Some evaporation may take place from the surface over time if vapor pressure is significant.

under pressure will jet from a breach in the container if that breach is below the surface level of the liquid, and large amounts will vaporize and fill the air around the release. If the container is breached in the space above the liquid, the gas will vent at high velocity, with the velocity slowing as the pressure drops. Examples of materials shipped and stored in pressurized containers are liquid anhydrous ammonia, ethylene, chlorine, vinyl chloride, and liquid petroleum gas (LPG).

CHEMICAL REACTIONS

Stability is the result of strong bonds between atoms. Bonds are formed when atoms exchange or share electrons. More energy is needed to break some bonds than others. It is generally harder to separate small atoms like hydrogen and oxygen from each other and to break bonds between atoms with relatively large differences in electrical charge, like hydrogen and fluoride. Heat is the form of energy most often available to break bonds; conversely, when new bonds are formed heat is released during the reaction.

Release of heat is the most serious hazard in many potentially dangerous reactions. The term exothermic refers to reactions that give off heat. Some reactions produce by-products that are hazardous, such as flammable hydrogen gas, irritating hydrogen chloride vapors, fire-supporting oxygen, or deadly phosgene gas. Some reactions occur so rapidly that fire or explosions can occur or may cause the rupture of containers.

Some of the reactive materials that can cause problems for hazmat responders are water-reactive materials, air-reactive materials, oxidizers, unstable materials, incompatible materials, and materials that polymerize. Polymerization under controlled conditions is a useful technology in the manufacture of plastics such as polystyrene, but if it occurs too rapidly in a spill or fire, it can result in an explosion.

Water-Reactive Materials

Water-reactive materials react with water, often violently, to release heat, a flammable or toxic gas, or a combination of these (Table 6.4). If there is a fire in which water is being used for extinguishment, the presence of water-reactive material can make the situation much more dangerous.

TABLE 6.4   Water-Reactive Materials

Material	Hazard of Reaction
Sulfuric acid	Heat
Potassium, sodium	Flammable hydrogen gas
Calcium carbide	Corrosive and flammable products
Aluminum chloride	Hydrochloric acid burns
Sodium peroxide	Oxygen and heat

Air-Reactive Materials

Some of the water-reactive materials mentioned above are also air-reactive and will ignite in air; potassium metal is one example. Diborane and some organic metal compounds, such as trimethylaluminum, also ignite when in contact with air. Another example is white phosphorus, which must be stored under water to prevent it from igniting.

Oxidizers

These materials, also called oxidizing agents, present special hazards because they react chemically with a large number of combustible organic materials, such as oils, greases, solvents, paper, cloth, and wood. Fire is often the result. Also, their reactions generate heat that may be absorbed by other materials, causing ignition. The oxidizers that contain oxygen may release that oxygen as they decompose, and help sustain a fire.

The halogens (fluorine, chlorine, and bromine) are powerful oxidizers and are strongly reactive. Another group of reactive oxidizing halogen compounds includes the hypochlorites, chlorites, and perchlorates.

Organic peroxides deserve special mention because they are very hazardous. In addition to being strong oxidizing agents, they are inherently chemically unstable. Most of them are sensitive to shock, friction, and heat and can decompose exothermally, releasing a great deal of heat. Examples are acetyl peroxide, benzoyl peroxide, cumene hydroperoxide, and peracetic acid. Because organic peroxides slowly decompose in storage, chemical inhibitors are added to prevent their decomposition.

Similarly, the inorganic peroxides sodium peroxide and potassium peroxide are sensitive to shock and are very reactive. They decompose in the same way that organic peroxides decompose.

Other oxidizers include ammonium nitrate, potassium dichromate, chromic acid, potassium permanganate, ammonium persulfate, and sodium nitrate. All of these decompose with heat, sometimes explosively, and release oxygen that can support combustion.

Unstable Materials and Polymerization

Materials designated as unstable are those that have a tendency to decompose all by themselves; they do not need to mix with other chemicals to react. They may generate heat or toxic gases or may burst into flame or explode as they generate flammable vapors. Unstable materials are often stored and shipped with inhibitors in the mixture to prevent decomposition. The organic and inorganic peroxides mentioned above are chemically unstable. There is another group of chemicals called ethers, some of which react with oxygen in the air to form organic peroxides, thus becoming sensitive to shock, friction, and heat. Examples of these peroxide-formers include ethyl ether, dioxane, and tetrahydrofuran (THF).

Another group of materials that is chemically unstable contains the monomers, or building blocks that form many types of polymers (resins, plastics, and synthetic rubber materials). Many plastics are made in this way: PVC, or polyvinyl chloride, is formed of polymerized vinyl chloride monomers. The polymers polystyrene, polyethylene, and polypropylene are widely used plastics.

Some of these materials can spontaneously polymerize, causing rupture of the container, often explosively. Their presence can compound a fire problem because some of them are sensitive to heat, light, and air. Some monomers like ethylene are fairly stable; others, such as butadiene and methyl methacrylate, are much more unstable. Chemical inhibitors are added to some of these before shipment to prevent unwanted polymerization and ensure safety in handling them. Most monomers are flammable, and many are toxic or irritating materials. These materials are not ordinary flammable materials, but they present multiple hazards to responders.

Incompatible Materials

No discussion of basic chemistry for first responders would be complete without some mention of incompatibility of chemicals. If two or more chemicals remain in contact with each other without an adverse reaction, they are considered to be compatible. Some materials when mixed with other materials can adversely affect human health and the environment in a variety of ways:

- Generation of heat
- Violent reaction
- Formation of toxic fumes or gases
- Formation of flammable gas
- Fire or explosion
- Release of toxic substances if they burn or explode

Some reactions proceed more readily in the presence of a catalyst, a chemical that is added to enhance a desired reaction.

Toxic Combustion Products

Toxic materials may form from reactions between chemicals during a fire and pose another potential hazard in addition to the numerous ones that firefighters already face. Formation of these toxic products depends on the nature of the material burning and the amount of oxygen present. Table 6.5 presents information about some dangerous products of combustion that may be encountered in a fire situation, according to the type of material involved in the fire.

TABLE 6.5   Toxic Combustion By-Products

Material	Toxic Products	Hazard
Organic materials	Carbon monoxide	Asphyxiation at low levels
		Brain damage
		Delayed brain toxicity
	Carbon dioxide	Oxygen deficiency
	Acrolein	Strong irritant to eyes and respiratory system
Materials containing chlorine	Hydrogen chloride	Damage to eyes, skin, and respiratory system
	Phosgene	Fatal by inhalation because of delayed pulmonary edema
Materials containing nitrogen	Ammonia	Pulmonary irritation, possibly fatal
		Flammable if confined
	Hydrogen cyanide	Respiratory arrest due to brain damage
	Nitrogen oxides	Acute respiratory illness
		Decreased resistance to infections
Materials containing phosphorus	Phosphine	Pulmonary edema, possibly fatal
Materials containing Sulfur	Sulfur dioxide	Constriction of breathing
	Hydrogen sulfide	Lack of oxygen to cells

RESEARCHING IDENTIFIED MATERIALS

Once a material has been identified by name, the next step is to use a reference source to completely assess the risks it presents. References include books and other printed material, telephone hotlines, and computer databases.

Material Safety Data Sheets

An employer is required by law to retain and make available to workers a Material Safety Data Sheet (MSDS) for each hazardous chemical to which workers may be exposed, and to send a copy of each, or a list of materials on-site, to the Local Emergency Planning Committee or its designated emergency response agency and the state Emergency Response Commission. At a fixed facility, for example, an electroplating shop, an MSDS for each chemical can be obtained when the chemical is ordered and filed for workers' use.

Although the format may vary, MSDSs are required to provide certain information. Examination of an MSDS will show that the sheet provides physical, chemical, and toxicological (health hazard) data that are useful to workers engaged in sampling, materials handling, spill control, and fire fighting and to persons responsible for choosing personal protective equipment. Figure 6.8 shows an example MSDS that was published by OSHA.

Material Identification

At the top of the first page of the MSDS, the name and synonyms of the material will be listed, as well as the trade names under which it may be sold or bought. The Chemical Abstract Service (CAS) number, a unique number that designates a specific chemical, can also be listed here. The CAS number can be used to cross-reference information found in other hazardous materials references.

Material Safety Data Sheet May be used to comply with OSHA's Hazard Communication Standard, 29 CFR 1910.1200. Standard must be consulted for specific requirements.	**U.S. Department of Labor** Occupational Safety and Health Administration (Non-Mandatory Form) Form Approved OMB No. 1218-0072
IDENTITY *(As Used on Label and List)*	*Note: Blank spaces are not permitted. If any item is not applicable, or no information is available, the space must be marked to indicate that.*

Section I

Manufacturer's Name	Emergency Telephone Number
Address *(Number, Street, City, State, and ZIP Code)*	Telephone Number for Information
	Date Prepared
	Signature of Preparer *(optional)*

Section II – Hazardous Ingredients/Identity Information

Hazardous Components (Specific Chemical Identity; Common Name(s))	OSHA PEL	ACGIH TLV	Other Limits Recommended	% *(optional)*

Section III – Physical/Chemical Characteristics

Boiling Point		Specific Gravity (H$_2$O = 1)	
Vapor Pressure (mmHg)		Melting Point	
Vapor Density (AIR = 1)		Evaporation Rate (Butyl Acetate = 1)	
Solubility in Water			
Appearance and Odor			

Section IV – Fire and Explosion Hazard Data

Flash Point (Method Used)		Flammable Limits	LEL	UEL
Extinguishing Media				
Special Fire Fighting Procedures				
Unusual Fire and Explosion Hazards				

(Reproduce locally)	OSHA 174, Sept. 1985

Figure 6.8   Material Safety Data Sheet.

Section V – Reactivity Data

Stability	Unstable		Conditions to Avoid
	Stable		

Incompatibility *(Materials to Avoid)*

Hazardous Decomposition or Byproducts

Hazardous Polymerization	May Occur		Conditions to Avoid
	Will Not Occur		

Section VI – Health Hazard Data

Route(s) of Entry:	Inhalation?	Skin?	Ingestion?

Health Hazards *(Acute and Chronic)*

Carcinogenicity:	NTP?	IARC Monographs?	OSHA Regulated?

Signs and Symptoms of Exposure

Medical Conditions
Generally Aggravated by Exposure

Emergency and First Aid Procedures

Section VII – Precautions for Safe Handling and Use

Steps to Be Taken in Case Material is Released or Spilled

Waste Disposal Method

Precautions to Be Taken in Handling and Storing

Other Precautions

Section VIII – Control Measures

Respiratory Protection *(Specify Type)*

Ventilation	Local Exhaust	Special
	Mechanical *(General)*	Other
Protective Gloves		Eye Protection

Other Protective Clothing or Equipment

Work/Hygienic Practices

Page 2   U.S.G.P.O.: 1986-491-529/45775

Figure 6.8 *(continued)*

Section I—Manufacturer's Information

The manufacturer's name and contact information or the name and contact information of the importer or other responsible party can be found at the top of the MSDS in Section I. The party listed here should also provide an emergency telephone number where personnel knowledgeable about the chemical and appropriate emergency procedures can be reached. The date the MSDS was finalized or prepared should be listed in this section. This piece of information is important so that workers and responders know that they are looking at the most up-to-date information on the chemical.

Section II—Hazardous Ingredients/Identity Information
This section lists the various components of the material and, if established, the allowable exposure limits. This section is included for manufactured products that are mixtures. If a reaction-inhibiting chemical is present in the mixture in a concentration of less than 10% by volume, the manufacturer is not required to identify the inhibitor by name, although some sheets do list inhibitors. Some inhibitors are toxic.

Many manufacturers classify chemicals into families or list the chemical formula. The chemical family is the general class of the chemical, such as acid, solvent, halogenated hydrocarbon, or organic amine. For simple substances, the manufacturer may provide the chemical formula. For example, the chemical formula for sulfuric acid is H_2SO_4.

Section III—Physical/Chemical Characteristics
This section lists chemical and physical properties of the substance as determined by laboratory testing. Only those tests applicable to the product will be shown, and these can vary from substance to substance. These properties we have already discussed, such as:

- Boiling point
- Vapor pressure
- Vapor density
- Solubility in water
- Specific gravity
- pH
- Evaporation rate

Others that we have not yet talked about include:

- Melting point: the temperature at which a solid substance changes to a liquid state. For mixtures, the melting range may be given.
- Appearance and odor: a brief description of the material under normal room temperature and atmospheric pressure. Noting whether the appearance of an "identified" chemical matches its description provides an additional indication of the accuracy of the identification.

Section IV—Fire and Explosion Hazard Data
This section describes properties such as flash point and flammable/explosive limits that should be considered when assessing an incident involving fire or the potential for ignition of the chemical.

This section also lists the most effective extinguishing medium to use if this product should become involved in a fire. Different flammable or combustible chemicals behave differently when burning. Therefore, the extinguishing medium must be selected for its ability to extinguish a fire without increasing the problems associated with the fire. Water, dry chemical, foam (AFFF), and CO_2 are some commonly used extinguishing media.

General fire fighting methods are not usually described but special or "exception to the rule" procedures may be listed.

Unusual fire and explosion hazards such as hazardous chemical reactions, changes in chemical composition, or by-products produced during a fire or high heat conditions will be described. Hazards associated with the application of extinguishing media will be mentioned if applicable.

If the substance has an autoignition temperature it will also be listed in this section. The autoignition temperature is the approximate lowest temperature at which a flammable vapor-air mixture will ignite spontaneously (i.e., without an ignition source such as a spark or flame).

Section V—Reactivity Data

This section describes any tendency of the material to undergo a chemical change and release energy. Chemical reactions may produce undesirable effects such as temperature increase and formation of toxic or corrosive by-products. The MSDS should describe these effects. Conditions that may cause a reaction, such as heating of the material or contact with other materials, should also be described.

Specific components of the Reactivity Data Section that should be addressed are:

- Stability
- Incompatibility
- Hazardous decomposition or by-products
- Hazardous polymerization

Stability. Stability is an expression of the ability of the material to remain unchanged and is the result of strong bonds between atoms.

Incompatibility. Incompatibility is an indication of the tendency of a material to react on contact with other materials.

Hazardous Decomposition. Hazardous decomposition is an indication of the relative hazards associated with decomposition of the material. For example, 1,1,1-trichloroethane may form phosgene gas in a fire. Phosgene is very toxic in small amounts; this explains the requirement for SCBA use in fighting a fire when this chemical is present. For other examples, see Table 6.5.

Hazardous Polymerization. Hazardous polymerization refers to a material's ability to undergo a reaction that is generally associated with the production of plastic substances. Basically, the individual molecules of the chemical (monomers) react with each other to produce a polymer chain ("poly" meaning many, "mer" referring to monomers). These reactions are usually exothermic (heat generating). The more heat generated once the reaction starts, the faster the reaction rate becomes. If this is not done in a controlled environment, the reaction can produce a lot of heat that may serve as a source of ignition and a lot of pressure inside a container or vessel that may increase to the point of valve or container failure. A well-known example of a monomer is styrene. Styrene molecules react with each other to form a polymer known as polystyrene or Styrofoam. The MSDS should tell you whether the possibility exists of a polymerization reaction and the conditions to avoid to prevent a polymerization reaction.

Section VI—Health Hazard Data
This section provides information on the ways in which the chemical may enter the body, its routes of entry, and the ways in which the chemical may harm you if you are overexposed. Routes of exposure and other information on health effects are discussed in detail in Chapter 7. The MSDS should list both acute and chronic effects of overexposure in addition to other toxicological information.

Section VII—Precautions for Safe Handling and Use
Spill, leak, and disposal information is provided describing how to properly contain and handle the material in the event of spills or leaks that may damage the environment. This may include recommended cleanup materials, equipment, and personal protective clothing. The manufacturer's recommended method for disposing of excess, spent, used, leaked, or spilled material will also be listed in this section. Unfortunately, because states' regulations differ, the section is usually not very detailed. Refer to your state regulations or environmental management office for specifics on disposal in your area.

This section should also list precautions to be taken in handling and storing. Usually manufacturers will provide information regarding hazards unique to the material and special measures for storage and/or handling that were not covered in other sections. Note that the MSDS is not always well organized and this information may be found in other sections.

This section may also provide information on the regulatory status of the chemical. Newer MSDSs have an addendum showing the Superfund Amendments and Reauthorization Act (SARA) reporting requirements and information complying with the Comprehensive Environmental Response, Compensation, and Liability Act (CERCLA) and the Resource Conservation and Recovery Act (RCRA) regulations.

Section VIII—Control Measures
Information on proper types of respiratory protection, ventilation, and chemical protective clothing will be reported in this section. Additional work/hygienic practices that may help workers limit exposure will also be listed.

NIOSH Pocket Guide to Chemical Hazards

NIOSH publishes the *Pocket Guide to Chemical Hazards*. This is a public domain document that is updated by NIOSH periodically and is available in both a bound format and an electronic version. This document provides a great deal of information important to responder protection. A copy of the NIOSH *Pocket Guide* can be acquired by contacting NIOSH Publications, 4676 Columbia Parkway, Cincinnati, Ohio 45226-1998 or by calling 1-800-35NIOSH (1-800-356-4674).

The electronic copy of the NIOSH *Pocket Guide* is easy to use and read, with fewer symbols and abbreviations to translate. There are also links to even more information in references such as the DOT *Emergency Response Guidebook* and other NIOSH and CDC resources. Of course, the drawback to the use of the electronic version in emergency response would be the availability of a computer while on the scene. That is why the rest of the discussion here is focused on the bound version of the *Guide*.

In the bound copy of the *Guide*, in order to provide a vast amount of information in a pocket-sized booklet, many specific symbols and abbreviations are used. To interpret these symbols and abbreviations, the glossary and codes located in the front of the *Guide* should be consulted by readers unfamiliar with the document.

Let us look at a specific example (methyl chloroform) found in Figure 6.9 from a bound copy of the NIOSH *Pocket Guide*. Be warned that chemical names are highly specific and can be misleading if the reader is careless. For example, two chemical names that differ only by a single number or letter represent two different chemicals that may have widely differing hazardous properties.

Column I—Chemical Name, Formula, Etc.

Chemicals are listed alphabetically in column I. If numbers precede the name, as in the case of 1,1,2-trichloroethane, they are ignored in alphabetization. If you cannot find the chemical you are looking for listed in the *Pocket Guide*, consult the synonym index found in the back of the *Guide*.

In our example, methyl chloroform, you will see the chemical name listed and directly underneath you will see the chemical formula CH_3CCl_3. The Cs represent carbon atoms. Note that there are two Cs, and therefore two carbon atoms: One has three hydrogen atoms attached (H_3), and the other attaches to three chlorine atoms (Cl_3).

The two numbers listed below the chemical formula are the identification numbers given to methyl chloroform by the Chemical Abstract Service (CAS) and the Registry of Toxic Effects of Chemical Substances (RTECS) published by NIOSH. These numbers could be used to access information from reference sources, in print or in a computer database.

The two numbers at the bottom of column I are the DOT identification number for methyl chloroform (2831) and the three-digit Emergency Guide number (160), which refers to orange section in the DOT *Emergency Response Guidebook*.

Column II—Synonyms

This column shows other names by which a chemical is known. Here we again see that methyl chloroform is also known as 1,1,1-trichloroethane, as well as some

Chemical name, structure/formula, CAS and RTECS Nos., and DOT ID and guide Nos.	Synonyms, trade names, and conversion factors	Exposure limits (TWA unless noted otherwise)	IDLH	Physical description	Chemical and Physical properties		Incompatibilities and reactivities	Measurement method
					MW, BP, SOL, Fl.P, IP, Sp. Gr, flammability	VP, FRZ, UEL, LEL		
Methyl chloroform CH_3CCl_3 71-55-6 KJ2975000 2831 160	Chloroethene; 1,1,1-Trichloroethane; 1,1,1-Trichloroethane (stabilized) (Chloroethanes) 1 ppm = 5.46 mg/m³	NIOSH C 350 ppm (1900 mg/m³) [15-min] See Appendix C OSHA 350 ppm (1900 mg/m³)	700 ppm	Colorless liquid with a mild, chloroform-like odor.	MW: 133.4 BP: 165°F Sol: 0.4% Fl.P: ? IP: 11.00 eV Sp.Gr.: 1.34 Combustible Liquid, but burns with difficulty.	VP: 100 mm FRZ: -23°F UEL: 12.5% LEL: 7.5%	Strong caustics; strong oxidizers; chemically-active metals such as zinc, aluminum, magnesium powders, sodium & potassium; water [Note: Reacts slowly with water to form hydrochloric acid.]	Char CS2; GC/FID; IV [#1003, Halogenated Hydrocarbons]

Personal Protection and sanitation	Recommendations for respirator selection – maximum concentration for use (MUC)	First aid (See Table 5)	Health hazards – exposure routes (ER), symptoms (SY), target organs (TO) (See Table 6)
Skin: Prevent skin contact Eyes: Prevent eye contact Wash skin: When contaminated Remove: when wet or contaminated Change: N.R.	NIOSH/OSHA 700 ppm: SA*/SCBAF §: SCBAF:PD,PP/SAF:PD,PP:ASCBA Escape: GMFOV(see page XV)/SCBAE	Eye: Irr Immed Skin: Soap wash prompt Breath: Resp support Swallow: Medical attention immed	ER: Inh, Ing, Con SY: Irrit eyes, skin; head, lass, CNS depres, poor equi; derm; card arrhy; liver damage TO: Eyes, Skin, CNS, CVS, liver

Source: U.S. Department of Health and Human Services, 2003. *NIOSH Pocket Guide to Chemical Hazards*. Washington, DC: U.S. Government Printing Office, p. 202-203.

Figure 6.9 *NIOSH Pocket Guide* information on 1,1,1-trichloroethane, listed by its synonym "methyl chloroform".

other synonyms. If this was not previously known and a response was made to a leaking container placarded 2831, you would immediately look this up in the DOT ERG. Beside number 2831 is the chemical name 1,1,1-trichloroethane, which is not listed in the NIOSH *Pocket Guide*. However, by looking up 1,1,1-trichloroethane in the synonym index in the back of the NIOSH *Pocket Guide*, you would be referred to the proper page number in the *Guide*. Turn to that page and look down the information in column II until you see "1,1,1-trichloroethane." Now look back at the first column and you see "methyl chloroform."

Column III—Exposure Limits

In column III, OSHA's legal Permissible Exposure Limit (PEL) and NIOSH's recommended limits are listed. OSHA's PEL is the maximum airborne concentration of the chemical to which unprotected exposure is allowed by law. If higher atmospheric concentrations of the chemical are present in the work area, respiratory protection must be utilized to reduce the concentration in the inhaled air to, or below, the PEL. For methyl chloroform, the exposure limit recommended by NIOSH is also shown. The abbreviation "C" listed before the NIOSH recommended limit of 350 ppm means it is a "ceiling limit," which is discussed in Chapter 9. If the abbreviation "Ca" is seen in this column or the next, it refers to a cancer-causing agent or carcinogen. This term and others are discussed in Chapter 7.

Column IV—IDLH Level

Column IV indicates concentrations of the chemical in air that are considered to be Immediately Dangerous to Life and Health (IDLH) and should never be inhaled. For some chemicals, the notation "Ca" is listed in this column. This designates a cancer-causing agent (carcinogen) and suggests that no exposure above the Permissible Exposure Limit should be permitted even though immediate death would not result. In fact, some scientists argue that there is no "permissible" safe exposure level for a carcinogen. Therefore, the IDLH levels for carcinogens are listed in brackets, indicating that they are thought to be hazardous at any level of exposure.

Column V—Physical Description

Column V provides a brief description of the appearance and odor of the substance. These descriptors can be used as clues for early identification, or for confirmation after identification. Purposely sniffing a chemical to determine odor should be avoided, because the detectable odor level may be higher than the safe breathing level.

Column VI—Chemical and Physical Properties

These are the same chemical and physical properties as listed in the Material Safety Data Sheet. For a discussion of these properties, see the previous sections of this chapter.

Column VII—Incompatibilities

Materials with which the chemical being researched may react are listed in column VII. The chemical should never be mixed or allowed to come into contact with any material listed in this column. The resulting chemical reaction could lead to fire, explosion, or generation of a toxic gas or vapor. For example, methyl chloroform, when mixed with sodium hydroxide (a strong caustic), forms three reactants, one of which is hydrogen, a flammable and potentially explosive gas. This column is of special importance to emergency responders, who are likely to encounter the product outside of its container.

Column VIII—Measurement Method

Column VIII provides information on suggested sampling and analysis methods used to determine the atmospheric concentration of the chemical in the work area. The abbreviations used are listed in tables near the front of the *Guide*.

Column IX—Personal Protection and Sanitation

Column IX provides recommendations for preventing or minimizing exposure to the chemical being researched. Translations of the terms and abbreviations used in this column for methyl chloroform tell us that we should:

- Wear protective clothing if we anticipate repeated or prolonged skin contact with the chemical.
- Wear protective goggles if there is a reasonable probability that the chemical may contact our eyes.
- Wash our skin promptly if the chemical gets on it.
- Remove promptly any nonimpervious clothing that becomes wet with the chemical.

Column X—Recommendations for Respirator Selection

Column X provides information on respiratory protection. The agency recommending respirator protection is identified at the top of the column. The abbreviations used are explained in Table 4 near the front of the *Guide*.

Column XI—First Aid

This column lists actions that should be taken immediately after accidental exposure to chemicals. Abbreviations used in this column are explained in Table 6 of the *Guide*. The following first aid procedures are recommended for exposure to methyl chloroform: irrigate (wash) the eyes or use soap to wash skin promptly to remove methyl chloroform; give artificial respiration to someone who has inhaled it and is not breathing; and seek medical attention immediately for someone who has swallowed any of the chemical.

Column XII—Health Hazards

This section provides information on how the chemical can enter the body and potential adverse health effects resulting from chemical exposure.

Exposure Routes (ER). The ER section lists the routes of entry by which chemicals may enter the body. Abbreviations used in this column refer to:

- Inh: inhalation (breathing in)
- Ing: ingestion (swallowing)
- Con: contact (with skin or eyes)
- Abs: absorption (through the skin into the blood vessels and to internal body tissues and organs)

Symptoms (Sy). Symptoms that may result from chemical exposure are listed in this section. Table 5 in the front of the NIOSH *Guide* explains the abbreviations used in this column. Methyl chloroform exposure may cause eye and skin irritation, headache, lassitude (slowing down), central nervous system not working well, poor equilibrium and balance, dermatitis (skin irritation), cardiac arrhythmia (irregular heartbeat), and chronic liver damage.

Target organs (TO). This section lists the organs of the body most likely to be affected by chemical exposure. Abbreviations used are explained in Table 5 of the *Guide.* Target organs for methyl chloroform are the eyes and skins if splashed and the central nervous system (primarily the brain), the cardiovascular system (heart and blood vessels), and the liver if the chemical is inhaled, absorbed, or swallowed.

Emergency Action Guides

The *Emergency Action Guides* are published by the American Association of Railroads. This document is made up of foldout pages in a loose-leaf binder for commonly shipped hazardous materials. The information given is a combination of all the information discussed previously. Purchase of the manual entitles the buyer to buy update pages as knowledge or regulations change.

CHRIS Manual

The *Chemical Hazard Response Information System* (CHRIS) *Manual* published by the U.S. Coast Guard is another large document containing information regarding hazards and emergency response procedures for hazardous materials. Materials are listed on separate sheets that are inserted into a binder. *Manual 2, Hazardous Chemical Data*, is the most relevant of the four volumes to emergency response.

Telephone Hotlines

A number of 24-hour manned telephone services will provide hazard information in an emergency. If possible, always consult the MSDS for an emergency phone number found in Section I. Other services such as the Chemical Transportation

Emergency Center (CHEMTREC), which is supported by the Chemical Manufacturers Association, are available 24/7. CHEMTREC can also provide conference calls with some manufacturers of the chemical in question. The number is 1-800-424-9300.

Several other services are available in the United States, Canada, Mexico, and Brazil. There is a list of these numbers on the back inside cover of the DOT ERG. There is also a specific number to be used when an incident involves a U.S. military shipment.

The Agency for Toxic Substance and Disease Registry (ATSDR), an agency of the Centers for Disease Control and Prevention (CDC), staffs a telephone line that provides information about toxicological effects of chemicals. ATSDR Emergency Response Teams are available 24 hours a day and are comprised of toxicologists, physicians, and other scientists available to assist during an emergency involving hazardous substances in the environment. On-site assistance can also be requested. The number is 1-404-498-0120.

The National Response Center and Terrorist Hotline (NRC), 1-800-424-8802, supplies assistance in identification, technical information, and initial response actions for oil and chemical spills, radiological events, and transportation-related incidents. Spills that are above the RQ should be reported to them.

The CDC is available for assistance in handling infectious disease-related incidents or suspected bioterrorism incidents. The number for the CDC 24-Hour Emergency Response Hotline is 770-488-7100. You should also contact your regional FBI and local and state public health agencies if you believe the incident is an act of bioterrorism.

Computer-Aided Management of Emergency Operations (CAMEO®)

The CAMEO® computer program system, as described in Chapter 3 of this text, was designed as a multifunctional and multipurpose computer program system with twelve informational modules and three software applications. CAMEO® contains a database of hazardous chemicals. Each chemical listing in the database includes chemical identification information, regulatory information, and a Response Information Data Sheet (RIDS). The RIDS is broken into several headings including:

- General description of the chemical
- Properties
- Health hazards
- First aid
- Fire hazards
- Fire fighting
- Protective clothing
- Non-fire response

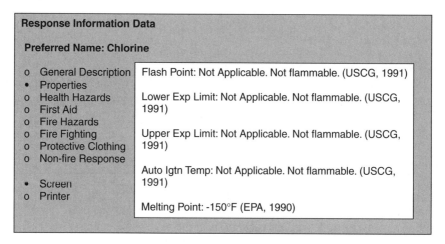

Figure 6.10   CAMEO® Response Information Data Sheet.

Figure 6.10 shows an example of the "Properties" screen, which provides information used to "estimate likely harm without intervention" and predict dispersion if the substance is released.

ASSESSMENT MODELS

The Street Smart Approach to Workplace Hazards

The "Street Smart Approach" was developed by Mike Callen of Fire Training Associates. It is a simple, commonsense approach whereby situations involving hazardous materials are classified as safe, unsafe, or dangerous. For our purposes, the approach will be broadened to include nonchemical hazards as well.

What is Safe, Unsafe, or Dangerous?
According to the Street Smart Approach, each of these categories is based on the likelihood of harm resulting to someone involved in the situation being assessed. Safe situations involve no significant likelihood of harm. Unsafe situations may be harmful but typically produce serious, permanent damage only if you have prolonged, repeated exposure to them. Dangerous situations are those likely to produce serious and/or permanent harm to you, even if you only have a limited exposure to them.

Types of Harm
Harm is an undesirable outcome. Examples include injury to our bodies, property damage, and environmental pollution. In the Street Smart Approach, we will consider harm to you as the number one concern as we review the various types

of harm we might encounter in during an incident involving hazardous materials. Harm can be due to many factors including:

- Chemical exposure
  - Contact with corrosive materials
  - Exposure to inhalation hazards
  - Oxygen deficiency
- Fire hazards
- Reactive materials
- Hazardous energy sources
- Other general safety hazards

Chemical Exposure. It would be easy to define what is "safe" if we could simply say that a safe situation is one in which no chemicals are involved. However, this is not realistic. In the modern world, chemicals are everywhere. We use them in our homes, and many of us use them in our hobbies. Because you are reading this book, chemicals are probably a concern for you. You may already know that the harm resulting from a given chemical is related to several factors, including:

- The dangerous properties of the chemical(s)
- The amount or concentration of the chemical(s) you are exposed to
- The length of time you are exposed

The higher the concentration of chemical you are exposed to, and the longer the time period of exposure, the greater the resulting harm. Another important factor is the route of entry: inhalation, ingestion, skin contact, or skin absorption.

Corrosives. Skin contact with acids illustrates the concept of harm. For example, you have probably been exposed to acetic acid in vinegar, as commonly used in various food products. Vinegar is a dilute, weak solution of acetic acid. Because of this, if you get vinegar on your skin or clothing, it's no big deal, even if it stays there for a long time. You can even ingest it without ill effects (unless you eat too many pickles and get indigestion).

As another example, consider a stronger, more concentrated acid: battery acid (aka sulfuric acid). If you have ever gotten it on your clothing or skin, you know that, unless it is removed or neutralized very quickly, it can destroy the material it contacts. If that material is your skin, you may be permanently injured as a result.

Inhalation Hazards and Published Exposure Limits. Consider inhalation exposure. The exposure limits you will be reading so much about in the upcoming chapters are functions of three factors: toxicity, concentration, and exposure time.

When you assess the hazards of a situation involving chemicals, keep in mind the differences between exposure limits. Remember that these limits are useless without good air surveillance information from the work area.

Oxygen Deficiency. One of the most basic facts of life is that people need adequate oxygen to stay alive and healthy. The air you normally breathe contains 20.9% oxygen. If you breathe air with significantly less oxygen, you will experience symptoms of oxygen deficiency. If an atmosphere you are breathing is severely oxygen deficient, you could be immobilized immediately and die very quickly.

Fire Hazards. Flammability and the properties that are used to predict materials' flammability have been discussed extensively earlier in this chapter. Because flammable and combustible liquids are very common in our society, statistically speaking, there is a very good chance that you might encounter these during a response. To assess the fire hazard of a substance, remember to assess properties such as flash point, explosive/flammable limits, and vapor pressure. Also keep in mind that, as with any atmospheric hazard, air surveillance is critical in assessment. Air surveillance is covered extensively in Chapter 9.

Reactive Materials, Flammable Solids, and Oxidizers. Some chemical reactions release enough heat to serve as an ignition source for any nearby fuel. This is why it is important to keep incompatible chemicals separated. Worse yet, some materials are inherently unstable and can break down spontaneously to release large amounts of heat. As discussed above, monomers may undergo violent polymerization. Merely moving shock-sensitive substances may trigger a violent reaction.

Flammable solid materials burn at a very high temperature and are difficult to extinguish. Some flammable solid materials will react with other substances to generate enough heat to ignite them. Worse still, some will do so merely through contact with water or air.

Oxidizers are substances that release lots of oxygen in chemical reactions. Extra oxygen in the atmosphere always increases the fire hazard. Some oxidizers also release lots of heat in addition to oxygen.

Any of these reactions can produce an explosive release of energy, so that we could actually class these materials in the energy hazard category. We could also class explosives in this category, because an explosion is simply a very rapid combustion reaction.

Hazardous Energy Sources. In addition to the heat energy of the fire hazards and reactive materials discussed above, you may encounter other types of hazardous energy. If you ever have to work around explosive materials, the hazardous energy potential they represent is obvious, because they function by a massive release of energy. Other energy hazards are not as obvious. Remember that any source of stored energy can injure you if released suddenly.

Any type of pressurized container poses a pressure hazard in addition to any other hazard associated with the contents. This should be obvious for materials stored in gas cylinders or other types of high-pressure containers. However, low-pressure containers, such as steel drums, can conceal a hazardous amount of pressure. Pneumatic or hydraulic systems can also pose a serious pressure hazard.

Electrical equipment and sources are probably the most common workplace energy source. Electricity poses a definite energy hazard.

Some incidents may contain energy hazards in the form of radioactive materials.

General Safety Hazards. While going about response activities, don't ignore the common, everyday sorts of hazards. Keep in mind that general safety hazards account for a much higher percentage of serious injuries than chemical hazards do.

Common hazards such as heavy equipment in operation, muddy site conditions, uneven terrain, and open excavations may be encountered. You may also contact unstable objects, such as improperly stacked containers or unshored, vertical trench walls. Even though such practices violate safety and health regulations, you may still encounter them.

Applying the Street Smart Approach

To apply the Street Smart Approach, size up the situation carefully. Try to categorize the hazards into one of the three categories.

Remember that Safe situations involve no significant likelihood of harm. Unsafe situations may be harmful, but they typically produce serious, permanent damage only if you have prolonged, repeated exposure to them. Dangerous situations are those likely to produce serious and/or permanent harm to you, even if you only have a limited exposure to them.

In using this approach, ask yourself first, "Is it dangerous?" then, "Is it unsafe?" and finally, "Is it safe?" In this way, you will be giving first priority to identifying the most hazardous situations.

The following guidelines may be helpful in using the approach. Keep in mind that although some hazardous conditions may be obvious, others can only be identified through air surveillance.

Is it Dangerous?

Dangerous situations may have atmospheric, chemical, and physical hazards present. Dangerous atmospheres may be:

- Toxic—containing vapors above IDLH concentrations
- Flammable—containing flammable gases or vapors at concentrations greater than 10% of the LEL, or combustible dust in concentrations thick enough to obscure vision at a distance of 5 feet or less
- Oxygen deficient—containing less than 19.5% oxygen
- Oxygen enriched—containing more than 23.5% oxygen

Hazardous chemicals may include:

- Corrosives that are not in proper containers
- Reactive materials, flammable solids, and oxidizers that are not properly stored or handled

Energy hazards may be represented by:

- Explosives that are not properly stored or handled
- Pressurized storage containers that have been damaged
- Bulging containers such as 55-gallon drums
- Electrical hazards
- Pneumatic and hydraulic systems
- Radioactive materials

General safety hazards may include:

- Poor walking/operating surfaces in high-traffic areas
- Traffic hazards, such as poorly regulated vehicle or heavy equipment traffic, and foot traffic and vehicle or equipment traffic in the same area
- Pits or excavations that are open and unguarded
- Unstable areas or objects such as unshored, steep trench walls or improperly stacked drums
- Machinery without proper guarding in place

Situations assumed dangerous until proven otherwise may include:

- Large spills or leaks
- Large containers with significant damage
- Visible vapor clouds
- Below-grade spills or leaks
- Confined space entries
- Chemical odors, unless odor threshold is known to be well below hazardous levels
- Biological indicators such as dead animals

Is it Unsafe?
Unsafe situations may have atmospheric hazards present but be absent of any other type of recognizable safety hazard.

Atmospheric conditions in this category include:

- Toxic gases or vapors—in excess of exposure limits but less than IDLH concentrations

- No oxygen-deficient areas—no atmospheres with less than 19.5% oxygen present
- No potentially flammable atmospheres—no flammable gases or vapors present at concentrations greater than 10% of the LEL and no combustible dust present in high enough concentrations to obscure vision
- No oxygen-enriched atmospheres—no atmospheres containing more than 23.5% oxygen

No hazardous chemicals unless properly stored and handled, no energy hazards, no significant general safety hazards, or no other obviously dangerous situation is present.

Is it Safe?
Safe situations are void of all hazardous conditions including hazardous atmospheres. Also, no hazardous chemicals unless properly stored and handled, no energy hazards, no significant general safety hazards, or no other obviously dangerous situation is present.

Using the Street Smart Approach for Hazard Assessment
Mike Callen's Street Smart Approach provides you with a different perspective for doing hazard assessment. Instead of an ironclad set of rules, think of this more as a general approach. Remember that this approach, like any other assessment method, is only useful if you actually use it on the job. Using it means using your eyes, ears, and, in some cases, your nose, and also all the accumulated knowledge and wisdom you have picked up in training. Also, remember that assessment and identification of hazardous conditions is worthwhile only if appropriate actions are taken in response to the hazards identified.

The General Hazardous Materials Behavior Model

The General Hazardous Materials Behavior Model (or GHBMO, pronounced "gebmo"), as developed by Ludwig Benner, can be used to relate events involved in an accidental release of hazardous materials to form a logical sequence of events (Table 6.6). This sequence of events begins with stress and damage to the hazardous material container(s) and ends with harm to people, property, and/or the environment. The GHBMO can thus be used to predict future events as an incident progresses and to identify options for intervening to prevent or minimize undesirable outcomes.

Events and Event Behaviors
The events making up the GHBMO sequence are shown in Table 6.6 and discussed below.

Stress Event. Hazardous materials are normally safely contained, as in tank cars, tank trucks, pipelines, steel drums, bottles, etc. However, containers involved

TABLE 6.6 General Hazardous Materials Behavior Model

Event	Event Behaviors	Event Interruption	
		Strategies	Tactics
Stress	Thermal Mechanical Chemical Radiation Biological	Influence applied stresses	Redirect impingement Shield stressed system Move stressed system

Breach	Disintegration
	Runaway cracking
	Attachments open
	Punctures, splits,
	  or tears
		Influence breach
		  size
			Cool contents
			Limit stress level
			Activate venting devices

Release	Detonation Violent failure Rapid relief Spill or leak	Influence quantity released	Reposition container Reduce pressure Contain breach
Engulfment	Cloud Plume Cone Stream Irregular deposits	Influence size of danger zone	Initiate controlled ignition Confine released material Dilute

Impingement	Short term
	Medium term
	Long term
		Influence
		  exposures
		  impinged
			Use shielding
			Evacuate

Harm	Thermal Radiation Asphyxiation Toxic Corrosive Biological Mechanical	Influence severity of harm	Decontaminate Increase distance from source Use shielding

Adapted from Benner, 1978 and Noll, Hilderbrand, and Yvorra, 1995.

in accidents are typically subjected to stresses that threaten the integrity of the container. The most common types of stress involved are as follows.

THERMAL STRESS results from extreme temperatures, such as those due to heat from flame impingement on containers (as evidenced by bulging containers, safety valve releases, etc.). Thermal stress may also be due to extreme cold related to cryogenics.

MECHANICAL STRESS results in physical damage, as when containers are impacted by other objects. Clues include gouges, punctures, and cracks.

CHEMICAL STRESS results from chemical reactions involving two or more substances. Examples include corrosion of metal and heat of polymerization.

Stress events can occur in combination. It is important during size up to identify stressed containers so that measures can be taken to protect personnel in the event of breaching and release. Also, it may be possible to identify actions that can prevent an impending container breach.

Breach Event. The breach event occurs when a container is stressed beyond the point of failure and opens up as a result. Basic types of breach events are as follows.

DISINTEGRATION involves total failure, resulting in the explosion or shattering of a container.

RUNAWAY CRACKING occurs when closed containers, such as drums or tank cars, fail suddenly as a small crack very quickly grows into a linear crack that circles the container. This type of total container failure is commonly associated with boiling-liquid expanding vapor explosions (BLEVEs) or BLEVE-type events.

FAILURE OF CONTAINER ATTACHMENTS occurs when devices such as valves, fusible plugs, etc. open up or break off of a container.

PUNCTURES are typically related to mechanical stresses that physically "punch" holes in the container.

SPLITS OR TEARS are elongated breaches that may occur at seams or welds on containers.

Response personnel should be aware of the extremely high hazards that disintegration or runaway cracking can pose to personnel at the incident scene. Control procedures designed to prevent or minimize the breach event should only be undertaken if they can be performed safely.

Release Event. Once a container is breached, the contents are able to escape as matter, energy, or a combination of both. The rate of release determines our ability to control the event and the likelihood of harm (including injury to responders) resulting. Types of releases are as follows.

DETONATION is an instantaneous explosive chemical reaction (e.g., involving dynamite, organic peroxides, etc.). No reaction time is allowed for response personnel to escape the affected area.

VIOLENT FAILURE involves a release time of less than one second (e.g., a deflagrating explosion). No reaction time is allowed for escape from the affected area.

This type of release is associated with overpressurization of closed containers and runaway cracking, as in BLEVE-type events, and can be deadly to response personnel or civilians in the area of engulfment.

RAPID RELIEF ranges from a few seconds to a few minutes and is associated with releases from pressurized containers because of punctures, splits, tears, damaged valves, or actuated pressure relief valves. Very limited reaction time for escape or a control measure is allowed.

SPILLS OR LEAKS involve release times varying from minutes to days, typically resulting from low-pressure flow through damaged valves or fittings, splits, tears, or punctures. The longer release times typically allow prolonged control measures to be initiated.

Engulfment Event. After hazardous materials and/or energy have been released from a container, they are free to migrate within the environment, thus engulfing a surrounding area. The rate and area of engulfment will depend on numerous factors (as described above in the section of this chapter on predicting dispersal of hazardous chemicals). The goal of hazardous materials control at this stage is to minimize the area of engulfment so as to limit exposures.

Impingement Event. As the engulfment event proceeds, the hazardous materials and/or energy released will impinge, or come into contact with, exposures. Exposures can include people (including response personnel), property, and the environment. Impingement may affect other hazmat containers, thus escalating the incident.

Short-term impingement occurs over a period of minutes to hours, such as when a plume of toxic gas is released during an incident. Medium-term impingement occurs over days, weeks, or months after an incident, such as when a stream is contaminated. Long-term impingement lasts for years or generations, such as when soil and groundwater aquifers are polluted.

Harm Event. The harm event refers to the effects of exposure resulting from impingement. Harm can be categorized into several types (Table 6.6).

The level of harm is determined by (1) rate and duration of release, (2) size of area of dispersion, and (3) toxicity of chemicals involved. The ultimate goal of response operations is the prevention or minimization of harm resulting from the incident.

Using the GHBMO in Incident Assessment

The goal in using the GHBMO is to determine to which stage of events an incident has progressed at the time of size up. On the basis of this, and other assessment considerations, response personnel can determine the probable course of future events and the harm likely to result without intervention (step two in the DECIDE process). Responders can then establish response objectives intended to prevent or minimize harm resulting from the incident (step three in the DECIDE process). Also, the GHBMO provides event interruption strategies and tactics (Table 6.6) that can be used to identify options to achieve the response objectives (step four of the DECIDE process).

Event Interruption Strategies and Tactics. Strategies and tactics for interrupting the GHBMO event sequence to prevent or minimize the resulting harm are discussed below.

INFLUENCE APPLIED STRESSES. In situations involving stress to containers, the ideal option would be to eliminate the stress before breaching of the container results. For example, a container under thermal stress due to flame impingement may be moved or shielded to terminate the impingement or hose streams might be used to cool the container.

However, it must be remembered that a stressed container may represent the greatest threat to responders because breaching and release could occur during this type of operation. In some situations, it may not be possible to intervene to influence applied stresses without undue risk to responders. In other situations, it may be possible to use remote control techniques, such as the use of unmanned hose monitors to cool stressed containers.

INFLUENCE BREACH SIZE. In situations in which breaching cannot be prevented, actions may be taken to minimize the breach size. For example, in some cases responders may activate venting devices to create a controlled release for pressure reduction to prevent total container failure. However, such procedures require careful assessment and decision-making and an expert knowledge of the containers involved.

When intervening at this stage, it is critical to know how the container involved is likely to breach, given the design of the container and the stresses involved. For unfamiliar containers, seek expert advice before proceeding. Surprises during intervention can have disastrous consequences.

INFLUENCE QUANTITY RELEASED. Once a container has been breached, it may be possible to intervene to minimize, or stop completely, the release of hazardous materials from the container. This requires the execution of advanced control techniques. Examples include repositioning a container to place a breach above the liquid-vapor interface and performing containment operations (such as plugging, patching, or installing specialized containment kits) to terminate the release at the source.

Remember that intervention at this stage typically requires that personnel work at the point of release, so exposure potential is high. As in all hazmat operations, safety of personnel must be given due consideration in determining control options.

INFLUENCE SIZE OF DANGER AREA. The goal of intervention after release has occurred is to minimize the area of engulfment so as to limit exposures. This can frequently be accomplished through basic control or confinement techniques. Examples of such techniques include diking, damming, and diverting spilled liquids and using fog patterns to control air releases.

INFLUENCE EXPOSURES IMPINGED. The degree of harm resulting from impingement is determined by the degree and duration of exposure. If engulfment of a given area cannot be prevented, responders may be able to minimize harm by removing or covering potential exposures before engulfment of the area.

INFLUENCE SEVERITY OF HARM. In situations in which impingement of exposures cannot be prevented or has already occurred before responders intervene, some actions may still be taken by responders to minimize the resulting harm. For example, emergency field decontamination of people contaminated by an accidental release may significantly reduce the extent of their injuries.

Safety Consideration and Assessment
As previously noted, it is important to be able to recognize stressed containers and breach events that have resulted (or are likely to result) from the stresses involved. This may allow an early interruption of the series of events as described in the GHBMO.

The recognition of stress or damage to containers is especially important in situations in which container breaching could occur while responders are located within potential areas of engulfment. Also, in some situations, the actions of response personnel (such as repositioning a damaged container) could trigger an impending breach and uncontrolled release of hazardous materials.

As a general rule, the earlier responders are able to intervene in this sequence of events the more effective the response operation will be in limiting harm. However, when sizing up a hazmat incident and selecting response procedures, remember that the safety of response personnel should always be given top priority. In some situations intervention may not be possible without undue risk to responders. In such cases, response personnel may be able to do nothing more than secure the area, remain at a safe distance, and take defensive actions as the incident runs its course.

SUMMARY

The dangers of actually or potentially released hazardous materials must be assessed before personnel enter the area and before response action decisions are made.

Being familiar with the information covered in this chapter can help achieve a sound chemical hazard assessment. There are many good sources of chemical information that can be utilized in planning for and responding to an emergency involving hazardous materials, only a few of which are discussed in this chapter. A complete assessment will also include information on health effects, air surveillance, and PPE selection, which are discussed in the following chapters.

7

HUMAN HEALTH EFFECTS

INTRODUCTION

Emergency responders are asked to take a great deal of risk when they go to the scene of a hazardous materials incident. This risk can be reduced by employing methods to limit the exposure of responders to toxic chemicals. Decisions regarding protective measures must be made in every incident; they can be made with more assurance if responders understand how toxic chemicals enter the body and cause harm.

In Portland, Connecticut, 47 employees of a manufacturing plant were taken to hospitals after being exposed to a plume of toxic vapors that were released from a tank of degreasing solution. On the same day, nine workers were taken to the hospital after inhaling a floor sealer being used in part of the building. A disaster drill scheduled for an area hospital was called off; it was felt the staff had enough practice with these simultaneous incidents.

In this chapter we present information about toxic chemical exposures and health effects. The goal of the chapter is to provide emergency responders with enough information to enable them to make good decisions for their own protection and the protection of the community.

TOXICITY

A toxic material is any substance that can harm your health. It may do this by burning your skin, irritating your throat and lungs, damaging the stomach lining,

Emergency Responder Training Manual for the Hazardous Materials Technician, Second Edition,
edited by Kenneth W. Oldfield
ISBN 0-471-21387-X   Copyright © 2005 John Wiley & Sons, Inc.

or impacting the cells in your body. Some toxins are produced naturally by bacteria and other living creatures such as snakes and insects; other toxic effects result from the actions of viruses or radioactivity. The toxic materials most likely to be encountered by emergency responders are chemicals, and our focus in this chapter is on the effects of toxic chemicals.

For a chemical to have an adverse effect on the human body, one must be exposed in such a way that the chemical gets onto or into the body. The effects that are produced are proportional to the dose the body receives and the duration of exposure. Doses administered experimentally in food or by injection are usually measured in milligrams per kilogram of body weight (mg/kg) of the test subject. In studies in which the dose is inhaled with air or is present in drinking water, the environmental concentration may be expressed as parts per million (ppm), meaning the number of parts of chemical in every million parts air or water, or as milligrams per cubic meter (mg/m^3), meaning the number of milligrams of chemical in a space of air that measures one meter on each side.

The relationship between the amount of chemical exposure (dose) and the body's biological response is called the dose-response relationship. In most cases, greater exposure will cause a greater response. It follows, then, that reducing or eliminating exposure can reduce or eliminate your body's reaction to the chemical exposure. A great deal of study has gone into providing people with information they can use to reduce or eliminate exposure to toxic chemicals.

LEARNING ABOUT TOXIC EFFECTS

One of the important things you learn in any job in which there may be exposure to toxic chemicals is that information is available about how to protect yourself from exposure. One key piece of information to use in choosing protective measures is the chemical's exposure limit. This is the maximum concentration of the chemical in air to which you can be exposed without experiencing any adverse health effects. These exposure limits are determined by toxicologists after conducting a number of different kinds of studies.

Toxicology can be thought of as the study of the beneficial and harmful interactions between chemicals and biological systems. A toxicologist is an individual with expertise in the nature of the adverse effects of agents on living organisms. They study the mechanisms of these interactions and assess the risk of harm to humans, other organisms, and the environment. They also recommend appropriate precautionary, protective, restrictive, and therapeutic measures to deal with these harmful effects. The following sections describe some types of toxicological studies that are used to learn about the harmful effect of chemicals.

Laboratory Studies with Animals

Controlled animal studies are the most common studies conducted with toxic chemicals. They were originally conducted to provide information about chemicals that had been found to have beneficial/medicinal effects. The studies determine the proper dosage as well as identifying the side effects these would have on humans. Animals are still used for this purpose, although regulations concerning animal testing have tightened. Animal studies have been expanded to include studies of workplace chemicals, to try to identify causal relationships between exposure to these agents and various forms of toxicity.

Mice, rats, rabbits, guinea pigs, dogs, monkeys, and other animals are used in laboratory studies, but mice and rats are usually the study animals of choice. Rodents are small, easy to keep, and relatively inexpensive to feed and have a life span of around three years. This last characteristic allows researchers to see a lifetime of exposure effects within a reasonably short period. Also, rodents of the same genetic strain can be purchased, allowing scientists to determine that variations in illness were produced by the variations in the dose of chemical and not by genetic differences in the study animals.

Permission to use experimental animals must be granted by the National Institutes of Health. It must be shown that the research is necessary for human health and that the animals will be humanely treated.

Types of Toxicity Studies

The general purpose of all toxicity studies is to identify the nature of health damage produced by a chemical agent and the range of doses over which damage is produced. There are many different types of toxicity studies (acute, subchronic, chronic), and each has a different and specific purpose.

Acute. Acute studies are the usual starting point for toxicological investigations. The study of the acute (single dose) toxicity of a chemical in animals is necessary to calculate the doses that will not be lethal to the animals used in long-term (chronic) studies. Another important aspect of the acute study is that it can help to identify the target organ affected in long-term (chronic) exposures. A detailed explanation of target organs is provided below in this chapter.

In acute studies the LD_{50}, the lethal dose for 50% of an exposed population, is estimated by toxicologists. Groups of test animals are given one exposure to the chemical, with the measured amount of the dose based on each animal's body weight. The animals are cared for in the normal way for 14 days, with notes kept on how many deaths occur in each dose group. The goal of the study is to determine the dose that is lethal to half the exposed animals in 14 days, the LD_{50}.

Table 7.1 shows how the results of an LD_{50} study might look. The LD_{50} in the study illustrated was found to be 8 mg/kg, because this dose was fatal to 50% of the animals who received it. Table 7.2 shows the LD_{50}s of some chemicals that may be encountered in emergency response.

LD_{50} studies give us only a relative idea of the dangers of the chemical tested. Because the objective of toxicological studies is not to find out how much exposure people can handle and suffer the loss of only half of the exposed population, the need for further study is obvious. So, once the LD_{50} for a chemical is known, further studies are conducted with new groups of animals to determine other, less immediately deadly effects of the same chemical. Through these types

TABLE 7.1   Results of a Hypothetical Lethal Dose Study[a]

Group	Number Tested	Oral Dose (mg/kg)	Number of Deaths
A	100	12	83
B	100	8	50
C	100	4	12
D	100	0	0

[a]$LD_{50} = 8$ mg/kg

TABLE 7.2 LD_{50} for Several Chemicals

Agent	Oral Rat (mg/kg)	Skin Rat (mg/kg)	Inhalation Rat (ppm)
Ammonia	350		2000
Aniline	440	1400	250
Benzene	3800		10,000
Carbon tetrachloride	2800	5070	4000
Chlorine			293
Creosote	725		
Ethyl alcohol	14,000		
Formaldehyde	800		250

of studies, toxicologists have revealed one of the basic principles of toxicology: Not all individuals who are exposed to the same dose of a chemical will respond in the same way.

In other types of acute studies, the chemical being studied may be administered to the animal in the same way as workers are expected to be exposed, by breathing, swallowing, or skin contact. The acute effects are noted and correlated to the dose received by each test group. At the end of the study, the range of doses expected to produce acute effects is known.

Subchronic. Subchronic studies look at exposure that occurs over a limited time period, whereas acute studies look at one-time exposures. These exposures could occur over days, weeks, or even months depending on the study design and the nature of the chemical being studied. Doses given to study animals are lower than those that have been shown in previous acute studies to produce effects.

Chronic. Chronic studies look at exposure that occurs over the entire lifetime of the test animals. Both subchronic and chronic studies will help toxicologists determine how the period of exposure (duration) affects the toxic response.

Common Terminology

The following are acronyms (new words formed by using first letters of words) and abbreviations (initials that don't make new words) that may be encountered in seeking toxicological information about a chemical.

No-Observed-Adverse-Effect Level (NOAEL). The purpose of both subchronic and chronic studies is not only to find the range of doses over which adverse effects occur but also to identify the dose at which these effects are not observed. This dose is the no-observed-adverse-effect level.

Toxic Concentration (TC). The toxic concentration describes an inhaled concentration that is expected to cause some type of adverse or toxic effect. TC_{LO} is the lowest inhaled concentration expected to cause an adverse or toxic effect. TC_{50} is the inhaled concentration expected to cause adverse or toxic effects in 50% of the exposed population.

Toxic Dose (TD). The toxic dose describes an ingested dose that is expected to cause some type of adverse effect. TD_{LO} is the lowest ingested dose expected to cause an adverse or toxic effect. TD_{50} is the ingested dose expected to cause adverse or toxic effects in 50% of the exposed population.

Lethal Concentration (LC). The lethal concentration describes an inhaled concentration that is expected to be lethal. LC_{LO} describes the lowest inhaled concentration expected to be lethal. LC_{50} is the inhaled concentration expected to be lethal to 50% of the exposed population.

Lethal Dose (LD). Lethal dose was discussed in detail above. LD_{LO} describes the lowest ingested dose expected to be lethal. LD_{50} is the ingested dose expected to be lethal to 50% of the exposed population.

As Low as Reasonably Achievable (ALARA). ALARA is a designation used when numerical limits cannot be set, such as for carcinogens and other highly toxic chemicals. Applying the ALARA concept means taking every reasonable precaution to reduce the exposure to as low as possible.

Correlating Animal Tests to People

Problems are encountered when we try to use information gained from animal experiments in calculating and setting safe exposure limits for humans. Obviously, humans are not laboratory rats. Because we do not look like rats, we can immediately assume that we are genetically different from rats. Our genes determine our bodies' abilities to handle toxic chemicals, so we may handle some of them in different ways than rats.

We are much bigger than rats and other laboratory animals. Doses are measured in such a way as to enable researchers to account for this size difference; oral doses are given to animals in milligrams or micrograms per kilogram of their body weight. The dose is adjusted to milligrams per kilogram and can be extrapolated to the weight of humans, whose weight is also measured in kilograms. Because this extrapolation for dose response does not take into account other differences besides weight, a safety factor is included in these calculations.

The conditions under which laboratory animals are tested are carefully controlled in a way that human living conditions are not. The rats do not smoke, drink, eat pesticide-laden fish, or drive to work in polluted air with no seat belts. They are protected from outside disease agents and other factors that might influence their response to the exposure. Because of these differences, additional safety factors are included in the calculations.

Effects must be measurable. All effects must be observable to be noted, whether visible (hair loss), countable (number of offspring), able to be weighed (liver shrinkage), or other clearly obvious symptoms of disease. It is impossible to get a verifiable answer from a rat to the questions, "How do you feel? Does anything hurt today?" Laboratory animals, therefore, are given larger doses than it probably would take to make them only mildly sick.

Almost all of these studies are conducted by using exposure to only one chemical at a time. It is very difficult to design a study to test the effects of exposure to more than one chemical at a time. In the real world, workers may be exposed to a combination of chemicals at the same time, or exposed to multiple chemicals at different times over a short time span. As we will see later, multiple exposures may lead to unexpected results.

Toxicity Information from Human Exposures

Epidemiology is the study of incidence, distribution, and control of diseases in humans. Epidemiologists define disease broadly and include accidents and other nontransmissible conditions. Sometimes these nontransmissible diseases are caused by chemical exposure. For example, mesothelioma (effect) is due to exposure to asbestos (cause).

Human exposure data on adverse health effects attributed to chemical exposure can potentially come from one of the following sources: self-reported symptoms in exposed people, case reports from doctors treating exposed people, and descriptive or analytic epidemiological studies. The first two sources are of limited use in determining cause-and-effect relationships because they are considered anecdotal.

Descriptive epidemiological studies look at identifying disease distribution and rate in human populations. Such studies often employ the correlational approach in which differences in disease rates in human populations are associated with differences in environmental exposure. Because of this, they are not usually helpful in identifying cause-and-effect relationships.

The last source, analytic epidemiological studies, is the primary source used for determining these causal relationships between specific chemical agents and human health effects.

Analytic Epidemiological Studies

Analytic epidemiological studies are observational, not experimental. There are three basic types of analytic epidemiological studies: cohort, case-control, or cross-sectional.

Cohort studies involve the evaluation of the disease experience of populations exposed to a particular chemical compared with similar, unexposed populations. In cohort studies, the exposed population (cohort) and the unexposed population (control) are followed for a period of time to find out who develops or dies from the disease in question.

Case-control studies compare people with a particular disease (cases) with similar people without the disease (controls) to evaluate their exposure history to a suspected agent.

Cross-sectional studies look at individuals irrespective of their disease or exposure status. They look at a cross section of the general population looking at disease and exposure status.

Epidemiologists use a number of different sources of information such as hospital records, insurance company records, and occupational safety and health records. They may also interview people who have worked with the chemical.

When enough people have been identified who have been exposed to the chemical, their health records are compared with the health records of an unexposed control group (case-control study). This unexposed control group should be as much like the study group as possible in sex, age, and other factors. If all other parameters are similar, and the only difference is exposure to a chemical, differences in health history may be attributed to the chemical.

Studies of exposure to some chemicals have shown that they cause or contribute to diseases in humans. Figure 7.1 shows the results of one study of the

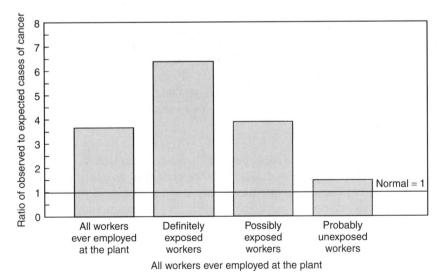

Figure 7.1 Incidence of cancer in a group of workers exposed to workplace chemicals.

relationship between chemical exposure and human disease. "Expected cases" are those cases of a certain cancer that would be expected to occur in this number of people, based on the percentage of the general population who get the disease. Only 1 person per 100,000 in the general population has the type of cancer studied. Of the group of workers who were definitely known to be exposed to the chemical in question, there were 6.48 cases of the cancer per 100,000 workers. This number of cases is statistically significant; it was found by statistical analysis to be much higher than can be explained by chance. The conclusion was drawn that the chemical exposure was responsible for the increased number of cases of cancer.

Several criteria must be met if a study of this kind is to be valid. Sometimes, when dealing with chemical exposures, these criteria are difficult to meet.

A study, particularly a cross-sectional study, must include large numbers of people. If the study group is not large, it cannot be demonstrated by statistical analysis that the effects were caused by the chemical and not by chance.

The exposures must be quantifiable. This is very difficult to do, as it requires that a number be placed on the air concentration, skin contact level, or amount swallowed of the chemical. If no one was doing this kind of monitoring, then these numbers are hard to estimate. You will remember that the dose-response relationship forces us to estimate the amount of the exposure to know that the response was due to the dose.

Groups of exposed subjects must be compared with groups of unexposed subjects who match them as closely as possible in terms of all other parameters. These parameters include sex, age, size, and lifestyle. Lifestyle includes many factors that are controlled by the person, not the researcher, and these almost always affect health and resistance to disease. Some of these factors are smoking, drinking, medications, and diet.

All of the above criteria must be met for epidemiological studies to produce valid results. Two other concerns exist that cannot be controlled. First, every person has a unique genetic makeup. Even identical twins have slightly different genes. These differences may result in different responses to the same chemicals. Also, human experimentation is illegal and immoral in present-day American society. We cannot put people in controlled, potentially harmful situations where researchers can control exposures and all other variables.

Learning from Disasters

Sometimes groups of people are exposed to chemicals accidentally. These exposures are called "experiments of nature" by the toxicologists and epidemiologists who study them. These are considered analytic epidemiological studies because the populations were exposed first and then they were tracked to determine the effects of the exposure. They share some of the characteristics with laboratory experiments: The chemical is known; many individuals are exposed; sometimes dose levels can be determined; and measurable results occur. Some of the most widely studied "experiments of nature" are mercury-induced disease in Minnemata, Japan, a deadly toluene diisocyanate release in Bhopal, India, and widespread dioxin contamination following a plant explosion in Sevaso, Italy. Surviving victims of these and other accidents have been extensively studied, some for as long as 30 years after exposure.

Table 7.3 provides data resulting from combining laboratory studies of rat exposures, epidemiological studies of human exposures, and follow-up studies on victims of an "experiment of nature."

Hexachlorobenzene is a chlorinated aromatic compound formerly used to prevent stored seed and grain from becoming moldy. From the data in Table 7.3 we can conclude that:

- Hexachlorobenzene is toxic to both rats and people.
- The effects are dose related.
- The chemical causes acute and chronic health effects in several organ systems.

Using Study Data to Predict Human Health Effects

Using data from laboratory and epidemiological studies, scientists from several organizations set legal or suggested safe exposure limits for humans. The

TABLE 7.3 Results of Studies of Hexachlorobenzene

Test Subject	Dose	Response
Animal Studies		
Rat	5 mg/kg	Half of the subjects died
Rat	0.08 mg/kg/day	Kidney damage in offspring
		No observable liver effects
Rat	0.4 mg/kg/day	Chronic liver damage
Rat	2 mg/kg/day	Acute liver damage
		Increased death in offspring
Ray	6 mg/kg/day	Liver cell tumors
Epidemiology studies		
Industrial workers	Not measured	Increased blood levels
Vegetable sprayers	Not measured	Blood levels 4–287 ppb
		Kidney damage
Hazwaste workers	Not measured	Increased blood levels
General population	0.2 µg/day	Fat retention 18–35 ng/g fat
		Retention for 15 years
Experiments of nature		
Humans in Turkey	Variable; ingested in bread over 4 years	100% death in children who drank milk of exposed mothers; 10% mortality in adults; liver disease; skin ulcers; neurological effects; short stature and small hands in adults exposed as children.

aim of setting these limits is to give potentially exposed people guidelines for protecting themselves from hazardous dose levels. Considerations in setting safe limits include the relative degree of danger presented by the chemical and the sort of study data (animal or human) on which the limits will be based. The application of exposure limits during a hazmat response are described in detail in Chapter 9.

Safety Factors

Safety factors are calculated into the current exposure limits. These safety factors take into account the severity of the hazard, the differences between lab animals and humans, and the differences between human individuals. The dose that produced no observable adverse effects (NOAEL) in animals is divided by 100 or 1000 for suggested safe exposure in humans; the magnitude of the safety factor depends on the seriousness of the potential response. A dose that was found to produce no observable effect in human studies is divided by a safety factor of 10. These safety factors are only general guidelines; researchers who are thoroughly familiar with the chemical use their judgment in setting limits.

Exposure limits are designed to protect most people, but they may not protect the most sensitive individual. It is known that certain individuals are more likely to have a negative response to chemical exposure than others, and these people may need more protection than provided by exposure limits. It is easy to understand differences in sensitivity to a chemical when we consider the differences we are aware of in allergic responses to agents such as plant pollens or poison ivy.

Different Limits Set by Different Agencies

Several agencies set human exposure limits; their criteria and suggestions for use of the limits they set may be different from those of other agencies. Safe exposure limits are useful only when understood and used in the context in which they are set.

The United States Occupational Safety and Health Administration (OSHA) sets legally enforceable workplace limits to protect workers who are exposed to a chemical on a regular basis in their daily workplace. It must be understood that even though OSHA limits cannot be legally exceeded in the workplace, they are not precise numerical cutoff numbers dividing safe from unsafe levels of exposure. OSHA encourages employers to keep workers' exposures as low as possible, even in atmospheres where levels are below published exposure limits.

The National Institute of Occupational Safety and Health (NIOSH) sets recommended limits, some of which are more conservative than the OSHA limits. They are often different interpretations based on the same experimental results.

The American Conference of Governmental Industrial Hygienists (ACGIH) also sets suggested limits. They are designed to be used in the practice of industrial hygiene. ACGIH limits were adopted by OSHA as the first legal standards. However, ACGIH now states that its limits are not intended for use as regulatory limits.

Although exposure limits are listed for most common airborne chemical hazards, their use is not the most accurate way to quantify human exposure. They only inform us about levels in the air and cannot tell us how these levels translate into the dose received inside the body at the place where dose leads to response. Monitoring would be most useful if it could detect internal dose before illness begins; usually this is not possible. Comparing listed safe exposure limits with measured concentrations in air is the most readily available means of judging chemical hazards.

SOURCES OF HEALTH EFFECT INFORMATION

A list of sources of information about specific chemicals, including their health effects, is provided in Chapter 6. For more extensive data concerning toxicological studies, the reader may consult one of the following sources:

- *ATSDR's Hazardous Substance Release/Health Effects Database.* These fact sheets are toxicological profiles for individual chemicals, published by the Agency for Toxic Substances and Disease Registry of the Centers for Disease Control and Prevention (CDC). They can be accessed for free on the following website. http://www.atsdr.cdc.gov/hazdat.html
- *RTECS (Registry of the Toxic Effects of Chemical Substances).* This multivolume reference is published by NIOSH and includes the data from thousands of published toxicology studies.

ROUTES OF ENTRY

Chemicals enter the human body by four major routes: inhalation (breathing), skin contact (contact), absorption through the skin (absorption), or ingestion (swallowing). Contact with a route of entry may cause harm at the site of contact or may lead to absorption by skin, respiratory surfaces, or the digestive tract, allowing harmful effects to occur at some distance from the site of entry. Once the route of entry of a chemical is known, protection from the chemical can be put in place.

The physical state of the chemical is important in determining the probable route of entry. Solids are unlikely to cause harm if they are in large pieces. It is only when they become finely divided by grinding, sanding, welding, or burning that they are small enough to be transported by air to the skin or lungs or to be carried to the mouth to be swallowed. Liquids may also get into the mouth or contact skin when they splash. Liquids that form aerosol mists or vapors can be inhaled. Gases are likely to be inhaled and may, under some conditions, condense on skin or other surfaces as liquids.

Contact with the Body Surface

Skin is an excellent protector against harmful agents in the natural environment. It covers the entire surface of the body and prevents entry to naturally occurring germs and toxins. Most of the chemicals used in industry today are man-made, and humans have not had time to develop natural protections against them. Two mechanisms may account for harmful effects—contact and absorption.

Skin Contact

Damage to the skin itself results from physical injury to skin cells by the chemical or by an allergic type response in the skin. Strong corrosives such as hydrochloric and sulfuric acids and bases like sodium hydroxide may burn skin severely. Mild to severe irritation, called dermatitis, can be caused by many halogenated solvents, which remove necessary natural oils. Some metal compounds, particularly those containing beryllium and nickel, initiate an allergic response in sensitive people. Skin diseases are the most common occupational diseases. Some skin diseases clear up quickly after exposure ceases, but others continue for a long time.

Skin Absorption

Many chemicals, especially solvents, permeate the outer, protective layer of skin and are absorbed into blood vessels in underlying layers. Chemicals known to cross skin at a significant rate are given a "skin notation" by their exposure limits. Examples of these are hexane, benzene, and trichloroethane. Skin cells contain fat, and solvents dissolve fat. This can lead to skin damage and entry into the blood transport system. Once in the blood, chemicals are carried throughout the body. Figure 7.2 shows the structure of skin and its underlying blood vessels.

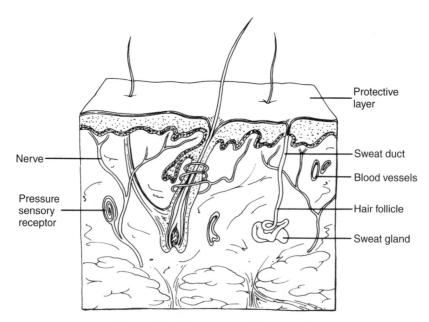

Figure 7.2 The structure of human skin.

Eye Contact

A small area of the body is not covered by skin but by the corneas of the eyes. This covering is even more vulnerable than skin, as eyes are always wet. This allows any water-soluble chemical to quickly dissolve, a prerequisite for crossing the body surface. Protection of eyes is essential, because they are a route of entry and because they are so delicate. Irreversible eye damage can begin within a few minutes of exposure to some chemicals. Contact lenses, which may trap and hold chemicals against the cornea, are not suggested, or approved by OSHA, for use in contaminated environments without eye protection.

Inhalation into the Respiratory System

All of the parts of the human respiratory system are vulnerable to inhaled toxic chemicals. If the chemical is in the form of dust or other large particles, it is likely to be stopped in the nose, throat, or upper airway and cause damage there. A study of career firefighters in Seattle, Washington, showed a higher incidence of upper respiratory cancer in this population, which inhales smoke, than in the non-firefighter population. Another group of chemicals known to cause upper respiratory problems are those that are very water soluble and dissolve out of inhaled air onto the damp surfaces there. Gases with high solubility, such as ammonia and hydrogen chloride, act on the upper respiratory tract within seconds and can cause fatal swelling. Moderately soluble gases like chlorine and sulfur dioxide cause both upper and lower respiratory tract distress within minutes. The low-solubility irritants like ozone and phosgene are more insidious; they may cause the lungs to fill with fluid 6 to 24 hours after exposure, without any irritant symptoms to serve as warnings.

Chemicals that reach the smaller branches of the bronchi or the far reaches of the lungs can cause a variety of health effects, ranging from damage of delicate membranes to lung cancer. Lung linings do not regenerate after traumatic injury, and damaged portions of a lung are no longer functional. The surface area of the lungs of an average 150-pound person has been estimated to be the size of a singles tennis court and the amount of air inhaled daily is around 430 cubic feet, so the potential for exposure to inhaled toxins is great. Figure 7.3 shows the structure of the human lower respiratory system.

The terminal pockets (called alveoli) inside the lungs are well adapted to perform their primary chore—the exchange of oxygen and carbon dioxide with blood inside tiny vessels surrounding these pockets. Many inhaled chemicals also cross these thin membranes and enter the blood, to be transported throughout the body.

Ingestion into the Digestive System

Ingested chemicals are swallowed. It is very unlikely that responders will knowingly ingest harmful chemicals during or after an emergency response, so attention to proper protection and personal habits can greatly limit exposure to chemicals by

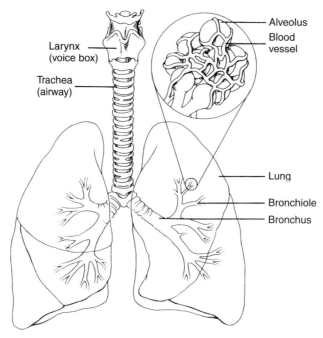

Figure 7.3 The human respiratory system brings air from outside the body into close contact with blood vessels surrounding the alveolus.

mouth. Avoiding splashes to the face will help, as will taking care to remove any contamination from the hands before touching the face or mouth. Hands should be washed before drinking, eating, or smoking. Some inhaled chemicals, particularly those in solid particulate form, may be brought up into the back of the mouth by mechanisms in the respiratory system that keep such particles from reaching the lungs. These are reflexively swallowed.

Chemicals that reach the stomach are not detoxified by the body's natural defense systems there, because these are designed primarily to kill bacteria. Swallowed chemicals can generally pass through to the small intestine, another body organ with an extremely large surface area (almost the size of a football field) and an excellent system for absorbing ingested materials. This organ is where almost all of your digested food is absorbed into the bloodstream for transport to cells throughout the body. Chemicals absorbed here are also available for transport. Figure 7.4 shows the structure of the human digestive system.

THE BODY'S RESPONSE TO CHEMICAL EXPOSURE

Responders are rightly concerned about how the chemicals they are exposed to will damage their health. However, it is important to remember that although the highly toxic man-made chemicals are relatively new, the human body

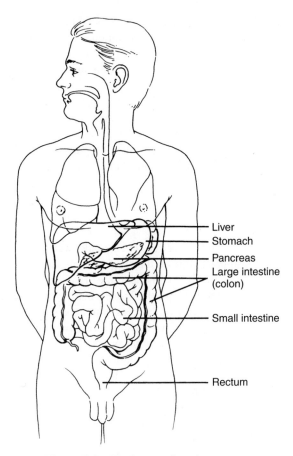

Figure 7.4   The human digestive system.

has long survived attacks by many toxic materials. There are a number of
defenses and mechanisms that provide some protection once a person has been
exposed—namely, excretion, storage, and metabolism. Although certainly not
universally effective, these defenses may serve to lessen the harm caused when
responders are exposed.

Blood travels everywhere. Most organs are not protected from materials carried
in the blood. All the cells in the body are near enough to the blood supply to
be reached by whatever the blood carries, because all cells need oxygen and
nutrients to remain functional. It is possible, however, to look up a chemical and
find that it has listed "target organs" where damage is most likely to occur. How
do these tissues happen to get singled out for harm?

Each cell is bounded by a membrane that surrounds its contents. Cell membranes are selective about what they let into cells, and different kinds of cells
let in different kinds of body chemicals. Ovary cells, for example, admit certain
kinds of hormones that regulate the daily business of the ovary. Muscle cells

admit calcium, and nerve cell membranes allow sodium and potassium to cross when messages are transmitted. The cell membrane around each of these cells has entry ports that are opened by a sort of lock and key system; the entering material is the key that fits the lock on the cell membrane.

Because the shape of the key depends on the chemistry of the entering compound, a toxic chemical that is similarly shaped can also open the lock. In this way, toxic chemicals get inside certain cells, and cause damage there, and are kept out of other cells. However, the absorption rate into the tissues and organs varies with different chemicals. This difference in uptake allows for chemicals to be acted on by the body's defenses before damage is done. These defenses are briefly described below.

Excretion

If the chemical is water soluble, or has been made water soluble by enzyme activities, it will be eliminated out of the blood as it goes through the kidney and will be excreted from the body in the urine. Even chemicals that are not water soluble may be excreted through the kidney if they are acted on by certain body chemicals. The urinary excretion process is more complex than this simple description sounds, but it is an effective means of removing some of the harmful chemicals before they can act on their target organ.

Other methods of excretion include exhalation of gases dissolved in the blood (consider the odor of alcohol in a drinker's exhaled breath or the function of the Breathalyzer to measure blood alcohol) and excretion of fat-soluble materials in mother's milk. Nursing infants have been harmed by exposure to pesticides like DDT and hexachlorobenzene in mother's milk. Some chemicals are treated by the liver and excreted in bile through the gastrointestinal tract.

Obviously, any chemical that is removed from the body by excretion before it can be absorbed into the target organ or system is no longer a threat to cause harm. Excretion, then, is one defense the body has against hazardous materials.

Storage

Chemicals may also bind to tissues, fluids, or parts of the body that are not necessarily affected in a harmful way. The body does this naturally with nutrients and minerals that are taken in as food so that they are available for energy and body activities throughout the day. This storage of chemicals can be a defense against harm as well.

The general mechanism is that the chemical enters the body by one of the routes of entry mentioned above. It is then carried throughout the body by blood or other body fluids, such as plasma. Equilibrium is established as some of it is concentrated in the storage tissue and thus taken out of the system, while some remains in the fluid. As the level in the body fluid begins to drop, the chemical then comes back out of storage. This is both helpful and potentially harmful. The immediate help is that much of the initially high dose of chemical is taken out of the system before it reaches the target organ to do damage, lessening the immediate effect. However, the chemical may persist and even accumulate with repeated exposure and pose long-term harm.

Fat tissue is a storage site for a number of chemicals such as DDT, polychlorinated biphenyls (PCB), and chlordane. Bone is a storage site for lead but is not harmed by it. Other tissues serve as storage depots to varying degrees.

Although it is often true that the storage depot is not harmed directly by the chemical, it is not always true. Sometimes the chemicals accumulate and are stored for extended periods in the very tissues that they damage, such as cadmium in the kidneys. Avoiding exposure is always the way to prevent harm. Nonetheless, storage may reduce the effect of a single high exposure that a responder may experience during a release.

Metabolism

Another possible response of the body to a chemical that enters is to react with it and change it into something else. Metabolism is the term used to describe the wide range of chemical reactions that occur between nutrients and other chemicals that enter the body and chemicals produced by the body called enzymes. The liver produces these enzymes, which are large protein molecules. Enzymes are made and preprogrammed to break molecules apart at certain specified spots. The result of the metabolic process is to break a complex compound down into simpler molecules, which often are water soluble. These water-soluble compounds are more easily excreted by the kidneys. This is also how the complex foods that we eat are broken down into the simpler nutrients and energy our body needs. The process may also act on toxic or nonnutrient chemicals as well.

If we think of nutrient molecules reaching the liver as a string of different-colored pop beads, we can think of enzymes as being programmed to separate the string between certain color combinations. For example, Enzyme A may always separate the string between red and yellow adjacent beads. If a toxic chemical containing this same combination reaches the liver in the bloodstream, Enzyme A, which cannot tell toxic chemicals from nutrients, breaks it between the red and yellow beads. There is no way the liver can control this enzymatic action; it happens automatically wherever red and yellow beads are connected.

The result of Enzyme A's action may be beneficial or it may be harmful to the body. If the broken sections are less toxic to cells as they float around the body than the original chemical would have been, the chemical has been broken down to the body's advantage. This is called detoxification. An example of this is when ethanol, a central nervous system (CNS) depressant, is changed into acetic acid, a water-soluble chemical that is removed by the kidneys.

Unfortunately, this is not always the case. If the broken bits combine in different ways with other bits to form chemicals more toxic than the original, the person will suffer more serious health effects than if Enzyme A had not done its job. The same enzymes that detoxify the ethanol mentioned above also react with methanol to form formaldehyde, a different chemical that is more dangerous than the methanol itself. This is called toxication.

Like storage, metabolism may either help or further harm a responder. Like storage and excretion, when detoxification is the result, it serves as another defense the body can mount against chemical exposure. However, when the defenses are overwhelmed or ineffective, the body can be harmed by the chemicals that enter the body and reach their target organs.

DAMAGE TO THE BODY

The human body is a complex system of diverse fluids and tissues that enter many chemical reactions to sustain vital processes. It should be no surprise then that foreign chemicals and pathogens (disease-causing organisms) can alter this system. The ways in which the alterations occur can be generalized into three types of actions:

- Inhibition of normal biochemical reactions or cell growth
- Stimulation of normal biochemical reactions or cell growth
- Alteration of genetic material

These actions may take place on a small, localized level with little or no significant impact on the exposed person's health. The body systems are able to overcome the damage or adapt to changes without any perceived illness or injury to the responder. On the other hand, some actions dramatically damage major organs or body systems to the point that the body cannot recover or adapt and the responder suffers a permanent effect, perhaps even death. Between these extremes is the full spectrum of health effects that a chemical may cause.

The dose-response relationship discussed above in this chapter determines the extent of damage to the body when responders are exposed. The toxicological studies mentioned previously point out that the nature of the health effect depends on the chemical and physical properties of the hazardous material. These factors combine with the personal factors of the responder to create the individual response.

Timing Terminology

The concentration of chemical in the air determines the amount of chemical available to enter the body of the responder. This important factor can be assessed by means of the air monitoring equipment and techniques described in Chapter 9. The other major factor in determining the dose of chemical that actually enters the body is the length of time or duration of exposure. The duration of exposure is described as being acute or chronic. An acute exposure is typically a brief exposure (hours or minutes) at a higher concentration. A chronic exposure occurs repeatedly over a longer period of time (weeks, months, years) to a lower concentration.

These terms are also sometimes used to describe the nature of the effects or damage. An acute effect is one that occurs at the time of an acute exposure or very soon thereafter. It is generally, although not always, detectable or sensed by the individual (e.g., nausea, skin irritation, dizziness). Although most acute effects are reversible or transient, some produce changes that the body cannot undo.

Chronic effects are those that result from repeated exposures and/or take a long time to appear after exposure occurs. Cancers and liver and lung diseases are examples of such long-term effects. Each chronic disease has a specific cause and name, but the idea is that it results from repeated exposures and takes a long time to become apparent.

Another term often used to describe an effect is "symptom." This term usually describes a relatively minor effect that should serve to warn of an acute over-exposure. Symptoms are usually transient effects that pass when the exposure stops. Responders should always be aware of the symptoms caused by the chemical involved in a release and be prepared to leave the hazard area immediately if they experience any of the symptoms.

Damage to Body Systems

Toxic chemicals are often classified by the part(s) of the body that they primarily affect. This is helpful in pointing to where the medical surveillance program should look for damage when an exposure to a chemical is suspected. Pulmonary toxins were discussed in the section on the inhalation route of entry, and chemicals that damage the skin were discussed in the section on the skin route of entry. The following sections describe other major body systems that may be damaged by chemicals, although there is not space in this text to describe all of them.

Genetotoxins

Each cell in the body contains genes or protein units that serve as a blueprint for the structure and function of that cell. Genetic material controls the hereditary makeup of the body and determines many of the characteristics of the body. Genes are made up of a series of proteins arranged much like beads on a string. The order of the beads (genes) on the string relays information to newly forming cells, telling them what shape, form, or function to accomplish. Beads are read three at a time, in a sort of three-letter code. If one bead is damaged, changed, or missing, the code no longer makes sense and cannot be used as a pattern for new cells. This type of change is called a mutation, and the chemical causing it is called a mutagen. The following example may help the reader to understand.

This three-letter word sentence makes sense to people who read English. If we remove a letter from one word, and our reader still reads the sentence three letters at a time, the sentence is no longer meaningful.

THE CAT ATE THE BIG FAT RAT

THE CAA TET HEB IGF ATR AT

This is not the only way genetic damage occurs; other kinds of damage also result in misinterpretation of the code. Caffeine and ultraviolet light are known mutagens, as are a few chemicals used in industry, such as hydroxy-lamine. Other industrial chemicals are under study to determine whether they are mutagenic.

Reproductive Toxins

Chemicals may damage the reproductive system of the body, affecting the ability to have children or causing effects in the children of the exposed person. Reproductive toxins may cause harm to the sperm cells in men or the egg cells in women or to the organs involved in sexual reproduction. The result is that couples are unable to become pregnant or the pregnancy ends before the child is born. 1,2-Dibromo-3-chloropropane (DBCP), methylmercury, and lindane are a few examples of such reproductive toxins.

A teratogen is a chemical that is known to cause damage in an unborn baby. The word teratogen comes from the Greek meaning "make a monster." Teratogen exposure should obviously be avoided by pregnant women, but some teratogens can influence embryos that are not yet conceived, by damaging eggs or sperm within the bodies of future parents. Because a woman already carries, at her birth, all the eggs she will ever produce, her exposure period lasts from birth until the last pregnancy.

Lead is a known teratogen, causing brain damage in offspring. A Supreme Court decision (*U.S. v. Johnson Controls*, 1991) determined that an employer cannot discriminate against women of child-bearing age in assigning workers to higher-paying, lead-exposed jobs. A number of other chemicals are known to be maternal teratogens, operating through exposure to the mother.

Recent research has found evidence for human sperm teratogenicity from exposure to vinyl chloride. Other studies indicate that children of male auto body workers exposed to a combination of hydrocarbons, metals, oils, and paints had a four- to eightfold increase in the incidence of kidney cancer. After the *Johnson Controls* discrimination suit was initiated, lead was found also to be a sperm teratogen.

Neurotoxins

Neurotoxic agents are toxic to the nervous system. They may damage the brain or spinal cord (called the central nervous system) or sensory or motor nerve communication in other parts of the body. One well-known neurotoxic effect is tremor of the limbs caused by exposure to several organic solvents; another results in a shuffling walk known as "Ginger Jake paralysis" because it was first identified in those who consumed contaminated ginger liquor imported from Jamaica in the 1930s.

Hepatotoxins

Hepatotoxic chemicals damage liver tissue. The liver is a large organ, and damage may not be detected until much of it is irreversibly harmed. It has a number of important duties in the body's functional operations. The liver is the primary, although not the only, site of metabolism or at least the production of metabolic enzymes. Hydrogen cyanide, formic acid, fluorine, and many chlorinated solvents are hepatotoxic.

Groups of similar chemicals are metabolized by the same enzyme, so exposure to two similar chemicals at once may result in a shortage of detoxifying enzymes. For example, ethyl and isopropyl alcohols are metabolized by the same enzyme that breaks down chlorinated organic chemicals like chloroform and carbon tetrachloride. Exposure to one of the alcohols together with chloroform or carbon tetrachloride leaves the emergency responder short of the necessary detoxifying enzyme and puts him at greater risk of harmful effects. This interaction is called synergism and is discussed below in this chapter. The ethanol exposure may be by vapor inhalation, skin absorption, or recent ingestion of alcoholic beverages.

Nephrotoxins

Nephrotoxic chemicals cause kidney damage. The kidney is the organ that controls the body's fluid volume and also is responsible for excreting much of the body's waste. Nephrotoxic chemicals therefore often cause a buildup of wastes in the body that result in harm to other systems in the body.

Certain workers—particularly auto assembly plant and metal work employees who clean machines—have a high risk of serious kidney disease. These workers have more than double the risk of developing kidney problems that require dialysis or transplants. Nephrotoxins include heavy metals (cadmium and mercury) and halogenated solvents (carbon tetrachloride, bromobenzene). Many cases of kidney damage are not diagnosed until a person has lost three-fourths of the function of the organ.

Immunotoxins

The body has a complex system to defend itself from infectious agents and to some extent search out and destroy neoplastic or cancer cells. The immune system functions by the actions of the spleen, lymph nodes, thymus, liver, and leukocytes (white blood cells). This system is designed to recognize the difference between the normal, desirable cells of the body and foreign cells and materials. Foreign substances act as antigens that cause the formation of protein substances called antibodies. Antibodies then spread through the body and work in several ways to destroy those foreign substances when they enter the body again. Because the body forms antibodies in response to encountering an antigen, an immune response may not occur at the time of the initial exposure to the chemical or agent. But after the antibody is formed and reproduced, it can produce a very strong reaction the next time the agent enters the body.

Immunotoxicity, or an undesirable effect on the immune system, generally takes one of three forms:

- Immunosuppression—depressed immunity to infection
- Autoimmunity—immune reaction to normal body substances
- Hypersensitivity—exaggerated immune response to an agent

Many chemicals have been shown to be immunotoxic, including benzene, PCBs, nickel, toluene diisocyanate (TDI), dioxin, and sulfur dioxide.

A very important form of immunotoxicity is sensitization. The typical progression of a sensitization reaction includes an initial exposure or repeated exposure in which there is little or no perceptible reaction. However, the antibody-forming process begins and intensifies until a future contact with the material causes a severe allergic reaction. A common form of sensitization is allergic contact dermatitis in which a person is repeatedly in contact with a metal like nickel or chromium. After a period of time, that person begins to experience severe rashes when exposed to even small amounts of the metal.

Allergic reactions to some materials can be severe, even fatal. Most people know of individuals who are "deathly" allergic to insect stings or other substances. These hypersensitive reactions often do not show up with the first exposure to a material. This makes it very hard to predict whether and when a person will have an allergic reaction. Therefore, it is wise to have even small allergic reactions evaluated by physicians familiar with allergic reactions.

Endocrine Toxins

A recent trend in environmental toxicology is the study of endocrine system disrupters. The endocrine system includes all of the organs in the body that produce and regulate hormones. It includes the pituitary, adrenal, thyroid, and parathyroid glands as well as the testes and ovaries. Hormones are chemicals produced by the body that control such activities as metabolism, cellular growth and chemical reactions, reproduction, and fetal (unborn child) development.

The full extent of the harm caused by environmental endocrine disrupters is not yet known. However, these agents have been implicated in effects as diverse as sexual effects, birth defects, hyperthyroidism (which leads to excessive weight loss), hypothyroidism (which leads to excessive weight gain), many forms of cancer, and myriad other health effects. Examples of endocrine disrupters include polychlorinated biphenyls (PCBs), DDT, parathion, lead, dioxin, and styrene. This promises to be a heavily studied issue in the future.

Exposures to Multiple Chemicals

Responders are often exposed to more than one chemical at a time. The chemicals may or may not cause damage to the same target organ. If they do, the combination of effects may be either additive or synergistic. In an additive effect, each chemical produces some effect and the total damage from all chemicals would be roughly equal to the sum of the effects. It may be represented by the equation:

$$2 + 2 = 4$$

However, in some cases, the mixed chemicals produce a much greater response then would be expected from simply adding the effects of all of the chemicals.

Such a response is called a synergistic response and might be represented by the equation:

$$2 + 2 = 8$$

A synergistic effect may result when one chemical activates or enables another chemical to produce a greater than expected effect. In other cases, the two chemicals may be metabolized by the same enzyme in the body. Exposure to one of the chemicals may tie up or consume the enzyme in the body and therefore allow the second chemical to pass unchanged through the body to the target organ. However it happens, synergism can produce a significant problem for responders who are exposed to multiple chemicals.

RECOGNIZING AND PREVENTING HEALTH EFFECTS

There are three primary methods that can be used to determine whether emergency responders have been exposed to toxic chemicals and to prevent continued exposure. These are recognizing the symptoms of exposure, monitoring exposure by biological monitoring, and medical evaluation through monitoring and surveillance.

Recognizing Symptoms

The first and obvious step in recognizing the symptoms of chemical exposure is to know what they are. In every hazardous materials incident, the chemical must be identified for a number of reasons, not the least of which is the gathering of human exposure data. The agency having jurisdiction should keep records regarding an individual's role and possible exposure in each incident. Once the chemical has been identified, part of the reference material that should be checked is the section describing symptoms of exposure. The Safety Officer should inform all responders and medical personnel what these symptoms are so that every one can watch for them. Responders should be informed during the postincident meeting what long-term symptoms they should watch for. Often, the accurate attribution of symptoms to their causes is the first indication of exposure to a chemical agent.

Because some physicians are poorly informed about chemical exposure, responders should be prepared to provide them with written information about the symptoms of such exposure and the target organs of incident chemicals.

Biological Monitoring

In Chapter 9, air monitoring techniques are described that can be used to measure the concentration of chemicals in the air that may enter the body by inhalation. However, air monitoring is not able to address entry of chemicals into the body by other routes such as skin absorption or ingestion. Biological monitoring is the process of using a body tissue, fluid, or exhaled air to measure how much of the chemical has entered the body.

The purpose of biological monitoring is not to look for disease or health effect but to measure exposure by looking at how much chemical has entered the body. An example would be to take a sample of blood and to have it analyzed to measure how much lead is in the blood. This amount could then be compared to the OSHA blood lead limit of 40 μg of lead per 100 μg of blood found in the Lead Standard (29 CFR 1910.1025). If the level in the blood is kept below this limit, most people should not experience damage from lead poisoning. OSHA has written a separate standard for a number of toxic chemicals, and each standard includes specifically required medical monitoring. Some of the other chemicals for which standards are written are asbestos, arsenic, cadmium, benzene, and vinyl chloride. These are found in 29 CFR 1910.1000–1101.

Another set of standards that can be used to evaluate biological monitoring results is the Biological Exposure Indices (BEIs) published by the American Conference of Governmental Industrial Hygienists (ACGIH). The BEIs are recommended (not regulatory) maximums of chemicals or their metabolites in blood, urine, or exhaled air. The BEI limits generally, although not always, represent the amount of chemical that would be expected to be in the body of someone exposed to the airborne chemical concentration limit called the Threshold Limit Value, which is described in Chapter 9. Again, the BEIs are intended to be used for monitoring exposure, not to look for health effect. If the BEI limit is exceeded in a responder's sample, he/she should be removed from further exposure and medically evaluated to look for signs of health effect.

Medical Surveillance and Monitoring

Medical surveillance is the regularly scheduled examination of any health parameters considered to be important to the job of an emergency responder. It consists of an initial physical exam to establish baseline health condition and annual exams to ensure that the responder remains healthy. The medical surveillance program is described in Chapter 2. Responders should be sure to tell their physician during the exams about any hazmat releases they may have responded to during the year to alert the physician to look for signs of damage.

Medical monitoring is a more incident-specific examination of a person who may have been exposed to a toxic agent. In a case where this agent is a chemical, monitoring should be specific to the action of the chemical and what is known about the way the body processes it. Medical monitoring may include taking blood pressure, respiration rate, and body temperature before and after entries during a response to look for changes in the responder's health caused by chemical exposure.

As part of the medical surveillance requirements of HAZWOPER, the employer must provide medical evaluation to any hazmat team member who is concerned about a possible exposure during an emergency incident. The purpose of medical surveillance and medical monitoring is to catch signs of health effects as early as possible in order to stop exposure and to provide medical care that may prevent serious harm.

SUMMARY

Most of the chemicals we find useful in our society have the potential to harm our health. This chapter reviewed how toxicologists study the mechanisms by which the chemicals cause harm to establish exposure limits and identify other ways to reduce or eliminate exposure. The routes of entry of chemicals into the body were reviewed as well as the ways in which the body responds to chemical exposures. The terminology used to describe toxic effects was introduced. Finally, the ways to recognize and prevent health effects were described.

8

PHYSICAL HAZARDS OF EMERGENCY RESPONSE

INTRODUCTION

Physical hazards may be present at an emergency response that can cause injury and must be considered and prevented if possible. The purpose of this chapter is to introduce the hazardous materials responder to typical physical and scene-related hazards and to discuss methods of preventing accidents. The physical hazards discussed in this chapter include temperature extremes such as heat and cold stress, noise, and radiation. The scene-related hazards discussed include the presence of hazardous energy, confined spaces, and musculoskeletal injuries. Finally, the prevention of accidents by developing and implementing job safety analysis and standard operating procedures is covered. The goal of this chapter is to help familiarize the reader with the possible types and controls of physical hazards that may be present at a hazardous materials incident.

PHYSICAL HAZARDS

Physical hazards are often a secondary thought of first responders following the consideration of the chemical hazard present in a hazardous materials incident. Physical hazards have the potential of adversely affecting the safety and health of hazardous material workers and should be considered during a hazardous materials response. The physical hazards discussed below include localized burns, heat stress, cold stress, radiation, and noise.

Emergency Responder Training Manual for the Hazardous Materials Technician, Second Edition, edited by Kenneth W. Oldfield
ISBN 0-471-21387-X   Copyright © 2005 John Wiley & Sons, Inc.

Localized Burns

Localized burns are a result of exposure to an energy source such as heat, chemicals, electrical current, and radiation from the sun. Burns from heat sources are the most common type of localized burn, yet burns from chemicals, electrical currents, and radiation from the sun can frequently occur in a workplace environment. The severity of the burn depends on the temperature of the material that caused the burn, the length of time the victim is exposed to the material, the location of the burn on the body, the size of the burn, and the victim's medical condition. Table 8.1 describes the different sources of materials that can cause localized burns.

Types

The three categories used to classify a type of burn are first, second, and third degree. A first-degree burn, also labeled a superficial burn, affects the top layer of the skin. The burn will be red, dry, and painful and will heal in five to six days. This type of burn is the least severe and will not leave a scar. A second-degree burn, also labeled a partial-thickness burn, affects the deeper layers of tissue. This type of burn will be red, and blisters may form that, if opened, will weep clear fluid. The area of the burn will be blotchy, painful, and swollen, and it will potentially leave a scar. A third-degree burn, also labeled a full-thickness burn, is very serious and affects all layers of the burn area. This type of burn will destroy all layers of the skin or tissues underneath the burn and may even destroy bones. The burn will look brown or blackish and may be either very painful or painless because of destruction of the nerve endings. This type of burn is considered critical, and immediate medical attention is needed.

First Aid

The steps involved in caring for a burn include first stopping the burning by putting flames out or removing the victim from the source of the burn. Next, cool the burned area with large amounts of water; do not use ice or ice water. If the burn is from a chemical, this may involve flushing the burn with enough water to clean the chemical off the skin. The final step is to cover the burn loosely with a dry, clean dressing. The covering will help keep air out of the burn, reduce pain, and help to prevent infection.

TABLE 8.1   Causes of Burns

Type of Burn	Causes or Source
Heat	Fires, explosions, hot surfaces
Chemical	Cleaners, lawn and garden sprays, paint removers, household bleach, industrial sources, laboratory sources
Electrical	Power lines, lightning, defective electrical equipment, unprotected electrical outlets
Radiation	Sun

Third-degree burns and even some second-degree burns require medical attention. The following is a list of burns that may be considered critical and must be treated immediately:

- Burns involving breathing difficulty
- Burns covering more than one body part
- Burns to the head, neck, hands, feet, or genitals
- Burns (other than a very minor one) on a child or an elderly person
- Burns resulting from chemicals, explosions, or electricity

Heat Stress

In the summer of 2003, Europe experienced a heat wave that was responsible for nearly 3000 deaths in France from illnesses related to heat stress. In the United States, approximately 175 deaths occur each year as a result of illness brought on by heat stress. Although stress from cold temperatures is sometimes encountered during hazmat response, heat stress is more prevalent, more dangerous, and more readily adapted to by a fit, acclimatized individual. Heat-related illnesses are sometimes serious and can be fatal; emergency responders should prepare for working in hot environments by learning the prevention, symptoms, and treatment of heat-induced illnesses.

Sources of Heat

Two sources of heat, environmental and metabolic, can cause an increase in body temperature. Both sources can be controlled to some extent during a hazardous materials response.

Environmental Heat. Environmental heat comes from sources outside the body. It may be present in the form of high ambient air temperature, infrared radiation from the sun or heat-generating equipment, or heat from a fire or chemical reaction. The impact of environmental heat can be different for each first responder and is dependent on the physical characteristics of the responder and on such environmental factors as ambient temperature, humidity, and wind. Procedures suggested by health professionals to reduce environmental heat during hazmat incidents include providing shade for rest breaks and selecting personal protective equipment that is not unnecessarily burdensome.

Metabolic Heat. Metabolic heat is produced by the body during cellular activities. It is increased when the body's workload increases. In work situations where tasks are repetitive and can be anticipated, the workload can be measured and the amount of metabolic heat produced predicted so that schedules can be adjusted in hot weather. The workload reduction is based on a formula that estimates total heat stress using combined environmental heat and predicted metabolic heat. Emergency responders do not work in highly predictable situations, and this sort of estimate would be difficult, if not impossible, for their duties. Responders can, however, implement good work practices, because metabolic heat production can be decreased by working more slowly and utilizing more human or mechanical help on heavy tasks.

Workers who are in good physical condition can perform tasks more efficiently with less effort. Because the energy expended in work efforts generates heat and adds to the metabolic heat load, physically fit workers are able to work harder and for longer periods with less metabolic heat production than workers who are not physically fit.

Thermoregulation

The human body has a thermoregulatory system that maintains body temperature at approximately 98.6° Fahrenheit. The "thermostat" is located deep within the brain, in an area called the hypothalamus. The hypothalamus receives input from specialized temperature-sensing cells throughout the body, and if temperature drops or rises, this part of the brain directs other organs to initiate processes to warm or cool the body. Two cooling responses occur when they are stimulated by the brain after an increase in body core temperature. These are vasodilation (enlargement of blood vessels) in the skin and sweating.

Vasodilation. Vasodilation is the enlargement of tiny blood vessels called capillaries. Enlargement of capillaries in the skin allows the volume of blood at the body surface to increase. The blood, having been warmed as it circulated through internal parts of the body, can now dissipate heat by radiation into the air. Heat radiation is effective whenever the temperature of the air is below 95° Fahrenheit.

Sweating. If surface vasodilation is not effective in lowering body temperature, the sweating response is stimulated. Sweat glands begin to produce sweat and deliver it to the skin surface via ducts where it evaporates, changing from liquid to vapor. Heat is used to change water into water vapor in the evaporative process; the heat is taken from the body. If the surrounding air is very humid, at or approaching saturation with water vapor, sweat will not evaporate and no cooling takes place. Because the message from the sensory cells to the brain still reads "Hot," sweat continues to be produced even though it does not evaporate.

Heat-Induced Illness

In situations where the body's thermoregulatory mechanisms are unable to reduce core temperature, illness will result. The illness may be heat rash, heat syncope (fainting), heat cramps, heat exhaustion, or heatstroke.

Heat Rash. Heat rash occurs when sweat is prevented from evaporating, usually by clothing. The outer skin layer of dead cells becomes saturated, swelling and clogging the sweat ducts. Sweat glands continue to produce (remember that temperature reduction is the only thing that will turn sweat production off), and sweat seeps out into the deep skin layers and causes irritation. Tiny red raised blisters appear on the surface of the skin, and prickling sensations are felt. Allowing the skin to dry under cool, dry conditions will stop heat rash. Because clothing is required for skin protection during emergency response, it may not be possible to prevent heat rash.

Heat Syncope. Heat syncope usually occurs when a person is standing still in a hot environment. Skin vasodilation results in an increase in the volume of blood in these vessels and, with blood also collecting in the legs and feet, the resulting shortage of blood to the brain may lead to dizziness and fainting. This is not a serious heat illness unless it is accompanied by symptoms of heat exhaustion; it can be prevented by simply walking around. Resting in a reclining position allows blood to circulate to the brain again, and full recovery is prompt and complete.

Heat Cramps. Heat cramps occur in working muscles that are deficient in sodium, potassium, calcium, or other electrolytes. Profuse sweating may deplete the body of electrolytes, especially in an unacclimatized person, causing electrolyte-deficient muscles to contract involuntarily. Treatment involves immediate infusion of intravenous electrolyte solutions. Orally administered electrolyte replacement drinks or salted water will probably resolve the problem quickly. To prevent muscle cramps, a balanced diet should be eaten and responders should acclimatize themselves to heat.

Heat Exhaustion. Heat exhaustion is a more serious state characterized by fatigue, nausea, headache, dizziness, pallor, and profuse sweating. It usually occurs in the setting of sustained exertion in hot conditions when an individual is dehydrated from insufficient water intake. Oral temperature may be elevated but is not in every case. Heat exhaustion can progress to heatstroke if the victim is not cooled. The individual suffering from heat exhaustion should rest in a cool place and drink water until urine volume indicates that body fluids have been replaced. He should not return to active work that day.

Heatstroke. Heatstroke is a serious condition requiring immediate medical treatment. The sweat resources have been depleted, so the skin is dry and hot but may still appear red because of vasodilation. Heatstroke quickly leads to collapse, delirium, and coma. Eighty percent of people who suffer full-blown heatstroke die, and half of the surviving twenty percent suffer brain damage.

The onset of heatstroke may be rapid, with very little warning to the victim that a crisis stage has been reached. Because the first symptoms include confusion and impaired judgment, the buddy system is vital in hot environments.

Heatstroke victims must be cooled immediately while waiting for emergency medical service personnel. Clothing should be removed and the victim continuously sprinkled with water or wrapped in a thin, wet sheet. Fanning to increase air movement and evaporation should be initiated. The patient should not be immersed in extremely cold water, since this may trigger vasoconstriction in the skin and shut down the one cooling mechanism still operating.

Prevention of Heat-Induced Illness

Reducing the Heat Stress. Working in hot environments cannot be totally prevented in hazmat response, but it can be reduced. Isolation of the worker from hot environments is not possible, because hazmat response is conducted at unpredictable and uncontrollable times. It is possible, however, to reduce the amount of time a responder spends in hot conditions by enforcing frequent rest breaks and to reduce metabolic heat by providing help with tasks that necessitate heavy work.

Reducing the Individual's Response to Heat Stress. Once environmental and metabolic heat have been reduced as much as possible by the means suggested above, the hazmat responder depends on physiological responses to cool his body. Beneficial responses can be enhanced, and dangerous responses reduced, by consideration and control of the factors that influence an individual's response to heat stress.

HYDRATION (providing water to body cells) is an important factor in one's response to heat stress. Drinking water must always be available to responders, and they should be reminded to drink water often, both on and off the job. It is possible to lose one liter of fluid each hour by sweating. Because the body's total fluid volume is only around 40 liters, it is easy to see why fluid must be replaced. If hypohydration (inadequate body fluid) exceeds one and a half to two percent of the body weight, tolerance of heat stress begins to deteriorate, heart rate and body temperature increase, and work capacity decreases. When hypohydration exceeds five percent, it may lead to collapse. Because the feeling of thirst is not an adequate guide for water replacement, responders working in heat should be encouraged to drink water every 15 to 20 minutes. The daily amount of hypohydration can be estimated by measuring body weight after work and comparing it to that morning's baseline weight; it should not exceed one and a half percent.

In the process of sweating, electrolytes are lost along with water. Electrolytes are minerals such as sodium, chlorine, potassium, and calcium. Sports and energy replacement drinks can be consumed to replace electrolytes; however, these types of drinks often contain large amounts of sugars that must be digested before the water can be absorbed. The best way to maintain a balance of water and electrolyte replacement is to drink water frequently and eat a diet rich in fruits and vegetables. At a hazardous waste site, workers should be provided copious amounts of different beverages that appeal to the workers and that do not promote dehydration. It is important to ensure that workers are drinking plenty of liquids when working in a hot environment.

One of the most important physiological factors causing hypohydration is alcohol and caffeine consumption. Both of these substances are diuretics that depress the body's production of a kidney hormone that prevents water loss. When the hormone is in short supply, more urine is produced and more water is lost from the body. Responders who consumed alcohol the night before are already on the way to hypohydration when they arrive at a hazmat incident.

ACCLIMATIZATION to hot conditions has been found to be effective in preventing heat-related illnesses. The process takes five to seven days, during which work hours in the hot environment are gradually increased. During seven days of acclimatization, heart rate and body temperature become lower while the individual is performing the same work. The major physiological mechanism involved in acclimatization is sweating; acclimatized persons sweat sooner and more profusely and produce more dilute sweat than unacclimatized individuals. Unacclimatized individuals lose four times the volume of electrolytes in a liter of sweat than acclimatized persons lose.

Acclimatization schedules may be difficult to adhere to, given the sporadic nature of emergency response. Industrial hazmat teams whose members will be called on to respond in hot areas should limit membership to individuals who can acclimatize. It is inappropriate to assign persons who work entirely in air-conditioned environments to active positions on hazmat brigades whose responses can be expected to include heat stress.

AGING results in a more sluggish response of the sweat glands that leads to less effective control of body temperature. One study of five years' accumulation of data on heatstroke in South African gold mines found a marked increase in heatstroke with increasing age of the workers. Men over 40 made up less than ten percent of the workers, but they accounted for fifty percent of the fatal and twenty-five percent of the nonfatal cases of heatstroke.

DRUG CONSUMPTION, whether prescribed or over-the-counter can interfere with thermoregulation. Almost any drug that affects central nervous system activity, cardiovascular reserve, or body hydration can potentially affect heat tolerance. Some over-the-counter medications, including those that lessen the symptoms of colds or allergies, fall into this group. A responder taking medications should seek the guidance of a physician in evaluating tolerance to heat.

Alcohol was mentioned above regarding its effect on hydration. Alcohol has been commonly associated with the occurrence of heatstroke. Some drugs other than alcohol that are used on social occasions have been implicated in cases of heat disorder, sometimes leading to death.

NONHEAT DISORDERS, such as degenerative diseases of the cardiovascular system and diabetes, pose an extra danger to responders when they are exposed to heat, particularly when a stress like hard work is imposed on the cardiovascular system.

PHYSICAL FITNESS, in the absence of disease or the use of medications or drugs (including alcohol), is the most effective way for an emergency responder to improve heat tolerance. All of the components of overall fitness are important; the two most critical for heat fitness are body composition and aerobic capacity.

It is well established that excess weight and obesity predispose individuals to heat disorders. Fat provides an insulating layer that prevents heat dissipation, adds weight requiring more energy to carry, and changes the body's surface-to-volume ratio, further reducing surface heat dissipation.

Trained individuals with relatively high aerobic capacity produce dilute sweat and sweat more profusely than untrained persons, thereby cooling their bodies better. They also generate less metabolic heat for a given amount of work because they can do it more easily and efficiently. Their cardiovascular system delivers oxygen to cells more efficiently, thereby allowing more blood to be used for cooling purposes.

INDIVIDUAL VARIATION affects acclimatization. Studies have been done to determine whether gender affects heat acclimatization. The results of these studies indicate that it does not: Men and women can acclimatize equally well if they meet the other heat acclimatization criteria. However, individuals vary in their heat stress tolerance. In all experimental studies of the responses of humans to hot environmental conditions, wide variation in responses has been observed. These variations are seen not only between different individuals but also to some extent within the same individual exposed to high heat stress on different occasions. Such variations are not totally understood.

Screening for Heat Tolerance. It would be helpful to know which individuals are likely to succumb to heat stress before making assignments to an emergency response brigade. This determination should be accomplished without placing hazmat team applicants in potentially harmful hot conditions.

The National Institute for Occupational Safety and Health (NIOSH) recommends that all individuals who will be required to work under hot conditions have a medical examination before assignment to the job. The examination should include:

- A comprehensive work and medical history, with special emphasis on information about any previous heat illnesses or heat intolerance
- A comprehensive physical examination that gives special attention to the skin, liver, and kidneys and the nervous, cardiovascular, and respiratory systems
- An assessment of the use of prescription, over-the-counter, or social drugs, including alcohol
- An assessment of obesity
- An assessment of the worker's ability to wear any protective clothing and equipment that might be required in the job and may add to heat stress

A physician will be able to use a medical examination to determine whether a person is in good medical condition to work in a hazardous environment. Along with the medical exam, physicians can use screening techniques that can further evaluate cardiovascular and respiratory fitness and can also offer advice on training techniques that can be used to improve the ability to withstand work in hot environments.

Monitoring for Hot Environments. Hot environments can produce both heat stress and heat strain. Heat stress is the external heat load on a body due to environmental conditions. Heat strain is the body's response to the external heat load.

Heat stress can be monitored by evaluating environmental conditions, and heat strain can be evaluated by medical monitoring.

ENVIRONMENTAL MONITORING for heat stress measures the external heat load on a person in a hot environment. Various methods can be used by safety and health professionals to assess heat stress, but the most common method used is the wet bulb globe temperature (WBGT). The WBGT is a calculated value based on air temperature, humidity, air velocity, and radiation heat from the sun. This value can then be compared to recommended values to help determine work and rest cycles for hazardous materials workers. Electronic instruments, as shown in Figure 8.1, are often used to collect data that are used to calculate a WBGT value.

During an emergency response, the task of measuring a WBGT value may be impractical, and the site safety personnel may rely on medical monitoring to ensure the safety of the responders. During prolonged work on a hazardous waste site, however, site safety personnel may use WBGT measurements to help determine the heat stress conditions as they change throughout the day, to ensure the safety of the hazardous waste site workers.

MEDICAL MONITORING for heat strain should be carried out during an emergency response. Several physiological parameters can be measured to judge an individual's response to heat. The two parameters that are easiest to monitor under incident conditions are heart rate and oral temperature. Baseline pulse and temperature should be taken at the beginning of every workday for comparison.

Table 8.2 provides suggested times for monitoring breaks. The environmental temperatures in the table were measured with a plain thermometer that does

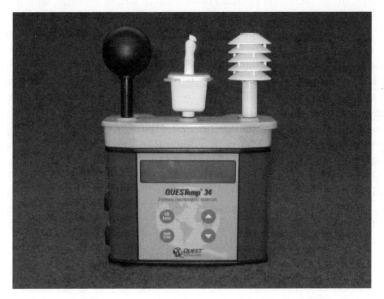

Figure 8.1   A WBGT instrument used to measure wet bulb globe temperatures.

TABLE 8.2   Suggested Frequency of Physiological Monitoring for Fit and Acclimatized Workers[a]

Adjusted Temperature[b]	Normal Work Ensemble[c]	Impermeable Ensemble
90°F (32.2°C) or above	After each 45 minutes of work	After each 15 minutes of work
87.5–90°F (30.8–32.2°C)	After each 60 minutes of work	After each 30 minutes of work
82.5–87.5°F (28.1–30.8°C)	After each 90 minutes of work	After each 60 minutes of work
77.5–82.5°F (25.3–28.1°C)	After each 120 minutes of work	After each 90 minutes of work
72.5–77.5°F (22.5–25.3°C)	After each 150 minutes of work	After each 120 minutes of work

[a] For work levels of 250 kilocalories/hour.
[b] Calculate the adjusted air temperature (ta adj) by using this equation: ta adj °F = ta°F + (13 × % sunshine). Measure air temperature (ta) with a standard mercury-in-glass thermometer, with the bulb shielded from radiant heat. Estimate percent sunshine by judging what percentage of the time the sun is not covered by clouds that are thick enough to produce a shadow. (100% sunshine = no cloud cover and a sharp, distinct shadow; 0% sunshine = no shadows.)
[c] A normal work ensemble consists of cotton coveralls or other cotton clothing with long sleeves and pants.
Source: NIOSH/OSHA/USCG/EPA. 1985. *Occupational Safety and Health Guidance Manual for Hazardous Waste Site Activities.* Washington, DC: U.S. Government Printing Office, p.8-2.

not take humidity into consideration. For responders not wearing impermeable ensembles, times should be adjusted downward in humid weather, with monitoring and rest occurring more frequently.

Temperature is the best indicator of response to heat stress. Deep-body temperature is the temperature of the body's innermost organs and is the most reliable indicator of body core temperature. It is taken by one of three ways: by monitoring rectal temperature, by inserting a wire thermometer into the nose or mouth and down the esophagus to heart level, or by swallowing a radio-telethermometer. The methods of monitoring deep-body temperature are not feasible in hazmat situations. The difficulty of taking a deep-body temperature makes taking oral temperature the preferred method of assessing temperature during a hazardous materials incident.

Oral temperature should be measured at the end of a work period, before drinking, by placing a clinical thermometer under the tongue. If the oral temperature exceeds 99.6°F, the next work period should be one-third shorter than the last and the rest period should stay at the same length. If the oral temperature still exceeds 99.6°F at the end of the next work period, the following work period should again be shortened by another third. Monitoring and adjustment of work periods should continue until the worker's initial temperature at the end of a work period is less than 99.6°F. No personnel should be permitted to wear semipermeable or impermeable clothing if oral temperature exceeds 100.6°F.

Heart rate provides an additional indication of heat strain. Heart rate can be detected at several pulse locations; the most easily accessible are the radial pulse on the inside surface of the wrist and the carotid pulse just on either side of the Adam's apple. Pulse should be taken during the first minute of seated rest after work and should not exceed 110 beats per minute (bpm). If it does exceed 110 bpm, the following work cycle should be shortened by one-third. Monitoring and shortening work cycles should continue until the responder's heart rate during the first minute of rest is lower than 110 bpm. A method that can be used to determine whether an individual is recovering from being overheated involves taking his pulse during the first and third minutes of seated rest. If the initial heart rate is over 110 bpm, but falls by 10 or more bpm by the third minute, one would conclude that the work is strenuous but the individual's thermoregulatory system is responding adequately to the stress.

Cold Stress

Just as emergency responders work in hot environmental conditions, they also respond to hazardous materials incidents when the temperature is low. Water is often present at an incident, in the form of rain or as part of the response, and wet conditions exacerbate cold injury. Responders may be injured by working in cold conditions and must take appropriate steps to reduce this hazard.

Conditions Leading to Injury

Three factors must be considered in assessing the danger of cold injury during a hazmat response: air temperature, wind velocity, and the presence of water in the work area.

Temperature. Although it is obvious that physiological damage from cold occurs more readily at low temperatures, it is impossible to define a temperature at which injury may begin. Assessment of the cold injury hazard must include consideration of the other important factors, wind velocity and the presence of water. Hypothermia has been known to occur in air temperatures as high as 65°F, or in water at 72°F, especially when the individual was fatigued.

Wind Velocity. The body responds to cold exposure by trying to slow heat loss. Heat loss by radiation from the body is greatest when the difference between body temperature and air temperature is great. Because wind constantly blows warmed air away from the skin's surface, the temperature difference and thus the heat loss is greatest when it is windy. Wind chill is used to describe the chilling effect of moving air in combination with low temperature. For example, an air temperature of 10°F with a wind velocity of 15 miles per hour (mph) is equivalent in chilling effect to still air at −18°F. As a general rule, the greatest incremental increase in wind chill occurs when a wind of 5 mph increases to 10 mph. Table 8.3 shows the cooling power of wind.

TABLE 8.3 Cooling Power of Wind on Exposed Flesh Expressed as an Equivalent Temperature Under Calm Conditions

Estimated Wind Speed (in mph)	Actual Temperature Reading (°F)											
	50	40	30	20	10	0	-10	-20	-30	-40	-50	-60
	Equivalent Chill Temperature (°F)											
Calm	50	40	30	20	10	0	-10	-20	-30	-40	-50	-60
5	48	37	27	16	6	-5	-15	-26	-36	-47	-57	-68
10	40	28	16	4	-9	-24	-33	-46	-58	-70	-83	-95
15	36	22	9	-5	-18	-32	-45	-58	-72	-85	-99	-112
20	32	18	4	-10	-25	-39	-53	-67	-82	-96	-110	-121
25	30	16	0	-15	-29	-44	-59	-74	-88	-104	-118	-133
30	28	13	-2	-18	-33	-48	-63	-79	-94	-109	-125	-140
35	27	11	-4	-20	-35	-51	-67	-82	-98	-113	-129	-145
40	26	10	-6	-21	-37	-53	-69	-85	-100	-116	-132	-148

(Wind speeds greater than 40 mph have little additional effect.)

LITTLE DANGER
Maximum danger of false sense of security

INCREASING DANGER
Danger from freezing of exposed flesh within 1 minute.

GREAT DANGER
Flesh may freeze within 30 seconds.

Trench foot and immersion foot may occur at any point on this chart.

Source: U.S. Army Research Institute of Environmental Medicine, Natick, MA.

The Presence of Water. When a person becomes wet, body heat is lost even faster; in fact, this is the reason why sweating is an effective means of heat dissipation in hot environments. Water conducts heat away from the body 240 times faster than still air does. An emergency responder whose clothing and skin are wet is at greater risk of cold injury than one who remains dry. Even inside boots and garments that are impervious to water, the body cools faster if the outer surface of the clothing is losing heat to water. A responder standing in water wearing waterproof boots can lose a great deal of heat from his feet, as the interior air space of the boot is cooled by heat loss through the boot material.

Another source of water on the skin is perspiration, if the responder has become overheated working inside protective clothing. Removal of the clothing exposes the wet skin to cold air, resulting in rapid cooling through the evaporation of sweat. A responder who was overheated can rapidly become chilled in this way.

Thermoregulation in Cold Environments

The same portion of the hypothalamus of the brain that sets cooling mechanisms in motion when core temperature rises also initiates thermoregulatory responses when core temperature drops. The two warming mechanisms are surface vaso-constriction and shivering.

Vasoconstriction. Because warmed blood loses heat at the surface of the body, reducing the flow of blood to the surface decreases heat loss. When body temperature drops, skin capillaries constrict; as they become smaller in diameter, they hold a smaller volume of blood. Heat is retained inside the body. One of the dangers of becoming exhausted in a cold environment is that the vasoconstrictive protective mechanism becomes overwhelmed, resulting in sudden vasodilation and rapid heat loss.

Shivering. Metabolic heat is produced by working cells; working muscle cells are set into motion by the thermoregulatory system when body temperature drops. Shivering is the involuntary contraction of muscle fibers. These contractions produce energy in the form of heat and help the body maintain its optimal temperature. Shivering is an obvious indicator that the body temperature has dropped.

Harmful Effects of Cold Stress

Responses to cold conditions range from discomfort to death. They include several stages of frostbite, may include trench foot under wet conditions, and can result in general hypothermia.

Frostbite. Frostbite is actual freezing of tissue due to exposure to extreme cold or contact with extremely cold objects. Wind chill can play an important role in accelerating frostbite. Frostbite is characterized by sudden blanching or whitening of the skin. If the tissues are cold, pale, and solid, deep frostbite has occurred; this is a serious injury that may lead to necrosis and loss of fingers, toes, or other affected parts.

Treatment of frostbite begins with slow, careful warming of the affected part. It should be immersed in water maintained at 102–105°F (comfortably warm to the inner surface of an unchilled forearm). Warming should be discontinued as soon as flushing indicates the return of blood. The injured part should be elevated after being warmed, and contact between the injured part and any surface except a sterile bandage should be prevented. The victim should not be allowed to walk on a frozen foot but should exercise a thawed part of the body by moving it around.

Trench Foot. Trench foot, also called immersion foot, is a condition resulting from long, continuous exposure to damp and cold while remaining relatively immobile. It can progress from a stage where the foot lacks an adequate blood supply to one in which gangrene occurs. Because emergency responders usually wear waterproof boots, and because they seldom stand still for long periods of time, they are unlikely to suffer from trench foot. The same condition can, however, affect the tip of the nose and ears under cold and wet conditions of immobility.

General Hypothermia. Individuals subjected to prolonged cold exposure and physical exertion are at risk from general hypothermia, the cooling of the entire body. Sweating and the fatigue of the vasoconstrictive response add to the hazard. Symptoms are usually exhibited in four stages: (1) shivering; (2) apathy, listlessness, sleepiness, and, sometimes, rapid cooling of the body to less than 95°F; (3) unconsciousness, glassy stare, slow pulse, and low respiratory rate; and (4) freezing of the extremities. Coma and death can result from general hypothermia if the cooling process is not reversed. The danger of hypothermia is increased by the consumption of sedatives and alcohol.

Prevention of Cold Injury
There is not a great deal a hazmat responder can do to acclimatize himself to working in cold temperatures. The body does not adapt physiologically to cold as well as it does to heat. Staying in good physical condition with an efficient cardiorespiratory system and not ingesting alcohol and other drugs will enhance an individual's thermoregulatory capability. Generalized prevention strategies focus on safe work cycles with adequate rest in a warm place and keeping a warm air layer around the body.

Work Rest Cycles. Work and rest cycles to prevent cold stress can be based on individuals' responses to the cold. When shivering begins, the body's temperature has dropped below optimum level. If shivering continues or grows stronger, an inability to warm the body is indicated and the worker is warned that it is time to move into a warmer location. Some sort of warm area should be provided at an incident site to enable chilled responders to warm up.

Protective Clothing. Many fabrics are available that greatly enhance body heat retention. The most efficient natural fibers for heat retention are silk and wool. However, synthetic fibers are available that have excellent insulating capabilities and offer the additional advantages of light weight and rapid drying. In addition to holding a warm layer of air next to the body, most synthetic insulating fabrics wick moisture away from the body to the outer surface of the garment. Garments made from these fabrics range in thickness from thin thermal underwear and glove liners to thick pile clothing. Many of these garments, which were developed for mountain climbing and other cold-weather outdoor sports, are extremely lightweight compared with natural fibers. They are most effective when worn under a tightly woven or otherwise wind-impervious outer garment.

It should be noted that at temperatures below 59°F the hands and fingers become insensitive long before cold injuries take place, thereby decreasing manual dexterity and increasing the risk of accidents. Hands and feet may be left with an inadequate blood supply and require especially well-insulated coverings when vasoconstriction in the extremities takes place in the body's effort to reduce heat loss. Sock liners of silk or polypropylene under double thicknesses of wool or polypropylene socks and boots that include insulating liners will help protect feet from cold. Glove liners and gloves with several separate layers provide the same protection for hands. Some of these can be worn under chemical-resistant gloves. Gloves that are tight or elastic should not be worn, as they inhibit blood flow when vasoconstriction has already occurred. The head and as much of the face as possible should always be covered with warm clothing, because these areas have many blood vessels and are locations where heat loss can be extensive.

Radiation

Radiation, also called electromagnetic radiation, is the transfer of energy through space and matter. The electromagnetic spectrum of radiation displays two distinct characteristics of wavelike and particlelike properties and can be described by frequency, wavelength, and photon energy. Wavelength is the distance between the ends of one complete cycle of a wave; frequency is the number of complete wave cycles that pass a point in space in one second; and photon energy is the electromagnetic energy present, expressed in electron volts (eV). The electromagnetic spectrum, as shown in Figure 8.2, is divided into two different categories of radiation: nonionizing and ionizing radiation.

Nonionizing Radiation
Nonionizing radiation has low photon energy, a long wavelength, and a low frequency. Nonionizing radiation does not have enough photon energy to ionize matter. The nonionizing radiation spectrum includes optical radiation (ultraviolet, visible, and infrared), laser, radio-frequency and microwave radiation, and extremely low-frequency radiation. Table 8.4 describes the different types, wavelength or frequency, biological and health effects, sources, and control of nonionizing radiation. Nonionizing radiation is measured with instruments appropriate to the type of radiation. Optical, radio-frequency, and extremely low-frequency radiation are measured with broadband instruments designed for each type. Complicated calculations are used to analyze potential laser hazards.

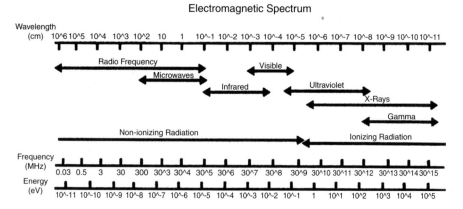

Figure 8.2 The electromagnetic spectrum.

Ionizing Radiation

Ionizing radiation is in the form of either electromagnetic waves or particulate energy that produces ions by interacting with matter. Ionization occurs when energy from a radiation source removes electrons from a neutral atom and results in the breaking up of an atom into charged components. A radiation source is called a radioactive isotope. Isotopes have the same number of protons, but the number of neutrons may vary, causing the atom to be either stable or radioactive. An isotope becomes radioactive if the nucleus will spontaneously change, releasing energy or nuclear particles in a process called radioactive decay. The nucleus will attempt to regain stability by shedding excess energy in the form of particles (alpha, beta, or neutrons) or electromagnetic radiation (gamma ray). Potassium-40, carbon-14, uranium-235, uranium-238, and tritium are naturally occurring radioactive isotopes found in the environment.

Types. Radiation can be in the form of a particle or an electromagnetic wave. Particle radiation includes alpha, beta, and neutrons. Electromagnetic wave radiation includes gamma and X ray.

ALPHA PARTICLES are the largest type of radiation particles, consisting of two protons, two neutrons, and zero electrons. Alpha radiation has a strong positive charge (+2) and causes ionization by pulling electrons from other atoms. Because of the relatively large size of the alpha particle, it does not travel very far in any type of material and it has limited penetrating power. Alpha radiation is not an external hazard to the human body; however, if an internal exposure occurs by inhalation or ingestion, cells and tissues of the body can be severely damaged.

BETA PARTICLES are smaller than alpha particles. They are very small electron-sized particles that are emitted from the nucleus. Beta particles have a negative charge and a lot of energy, and they travel at very high speeds. Beta particles

TABLE 8.4 Nonionizing Radiation

Types of Nonionizing Radiation	Wavelength (λ) or Frequency (f)	Biological and Health Effects	Sources	Control
Optical Radiation Ultraviolet (UV) Visible Infrared (IR)	(λ) 100 nm–1 mm (λ) 100 nm–400 nm (λ) 400 nm–770 nm (λ) 770 nm–1 mm	Skin, eyes, and immune system. UV can be beneficial in Vitamin D synthesis.	Sunlight, arc lamps, germicidal lamps, mercury-HID lamps, industrial IR sources, metal halide UV-A lamps, sunlamps, and welding arcs	Protect the skin and eyes. Isolation or enclosure of industrial source Sunblocks, protective eyewear, and clothing
Laser—Light Amplification by the Stimulated Emission of Radiation	Range can be from (λ) 180 nm to 1 mm but is usually a single λ or color.	Skin and eyes. Retinal hazard if λ is between 400 and 1400 nm, including behavior effects, temporary impairment of vision, and scarring resulting in permanent visual impairment. Can be reflected from or transmitted into skin.	Medicine, dentistry, science, industry, communication, construction, education, entertainment, criminal justice, military applications	Necessary for class 3b and 4 lasers that have open-beam paths and limited open-beam paths
Radio frequency Microwave Radio waves	(f) 300 GHz–3 kHz (f) 300 GHz–300 MHz (f) 300 MHz–3 kHz	Human data is limited and presents no clear trends. Animal studies show effects in nervous, neuroendocrine, reproductive, immune, and sensory systems.	Natural such as sun, galaxies, lighting, human body. Man-made radiators such as antennas, leakage from improper maintenance	Shielding, bonding, and grounding Control duration of exposure, increase distance to exposure, restrict access. Protective footwear, clothing, and gloves
Extremely low frequency	(f) 3000 Hz–3 Hz	Little consensus on biological and health effects. Effects on visual, nervous systems, and melatonin have been noted. Studies on effects of cancer are inconclusive.	Generators, power lines, power plants, high-amperage equipment.	Shielding, separation, cancellation of source, prudent avoidance

react with electron clouds of other atoms by pushing out electrons from their orbit around the nucleus. The high energy of beta particles allows them to travel up to 10 feet in air and also allows them to penetrate thin materials such as paper and human skin. Beta radiation is an external as well as internal hazard to human health.

NEUTRON PARTICLES emitted at high speeds from the nucleus produce neutron radiation. Neutron particles ionize atoms by direct collisions with electrons that result in the formation of a thermal neutron. A thermal neutron is then capable of neutron activation, which "turns" an atom into a radioactive isotope. Neutron radiation is considered a whole body hazard and has a slightly high penetrating power.

GAMMA AND X RAYS are electromagnetic rays with very short wavelengths that are made up of pure energy photons. Gamma and X rays travel at the speed of light and interact with atoms by colliding with electrons and knocking them out of their atoms orbit. Gamma and X-ray radiation penetrate most materials very well and can only be blocked by very dense materials. Because of their penetrating power, gamma and X-ray radiation are considered a whole body hazard; all the organs in the human body can be penetrated and damaged.

Radioactive Contamination. Radioactive contamination occurs when a material that contains radioactive atoms is deposited in any undesirable place, such as skin, clothing, or soil. Radioactive material is material that contains radioactive atoms and is not considered contamination as long as it is properly contained. There are three types of radioactive contamination:

- Fixed—contamination that cannot be easily removed from surfaces and cannot be removed by casual contact
- Removable—contamination that can be removed from surfaces easily; any object that comes into contact by casual contact, brushing, wiping, or washing with it may become contaminated or may transfer it. Air movement across the contamination may cause it to become airborne.
- Airborne—contamination that is suspended in air

Recognition of Sources. Radiation comes from sources of radioactive isotopes. Radioactive isotopes come from natural and man-made sources and nuclear reactors. Naturally occurring sources of radiation include cosmic radiation, sources in the earth's crust, sources in the human body, and radon. Radon gas, an alpha emitter, is given off by uranium ore and can be found in the soil and air. Man-made sources of radiation include tobacco products, medical radiation sources, building materials, and the domestic water supply. Other man-made sources that contribute a small amount of radiation are consumer products such as smoke detectors and industrial sources such as welding rods and X-ray machines for baggage inspection. The third major source of ionizing radiation is from nuclear reactors. Table 8.5 lists the annual radiation doses from natural and man-made

TABLE 8.5 Annual Radiation Doses from Natural and Man-Made Sources

Natural Background Source	mrem/yr
Cosmic radiation	26
Terrestrial (Earth's crust)	28
Internal sources (body)	40
Radon	200

Man-Made Sources	mrem/yr
Smoking (tobacco products)	1300
Medical X rays	40
Medical diagnosis and therapy	14
Building materials	7
Domestic water supply	5

sources. The total average dose to an individual from natural and man-made sources of radiation is approximately 360 mrem/yr.

Measurement Devices. Ionizing radiation can be either detected or measured to determine the total amount of radiation exposure received. Survey meters are used to detect whether radiation is present and at what rate. Dosimeters measure the total exposure or accumulated dose of radiation received by a person. During a hazardous materials incident, it may be necessary to detect the presence of radiation in an area and to monitor a responder for a possible accumulated dose of radiation.

SURVEY METERS are used to detect the interaction of radiation with matter by measuring the intensity of a field of radiation, called the exposure rate. Two common types of survey meters (also called rate meters) are the ion chamber and the Geiger–Mueller meters. The ion chamber (Fig. 8.3) measures the exposure rate due to gamma radiation; it is not able to measure exposures from alpha or beta radiation. The presence of gamma radiation causes ionization of gas in the detector to occur. This ionization causes a split between the positively charged gas molecules and the negatively charged electrons from the gas. The charged molecules and electrons are collected on electrodes and create an electrical current that can be measured. The electrical current is proportional to the gamma dose rate, and the exposure dose is displayed in roentgens per hour (R/h). If radiation contamination is suspected, a portable ion chamber should be used by responders on initial entry to determine whether large levels of gamma radiation are present. The ion chamber is not accurate in low-level radiation fields, and it will not measure normal background radiation.

The Geiger–Mueller meter (Fig. 8.4) measures the exposure rate due to alpha, beta, and gamma radiation. This type of survey meter has a Geiger tube (or G-M

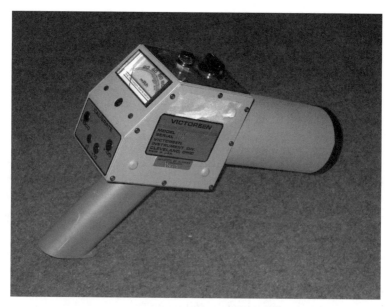

Figure 8.3   A portable ion chamber survey meter for measuring gamma radiation.

Figure 8.4   A Geiger–Mueller survey meter for measuring alpha, beta, and gamma radiation.

tube) that is composed of gas and a high-voltage electrode. Alpha, beta, or gamma radiation enters the tube and causes ionization to occur, which causes electrons to become free and accelerate toward the high-voltage electrode. The accelerating electrons then ionize gas molecules that free more electrons to accelerate toward the high-voltage electrode. This series of ionization of electrons results in short, intense pulses of current that are counted by the instrument. The number of pulses per second is an indication of the intensity of the radiation field. Alpha and beta sources display an exposure rate of counts per second (cps), counts per minute (cpm), or disintegrations per minute (dpm), and gamma sources display an exposure rate of milliroentgens per hour (mR/h). The Geiger–Mueller meter is not good to use for measuring high-exposure areas, but it can measure small areas of radiation contamination.

It is possible for a source of radiation to emit more than one type of radiation. Shielding techniques can be used to determine whether the source is alpha, beta, or gamma radiation. Various types of Geiger–Mueller meter probes can be used to help determine the type of radiation present. One type of probe blocks alpha radiation from the instrument and uses a sliding metal cover to distinguish between beta and gamma radiation. Another type of probe blocks both alpha and beta radiation and only responds to gamma radiation. The pancake probe (Fig. 8.5), made with a thin mica window, is a type of Geiger–Mueller probe used to detect low levels of radiation. This type of probe is shaped like a magnifying glass and is able to detect small amounts of alpha, beta, and gamma radiation.

Survey meters have been developed that are small, pagerlike detectors that can detect the presence of gamma and/or neutron radiation. These meters are designed

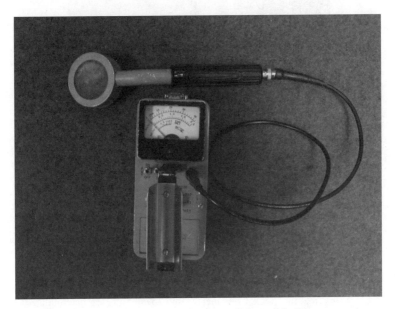

Figure 8.5   A pancake probe attached to a Geiger–Mueller survey meter.

to detect low levels of radiation by using a cesium iodide (for gamma) or a lithium iodide (for neutron) scintillator. This type of detector allows a personal monitor to be worn that measures dosage rate and might be useful in detecting low levels of gamma or neutron radiation before they cause a larger-scale problem. This type of monitoring device could be helpful for groups such as customs and border patrols, law enforcement, security officers, hazmat teams, and fire departments.

Various types and brands of the above-mentioned survey instruments are available from many manufacturers. The instruments require periodic calibration checks and daily background and operability checks. Site safety personnel should be knowledgeable in the selection, use, operation, and care of radiation survey instruments to obtain useful and meaningful readings.

DOSIMETERS are used to measure the total exposure or accumulated dose of radiation received by a person. Personal dosimeters may be read directly to provide an immediate assessment of exposure dose, or they may require processing to determine exposure dose.

The self-reading pocket dosimeter (Fig. 8.6) measures the total exposure or accumulated dose during an exposure to gamma radiation only and allows the results to be read immediately after the exposure is over. The instrument initially

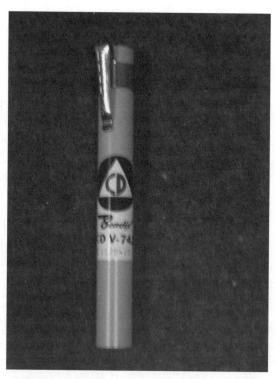

Figure 8.6 A self-reading pocket dosimeter used to measure an accumulated dose of exposure to gamma radiation.

is "charged" and shows a hairline at a specific point on the scale that is recorded before the exposure begins. When the dosimeter is exposed to radiation, ionization takes place inside the instrument, causing it to lose its charge, which moves the hairline indicator up the scale. The scale on the dosimeter is then read, and the dose received by the worker is recorded. The advantage of the self-reading pocket dosimeter is that the exposure dose result can be read immediately; however, this is not a permanent record and the dosimeter can be broken easily in the field.

Thermoluminescent dosimeters (TLDs) (Fig. 8.7) are used to assess an exposure dose that is determined after the detector has been processed in a laboratory. Results from a TLD are often used to determine the legal dose of record because they are the most accurate indicator of an exposure dose. The TLD can be used to measure exposures to high-energy beta, gamma, X-ray, and neutron radiation. On exposure, ionizing radiation transfers energy to electrons in the atoms of lithium fluoride chips inside the detector. During processing, the lithium fluoride chips are heated to 300°C, which allows light to be given off. The light emitted is proportional to the energy transferred by the radiation and is used to calculate the exposure dose. The advantage of the TLD is that it is rugged, reliable, and allows a wide range of doses to be measured. The disadvantage of the TLD is that it needs to be sent to a lab for processing, and after the TLD is read, the original record of exposure is lost.

The film badge (Fig. 8.8) is another type of personal dosimeter that needs processing before the exposure dose can be read. The badge is composed of X-ray film that contains either silver bromide or silver iodide. When the badge is developed it displays a curve that can be read and compared to calibrated film

Figure 8.7 A thermoluminescent dosimeter used to measure an accumulated dose of exposure to beta, gamma, X-ray, and neutron radiation.

Figure 8.8 A film badge used to measure an exposure dose of radiation.

standards. The film badge can be read as many times as desired, and the type of radiation that is exposed to it can be determined. Film badges can thus be used for any type of radiation and can also be a permanent record of an exposure dose.

Measurement. Radiation is measured in three different ways, by determining the source activity, the exposure, and the dose. The measurement we will focus on for this text is the dose. The dose of radiation can be measured in roentgen-equivalent-man (REM) or radiation absorbed dose (RAD). The emergency dose limits for rescue and recovery operations are given in terms of REM, the most common unit used for measuring dose equivalence. The REM pertains to the human body, and it applies to all types of radiation. The REM also takes into account the energy absorbed (dose) and the biological effect on the body due to different types of radiation.

Exposure Limits. Occupational exposure guidelines and exposure levels for radiation in routine conditions are given in various sources. However, any exposure to ionizing radiation has a probability of causing adverse health effects and radiation exposure should be limited. Situations may arise in which emergency responders are needed to rescue and control an area where high levels of radiation are present. The Environmental Protection Agency (EPA) has established dose guidelines for rescue and recovery operations (Table 8.6). The guidelines include the following principles:

- No individual shall be required to perform a rescue action that might involve substantial personal risk.

TABLE 8.6   EPA Dose Guidelines for Rescue and Recovery Operations

Dose Limit (whole body)*	Emergency Action Dose Guidelines Activity Performed
5 rem	All activities
10 rem	Protecting property
25 rem	Lifesaving or protection of large populations
>25 rem	Lifesaving or protection of large populations, only by volunteers who understand the risks.

*The whole body includes the top of the head down to just below the elbow and just below the knee.

- Personnel shall be trained to a level appropriate for the hazards in the area and any required controls.
- Personnel should be briefed beforehand on the known or anticipated hazards to which they could be subjected.

Control. There is considered to be no safe level of exposure to ionizing radiation; thus the most common approach to controlling ionizing radiation is by using the ALARA principle. The ALARA principle (as low as reasonably achievable) strives to reduce radiation exposure to the lowest possible level. The three components of the ALARA principal include time, distance, and shielding. A reduction in the amount of time spent around ionizing radiation decreases the exposure dose received and thus decreases the chance of having an adverse health effect from an exposure. An increase in the distance from a radiation source would also decrease the dose to an individual because radiation has a limited distance that it can travel from its source. The third principle, shielding, is often used in an occupational setting to control the exposure to ionizing radiation. Alpha particles are easily shielded with material as thin as a piece of paper, as long as there is no exposure by ingestion or inhalation. Beta particles can be shielded by use of materials such as plastic, glass, or metal foil. Gamma and X-ray radiation are harder to shield because of their high penetrating power. Gamma and X-ray radiation can be shielded by using highly dense materials such as lead, steel, and concrete.

Noise

Exposure to noise can be harmful when it is loud for a prolonged period of time or when there is a sudden intense acoustic event such as a gunshot, explosion, or fireworks. The harmful effects of noise exposure include temporary hearing loss due to short-term exposure to loud noises or permanent hearing loss due to long-term exposures to noise. Exposure to sudden intense noises can cause a sensation of ringing in the ears called tinnitus or acoustic trauma that can lead to a temporary or permanent hearing loss. During a hazardous materials incident, noise exposures can startle workers and drown out communication, both of which can contribute to the occurrence of accidents.

Sound Basics

Sound is any pressure variation in air, water, or various other media that the ear can detect. Sound travels in the form of waves, and the speed of sound is dependent on the elasticity and density of the medium. The speed of sound in air at sea level is 344 meters per second (m/s); however, in solids and liquids that are denser, sound travels faster. The speed of sound in water is 1500 m/s, and the speed of sound in steel is 5000 m/s.

Frequency. Frequency is the number of pressure variations over time and is measured in hertz (Hz), or cycles per second. Humans are able to hear frequencies ranging from 20 to 20,000 Hz. As we age, the ability to hear high-frequency sound decreases in a process called presbycusis. The average adult has difficulty hearing sounds above 14,000 Hz. The frequency composition of a sound heard by humans can be described by its spectrum. A narrow frequency spectrum can be harmful to hearing and a high-frequency spectrum tends to be annoying to humans.

Loudness. Loudness is dependent on sound pressure and frequency. The human ear is more sensitive to high-frequency sounds then it is to low-frequency sounds. Two different frequencies could have the same sound pressure level measurement, but the higher-frequency sound is often perceived to be louder. Loudness also varies with factors such as the person's health, the characteristics of the sound, and the person's attitude toward the generated sound.

Hearing. To be able to perceive and understand sound it has to be translated into hearing. The human hearing mechanism gathers, transmits, and perceives sound by using the outer ear, middle ear, inner ear, nerves, and brain. Sound pressure waves are first concentrated and converted into mechanical vibrations in the outer ear and eardrum. Bones in the middle ear amplify these vibrations and transmit them to the inner ear or cochlea. Hair cells in the cochlea respond to the mechanical vibrations and activate nerves that turn the sound into an electrochemical response in the inner ear. Vibrations of different frequencies will excite the hair cells at different points along the cochlea, thus enabling us to perceive different pitches of sound. The auditory nerve then transmits the activity to the brain, where the signal is interpreted into sound. The hair cells are very important in the process of translating sound into hearing, and as we get older or experience excessive noise exposures, those cells are damaged and no longer usable.

Sound Measurement

Equipment. Sound is measured by using many types of instruments. The three most commonly used instruments are the sound level meter, the octave band analyzer, and the noise dosimeter. A direct reading instrument, the sound level meter (Fig. 8.9) measures sound pressure variation in the air. The octave band analyzer is used to determine the distribution of frequencies in a noise pattern. The noise dosimeter (Fig. 8.10) is an instrument that can measure sound level pressures throughout the entire work shift to determine a person's accumulated noise exposure dose.

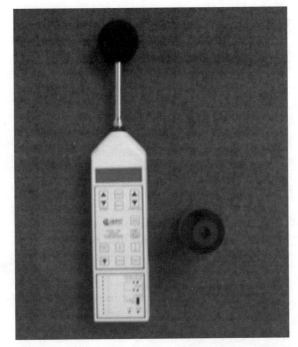

Figure 8.9 A sound level meter used to measure noise levels in decibels.

Figure 8.10 A noise dosimeter is used to measure an accumulated noise exposure dose.

TABLE 8.7   Sound Pressure Level Values for Some
Typical Sounds

Overall Sound Pressure Level	Example
0	Threshold of hearing
10	Sound testing chamber
20	Studio for sound pictures
30	Soft whisper (5 ft)
40	Quiet office; audiometric testing booth
50	Average residence; large office
60	Conversational speech (3 ft)
70	Freight train (100 ft)
80	Very noisy restaurant
90	Subway; printing press plant
100	Looms in textile mill; electric furnace area
110	Accelerating motorcycle at 1 meter
120	Hydraulic press; 50-HP siren (100 ft)
140	Threshold of pain; jet plane
180	Rocket-launching pad

Adapted from: B.A. Plog, J. Niland, and P.J. Quinlan, ed., *Fundamentals of Industrial Hygiene*, National Safety Council, Itasca, IL, Fourth Edition, 1996, p. 203.

Units. Sound pressure level measurement involves a complicated calculation that integrates pressure fluctuations of sound and reports these as decibels (dB). The decibel is a dimensionless unit that expresses the ratio of a logarithm of a measured quantity to a reference quantity. The typical range of sound pressure levels in decibels is 0 (threshold of hearing) to 140 (threshold of pain). Table 8.7 describes various sound pressure levels for some typical noise sources.

OSHA's Hearing Conservation Standard

The Occupational Safety and Health Administration (OSHA) was created in 1970 to promulgate and enforce occupational safety and health standards, whereas the National Institute for Occupational Safety and Health (NIOSH) was created to conduct research and to recommend new occupational safety and health standards. In 1972, NIOSH developed the document named *Criteria for a Recommended Standard: Occupational Exposure to Noise*. Based on the NIOSH criteria document, OSHA published the Hearing Conservation Standard (29 CFR 1910.95), the rules for controlling occupational noise exposure, in the *Federal Register* on October 24, 1974.

The standard requires employers to reduce noise exposure levels to below 90 dBA, for an eight-hour time-weighted average, by means of engineering and administrative controls or, if necessary, by personal protective equipment. The OSHA Hearing Conservation Standard sets the permissible exposure limit for occupational noise at 90 dBA as an eight-hour time-weighted average. A hearing conservation program is required if an eight-hour time-weighted average exposure is 85 dBA or greater.

The exposure limit of 90 dBA set by OSHA is for an eight-hour time-weighted average per day. The exposure limit is based on the principal that if the sound pressure level increases by five decibels, then the permissible duration time is decreased by fifty percent. For example, if the sound pressure level is 95 dBA, then the amount of time the noise exposure could occur is four hours per day. The important point to remember is that as the sound pressure increases, the allowable exposure time decreases.

Hearing Conservation Program

A hearing conservation program is designed to protect workers from hearing impairment that can result from noise exposure while at work. The hearing conservation program consists of noise monitoring, audiometric testing, hearing protectors, training, and recordkeeping.

Noise Monitoring. Employers are required to monitor workers for noise exposure. The employer must identify workers who are exposed to a noise level at or above 85 dBA and is required to notify the workers of the noise monitoring results. Additional noise monitoring is required if there is a change in exposure conditions for a previously monitored employee or if a new employee is moved into an area of the worksite that has a noise level of 85 dBA or greater.

Audiometric Testing. The Hearing Conservation Program requires audiometric testing to be done as a baseline test and then annually for workers who are exposed to noise levels of 85 dBA or greater. A baseline audiogram is done within six months of an employee working in an environment with a noise level of 85 dBA or greater and is used to compare future audiograms to determine whether there has been any loss of hearing ability.

Hearing Protectors. Hearing protectors are required to be available to all workers who are exposed to noise levels at or above 85 dBA. The employer is required to provide the hearing protection and will ensure that the worker is shown how to use it properly.

Training and Recordkeeping. Workers exposed to noise levels of 85 dBA and greater are required to be trained annually in the areas of:

- The effects of noise
- The purpose, advantages, disadvantages, and attenuation characteristics of various types of hearing protectors
- The selection, fitting, and care of protectors
- The purpose and procedures of audiometric testing

Records of noise monitoring and audiograms must be kept by the employer. Noise exposure results must be kept for at least two years, and audiogram results must be kept for the duration of the worker's employment.

Control of Sound

Sound control efforts can be applied at any of three points: the source, the path, or the receiver. Controlling noise at the source is the most desirable and can be accomplished by modifying equipment or by introducing noise-reduction measures at the design stage of new equipment. Blocking the path of noise can reduce the exposure to the receiver by using shielding or enclosing the source, increasing the distance between the source and the receiver, and also by placing acoustical material on walls, ceilings, and floors to absorb rather than reflect sound waves. The third point of noise to control is the exposure to the receiver, specifically the human ear. The ultimate goal of noise control is to reduce noise exposure to the human ear to levels at which no injury to the ear or loss of hearing occurs.

Noise Control Measures. The preferred method of controlling any kind of workplace hazard is by the use of engineering controls. If a hazardous condition can be designed out of the workplace, this eliminates any potential for the hazard to cause an injury or illness. Situations arise when engineering controls are not feasible or take a long time to develop, and the next type of control that is preferred is administrative followed by the least desirable control, personal protective equipment.

Engineering Controls. In theory, it is desirable and easier to design machinery that is less noisy than to fix existing machinery. This cannot always be done, and a few steps can be taken to reduce the noise output from old machinery. Proper maintenance, substitution of machine parts, substitution of process, and a reduction in the force of vibrating surfaces are engineering controls that can help reduce a noise source. Other types of engineering controls include enclosures such as a soundproof booth that can either prevent noise from getting inside or keep noise from getting outside.

Administrative Controls. The second most preferred method of controls is administrative controls. The main purpose of administrative controls is to reduce the employee's overall time of exposure to noise. Methods of reducing time of exposure can include changing the production schedule, job rotation, scheduling machine operating times, or transfer of susceptible employees to other locations.

Personal Protective Equipment. The final type of noise control, and the most frequently used, is personal protective equipment. Personal protective equipment is only meant to be used while engineering controls are implemented. Types of hearing protective devices include inserts or ear plugs, canal caps, and earmuffs. Hearing protectors do nothing to reduce or eliminate the noise hazard, and employees must be trained in the proper use and care of their hearing protective devices.

More information about personal protective equipment for noise is provided in Chapter 12, Personal Protective Equipment.

SCENE-RELATED HAZARDS

Various types of hazards that may cause accidents to occur are common at hazardous materials response scenes. The types of accidents that may occur depend on the scene conditions, the operation performed, and the equipment used. It is impossible to list all potential accidents that might occur; however, a useful approach is to classify accident types into the following general categories:

- Falls to the same level on which working or to a lower level
- Struck by or against objects in motion, projectiles, or stationary equipment
- Caught in, on, or between rotating equipment parts, projecting parts from stationary equipment, moving equipment and stationary objects, or pinch points on equipment in operation
- Contact with charged electrical equipment
- Exposure to chemicals, toxic atmospheres, noise, extreme heat or cold
- Overexertion through lifting, pulling, pushing, or fatigue

Vehicles and Heavy Equipment

Vehicles such as fire engines, hazmat trucks, ambulances, and cars can all pose a danger to emergency response personnel, victims of a hazardous materials incident, and by-standers. Care should be taken to avoid injuries occurring from the arrival on-scene of any vehicles. Roads and pedestrian pathways that are near a hazardous materials scene should be blocked off and secured by emergency personnel.

Heavy equipment that may be present at a hazardous materials incident may include bulldozers, backhoes, forklifts, and powered hand tools. Various kinds of industrial equipment may be present to help stop a release or help clean up a hazardous materials spill. Caution should be taken to avoid contact with both aboveground and belowground power lines, utility lines, sewers, and public works systems.

Falls from vehicles, heavy equipment, and slippery areas where a spill may be present can easily occur, especially if the responder is outfitted in cumbersome PPE such as Level A or B gear. Procedures such as using a ladder to climb up onto a railcar (Fig. 8.11) instead of using the railcar-mounted ladder should be followed to prevent injuries resulting from falling to the ground.

Hazardous Energy

Hazardous energy can be released from industrial equipment and tools that use energy. The equipment used at a hazardous materials incident has the potential to injure workers by releasing electrical, mechanical, hydraulic, pneumatic, chemical, or thermal energy. The primary OSHA standard covering the control

Figure 8.11 A ladder should be used to climb up onto a railcar.

of hazardous energy sources is 29 CFR 1910.147: The Control of Hazardous Energy (lockout/tagout).

Electrical energy may be used in a number of ways during a hazardous materials incident including the use of equipment ranging from small power tools to electrical motors and from pumps to large pieces of remediation equipment. Electrical hazards can lead to five kinds of accidents:

(1) Shock or electrocution
(2) Ignition of flammable or explosive materials
(3) Overheating, causing burns to people or equipment
(4) Electrical explosions
(5) Activation of equipment at the wrong time

Electrical shock is very painful, and sometimes deadly. The following is a list of factors that influence how severe a shock will be:

• The voltage involved: As little as 30 V is enough to push current through your body.

- The amount of current available from the source: Lighting and equipment circuits have enough current to produce a fatal shock.
- The path the current takes through your body: A current path through your heart is more dangerous than a path that does not pass through your heart.
- The resistance of your skin: Wet skin has less resistance and will contribute to a stronger shock than dry skin.

Lockout/Tagout

The lockout/tagout procedure can be used to avoid accidents occurring from exposure to hazardous energy. The lockout/tagout procedure is a method of controlling energy during service and maintenance of equipment, tools, pipes, and any other places where a worker may be injured by the flow of electricity or product. Energy types to which this standard applies include electrical, mechanical, hydraulic, pneumatic, chemical, or thermal energy. Pipes carrying hazardous chemicals are covered by this standard.

Lockout. A lockout is a method of using a locking device (Fig. 8.12) to keep equipment from unexpectedly being set in motion and endangering workers. The steps involved in locking out equipment include:

- A disconnect switch, circuit breaker, valve, or other energy-isolating mechanism is put in the safe or off position.
- A device is often placed over the energy-isolating mechanism to hold it in the safe position.
- A lock is attached, so that the equipment cannot be energized. Use a separate lock for each person who works on the equipment.

Tagout. In a tagout procedure, the energy-isolating device is placed in the safe position and a written warning, or tag (Fig. 8.12), is attached to it. Tagout should not be used on equipment that can have a lockout; lockout is preferred. Tagout is allowed if the employer can clearly demonstrate that it renders the equipment completely safe. Some old equipment may be made in such a way that a lockout cannot be applied. Equipment bought or modified after January 2, 1990 must be able to be locked out.

Figure 8.13 shows an example of a process that has been properly locked out by using these steps that must be followed when applying the lockout/tagout procedure:

(1) Prepare for shutdown by knowing the type and magnitude of energy, its hazards, and the method or means to control it.
(2) Shut down the machine or equipment, using procedures established by the site or location safety and health program.
(3) Isolate the machine or equipment with a physical block or device.
(4) Apply the lockout or tagout device to hold the block in place.

Figure 8.12   An example of a lock and a tag that can be used to control the release of hazardous energy.

Figure 8.13   An example of a lock and tag in place to control the release of hazardous energy.

(5) Relieve any stored or residual energy.

(6) Verify that isolation and de-energization have been done.

Excavations

Excavations are sometimes necessary to uncover buried drums, underground containers of hazardous materials, or utility pipes or for various other reasons. Excavation work can be difficult and dangerous, and excavation accidents have resulted in many injuries and deaths. The OSHA excavation standard is included in Subpart P of OSHA's Construction Standards (29 CFR 1926.650, 651, and 652).

Hazards

An excavation is any man-made cut, cavity, trench, or depression in the Earth's surface formed by earth removal. A trench is a narrow excavation in which the depth is greater than the width and the width is usually less than 15 feet. Hazards of excavations include:

- Heavy equipment in operation
- Contact with energy or materials in buried wires or pipes
- Electrical or fire hazards
- Slips and falls
- Falling objects
- Cave-in of trench walls
- Oxygen-deficient atmospheres
- Accumulations of toxic and/or explosive gases or vapors

General Precautions

Important precautions must be taken before excavations are made:

- Check the area to be excavated for underground pipes and wires. Consult with utility companies as needed to locate buried lines. Use remote-sensing equipment, if needed.
- Determine site conditions, such as soil composition, with special emphasis on conditions that may lead to cave-in. This can be done with soil maps and soil sampling.
- Follow an excavation safety plan for dealing with excavation hazards. Modify or update the plan as required by changing site conditions.
- Train all involved employees in safe excavation procedures. Training should cover factors such as:
  - Utility line locations
  - Cave-in prevention measures
  - Recognition of conditions that may cause cave-in
  - Clues to impending cave-in (such as tension cracks, bulging walls, etc.)

Avoiding a Cave-In

The occurrence of a cave-in is the major concern for workers in excavation operations. Responders should learn the conditions that could lead to a cave-in and take precautions to prevent such a cave-in from occurring. A cave-in is likely if the soil is unstable, has high water content, or has been subjected to freeze-thaw or frost heaving. Other dangerous conditions to consider are loading of trench walls with heavy equipment, supplies, structures, or dirt piles; vibration caused by equipment operating near excavations; overly deep trenches or high trench walls; and trench walls that are too steep.

Cave-in can be prevented by sloping the walls of the excavation at a stable angle. Three different slope angles of the walls, A, B, and C (Fig. 8.14) can be used and are determined on the basis of the type of soil at the excavation site. Material that would slide into the excavation during a cave-in is removed as the excavation is made. Benching, the process of cutting walls into a series of steps, can also be done to prevent a cave-in.

Vertical trench walls over five feet deep are not allowed by OSHA unless the wall is solid rock or a restraining system is used. Use of a restraining system (Fig. 8.15) can prevent a cave-in by holding back material in the trench walls. A shoring system

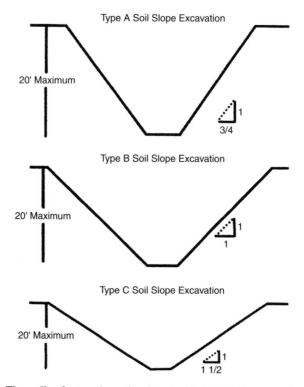

Figure 8.14   The walls of a trench can be sloped with three different angles (A–C) based on the soil type at the excavation site.

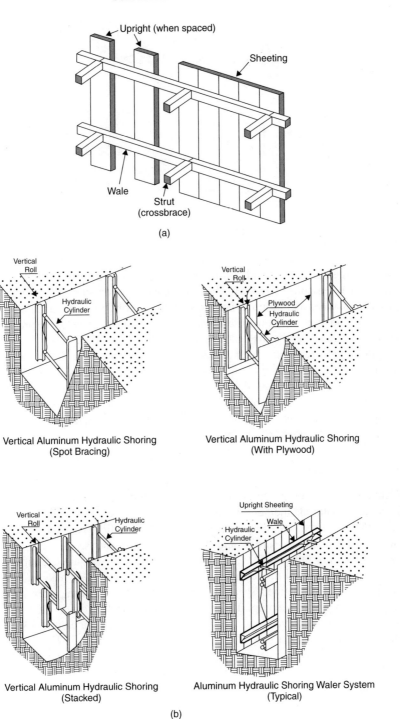

Figure 8.15 Restraining systems consist of timber shoring (A), aluminum hydraulic shoring (B), or a prefabricated trench box (C).

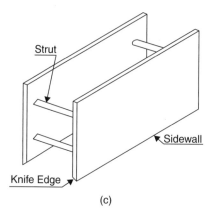

(c)

Figure 8.15   (*continued*)

and a moveable trench shield or box can be used as a restraining system. A shoring system may be constructed with timber components on-site or may consist of a premanufactured aluminum hydraulic system. Shoring must be installed from the top down and removed from the bottom up. Trench boxes or shields are prefabricated enclosures that can be moved along an excavation as needed.

Additional precautions (e.g., added shoring bracing or a flatter slope angle) should be utilized whenever site conditions indicate that they are needed to prevent cave-ins. Excavations should be inspected by a competent person daily and after any event (such as rainfall) that may increase the likelihood of a cave-in. If inspection indicates the potential for a cave-in, all work should cease until appropriate precautionary measures are taken.

Confined Spaces

Confined spaces are another type of work hazard that may be present during a hazardous materials incident. From 1980 through 1989, 585 fatal confined-space incidents occurred, resulting in a total of 670 fatalities. NIOSH estimates that 60% of the fatalities were would-be rescuers. In response to these grim numbers, OSHA promulgated 29 CFR 1910.146, the Permit-Required Confined Space Standard, which went into effect in April, 1993. The standard has clear and strict requirements for confined space definition, hazard assessment, permit program, key personnel, and training. This section is an introduction to what a confined space is. If an employee's job requires regular entry into a confined space, then more training is necessary.

Definition
According to the OSHA confined space standard, a confined space must have all of these characteristics:

- Large enough that an employee can enter it and perform assigned work

- Has limited or restricted means for entry or exit
- Is not designed for continuous employee occupancy

Some confined spaces do not pose a hazard to persons entering them; however, there are other enclosed spaces that can be deadly. OSHA considers these spaces to be permit-required confined spaces. A space does not have to be completely enclosed to be considered a permit-required confined space. Pits and open-topped vats fit the OSHA description because they may have limited means of entry and exit, are not intended for occupancy, and can have hazardous atmospheres. A permit-required confined space is a confined space that has one or more of the following characteristics:

- It contains or has the potential to contain a hazardous atmosphere:
  - Flammable gas, vapor, or mist in excess of 10% of its lower flammable limit (LFL)
  - Airborne combustible dust at a concentration that obscures vision at a distance of five feet or less (considered potentially flammable)
  - Oxygen concentration of less than 19.5% or more than 23.5%
  - Hazardous substances in excess of OSHA's permissible exposure limits
  - Any other atmospheric condition that is immediately dangerous to life and health
- It contains a material, like water or grain, that could engulf an entrant.
- It is built so that it could trap or asphyxiate an entrant in a narrow bottom.
- It contains any other recognized serious safety or health hazard.

The confined space standard has requirements for entry and permitting and even defines the key personnel who are necessary for an entry into a confined space. The confined space standard assigns responsibility to each of the key personnel in the entry process with the intent that the entry will be made safely and with no complications.

Key Personnel
The key personnel involved in a confined space entry are the entry supervisor, the attendant, and the entrant. The entry supervisor is in charge of the entry and must ensure that the conditions of the permit have been satisfied. The entry takes place under the direction and supervision of the entry supervisor. The attendant remains just outside the confined space, in a position to communicate with and continuously monitor entrants. The attendant must recognize immediately when entrants need help and summon rescue personnel if needed. The authorized entrant must understand the hazards in the confined space, be trained in the proper use of all equipment, and communicate with the attendant.

Rescue
Arrangements with rescue personnel must be made before an entry can occur. The entry supervisor must ensure that the communication mechanism for summoning

rescue personnel works and that the rescue service or equipment is available and functional. More than half of the fatalities in confined spaces are untrained rescuers. The confined space standard allows a choice of two options for rescue:

- The employer can select, train, and equip a company rescue team. An on-site team will be available for rescue quickly if needed.
- The employer can choose to call an outside contractor or agency (such as the fire department) when rescue is needed. If this second option is selected, the employer must be sure that the outside team is properly trained and equipped and can respond in a timely manner. A written agreement should be signed that describes the arrangements that have been agreed upon.

A hazardous materials incident may require responders to enter a confined space whether acting as an entrant, an attendant, or a supervisor of confined space entry. The responder should know the requirements of the standard and be able to evaluate the space to ensure that permit conditions are met. The fatality risk for entry and rescue is high and can be controlled only by strict adherence to the requirements of the standard.

MUSCULOSKELETAL INJURIES

Ergonomics is the science of fitting workplace conditions and job demands to the capabilities of the working population. Musculoskeletal disorders are just one aspect of ergonomics. Musculoskeletal disorders are a group of conditions that involve the nerves, tendons, muscles, and supporting structures, including disks located in the vertebrae of the back. NIOSH defines musculoskeletal disorders with the following conditions:

- Disorders of the muscles, nerves, tendons, ligaments, joints, cartilage, or spinal disks
- Disorders that are not typically the result of any instantaneous or acute event, such as a slip, trip, or fall, but reflect a more gradual or chronic development
- Disorders diagnosed by a medical history, physical examination, or other medical tests that can range in severity from mild and intermittent to debilitating and chronic
- Disorders with several distinct features, such as carpal tunnel syndrome, as well as disorders defined primarily by the location of the pain (e.g., low back)

Upper Extremity Disorders

Musculoskeletal disorders of the upper extremities include carpal tunnel syndrome, wrist tendonitis, and rotator cuff tendonitis. The upper extremities include the neck, shoulders, elbows, hands, wrists, and fingers. The risk factors associated with upper extremity disorders include repetition, force, posture, static loads, mechanical stress, low temperatures, and vibration. Controls that can be implemented to help reduce the occurrence of musculoskeletal disorders include improving the fit and reach of the workstation, work organization, and wearing gloves to protect against vibration and cold.

Lower Back Disorders

Musculoskeletal disorders of the lower back are common and very costly. In the United States, back disorders account for 27% of all nonfatal occupational injuries and illnesses involving days away from work. Lower back disorders can be caused by overexertion activities such as lifting improperly or carrying too much weight. The risk factors that lead to lower back disorders include posture, frequency, repetitive handling, static work positions, improperly placed handles, asymmetric handling, space confinement, and wearing personal protection equipment. NIOSH has developed an equation that can be used to determine the Recommended Weight Limit (RWL) for a particular job task. This can be calculated for work environments that do not change very much, such as in a manufacturing environment. In a hazardous materials incident, the best way to ensure that lower back injuries do not occur is to be trained in proper lifting techniques and container handling as discussed in Chapter 15, Advanced Hazardous Materials Control.

PREVENTING ACCIDENTS

An accident is any unexpected event that interrupts the work process and may result in injury, illness, or property damage to the extent that it causes loss. An accident cannot always be predicted or prevented during a hazardous materials incident; however, techniques such as using a job safety analysis and a standard operating procedure can help reduce the occurrence of accidents.

Job Safety Analysis

Job safety analysis (JSA) is used by safety personnel to review job methods and uncover hazards. Once the safety and health hazards are known, the proper controls can be developed. The JSA breaks the job into steps and identifies hazards leading to the recommended action or procedure. The JSA is done by using the following five step procedure:

(1) Select the job to be analyzed.
(2) Break the job down into individual steps.
(3) Identify the hazards and potential accidents of each step.
(4) Develop safety precautions that can be used to prevent accidents.
(5) Put the safety precautions into effect in the workplace.

Table 8.8 is an example of a JSA worksheet that can be completed to analyze a specific job task. A JSA is a very useful tool that can identify hazards in most

TABLE 8.8 Example of a Job Safety Analysis Work Sheet

Title of Job/Operation _____ Date _____ No. _____

Position/Title(s) of Person(s) Who Does Job _____ Name of Employee Observed _____

Department _____ Analysis Made By _____ Analysis Approved By _____

Sequence of Basic Job Steps	Potential Accidents or Hazards	Recommended Safe Job Procedures

1. Struck By (SB)
2. Struck Against (SA)
3. Contacted By (CB)
4. Contact With (CW)

5. Caught On (CO)
6. Caught in (CI)
7. Caught Between (CBT)
8. Fall—Same Level (FS)

9. Fall to Below (FB)
10. Overexertion (OE)
11. Exposure (E)

322

work environments; however, the unpredictable nature of a hazardous material incident may make creating and following a JSA challenging.

Standard Operating Procedures

Standard Operating Procedures (SOPs) are intended to provide uniform instructions for accomplishing a specific task. SOPs identify what should be done, how it should be done, and who should do it. SOPs effectively communicate proper procedures for a variety of tasks to other workers in an organization, and also to emergency responders who may be required to respond at an organization.

An example of an incident that illustrates the importance of developing and using effective SOPs occurred at a natural gas distribution facility several years ago in a semirural area of Kentucky. At 11 o'clock one weekday morning, an explosion occurred at the facility, killing the entire crew, including the maintenance personnel, an engineer, a computer specialist, and the plant manager. The fire department arrived to find the site devastated, with no apparent survivors. Because the SOPs for the facility were all tied to the computer system, which did not survive the explosion, no one knew exactly what to do to shut down the system. After several hours of searching for an individual with knowledge of the facility, a night watchman was brought in to assist in locating the shutoff valves. If there is any possibility that this type of accident could occur, then an SOP should be developed to notify the other shift personnel for assistance. The several hours of delay could have been crucial in preventing further mitigation of the gas.

Emergency response personnel generally rely on verbal safety instructions and use existing SOPs until, time permitting, the plan can be modified to fit the site-specific situation. The OSHA HAZWOPER Standard (29 CFR 1910.120) states that the first responder trained to the operations level, at a minimum, shall have an understanding of relevant SOPs. Federal regulations require facilities that use or store hazardous materials to provide Local Emergency Planning Committees (LEPCs) with the information needed to prepare and maintain emergency plans. Those facilities that store quantities of extremely hazardous substances in excess of their designated threshold planning quantities are required to appoint a facility emergency coordinator to assist the LEPC in its planning efforts.

Other facilities may also be required by State Emergency Response Committees (SERCs) to participate in the planning process under Title III of SARA. In addition, hazardous material technicians and specialists are required to meet all requirements of the preceding levels; therefore, each responder must have a working knowledge of all relevant SOPs applicable to an emergency response. Chapter 3 provides a thorough discussion of emergency planning requirements.

SOPs for handling various types of incidents should be a part of emergency plans and should include the roles and responsibilities of responders, lists of available equipment, evacuation procedures, and decontamination capabilities, among others. A good plan will include accurate information about the identity, location, and characteristics of hazardous materials, related processes, and all response capabilities available. Maintaining an up-to-date plan is of prime importance. When corrections, additions, or changes are made, they should be recorded in a simple bookkeeping style so that all plan users will be sure that they are using a current plan. Not only must a variety of technical tasks be conducted efficiently to mitigate an incident, but they must be accomplished in a manner that protects the worker. For SOPs to be effective, they must be:

- Written in advance, because it is impossible to develop and write safe, practical procedures under stress while responding to an incident
- Based on the best available information, operational principles, and technical guidance
- Field-tested, reviewed, and revised when appropriate by competent safety professionals
- Concise, understandable, feasible, and appropriate
- Distributed to and understood by all personnel involved in response activities
- A part of the regular training of response personnel that is required by the applicable EPA standards and OSHA Standard 29 CFR 1910.120

Emergency response personnel along with site safety and health personnel can greatly increase their ability to prevent accidents by following SOPs. Appropriate equipment and trained personnel, combined with SOPs, can help reduce the possibility of harm to response workers.

SUMMARY

Physical hazards of emergency response can cause injuries and must be considered and prevented if possible. The physical hazards discussed in this chapter included temperature extremes such as heat and cold stress, noise, and radiation. Scene-related hazards of hazardous energy, confined spaces, and musculoskeletal injuries are dangerous when encountered during a hazardous materials incident and, if recognized, can be properly controlled or prevented. The tasks of completing a job safety analysis and standard operating procedure can help emergency response personnel recognize the presence of physical hazards and can increase their ability to prevent accidents from occurring.

9

AIR SURVEILLANCE

INTRODUCTION

The senses of the human body are not appropriate for locating and measuring the presence of hazardous materials in air (Fig. 9.1). Although some chemicals may briefly form a visible cloud, most are invisible when airborne even at very high concentrations. Likewise, our noses may detect some chemicals but they cannot accurately estimate the concentration. More importantly, many highly toxic, flammable, or otherwise dangerous chemicals have no odor and cannot be detected by our noses. The hazardous materials technician must use air surveillance equipment and procedures to assess the threat of airborne hazardous materials.

Environmental professionals responded to a call by a hotel chain about complaints of odors by hotel personnel and guests at one of their facilities. The hotel and surrounding development had been built on the site of a closed petroleum refinery approximately three years earlier. Complaints of odors from surrounding areas received local media attention, and the hotel management took seriously similar complaints among their employees and guests. The complaints followed a period of seasonal rain.

The responders surveyed the lodging suites with a combustible gas indicator (CGI) and a flame ionization detector (FID), which detects a variety of organic compounds. Although most of the individual suites did not show detectable contaminants, peak readings of over 1000 units on the FID (calibrated to methane)

Emergency Responder Training Manual for the Hazardous Materials Technician, Second Edition, edited by Kenneth W. Oldfield
ISBN 0-471-21387-X   Copyright © 2005 John Wiley & Sons, Inc.

Figure 9.1 The human senses are not safe and appropriate for detecting hazardous chemicals.

and 2–3% of the lower explosive limit (LEL) on the CGI were taken in a few locations, especially near plumbing and other conduits to the ground. Occupants were moved out of the suites in which contaminants were detected, and periodic monitoring was conducted in all suites.

Air samples, using charcoal tubes, were taken from the units with the highest survey readings and from an area considered to represent background conditions. These samples were sent to a laboratory that had agreed to provide 24- to 48-hour turnaround for analytical results. The samples were analyzed for common petroleum constituents, and the total contaminants in any of the samples did not exceed 1 ppm. However, readings of the FID taken concurrently with the air sampling had ranged from 10 to 100 ppm. Colorimetric indicator tubes for the above contaminants exposed during the air sampling failed to produce a color change, indicating that no contaminant was detected. Both the FID and the air sampling equipment were calibrated before and after use.

The discrepancy in readings and analytical data was later attributed to methane gas emissions that were forced out of the soil by the rising groundwater level after the rain. Methane gas is a by-product of the decomposition of hydrocarbons and would be detected by the FID, but would not adsorb to the charcoal in the air sampling tube. Gas bag samples collected later from an apartment near the site and analyzed in a laboratory yielded methane levels of 4000 ppm.

The incident illustrates the importance of understanding the use of air surveillance during a response. For example, relying on a single means of monitoring airborne hazards may result in misinterpretation of data or insufficient information to make sound decisions during a response. Therefore, a hazmat responder should be as familiar as possible with general principles and equipment of air surveillance in order to make quick, sound decisions during an incident. This chapter describes the principles and equipment used in air surveillance.

Technology continues to advance, and today's air surveillance equipment is much smaller and lighter than equipment available "in the old days." This equipment is also more sophisticated and reliable than ever before. Instruments are able to measure very small concentrations and provide important information rapidly. However, emergency responders must understand both the capabilities and the limitations of the available equipment to effectively incorporate air surveillance into their hazard assessment process. For instance, each type of sensor used in direct-reading instruments (DRIs) will respond to certain chemicals, but not all chemicals. Therefore, responders may not be able to measure some of the hazards at the scene.

This chapter introduces hazmat technicians to the equipment and procedures available to measure the airborne hazards at a release. Equipment is described in general terms and not by brand because the technology and features change so rapidly. Responders can keep current on the features of all available equipment through magazines and catalogs or by contacting equipment sales representatives.

In this chapter we can only introduce the field of hazardous materials air monitoring and acquaint the responder with basic principles. Maslansky (1993), Ness (1991), ACGIH (2001), and Hawley (2002) have provided excellent books that can give a more detailed discussion if it is needed. This chapter will address the following:

- General response procedures
- Air sampling with laboratory analysis
- Air monitoring with direct-reading instruments (DRIs)
- Interpreting and applying air surveillance results

A general approach to hazardous materials air surveillance is described first. These procedures apply to monitoring any type of chemical hazard at any scene. The operation and interpretation of kinds of air surveillance equipment is explained next. Each type of equipment has unique capabilities, limitations, and applications that determine whether and how it can be used in a hazardous materials response. Finally, the chapter discusses how to use the data and information gathered in air surveillance to assess the hazard to responders and to the public, including the issues that can result in misinterpretation of the readings.

GENERAL AIR SURVEILLANCE STRATEGIES

In an industrial setting, the identity of the hazardous material may be relatively easy to determine because the list of possibilities is limited. In a transportation incident, virtually any chemical might be transported through a jurisdiction and the identity of the released material may be more difficult to determine initially. Nonetheless, the various types of equipment available for air surveillance have limitations as to what they can measure. Therefore, before air surveillance is initiated, preliminary information gathered from visual observations, knowledge of the materials stored in the location of the release, container labels, or other clues should be used to plan the air surveillance activities. If possible, witnesses to the release, plant managers or engineers, or other individuals with information about the identity and specific hazards of the released material (e.g., chemical supplier representatives) should be interviewed. If the chemical or physical hazard can be identified before monitoring begins, valuable time can be saved and the proper equipment can be chosen. If the hazard cannot be identified exactly, preliminary information such as the physical state and approximate size of the release is still useful in selecting the appropriate level of protection for personnel entering the site to gather more information.

The first decision to be made in planning the air surveillance strategy is whether air monitoring should be attempted. Air surveillance involves time and effort that must be invested wisely during a response. In some cases, the monitoring equipment available to the hazmat team is not appropriate for the hazard (e.g., the instrument will not detect the hazard). For instance, if perchloroethylene, a noncombustible liquid, is spilled in an open area and the only monitoring equipment available to the team is a combustible gas meter, the time and effort invested in air surveillance could be better used to control the release. Also, if the released chemical is identified and judged unlikely to generate high concentrations of vapors, it may be decided that air monitoring is not necessary. This decision must only be made by professionals with a thorough understanding of the chemical and physical properties of all of the chemicals involved.

Once the decision to monitor for airborne hazards has been reached, the strategy for monitoring must be established. Preliminary information about the nature of the hazard involved in the incident should be reviewed to determine what information is missing in order to effectively evaluate the potential risk to the health and safety of the responders and the public. Objectives should be established that provide the specific information needed.

General monitoring procedures should be addressed in the standard operating procedures (SOPs) for the response team before the incident occurs. During the incident, the need is to determine how to adapt the SOPs to the specific conditions of the site. The air surveillance strategy may involve the types of monitoring discussed in the following sections.

Perimeter and Background Survey

As discussed in Chapter 11, the first responder to a release should establish an isolation perimeter large enough to include the hazard area and an additional buffer zone. Before response personnel enter the hazard area, an inspection of the hazard area should be conducted from the isolation perimeter to gather information about the identity and extent of the hazard (Fig. 9.2). This survey should use air monitoring equipment to detect air contaminants in two specific areas—upwind and downwind of the hazard area.

Upwind monitoring will provide data on background levels of contaminants from sources away from the site of the incident, such as emissions from other processes in the plant or sources located on property nearby. Background levels of contamination may be compared with readings taken on-site to determine what portion of the air contamination originated with the released material.

Downwind monitoring can be used to determine whether the isolation perimeter contains all of the hazard area. In other words, if air contamination greater than background levels is detected downwind of the hazard area, the isolation perimeter should be expanded to prevent exposure to unprotected people.

Figure 9.2   Responders should begin air surveillance at the boundary of the hazard area.

Air monitoring around the entire perimeter may not be necessary because most of the chemical contamination will be carried by the wind in a plume. The upwind and downwind monitoring should be sufficient to detect the plume. However, the release areas closest to the perimeter should be monitored regardless of the wind direction. Also, if radiation hazards are suspected, the entire perimeter should be monitored because radioactive emissions are affected only minimally by air movement. Perimeter monitoring should be included in the periodic monitoring described below in this chapter.

Initial Entry

To monitor conditions in the actual hazard area, an initial entry must be made. The objectives must be clearly outlined and communicated before the entry so that the team can efficiently conduct the survey and leave the hazard area quickly. The entry objectives for air surveillance may include:

- Establishing that airborne hazards exist or potentially exist at the site
- Locating and delineating areas of high air concentrations of the released material(s)
- Verifying preliminary or existing information with respect to the nature of the release
- Establishing boundaries for the site control zones based on visual observations of the current location and potential movement of the released materials
- Collecting information related to the specific protective measures and equipment required for response personnel
- Collecting information useful in choosing response actions

Preliminary information should be used to determine what specific hazards to monitor during the entry. If the identity of the hazardous material is known, DRIs or colorimetric indicators capable of monitoring the material should be selected. If the material cannot be specifically identified, the hazard it represents can be classified in one or more of the following groups:

- Combustible gases or vapors
- Oxygen deficiency or enrichment
- Toxic gases, vapors, or particulates
- Radioactivity

Air monitoring equipment capable of monitoring one or more of these hazards may be selected to provide general, initial detection of the possible hazards.

The initial entry should be a brief survey of the hazard area to identify and outline areas of high air concentrations or other immediately dangerous to life and health (IDLH) conditions such as confined spaces or potentially explosive environments. The team should note any visual indicators of the existence of a hazard such as pooled liquids, discoloration or degradation of the ground surface, condensed vapor clouds, peeling or staining of painted surfaces, and the location of any victims.

If the approximate location of the release is known, the entry team should approach from the upwind direction and monitor the air on a continuous basis (Fig. 9.3). Because chemical air contaminants are typically dispersed by wind, approaching the incident from upwind should allow the team to approach the release without entering the area of highest air contamination. Continuous monitoring is especially important if radiation hazards are suspected to be present, because the release of radioactivity is not significantly affected by wind direction.

Site conditions may be used to prioritize the hazards to be monitored if the chemical has not been identified. For instance, in areas with restricted air movement, the presence of an explosive and/or oxygen-deficient environment may pose the most immediate threat to the entry team. Because the entry team should be using at minimum Level B protection if the hazards are unidentified, the investigation of toxic atmospheres might be conducted after these hazards have been assessed and determined not to pose a life-threatening situation. On the other hand, if the area around the release is open and seems to be well ventilated, the accumulation of sufficient explosive vapors may be less likely and monitoring may focus on the comparatively lower-concentration toxic atmospheres.

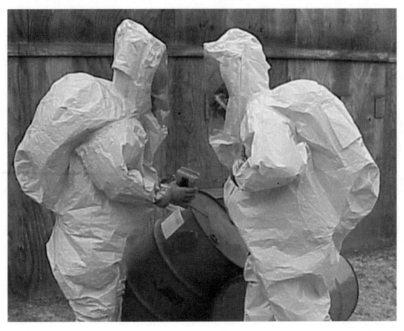

Figure 9.3   During the initial entry, responders should cautiously approach the release area from the upwind direction.

The actual use of monitoring equipment should follow general guidelines set forth in the team's SOPs. It is important to remember that some air contaminants are heavier than air and may accumulate near the ground. Air monitoring should be conducted near the ground surface and at the level of the breathing zone. Monitoring should emphasize areas where air movement is reduced; such as diked areas, ditches and low areas, buildings, or confined spaces, because vapors tend to accumulate in these areas. If radiation is being monitored, any debris or other materials that may be covering (i.e., shielding) the radioactive source should be monitored without being moved, as this may expose the team to unshielded radioactivity.

The team should periodically test that the instrument is functioning properly by using procedures recommended by the manufacturer. These procedures may include visual indicators such as lights or audible signals that sound at regular intervals or special alarms that sound in the event of instrument failure. Alarm sounds or lights that warn of hazardous conditions should always be functional and adjusted to the appropriate limits for the hazard of concern (e.g., below 19.5% oxygen). Alarms are discussed below in this chapter.

Periodic Monitoring

Air surveillance is an ongoing activity that does not end with completion of the initial entry survey. Even if the air concentrations in the hazard area are well defined, the changing nature of emergency response activities and site conditions makes periodic monitoring necessary. The objective of this type of monitoring is to detect changes that may positively (e.g., consistently lower concentrations allowing a downgrade to a lower protection level for responders) or negatively (e.g., detection of air contaminants leaving the hazard area, leading to an enlarging of the hazard zones) affect the emergency response activities. Some factors that may cause a change in air concentration during an emergency response include:

- Response actions such as accidental releases while moving drums or containers, handling spilled materials, or conducting containment or confinement procedures
- Changes in weather conditions such as rain, decreased cloud cover leading to hotter conditions, changes in wind direction and speed, or possibly sudden changes in atmospheric pressure
- Accumulation of spilled material because of continued release before containment procedures can be completed

Periodic monitoring may take the form of regular surveys of all or part of the site similar to the initial entry survey or may be performed as area monitoring at a fixed position near the area of highest hazard. Fixed-area monitoring is not practical if only one instrument is available and a large hazard area may be involved. Survey-style monitoring dedicates at least one person to conducting the survey at the expense of other activities in the response. The frequency and

Figure 9.4   The incident commander should include air monitoring data in the decision to terminate an emergency response.

method of monitoring should be site-specific and based on the degree of hazard present and the potential for changing conditions to affect the air concentrations of contaminants.

Termination Monitoring

The Incident Commander should include air monitoring data in the criteria for terminating response activities (Fig. 9.4). Air contamination may remain in the hazard area even after the source of the contaminant has been contained or confined. Air monitoring should be performed throughout the hazard area to confirm that all spilled material or sources of contamination have been contained. Also, before incident termination, air contamination should be reduced to background levels. At that time, normal traffic (non-response personnel and vehicles) is allowed into the response area. This is especially true of incidents in confined spaces or areas of poor ventilation where hazardous gases or vapors may accumulate.

AIR SAMPLING WITH LABORATORY ANALYSIS

Air sampling involves the collection of the contaminant from the air onto an appropriate sample medium (e.g., activated charcoal, liquid media in an impinger, filters) that is then sent to a laboratory. The contaminant remains associated with the medium until it is removed at the laboratory for analysis. The result of this analysis is the total mass (or volume) of contaminant divided by the total volume of air sampled and is reported as the average air concentration for the sample period. Laboratory analytical equipment can provide more accurate qualitative and quantitative analysis of air contaminants than most field air monitoring equipment.

The unpredictable nature and short duration of hazardous materials incidents makes the use of air sampling difficult. Responders must plan to collect the air samples by keeping some amount of the equipment and media described below on hand. Personnel responsible for air monitoring should be able to deploy the equipment rapidly. A qualified laboratory should be available to provide rush analysis. Laboratories are usually very helpful in providing information necessary to plan sampling events.

Sample Period

The length of time over which the sample is to be collected (sample period) will largely be determined by (a) the requirements of the analytical method and (b) the type of exposure concern. Most analytical equipment and methods require that a minimum amount of contaminant be present on the sample medium (minimum detection limit) to be detected. The amount of contaminant deposited on the sample medium is a function of the concentration of the contaminant in the air and the amount of air sampled. Therefore, sampling a low-concentration atmosphere will require a longer sample period to collect enough contaminant for analysis.

The other consideration for the sample period is the type of exposure of concern. If the suspected contaminant produces short-term (acute) effects after brief exposure to a high concentration, a short sample period should be used at times of expected highest exposure to minimize the effect of averaging over time. On the other hand, the contaminant may cause long-term or cumulative effects after longer exposure to lower concentrations. In this case, a long sample period would provide data on the average exposure for a longer period of time, perhaps the entire life of the incident. Planning and training before an incident occurs are necessary to make good decisions about the type of sampling to be used during an incident.

Sampling Systems

Two types of sampling systems are used for the collection of air samples: active samplers, which draw contaminated air through a sample medium by means of a pump, or passive samplers, which rely on natural forces such as diffusion or permeation to collect samples. These systems are described in the following sections.

Active Samplers

Active sampling systems mechanically draw air through a sampling medium that collects the contaminant. In this way, the contaminant contained in a large

Figure 9.5 The active sampling train includes a pump, tubing, and sample medium.

volume of air is concentrated onto the medium, which is then analyzed in the laboratory. An active air sampling system (Fig. 9.5) typically consists of the following components:

- Sampling pump to move the air
- Inert tubing to carry the air
- Sampling medium that collects the contaminant

Active sampling systems typically rely on battery-powered pumps to draw air through the sampling medium or into a sampler container. Most pumps now have some means of adjusting the air flow rate, which may be specified by the analytical method. Flow rates may vary from a few cubic centimeters per minute (cc/min) to over 10 liters per minute (10,000 cc/min). Pumps with the broadest range of flow rates will offer the most flexibility for air surveillance of a variety of contaminants.

The sampling pump must be calibrated before and after each use to ensure a constant flow rate. Unlike DRIs, active air samplers concentrate the contaminant on a sample medium for later analysis. To get a concentration value, the amount of contaminant analyzed by the laboratory must be divided by the total volume of air sampled. Therefore, it is essential that the flow rate be kept constant during the sample period so that the total volume of air can be calculated. Calibration of the pump flow rate is the means of ensuring that the flow rate is right and has not changed over the sample period. Because sample media and tubing may affect the air flow rate, calibration should be performed with the whole sampling system intact.

Passive Samplers

Passive samplers or dosimeters are becoming increasingly popular as alternatives to active sampling methods. These samplers have no moving parts and therefore are relatively easy to use and do not require calibration or maintenance. Some may be read directly, similar to colorimetric tubes, and others must be sent to a laboratory for analysis.

Passive samplers may be classified as either diffusion- or permeation-type samplers (Fig. 9.6) Diffusion samplers have a sorbent medium that is separated from the contaminated air by a grid section. This grid creates a layer of stagnant air between the contaminated air (high concentration) and the sorbent material (low concentration), forming a concentration gradient. If given the opportunity, chemicals will move from areas of high concentration to areas of low concentration by natural molecular forces. Thus the air contaminant will naturally move to the sorbent medium from the air.

Permeation samplers take advantage of similar natural forces, but the sample medium is covered by a membrane. The membrane may be permeated (passed through) by certain chemicals and not by others, thus screening out unwanted contaminants. As the chemical permeates through the membrane, it is collected by the sorbent medium on the other side.

Figure 9.6 Passive samplers may be either diffusion (left) or permeation (two on right) designs.

Sampling Media

Air sampling methods for gaseous and vapor contaminants make use of a variety of sample collection media, including solids, liquids, long-duration colorimetric tubes, and sampling bags (Fig. 9.7). When solid sorbents are used, contaminants in the air adsorb to the solid medium itself or to a chemical coating on the medium. Two widely used solid sorbents are activated charcoal and silica gel. Other solid media include porous polymers such as Tenax and Chromosorb and specialty sorbents for unique uses. The sampling and analytical method will specify the medium to be used.

Sampling with a liquid medium typically involves drawing contaminated air through a liquid that absorbs the contaminant. The liquid is contained in an impinger or a bubbler that allows for contact between the contaminant, which is either reactive or soluble, and the liquid reagent. The liquid is then sent to a laboratory for analysis.

Airborne particulates, including liquid and solid aerosols, are typically sampled with filter media. The two most common types of filters are fiber mesh and membrane (thin polymer membranes or sheets with tiny holes). As contaminated air is drawn through the filter medium, the particulates are impacted and trapped on the filter, which is sent to the laboratory.

In some cases it may be desirable to collect samples of the air in a gas bag for analysis outside the hot zone (Fig. 9.8). This may be the case when an unidentified contaminant is present in high concentrations. The use of multiple

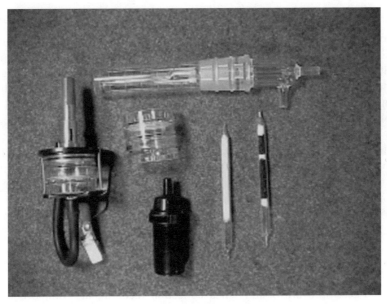

Figure 9.7 Different types of sampling media are used to collect different kinds of chemicals.

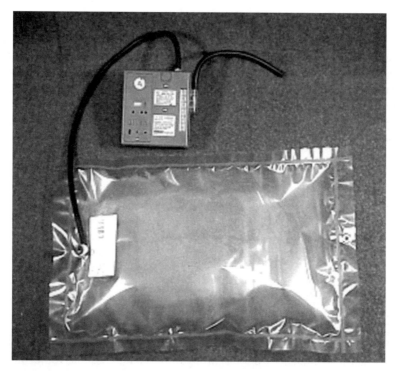

Figure 9.8 Air from the hazard area can be pumped into gas bags for analysis outside the hot zone.

colorimetric indicator tubes takes time and could easily be conducted by personnel outside of the hot zone, where a lower level of protection is appropriate. Analysis on-site by gas chromatography may also be desired. Shipment of the bag off-site to a laboratory may not be practical because (a) the bag may leak, (b) the chemical may permeate through the wall of the bag, or (c) the chemical may simply degrade over the time required for shipment. Therefore, any analysis must be conducted on-site.

The sample can be collected by attaching inert tubing to the exhaust port of a sampling pump and to the inlet of the gas bag and placing the pump inlet in the area to be sampled. This procedure draws contaminated air through the pump, so only pumps certified inherently safe should be used if explosive gases or vapors may be present. Also, the potential for incompatibility between the contaminant and the materials used in the pump must be considered.

Laboratory Analysis

Once air has been drawn through a collection medium, the sample container is sealed and sent to an analytical laboratory. Gas or vapors are then desorbed (removed) from the collection medium and run through analytical equipment (often a gas chromatograph with appropriate detector) for analysis. The analysis may be both qualitative (what is present) and quantitative (how much is present), depending on whether the contaminant has been identified.

The results are usually reported in mass units (milligrams, mg, or micrograms, μg) or in volume units (microliters, μl). If the total amount detected is divided by the total volume of air sampled (expressed as liters or cubic meters), an average concentration for the sample period is obtained. The units often used for air concentrations are milligrams of contaminant per cubic meter of air (mg/m^3). This is a mass (mg)-to-volume (m^3) relationship and is not the same as parts of contaminant per million parts of air (ppm), which is a volume-to-volume relationship. Results given as mg/m^3 may be converted to ppm based on the molecular weight of the contaminant and the temperature and atmospheric pressure during the sampling. Laboratory personnel can make this conversion if it is desired.

Samples of particulates may be analyzed in two ways. First, the exposed filter media is weighed on a very sensitive scale and that weight is compared to the preexposure weight of the filter. The added mass is the mass of the particulate in the air that was sampled. The sample mass is divided by the total volume of air sampled to obtain an average concentration. This is called gravimetric analysis and is usually used for dust particles.

The other particulate analysis involves fiber counting. A filter that has been exposed to contaminated air is turned clear, and the fibers on a portion of the filter are counted under a microscope. With mathematical formulas, the number of fibers per cubic centimeter of air is calculated. This type of analysis is most commonly used for determining airborne asbestos concentrations.

Using Air Sampling Data

The point of air surveillance is to gather information that can be used to make decisions during an emergency response. Air monitoring with direct-reading instruments provides immediate information, but the information is usually not chemical specific and the accuracy of these instruments varies. Air sampling data provide concentration data that are chemical specific and accurate at lower levels.

If the samples can be analyzed during the response, the data may be compared directly to the appropriate criteria to make decisions about protective measures. For instance, evacuation decisions can be made by comparing concentration data of a specific chemical to the *Emergency Response Planning Guide* for the chemical as discussed below in this chapter. Selection of proper protective equipment for emergency response then can be based on the specific hazardous characteristics and known concentrations of the released material. Potential health effects can be predicted and monitored if the chemical identity and exposure data are available.

Even if laboratory results are obtained too late for use during the incident, the data may be useful in clean up efforts to select the appropriate protective equipment based on actual exposure data. Also, the data may be used to help prepare for future responses to similar incidents. Emergency responders could estimate air concentrations during a future incident involving the same material based on air sampling results obtained during the initial incident. To make the comparison, sufficient information about site conditions must be reported to allow knowledgeable individuals to assess the impact of differing conditions (e.g., higher temperature, larger quantity release, higher humidity) on the concentration.

AIR MONITORING WITH DIRECT-READING INSTRUMENTS

Because the physical properties of the chemical are interacting with the changing environmental conditions, the chemical concentration in the air changes constantly. This variability means that responders need equipment that gives immediate information but also responds to changes in the concentration. Direct-reading instruments (DRIs) provide such real-time measurement. A discussion of DRIs and their use follows.

Basic Operation of DRIs

Many manufacturers develop and sell DRIs. As in other technologies, these instruments are becoming smaller, lighter, smarter, faster, and, in some cases, less expensive. Although the packaging and features may vary, all instruments have at least these components:

- Sensor(s)
- Circuitry
- Battery power
- Display
- Audible and visual alarm

These are the basic elements of any DRI (Fig. 9.9). There is a chemical or physical interaction between the sensor and the chemical in the air that produces or affects an electronic current or signal. The signal is converted by circuitry and microprocessors in the instrument to an appropriate displayed reading. The DRI typically is able to store some form of the information.

Sensors are designed to respond to a particular chemical or to a broad range of chemicals with a common characteristic, such as flammability. Thanks to miniaturization of components, instruments may include as many as five different sensors installed in a single meter (Fig. 9.10). These multigas meters provide readings from all of the sensors at the same time. If the alarm point of any sensor is crossed, the alarm sounds and the dangerous condition is highlighted in the display.

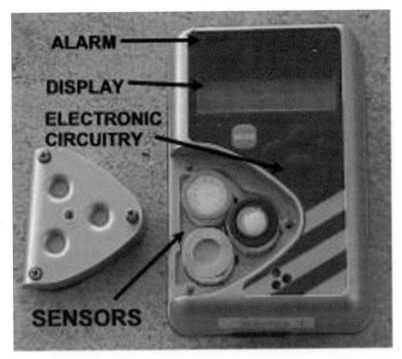

Figure 9.9   The basic elements of a direct-reading instrument.

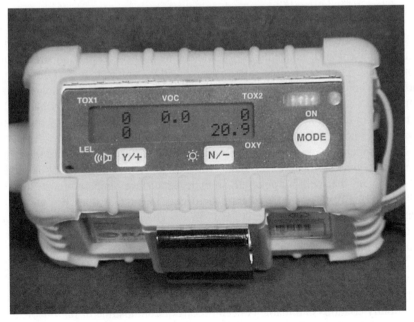

Figure 9.10   Multigas meters can contain as many as five or six sensors.

Common Features of DRIs

Although the DRIs to be discussed perform unique functions, there are a few characteristics or features common to almost all of them.

Intrinsic Safety

DRIs must be safe to use, even in hazardous environments. Because an instrument contains electronics and possibly other sources of heat, it must be constructed in such a way that it will not cause ignition of an explosive atmosphere. All potential sources of heat must be sealed away from exposure to the flammable atmosphere outside the instrument.

The National Electrical Code of the National Fire Protection Association describes minimum criteria for an instrument to be considered "intrinsically safe." Instruments are typically tested by Underwriters' Laboratory (UL) or Factory Mutual (FM) and must be marked as to the hazardous atmosphere for which they are certified. The code classifies hazardous atmospheres by class, group, division, and temperature code. Class and group are used to describe the type of flammable material present as follows:

- Class I is flammable vapors and gases and is further divided into groups A, B, C, and D based on similar flammability characteristics. Examples include gasoline and hydrogen.
- Class II is combustible dusts and is divided into groups E, F, and G. Examples include coal, grain, or metals such as magnesium.
- Class III is ignitable fibers such as cotton.

Divisions are used to describe the likelihood that the flammable contaminant will be present in a concentration sufficient to pose an explosion or combustion hazard. Division I atmospheres are considered most likely to contain the hazardous substance in flammable concentrations. Division II atmospheres have flammable or combustible substances present, but they are typically handled or contained in closed systems that are not likely to generate hazardous concentrations under normal conditions.

The temperature code represents the allowable surface temperature the equipment may generate. The codes range from T1 to T6. A code of T1 is applied to equipment that will not generate an exposed surface temperature greater than 450°C (842°F), whereas T6 is applied to equipment that does not generate a surface temperature greater that 85°C (185°F). The greater the number code, the cooler the equipment and therefore the less likely it will serve as a source of ignition to a flammable atmosphere.

A typical marking on an instrument is that it is "intrinsically safe for Class I, Division I, Groups ABCD, and Temperature Code T3 as approved by UL" (Fig. 9.11). This means that the instrument can be used in an atmosphere that potentially contains flammable concentrations of combustible or flammable gases or vapors. Approval of an instrument for use in one hazard class does not mean

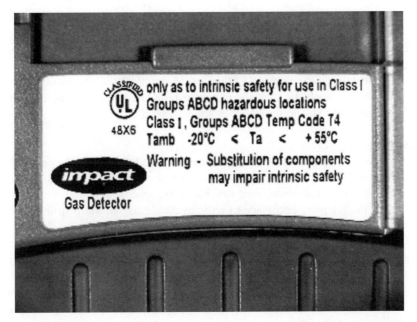

only as to intrinsic safety for use in Class I
Groups ABCD hazardous locations
Class I, Groups ABCD Temp Code T4
Tamb -20°C < Ta < +55°C

Warning - Substitution of components
may impair intrinsic safety

48X6

impact

Gas Detector

Figure 9.11 Information about the intrinsic safety of air monitoring equipment is always printed on a label on the device.

it can be used in all hazard classes. This approval assumes that the instrument will be used according to the manufacturer's directions and that it will not be modified by the user.

Alarms

Air monitoring may be one of several tasks assigned to an entry team, and their attention may not always be on the instrument. DRIs are equipped with audible and visual alarms to alert the user to a dangerous condition. A loud tone and a flashing light activate whenever a reading crosses a preset alarm point. The user can adjust these alarm points, but typically they correspond to a regulatory or action limit for the hazard being monitored.

Alarms are either "latching" or "nonlatching." A latching alarm locks on if an alarm condition is reached, and the user must push a button to unlock or turn off the alarm. Nonlatching alarms sound while the instrument readings are in alarm conditions but turn off automatically when readings return to normal. Latching alarms ensure that responders are made aware of temporary excursions into alarm conditions. These excursions, however brief, should serve as warnings of the potential for hazardous atmospheres, and they may be missed if nonlatching alarms are used.

Besides alerting to a real-time, immediately dangerous atmosphere, instruments with sensors measuring exposure to gases or vapors that are toxic also have averaging alarms. The instrument integrates the concentration readings over time to keep a "running average" of exposure. At some point, the 8-hour time-weighted average (TWA) or the 15-minute short-term exposure limit (STEL) of that particular gas may be exceeded. In other words, a worker may have spent a long enough portion of time at such a high concentration that even if for the rest of the appropriate time period (8 hours or 15 minutes) the exposure was zero, the TWA result would be over the exposure limit. At that point, the instrument would alarm and indicate the gas that was exceeded.

Data Logging

Data logging is the ability of an instrument to record the electronic output from the sensor and store it in memory for later printout or downloading to a computer. This function is usually integrated into the instrument design, and the manufacturer sells the adapters necessary to retrieve the data from the instrument as an optional feature.

Data logging in an instrument typically operates in one or all of three modes. First, it may be programmed to record the readings for each sensor at specific intervals of time. Another mode is to record only readings that exceed preset levels or alarm points. Third, the data logger may record readings only when started manually by the user. For many instruments, the user can choose the desired mode in the field.

Portability and Ease of Operation

Instruments should be easy to operate and should have a display that makes it easy to read the results. Instruments that are difficult to operate likely will require too much of the responder's attention, perhaps causing him to miss important visible information. Any buttons, switches, and knobs should be easy to find and operate, even while wearing gloves.

A bulky instrument may limit the responder's actions during an entry. A small, compact device is preferable, especially if the responder will be carrying other items. The equipment must also be durable. The instrument must be able to withstand a certain amount of abuse and still function properly.

Selectivity

Selectivity is the ability of an instrument to monitor one chemical or group of chemicals and ignore others. It is useful when only a few chemicals may be present or when one specific chemical poses a higher hazard than others do. The user can focus on a particular hazard with a greater degree of confidence; however, the inability to detect other chemicals is a potential limitation if multiple hazards may be present. Responders must also be aware of interference by other chemicals that may cause false positive responses by the instrument.

Sensitivity and Operating Range

An important feature of an instrument is its sensitivity. This is the lowest concentration of chemical in air that will cause a response by the instrument. The full range of concentrations that can be measured by an instrument is called its operating range. The sensitivity and operating range requirements for an instrument will depend on what hazard you want to monitor.

For example, a combustible gas meter that measures in the range of percentage (parts per hundred) would not be appropriate for monitoring toxic gas or vapors. This is because most gases and vapors are toxic at concentrations in the range of parts per million (1% equals 10,000 ppm). In other words, a concentration of the gas that is high enough to be detected by the combustible gas meter would be much higher than the toxic exposure limit, perhaps higher than the IDLH level of that chemical. On the other hand, an instrument designed to monitor toxic exposures may not have a sufficient upper range to monitor the flammability of a gas.

Some instruments are capable of reporting measurements in more than one range. For instance, some CGIs display readings as percentage of LEL or percentage (by volume) of methane. Some older instruments that use an analog (needle) display change the range of the scale by a switch on the instrument. The user must be sure he knows what measurement units the display represents.

Multigas Meters

Multigas meters conveniently consolidate several sensors into a single instrument. However, not all meters have multiple sensors and some of the sensors can be found in stand-alone instruments. The following sections describe the main types of sensors used to measure hazardous materials—combustible gas sensors, oxygen sensors, chemical-specific sensors, and broadband sensors for toxic gases and vapors. Special sensors for chemical or biological weapons are also described briefly.

Measuring Combustible Gases and Vapors

Combustible or flammable gases and vapors are measured with instruments that commonly are called combustible gas indicators (CGIs). These instruments warn of the presence of combustible or flammable gases or vapors at concentrations lower than the lower explosive limit (Fig. 9.12), which is explained in Chapter 6. Several types of sensors may be used in CGIs, and this section looks at the most common types.

Catalytic Sensor. Most CGIs use a form of catalytic sensor to detect the presence of a potentially flammable atmosphere. The sensor chamber contains two filaments that carry an identical electrical current that heats the filaments. In the chamber, the two filaments are simultaneously exposed to air from the environment. One of these filaments is treated with a catalyst that will burn flammable gas even at low levels, and this further heats the filament and decreases the current it carries. The other filament is untreated and will not burn the gas, so its current is unchanged. The difference in electrical current between these two filaments, measured by a bridge circuit, indicates how much flammable vapor is in the air.

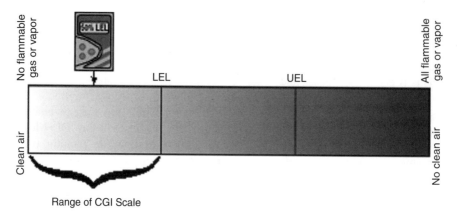

Figure 9.12 Combustible gas sensors measure flammable gases and vapors at concentrations lower than the lower explosive limit (LEL).

CGI Readings. CGIs display the sensor response as a percentage of the lower explosive limit (percent LEL) of their calibration gas. As discussed in Chapter 6, LELs differ from one flammable material to another. Because each CGI manufacturer recommends a calibration gas for its instrument and CGIs are not able to distinguish one flammable gas from another, the displayed percentage of the LEL is relevant only to the calibration gas. If flammable gases or vapors other than the calibration gas are being measured, the readings do not tell you the true percentage of the LEL of the actual gas present. This is discussed further in the section on relative response below in this chapter.

Atmospheres that provide any response on the CGI must be treated carefully. OSHA considers the atmosphere hazardous if the reading exceeds 10% of LEL. This conservative limit is appropriate because if there is enough flammable material present to give a reading of 10% LEL where and when the testing is performed, the concentration may be higher in another location or at a different time. It also allows for the difference in response and LEL described in the previous paragraph.

Limitations and Considerations. CGIs do not respond to combustible dusts at all, even in high concentrations. In fact, there is no DRI available that is able to discriminate and measure flammable or combustible solids. The OSHA confined space standard states that the LEL of a flammable dust has probably been exceeded if it is difficult to see something within five feet [29 CFR 1910.146(b)(2)]. Responses to releases of flammable solids must be made with caution because the hazard cannot be measured.

The catalytic sensors require adequate oxygen to function properly. Multigas instruments with oxygen sensors have the advantage of being able to confirm that there is adequate oxygen for the proper operation of the combustible gas sensor.

Combustible gas meters that do not have an oxygen sensor included should be used only if it is confirmed that they do not require oxygen.

Certain chemicals, such as corrosives and leaded gasoline, can "poison" CGI sensors. Halogenated compounds exposed to high temperatures can degrade to corrosive materials, as can other compounds. This damage may be permanent and may not be immediately apparent when it occurs. It is a good practice to check the function and calibration of a CGI after exposure to a high concentration of any material.

Oxygen Sensors

Oxygen sensors are basic electrochemical sensors. Oxygen permeates from the atmosphere through a barrier into the sensor and interacts with an electrolyte solution and electrodes to create an electrical current. Specifically, oxygen reacts with a sensing electrode (usually of lead or zinc) in potassium hydroxide (electrolyte) solution, releasing electrons. These electrons migrate to a counter electrode (usually gold or platinum), and an electrical current proportional to the amount of oxygen is produced.

Interpreting Readings. Oxygen meters display readings as percent oxygen (by volume) in air. Normal air has about 20.9% oxygen and about 79% nitrogen, with trace amounts of other gases. OSHA treats as hazardous atmospheres with an oxygen concentration below 19.5% or above 23.5% in air.

The above normal concentration of oxygen does not vary across the Earth's entire surface, even at higher elevations. What *does* change at higher elevations is the total atmospheric pressure and, therefore, the partial pressure of oxygen. This lower partial pressure affects the movement of oxygen into the blood in the lungs and results in the perception that there is less oxygen. However, the actual volume-to-volume ratio of oxygen to nitrogen is the same unless something unusual acts to add or displace oxygen. This constant oxygen level means that any fluctuation in the oxygen reading of a meter must be considered carefully, as the following discussion demonstrates.

Besides simply measuring the oxygen concentration in the atmosphere, an oxygen meter may warn of the presence of other hazards. For instance, say the oxygen meter reads 19.9%—down 1%, but still above the minimum allowable oxygen level in air. Because oxygen is about 1/5 (20.9%) of the air, a drop of 1% in the oxygen reading may mean that there is 5% of some gas or vapor in the air. Now consider that a 1% concentration equals 10,000 ppm. A drop of 1% oxygen may not be an oxygen-deficiency problem, but it could warn that there is 50,000 ppm of some chemical in the air. That's likely to be a dangerous concentration of any chemical!

Limitations and Considerations. Although oxygen meters are pretty simple to operate and read, there are some things to keep in mind. First, the electrolyte in the sensor is typically a base (alkaline) solution that can be neutralized when exposed to acid gases, such as carbon dioxide in exhaled air. Other chemicals may also poison the sensor, and these would be indicated by the instrument manufacturer. Oxidizers may cause an artificially high reading. If the instrument is used in a response involving the release of such chemicals, it should be calibrated afterwards to ensure that it is still functioning properly.

Excess moisture and humidity also can affect the permeation of oxygen through the protective membrane of the sensor. Temperature extremes may also affect the electrochemical reaction. However, instruments are typically able to operate in a range of 32°F to 104°F because they are temperature compensated. Measuring in temperatures outside this range may require special procedures, such as recalibrating the instrument at the temperature at which it will be used.

A change in the partial pressure of oxygen can affect the permeation of oxygen through the protective barrier of the sensor. This will lead to low readings, even if the actual concentration is normal. Therefore, oxygen sensors should be calibrated at the elevation or pressure conditions at which they will be used.

Chemical-Specific Toxic Sensors
Multigas instruments almost always include the combustible gas and oxygen sensors mentioned above and add at least one and as many as four "toxic" sensors. These sensors are designed to respond selectively to a single chemical at concentrations in the ppm range. The most common examples of these are carbon monoxide and hydrogen sulfide, but others include:

- Ammonia
- Chlorine
- Formaldehyde
- Carbon dioxide
- Hydrogen cyanide
- Sulfur dioxide

These chemical-specific sensors typically operate as electrochemical sensors like the oxygen sensor described above. The electrolyte solution, electrodes, and operating conditions are changed to make the sensor selective for a particular chemical, but the general operation is very similar.

The instrument displays the measurement of the chemical in units of ppm. Because the sensor theoretically responds only to the chemical of interest, any reading can be compared directly to an exposure limit for that chemical. Most instruments with these sensors store the concentration data as peak readings and as "rolling averages" for comparison to either a short-term exposure limit (STEL) or an 8-hour time-weighted average (TWA). These limits are described below in this chapter.

Limitations and Considerations. Because these sensors rely on a chemical interaction to detect and measure the chemicals, a number of factors can impact the operation and accuracy of these devices. For instance, a sensor designed to measure one particular chemical will usually also respond to the presence of some other chemicals (called interferents) that have similar properties. The instrument manufacturer should identify known interferences in the manual for the instrument.

Like the oxygen sensor, these electrochemical sensors may be affected by extreme environmental conditions, such as temperature, humidity, and atmospheric pressure. The effect of these conditions is usually negligible within a specified operating range, but responders should become familiar with how the equipment operates in all the conditions they expect to encounter. Even under normal operating conditions, these sensors have a limited useful service life (typically 1–2 years) and must be replaced. This life may be significantly shortened by exposure to some chemicals that damage or consume the sensor.

Broadband Toxic Sensors

Another type of sensor included in some multigas meters is the broadband toxic sensor. The name derives from the fact that these sensors respond to a broad range of different chemicals without discriminating one from another. Their results are often displayed in units of ppm. This can be confusing because the instrument responds differently to different chemicals but is only calibrated to one chemical. Therefore, the readings as ppm would only be accurate if the instrument was measuring the same gas used for calibration. This is discussed in the section on relative response below in this chapter. Nonetheless, the readings of broadband sensors can be useful in alerting the responder to the presence of detectable chemicals and perhaps in locating a leak or source of a chemical release. Some common broadband sensors are discussed below.

Thermal Conductivity Sensors. Thermal conductivity sensors also have two filaments like the catalytic sensor of the CGI; however, only one is exposed to the contaminated air. The other is sealed away from the air. The filaments are heated to a high temperature, and they carry an identical electrical current. Gases in the sample air passing over the exposed filament may either cool it and increase its electrical current or heat it and decrease its current, while the unexposed wire is not affected. Once again, the difference in current between the two filaments is measured by a bridge circuit to tell how much vapor is present.

The thermal conductivity sensor is not selective for particular gases or vapors and does not respond as well to low concentrations. It is used primarily to measure high concentrations of methane and some other gases. The results are reported as percent gas in air.

Metallic Oxide Semiconductor. Metallic oxide semiconductor (MOS) sensors are a type of solid-state sensors. The sensor consists of two electrodes and a heating element imbedded in a ceramic bead that is coated with a metal oxide semiconductor such as tin oxide. The sensor surface is heated to a high temperature, and the semiconductor material absorbs atmospheric oxygen, establishing a baseline electrical conductivity through the sensor. As a combustible gas contacts the bead and reacts with the absorbed oxygen, this changes the conductivity of the sensor. As the gas leaves, the conductivity returns to normal. The change in conductivity indicates the amount of gas or vapor present.

Survey Instruments

All or some of the kinds of sensors described to this point are commonly found in multigas instruments. In addition to these multigas instruments, there are instruments that use somewhat more advanced sensor technologies to measure chemicals at smaller concentrations. These survey instruments often represent a durable field version of technologies used in analytical laboratories. The most common of these are:

- Photoionization detector (PID)
- Flame ionization detector (FID)
- Infrared spectrophotometer (IRSpec)

These instruments all respond to gases or vapors in concentrations as small as 1 ppm or less. None of them is able to identify an unknown chemical or to separate and measure each chemical in a mixture. Nonetheless, their ability to respond to chemicals in very low concentrations makes them useful in some hazmat response situations. They are described in the following sections.

Photoionization Detector

PIDs are survey instruments that use ultraviolet (UV) light to strip electrons off the gas or vapor molecules in the air, thus forming ions (charged particles). The ions migrate to appropriately charged electrodes and create an electric current that is proportional to the amount of chemical in the air. The PID will detect many organic and some inorganic compounds at concentrations as low as one part per billion (ppb) and up to 2000 ppm.

The ionization potential (IP) of a chemical is the amount of energy necessary to ionize it (remove one or more electrons.) The IP, expressed in units called electron volts (eV), is different for different chemicals. The UV lamp in the PID must be stronger (have more eV) than the IP for the chemical(s) for the instrument to respond. Most PIDs use a 10.6-eV lamp, although other strength lamps, such as 9.8 eV and 11.7 eV, are available.

Although PIDs are not selective (i.e., they will detect a wide variety of chemicals), some selectivity may be achieved on the basis of differing ionization potentials. As mentioned above, the IP differs for various chemicals. For example, the IP for 1,1,1-trichloroethane is 11 eV. If a 9.8-eV lamp is used with the PID, most of the 1,1,1-trichloroethane present in the environment will not be detected, whereas a chemical with an IP of 9.7 eV or less would be detected. Thus the PID becomes selective for chemicals with an IP less than the strength of the UV lamp used. (Some ionization does occur at strengths less than the IP of a compound but is usually a very small fraction of that which occurs at or above the IP.)

High humidity and high electromagnetic energy may interfere with the PID's response. If the UV lamp is dirty, the UV light will be blocked and unable to ionize the chemical, causing the instrument to give an inaccurate reading. Charged particles in the air other than the ions of the contaminant (e.g., dusts, particulates from diesel engines) will collect on the electrode and cause a false reading.

Flame Ionization Detector

The FID uses a hydrogen flame to burn the contaminant, resulting in the release of charged ions. Like the PID, the ions are collected on electrodes and cause an electric current "signal." The FID will ionize organic compounds with an IP of 15.4 eV or less. It is not significantly affected by humidity and can measure over a wide range of contaminant concentrations.

The FID detects only organic compounds. Because the flame requires oxygen, the unit may not function in oxygen-deficient atmospheres. Impurities in the hydrogen used as fuel will be ionized and detected just like contaminants in the air, causing a false reading. Like the PID, other charged particles may collect on the electrode, interfering with the instrument readout.

Infrared Spectrophotometer

Molecules of most organic and some inorganic chemicals absorb infrared light energy that strikes them. Different molecular structures will absorb infrared light with different wavelengths. The IR sensor takes advantage of the fact that chemicals will absorb infrared light at certain discreet wavelengths. Therefore, the IRSpec can be fairly selective (i.e., monitor one chemical at a time) on the basis of the wavelength(s) of infrared light used. The spectrophotometer consists of a sample chamber through which infrared light is projected. The wavelength of the infrared light can be adjusted, allowing the instrument to scan a sample with different wavelengths. Some of the light in the sample chamber is absorbed by the contaminant molecules, and the remainder is transmitted. The instrument determines the amount of infrared light absorbed by the chemical at each wavelength and uses data provided by the manufacturer to tentatively identify the material and calculate the concentration present in ppm.

Special Sensors for Chemical and Biological WMD

The types of agents that may be incorporated in a weapon of mass destruction (WMD) are described in Chapter 10. One of the goals of terrorists employing a WMD is to cause fear and disruption beyond the immediate area of the device by making people believe that the agent has been spread over the widest possible area. Therefore, good detection equipment can help responders minimize the chaotic effect of such an incident by clearly delineating the true threat area.

The relatively new field of commercially available WMD detectors is scrambling to develop the necessary detection technologies. The following are some of the newer sensor types available at the time this book was being written. Because of the specialized nature of some of this equipment, brand or model names of some devices are associated with the sensor types. The relative effectiveness and value of these technologies in actual incidents is still to be determined. Also, newer technology may be available by the time the reader reads this text.

Biological Agent Detectors

Many emerging and existing technologies are being applied to the problem of identification of biological agents. Many are either so technologically advanced or expensive that they are not likely to be used by most hazmat responders. The two most common technologies at the time of this writing are described here. Neither of these technologies is an air monitoring instrument, but they may be used as part of the hazard assessment process.

Immunoassay Technology. Bio Threat Alert (BTA) Strips and Sensitive Membrane Antigen Rapid Test (SMART) tickets are examples of immunoassay tests. These test kits initiate an antigen-antibody reaction in the presence of biological agents. The test strip contains an antibody that will react with proteins (antigens) unique to biological agents. This reaction creates a color change on the test strip, which is then read to indicate the presence or absence of the biological agent.

Polymerase Chain Reaction (PCR). PCR devices detect the presence of genetic material from viral or bacterial agents. The term PCR refers to the process by which the device concentrates and amplifies the DNA from a sample material. The genetic material is then "tagged" with a material, such as fluorescent dye, that is read by an optical reader.

Chemical Agent Detectors

Chemical warfare agents are similar to traditional hazardous materials, although their toxicity is higher. The sensors and instruments described above (PID, FID, IRSpec, electrochemical) have shown varying usefulness in the detection and measurement of chemical agents. The following are technologies specifically applied to the detection of chemical agents.

Ion Mobility Spectrometry. This detector is based on the fact that ions (charged particles) from different chemicals will move through an electrical field at different rates. The detector uses a Ni^{63} (nickel 63) or Am^{241} (americium 241) source of beta radiation to ionize the air sample and thus the chemical agents. The sample is then injected into a "drift tube" that has a slight electrical charge gradient. The gradient draws the ion toward a collecting electrode that generates a charge on ion impact. The travel time down the tube identifies the chemical species, and the amount of charge generated indicates the relative concentration. Examples of this type of instrument include the APD2000, the chemical agent monitor (CAM), and the M8A1/M43A1 military detectors. The detector measures both nerve and blister agents but is vulnerable to interferents and false positives.

Flame Photometry. Flame photometry devices burn the air sample in a hydrogen flame. Nerve agents contain phosphorus and sulfur, and blister agents contain sulfur. These elements emit light of characteristic wavelengths when heated in the hydrogen flame. A photometric sensor measures the light emitted at the appropriate wavelengths to indicate the amount of chemical agents present in the sample. The MiniCAM and AP2C are examples of stand-alone detectors. However, the detector is susceptible to interference from chemicals that contain phosphorus and sulfur but are not chemical weapons agents. Therefore, these detectors are also coupled with a gas chromatograph (discussed below) to separate the constituents of an air sample to be analyzed independently.

Surface Acoustic Wave (SAW). These detectors use piezoelectric crystals that are coated with films that absorb chemical agents. The crystals vibrate with a characteristic frequency. Chemical agents absorb onto the surface, changing this resonant frequency. The instrument uses multiple crystals that are coated with different films and therefore absorb different chemical types simultaneously. The pattern of responses of the crystals when exposed to a particular chemical agent is stored in a microprocessor. As an air sample is introduced into the device, the sample pattern is compared to the stored patterns to look for a match that would indicate the presence of a chemical agent. JCAD, SAW MiniCAD mkII, and Chemical Warfare (CW) Sentry are examples of this technology. They will detect nerve and blister agents but have slow response times and are very sensitive to interferents.

Gas Chromatography

Identifying components of an unknown mixture is a difficult job at best. One technology that may be used to help in this process is gas chromatography (GC). A GC is not a detection device but is useful in separating out components of a mixture or, in some cases, in identifying an unknown compound. Some instrument manufacturers have coupled this technology with detectors such as the PID or FID in field instruments. These instruments can operate in the survey mode (detector only) or the GC mode (GC and detector). Because of the time requirements of operating in the GC mode, these instruments cannot simultaneously survey an area and separate the components of a mixture with the GC. Nonetheless, the GC can be a valuable tool for identifying and quantifying the components of a mixture.

The GC is essentially a long, thin tube (usually coiled) called a column that contains a medium. Contaminants are introduced into the column in a carrier gas. Each contaminant adsorbs to the medium with a different strength of attraction and subsequently is released to flow out the other end of the column into a detector (e.g., FID, PID, atomic absorption, mass spectroscopy). The detector is usually attached to a chart recorder that records the concentration detected over time. Based on the strength of the attraction, the column temperature, and properties of the contaminants, all of a particular chemical or group of chemicals will release from the medium and leave the column at the same time and this time will be different than that of other chemicals.

The length of time from injection of the chemical into the column until it exits the column and is detected is called its retention time. The retention time for a chemical at specific column conditions is constant and will vary from chemical to chemical. The temperature of a column greatly affects the retention time of a chemical. Put simply, the individual chemicals in a mixture injected in one end of a GC column will exit the column at different retention times, but all of each chemical will come out at approximately the same retention time. The chart recorder will show a peak for each chemical at the appropriate retention time. The area under the peak is proportional to the amount of the contaminant present.

The practical use of this technology is that if column conditions (temperature, medium type, carrier gas flow rate, etc.) are kept constant and the retention times of individual contaminants in a mixture under those conditions are known (two very big "ifs"), the quantity of each chemical can be determined as it is detected

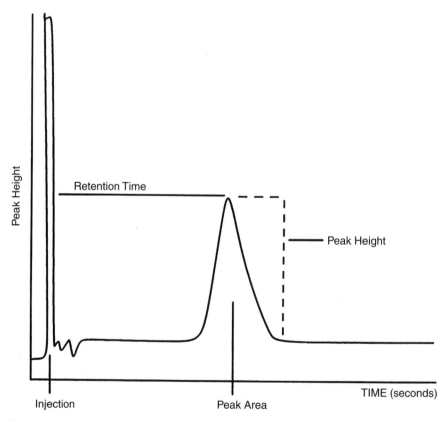

Figure 9.13 The amount of chemical measured by a gas chromatograph (peak area) is a function of the detector response (peak height) and the time of response. The retention time can help identify the material.

by the detector and, based on its retention time, it can be identified. Figure 9.13 shows a chromatogram.

The technology is useful but does have limitations and drawbacks. Obviously, someone using GCs must be well trained. Many things must be controlled. Calibration standards of each chemical to be detected must be run through the GC under specific conditions to confirm the retention times and response of the detector to the chemical. If your response team has a large variety of potential contaminants and can afford the equipment and training costs to operate a field GC, it can provide you with valuable information about specific contaminants quickly.

Instrument Calibration

The components of direct-reading instruments, such as sensors, circuitry, or moving parts may begin to malfunction or wear out with use or over time. This may affect the accuracy of the instrument in a way that is subtle and goes unnoticed. This could have the potentially disastrous effect of causing the instrument to give false readings to responders, who then base response decisions on bad information. Therefore, checking the calibration of the instrument is of critical importance.

Instrument calibration, then, performs two functions. First, calibration determines whether the instrument is even operational. Almost all sensors (except oxygen) would be expected to read zero in normal atmospheres. However, a malfunction in the instrument may also cause it to read zero. Therefore, the fact that an instrument reads zero when turned on in clean air does not confirm that it is functioning properly. It must be exposed to some atmosphere that would cause a response to confirm that it will respond at all.

Second, calibration ensures that the instrument will also read accurately. The calibration process compares the instrument reading to a known concentration of an appropriate gas to be sure that subtle changes in the instrument function have not caused even minor errors in its response. Calibrating the instrument involves adjusting the electronics that interpret and display the signal from the sensor. Adjusting the instrument display to agree with a known concentration of a chemical gives confidence that the instrument will measure a hazardous atmosphere accurately.

Calibration Procedures

Zeroing. Zeroing the instrument involves setting a baseline response from the sensor when the instrument is in normal air. The instrument can be zeroed to air in a clean atmosphere or by connecting it to a gas cylinder of purified air. The purpose of this is to eliminate the effects of electronic noise and other factors not related to the actual measurement of a contaminated atmosphere.

Bump Test. A calibration check or "bump test" involves simply exposing the instrument to an atmosphere that will cause a response from the sensor. In a strict sense, it is not critical that the exact concentration be known as the purpose is simply to cause any reading greater than the baseline reading. The test is most useful when the instrument is exposed to a known concentration and the reading is compared to that concentration. A bump test differs from the full calibration discussed below in that no adjustment to the reading is made. The user simply confirms that the instrument will react in a hazardous atmosphere.

Span Adjustment. A full calibration, or span adjustment, involves connecting the instrument to a source of a known concentration of a calibration gas and adjusting the instrument's displayed reading to match that concentration. To prevent accidental calibration adjustments, the instrument generally requires the user to enter a special calibration mode before the reading can be adjusted. Once in the calibration mode, the user attaches the instrument to a calibration cylinder, allows the reading to stabilize, and then presses appropriate buttons on the instrument to change the reading to match the known concentration.

"One-Button" or Auto Calibration. Some instruments have been designed to automate the calibration process. The user simply connects the instrument to the source of calibration gas and presses a button. The instrument waits for the signal from the sensor(s) to stabilize and then resets the display to match the preprogrammed readings. This greatly simplifies the calibration process. However, the user must use a calibration gas exactly matching the preprogrammed settings.

Calibration Gases

The instrument manufacturer specifies a calibration gas for each instrument. A chemical-specific instrument uses a known concentration of that chemical for calibration. The concentration used is usually lower than the exposure limit for the chemical so that the person calibrating the instrument is not overexposed during the calibration.

For combustible gas sensors or survey instruments that respond to different chemicals, the calibration gas chosen is one for which the instrument has medium sensitivity. This will enable the instrument to respond to other gases to which it is more or less sensitive. The properties of other flammable gases may cause more or less response than the same amount of the calibration gas. For example, many CGIs use pentane as their calibration gas. If a CGI calibrated to a pentane standard is used to measure xylene, another flammable material, it will give a lower reading than the concentration that is actually there. The reading can be converted by means of a relative response factor described below in this chapter.

A survey instrument can be calibrated to directly and accurately read almost any gas that will cause its sensor to respond. If responders expect to respond to one chemical and it has been identified, they can use a known concentration of that chemical to calibrate an instrument. Once calibrated to that chemical, the instrument will display the actual concentration of that chemical when measuring the air in the space. However, if the instrument is used to measure another chemical, the reading will not be accurate and cannot be converted with the manufacturer's response factors.

Calibration gases are most often provided in premixed compressed gas cylinders. The gas flow is controlled by a regulator. If the instrument uses a pump to move air into the sensor chamber, the flow rate of the regulator must match the flow rate of the pump. Purchasing the calibration kit recommended or sold by the instrument manufacturer will insure compatibility.

Frequency of Calibration

Instrument sensor technology is constantly improving, and today's sensors are much more stable in their operation than in the past. Nonetheless, because many factors can affect instrument accuracy, instruments should be calibrated before each day's use. Calibration checks are simple to perform, and important response decisions often depend on air monitoring information. Responders should include calibration of detectors in their standard response procedures.

The International Safety Equipment Association (ISEA), a nonprofit professional organization for safety equipment manufacturers, issued a statement on November 2, 2002 outlining recommendations for the frequency of instrument calibration. The recommendation is for instruments to undergo a bump test or full calibration daily, more frequently if environmental conditions that may affect instrument performance are suspected to be present. It allows for a less frequent calibration (up to once every 30 days) if a test period of at least 10 days of use in the intended atmosphere and under all exposure conditions shows that it is not necessary to make adjustments (http://www.safetyequipment.org/calibration.PDF).

Calibration checks can be made in a laboratory or office environment before deployment in a response. Instruments have a range of conditions of temperature, pressure, and humidity in which they operate properly. In extreme conditions or at significantly different altitudes, the calibration should be under conditions similar to those of the response scene in which the instrument will be used. Periodic checks should be made of equipment that is infrequently used and kept in storage to be sure it will function when it is needed.

Calibration can also be done after each use to confirm that the instrument continued to function throughout the response. It may also identify repairs that may be necessary before the next use.

INTERPRETING AIR MONITORING RESULTS

The appropriate interpretation criteria for each type of monitoring device were discussed above. A user can compare the readings in most situations directly to the action level or exposure limit to determine whether a hazardous atmosphere exists. There are circumstances in which interpretation of the results is more challenging. Some of these circumstances are discussed below.

Relative Response

Sensors that respond to many different chemicals are calibrated to a single calibration gas. When the instrument is used to measure a chemical other than that
gas, the reading that is displayed is not necessarily the actual concentration of
the chemical in the air. Although inconvenient, this is not an insurmountable
problem if the chemical that has been released has been identified. Manufacturers of such instruments usually have determined relative response factors for
many common chemicals. These factors are applied to the reading to calculate
the actual concentration of the chemical from the reading.

Figure 9.14 shows a situation in which the responder is using an instrument
properly calibrated to isobutylene, the calibration gas recommended by the manufacturer. The manufacturer has included a list of response factors in the operating
instructions for the instrument. For n-hexane, the response factor is 6.2 and readings of the instrument measuring hexane only should be multiplied by this factor.
A reading of 10 ppm means that the actual concentration of hexane is 62 ppm
(6.2 × 10 ppm). In the same way, an actual 100 ppm concentration of hexane
would be expected to cause a reading of about 16 ppm on the instrument.

Responders can only apply a relative response factor to a situation in which
there is just one chemical in the air and it has been identified. Obviously, it
is impossible to choose the proper response factor if the chemical has not been
identified. If a mixture of two or more chemicals is known to be in the atmosphere

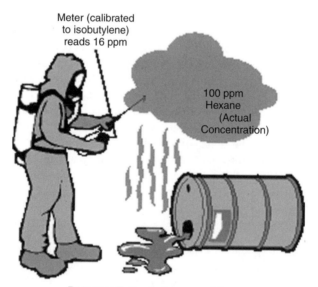

Meter (calibrated
to isobutylene)
reads 16 ppm

100 ppm
Hexane
(Actual
Concentration)

Response Factor for hexane = 6.2

Figure 9.14   The actual concentration of chemical in the air can be estimated by multiplying the reading displayed on the instrument by the relative response factor from the
manufacturer.

being measured, the reading could not be multiplied by more than one response factor. Readings of atmospheres containing unidentified chemicals and mixtures cannot be converted to an actual concentration(s) of the chemical(s) present.

Unidentified Contaminants

Atmospheres that are known or suspected to contain a chemical that has not been identified are very difficult to assess. Emergency responders should only make an entry into a hazard area that is suspected of containing an unidentified hazardous material if the incident commander deems it vital to gather information or observations and the safety officer agrees that conditions are unlikely to be an immediate danger to the responder. Such an entry would still be made with the highest level of protection and using a survey instrument, such as a multigas instrument with combustible gas and oxygen sensors that would warn of any fire or explosion hazard.

For instance, if a CGI gives no reading along the perimeter of a hazard area that has no obvious dust present, it should be safe to slowly approach the hazard area from upwind while continuing to monitor for flammable atmospheres. Oxygen content also can be measured in any atmosphere. If abnormal oxygen and flammability are ruled out, that leaves only the possibility of a toxic atmosphere. It may still be possible to enter such a space with the highest level of respiratory protection and adequate skin protection. Such an entry would be made only if it were necessary to make a rescue or to gather information.

Mixtures

When an atmosphere contains a mixture of chemicals that cannot be measured separately, it is difficult to assess the hazard to the responder accurately. Once again, measuring the oxygen atmosphere and the flammability provides some information. Measuring any known chemicals with chemical-specific instruments or detector tubes will also be helpful. One is still left with addressing the toxicity of the remaining mixture.

One approach to this would be to use a survey instrument such as a PID to assess a total reading for the mixture. If the chemicals that make up the mixture are known and all can be measured by a PID, the responder can identify the chemical with the highest response factor and multiply the reading by that factor. The result then could be compared to the lowest exposure limit value of all of the chemicals present. This would be a very conservative approach, and response decisions made based on this interpretation might be unnecessarily restrictive.

COLORIMETRIC INDICATORS

Colorimetric indicators provide a means of quantifying air contamination with a reasonable degree of selectivity. Three types of indicators may be used: liquid reagents, chemically treated papers, and glass tubes containing chemically treated solids. The principle of operation is that a contaminant in air reacts with a chemical reagent to cause a change in color, which is proportional to the amount of contaminant in the air. Liquid reagent devices are available that produce color changes that are fairly easy to observe. But handling liquids in the field may be awkward and inconvenient. Chemically treated papers are easy to use but typically do not have a means of controlling the volume of air contacting the paper. Therefore, accuracy may be affected by factors such as the amount of air movement over the paper.

Detector Tubes

Because of their ease of use and quick results, the glass tubes containing chemically treated solids—known as colorimetric or detector tubes—are widely popular. They typically consist of a glass tube containing a granular carrier solid that has been impregnated with a specific chemical reagent. As contaminated air passes through the glass tube, the contaminant reacts with the reagent on the carrier solid to produce a color stain. The air is drawn through the tube by either a hand pump or a battery-powered pump. Thus the volume of air sampled can be controlled. It should be noted that detector tubes are designed and calibrated to be used exclusively with the manufacturer's pump. Using a detector tube with another manufacturer's pump, even if the pump volume is the same, can lead to inaccurate results.

The tubes may be specific for a certain gas or vapor or may detect groups of chemicals, such as alcohols or aromatic hydrocarbons. Tubes that are designed to detect one chemical will usually react with certain other chemicals (known as interferences) to produce a similar color change. The manufacturer will provide a list of known interferences. In general, detector tubes provide an opportunity to make a qualitative and quantitative determination of the presence of a particular chemical.

Detector tube manufacturers often package sets of tubes as "hazmat kits" that can be used to try to identify or classify a released hazardous material. The kits include tubes that directly respond to commonly released chemicals and other tubes that have a broad range of response or many cross-sensitivities. The latter tubes provide a quick screening of the presence or absence of a wide group of chemicals and, based on the qualitative response, direct the user to other specific tubes or tests.

One variation on the broad range tube is the tube that includes four or five sections with different reagents. Each section responds to different contaminants, and a single tube is then able to screen for multiple contaminant families. A color chart representing different patterns of response to the sections is used to determine further specific tests that may specifically identify the contaminant.

Operation of Detector Tubes

The tubes may operate in one of three ways (Fig. 9.15). In the first type, the length of stain tubes (A), a set number of pump strokes (volumes) of air is

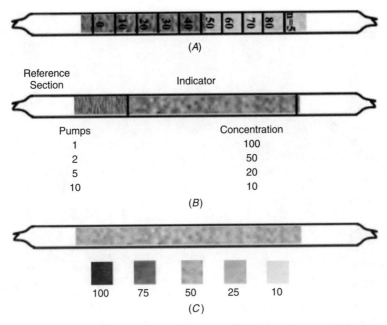

Figure 9.15 Detector tubes display chemical measurements by one of three methods (A–C).

drawn through the tube and the length of stain is compared to a calibration scale, typically printed on the tube, to determine the concentration. For a set number of pump strokes, a high concentration would cause a longer stain. In another tube type (B), the pump is operated until the degree of color change matches a reference section inside the tube. The number of pump strokes (i.e., the volume of air) required to reach this full stain is compared with a chart to determine the concentration of the contaminant in air. In this case, a high concentration of contaminant in air would require fewer strokes to reach full stain. In the third type of tube (C), a predetermined number of pump strokes of air are drawn through the tube and the degree or tint of the color change is compared with a chart to determine the concentration. For these tubes a high concentration would cause a deeper or darker color change after a set number of pump strokes. It is critical that the operator of the tube be familiar with the manufacturer's directions and knows which mode of operation is used.

Two kinds of pump are used to take detector tube samples (Fig. 9.16). A detector tube, with both ends broken open, is inserted into the inlet for the pump. The bellows pump has a collapsible rubber bellows that is compressed by the user. As the bellows expands, air is drawn through the tube. Each complete stroke draws 100 ml of air, and the number of strokes determines the volume of air sampled.

The other type of pump used to take detector tube samples is a piston pump. An opened tube is inserted into the pump inlet. A spring-loaded handle is pulled

Figure 9.16 Air is moved through the detector tube by means of bellows (top two) or piston (bottom) pumps. The same brand of pump and tubes must be used.

on the pump. This action draws a piston to expand the cylinder inside the pump body. The air to fill this expanding cylinder is drawn through the tube. It is the reverse action of a bicycle tire pump.

The tube manufacturer's pump must be used to draw air through the tube. Using a detector tube with another manufacturer's pump, even if the pump volume is similar, can lead to inaccurate results.

Sample pumps must be checked for leaks to ensure that the appropriate volume of air is drawn through the tube. Also, the pump must be allowed to fully complete every pump stroke. Incomplete strokes or leaks in the system will cause less than the appropriate volume of sample air to pass through the tube, potentially resulting in a reading lower than the concentration actually present.

Limitations of Detector Tubes

Some limitations must be considered when detector tubes are used. Detector tube systems have rather poor accuracy, with errors ranging from 25% to 50% for many tubes. The National Institute for Occupational Safety and Health (NIOSH) tested and certified detector tubes at one time but has since discontinued the practice. Manufacturers generally provide accuracy information with the instructions.

Because a chemical reaction is involved, detector tube accuracy may be affected by such factors as temperature, humidity, and atmospheric pressure. Where temperature will significantly affect the performance of the tube, the manufacturer will include compensation factors in the instructions. Colder temperatures will usually slow down the reaction, so if detector tubes are to be used in cold weather, they should be stored in a warm place and carried next to the body. High temperatures may also affect the rate of the chemical reaction and may reduce the shelf life as described below.

One source of error in the use of colorimetric tubes is the visual interpretation of the length or degree of color change. The leading edge of the stain may be uneven or may be lighter than the rest of the stain, calling for a judgment on the part of the operator as to what constitutes the end of the stain. The same difficulty applies to judging the degree or tint of color change, even when comparison charts are provided. When in doubt, it is advisable to use the most conservative (i.e., highest) reading so that more protection is provided to the responder and to the public.

Detector tubes have a specific shelf life. Chemical reagents will deteriorate over time, even if the tube is not opened and exposed to air. Also, high temperatures may cause degradation of the reagent. The manufacturer will stamp an expiration date on each pack of tubes. Storing the tubes in a refrigerator may maintain or extend the shelf life, but expired tubes should not be used.

Chip Measurement System (CMS)

Dräger Safety Company's CMS® automates the colorimetric measurement process. The system uses a card (chip) with ten capillary tubes containing liquid reagent that reacts with specific chemicals or groups of chemicals. The card is inserted into the analyzer instrument that automates both the sampling and analysis process. The analyzer opens the capillary tube and pumps air across the chip at a constant flow rate that allows a consistent reaction to occur between contaminant in the air and the reagent. An optical reader in the analyzer detects and measures the slightest resulting color change, and the instrument reports the result on a digital display.

The analyzer's automated process and optical analyzer increase the accuracy of this colorimetric indicator system. The manufacturer reports error ranges of $\pm 4-7\%$ for many of the chemicals measured. Dräger can provide a handbook describing the CMS® operation and a list of the available chips.

Papers and Tapes

A basic application of the colorimetric measurement process is the chemical indicator tape or paper strip. The military has used tickets (M8 and M9) to verify the presence of chemical agents for years. Similar kits have been developed for industrial and hazardous materials applications. These tests are qualitative in nature, meaning that they give a color change if the material is present but do not give an estimate of the concentration. Some are intended to respond to airborne chemicals, whereas others indicate surface or liquid contamination.

EXPOSURE LIMITS

How much chemical in the air is too much? For carbon monoxide in a residence, any amount is a sign of a potential problem. However, industrial facilities and other areas where chemicals are used and stored regularly may have some level of these chemicals in air. It is generally accepted that the body can tolerate some exposure to these chemicals. However, there is a level of exposure at which health effects begin to occur.

Exposure limits are set to act as indicators of what exposure is too much. A number of different agencies, both government and professional, publish exposure limits. These are simply maximum acceptable concentrations of chemicals in the air over a given period of time. There are a number of terms and abbreviations related to which agency publishes the limits and over how much time they are applied.

Occupational Exposure Limits

Most of the limits that have been developed are intended for normal workplace exposures, not emergencies. These have been developed through extensive toxicological studies like those discussed in Chapter 7. The exposure situation they typically address is that of a worker who works in the same job or type of work from day to day for an extended period of time. Because the exposure pattern is expected to be repeated each day, the chance or likelihood of the worker accumulating a substantial dose is a real concern. Therefore, the allowable exposures are set very low to ensure that the amount of chemical in the body is kept low over time.

Who Sets the Limits?

Occupational exposure limits are set by four groups. Each has its own name for its set of limits. The key distinction between these is that OSHA's limit carries the force of law. The terms are:

- **PEL**—Permissible Exposure Limit—OSHA's *legally enforceable* regulatory limit
- **TLV**—Threshold Limit Value—*recommended* limit set by the American Conference of Governmental Industrial Hygienists (ACGIH)
- **REL**—Recommended Exposure Limit—*recommended* limit set by the National Institute for Occupational Safety and Health (NIOSH)
- **WEEL**—Workplace Environmental Exposure Limit—*recommended* limit set by the American Industrial Hygiene Association (AIHA)

Each of these agencies reviews toxicology information and sets the limits. If workers' exposures (what they actually breathe) are kept below these concentrations, most will not suffer any health effects. It is important to remember that some people are more sensitive to chemicals and may suffer effects, even at the lower levels.

Limits for How Long?

Each of the sets of limits listed above includes chemicals that cause acute effects, chronic effects, or both. Therefore, limits of exposure must take into account the duration of exposure. The following terms are used to describe the time factor of exposure:

- **TWA**—Time-weighted average—This limit is an average concentration over a period of time, usually 8 hours. It is intended to protect the worker against long-term (chronic) health effects.
- **STEL**—Short-term exposure limit—This limit is an average concentration for a 15-minute period. It is intended to protect against short-term (acute) health effects.
- **C**—Ceiling limit—This limit is the concentration of a contaminant in air that should never be exceeded.

The above exposure limits are established to protect workers in normal occupational settings. These limits are set very low because workers may be exposed to the chemicals repeatedly for months or even years. This means, however, that the limits are usually many times lower than the levels needed to protect emergency response personnel for brief, isolated exposures. In many cases, it would be unnecessarily conservative to compare emergency exposures to the PEL or TLV for a chemical.

Another problem with using occupational exposure limits in emergencies is when exposure occurs to people in the community. The general public is more likely to include people such as the very young, the elderly, or less healthy individuals. In this case, exposure limits intended to protect "healthy workers" may not be appropriate.

Emergency Exposure Limits

Exposure limits are established for use in emergency response situations. These limits take into account the problems mentioned above. Because emergencies involve so many potential problems, the limits are only estimates and are not intended to draw clear lines between "safe" and "unsafe." They do provide guidance that responders can use with good judgment to make decisions.

Immediately Dangerous to Life and Health (IDLH)
This limit is established by NIOSH for chemicals in the workplace. People exposed to concentrations higher than the IDLH without respiratory protection may suffer an effect within 30 minutes that makes them unable to rescue themselves. NIOSH intends that only the highest level of respiratory protection be used in exposures above the IDLH level.

Emergency Exposure Guidance Level (EEGL)
The Committee on Toxicology of the National Research Council established the EEGL and its companion, the Short-Term Public Emergency Guidance Level (SPEGL). The EEGL was established for about 70 substances and was intended as a maximum level to which young, healthy military personnel could be exposed

to during work in a single instance of a chemical release. The SPEGL was developed as a maximum level of exposure for the general public, but only for four substances. The same committee has published guidelines for how to develop Community Emergency Exposure Levels based on toxicological data but has not published any further lists of actual exposure levels.

Emergency Response Planning Guidelines (ERPG)

The American Industrial Hygiene Association publishes these limits (AIHA, 2002), which are actually a series of three limits for each chemical. The limits, ERPG-1, ERPG-2, and ERPG-3, represent estimates of progressive levels at which certain types of health effects would occur. They are defined as follows:

- **ERPG-1.** The maximum airborne concentration below which it is believed that nearly all individuals could be exposed for up to 1 hour without experiencing other than mild, transient adverse health effects or perceiving clearly defined, objectionable odor

- **ERPG-2.** The maximum airborne concentration below which it is believed that nearly all individuals could be exposed for up to 1 hour without experiencing or developing irreversible or other serious health effects or symptoms that could impair an individual's ability to take protective action.

- **ERPG-3.** The maximum airborne concentration below which it is believed that nearly all individuals could be exposed for up to 1 hour without experiencing or developing life-threatening health effects.

An example of these levels is methanol, with the following values:

- ERPG-1 = 200 ppm
- ERPG-2 = 1000 ppm
- ERPG-3 = 5000 ppm

Carbon monoxide is given these values:

- ERPG-1 = 200 ppm
- ERPG-2 = 350 ppm
- ERPG-3 = 500 ppm

AIHA provides full documentation for the guidelines that includes summaries of the animal toxicity data and human experience that is used to set the levels. These limits are peer reviewed and based on extensive toxicological studies. There are 100 chemicals in the 2002 edition of the guides (AIHA, 2002).

Temporary Emergency Exposure Limits (TEELs)

When one considers that only 100 ERPGs have been established, it should be clear that many chemicals have no such emergency limits yet. The Department of Energy (DOE) Emergency Management Advisory Committee developed a set of emergency guidance limits that could be used until ERPGs for other chemicals could be developed. The TEELs are not based on direct peer-reviewed studies of each chemical like the ERPGs but are values derived by mathematical calculations incorporating occupational exposure limits (PELs, TLVs, etc.) and toxicity data such as LD_{50}, TC_{LO}, TD_{LO}, and others (see Chapter 7). Because each limit is not studied as thoroughly as the ERPGs are, the TEELs are not considered as reliable. They are used by DOE facilities for emergency planning purposes for a given chemical only until an ERPG is developed for that chemical. There were over 1400 TEELs listed at the time this book was being written.

The TEELs follow a structure very similar to that the ERPGs:

- **TEEL-0**—The threshold level below which most people experience no appreciable risk of health effects
- **TEEL-1**—The maximum concentration in air below which it is believed nearly all individuals could be exposed without experiencing other than mild, transient adverse health effects or perceiving a clearly defined, objectionable odor
- **TEEL-2**—The maximum concentration in air below which it is believed nearly all individuals could be exposed without experiencing or developing irreversible or other serious health effects or symptoms that could impair their abilities to take protective action
- **TEEL-3**—The maximum concentration in air below which it is believed nearly all individuals could be exposed without experiencing or developing life-threatening health effects

The DOE's guidelines for using these limits indicate that they should be compared to 15-minute time-weighted exposure measurements. They would help emergency responders make decisions about the actions necessary to protect the public when the limits are applied to measurements made in a particular area.

Applying Exposure Limits

Emergency responders must protect the public from exposure to chemicals that can harm their health. The air monitoring data that are generated during the hazard and risk assessment process tell what the actual air concentrations are. Exposure limits are the estimates of what levels would be safe for most people. Whether or not these concentrations can cause harm to people depends on a number of factors, including their age and health.

The AIHA ERPG values are not intended to be used as exposure limits during a response, according to the preface of the guidelines, because they have not had the typical safety factors applied to the data. Like other exposure guidelines and limits, they will not be sufficiently protective for some extremely sensitive or vulnerable members of the population. They are intended as guidance for the planning process as estimates of the concentrations at which observable effects of progressive severity would be observed. Responders who want to use these values during a response should consult with knowledgeable industrial hygienists or other safety and health professionals to determine how the guidelines can be applied to the specific incident conditions.

None of the exposure limits is intended to be a fine line between "safe" and "unsafe" conditions. Where exposures are measured and found to be close to the limit, it is safest to assume that with a change in location or time the exposure will exceed the limit. Therefore, judgment and caution should always accompany the decisions responders make by using exposure limits. Although this may seem imprecise, having the exposure limit for a released chemical will give the incident commander a frame of reference when making response decisions.

SUMMARY

To determine whether there is too much chemical in the air at a hazmat release, responders must use and understand air surveillance equipment and procedures. Most information will come from air monitoring with direct-reading instruments. However, air sampling followed by laboratory analysis may provide more detailed and specific information that may be of use after the response. In any case, the proper use of air surveillance during a response will require prior training and practice.

10

TERRORISM AND WEAPONS OF MASS DESTRUCTION

Chemical, biological, and nuclear weapons have been used for hundreds of years to achieve military objectives. In recent years, terrorists have unleashed some of the same types of weapons on civilian populations, and intelligence information indicates that more such attacks are likely in the future. The use of weapons of mass destruction (WMDs) by terrorists presents significant new threats and new challenges for emergency responders. In this chapter, we examine terrorism, WMDs, and considerations for responding to those types of incidents. Several of the chapters that follow will build on the information presented here to provide additional information relating to WMD incidents.

TERRORISM AND WEAPONS OF MASS DESTRUCTION DEFINED

In 28 CFR Section 0.85, which provides operational guidelines for the Federal Bureau of Investigation, terrorism is defined as "the unlawful use of force and violence against persons or property to intimidate or coerce a government, the civilian population, or any segment thereof, in the furtherance of political or social objectives." This definition includes three key elements: (1) force is used illegally; (2) the use of force is intended to intimidate or coerce; and (3) the ultimate purpose is political or social in nature. The third element of the definition is the key distinction between acts of terrorism and more typical criminal uses of force such as armed robbery.

Emergency Responder Training Manual for the Hazardous Materials Technician, Second Edition, edited by Kenneth W. Oldfield
ISBN 0-471-21387-X   Copyright © 2005 John Wiley & Sons, Inc.

A weapon of mass destruction is defined under Title 18 of the United States Code as:

(1) Any explosive, incendiary, or poison gas, bomb, grenade, rocket having a propellant charge of more than four ounces, missile having an explosive or incendiary charge of more than one-quarter ounce, or mine or device similar to the above

(2) Any weapon or device that is designed or intended to cause death or serious bodily injury through the release, dissemination, or impact of toxic or poisonous chemicals or their precursors

(3) Any weapon involving a disease organism

(4) Any weapon that is designed to release radiation or radioactivity at a level dangerous to human life

From the perspective of the hazmat emergency response, we can think of a terrorist attack involving WMDs as a hazardous materials incident that is intentionally perpetrated to create a mass casualty incident in order to achieve political or social objectives. This means that training and procedures for hazardous materials emergency response provide a good foundation for preparing to respond to WMD incidents. At the same time, we must be aware of critical differences between the two types of events. WMD response is a specialized type of operation with specialized training and logistical requirements beyond the scope of regular hazmat operations. If we fail to make this distinction and approach a WMD incident as we would a traditional hazmat accident, we may quickly become part of the problem.

THE ROLE OF WMDs IN TERRORISM INCIDENTS

Although hazardous materials have been used as weapons of mass destruction for generations, recent events have focused our attention on this threat as never before. Let's examine several such events and some of the insights gained from them.

The Tokyo Subway Sarin Attack

In March of 1995, members of a cult known as Aum Shinrikyo released sarin, a lethal chemical warfare nerve agent, within a subway in Tokyo, Japan. As a result of the release 12 people died and 5500 injuries were reported. Of the reported injuries, 1000 were emergency responders. The Tokyo attack drew world attention to the threat of chemical warfare agents used in terrorist attacks on civilian populations.

The sarin used in the attack was reportedly a "homemade" version with only around 40 percent of the lethality of the military-grade agent. The attackers also lacked a means of effectively dispersing the agent. They simply carried the sarin into the subway in plastic bags and ripped the bags open with sharpened umbrella tips to release the agent. Were it not for these factors, the number of casualties resulting from the attack would have been much higher.

After the event, it was reported that only approximately a fifth of the 5500 injury cases initially reported were actually injured in the incident. The other four-fifths were "worried well" individuals who swamped the medical system by self-reporting simply because they were near the event when it occurred or because of psychosomatic symptoms.

The Bombing of the Murrah Federal Building in Oklahoma City

In April of 1995, a truck bomb was detonated in front of the Alfred P. Murrah Federal Building, a nine-story structure of concrete and steel in Oklahoma City, Oklahoma. Approximately 170 people died as a result of the attack. The explosive used was a mixture of two hazardous materials: ammonium nitrate, an oxidizer commonly used as an agricultural fertilizer, and fuel oil, a combustible liquid commonly used as motor vehicle fuel. When properly combined, the ammonium nitrate and fuel oil mixture is referred to as "ANFO," a common blasting agent used in mining and construction. Forty-eight hundred pounds of ANFO was used in the attack.

Timothy McVeigh, a lifelong U.S. citizen and decorated veteran of the Persian Gulf War, and Terry Nichols, who served with McVeigh in the U.S. military,

Figure 10.1 Terrorism involves the unlawful use of force for intimidation or coercion to achieve political or social objectives.

were tried and convicted for the attack on the Murrah building. McVeigh, the "trigger man" of the attack, has since been executed. The motivation for the attack was reportedly revenge for the deaths of David Koresh and his followers that occurred two years earlier during the siege by federal agents of the Branch Davidian compound near Waco, Texas.

The attack on the Murrah building reminded the United States of two significant facts about terrorism. Terrorist attacks on U.S. targets can have domestic as well as foreign origins. Commonly available substances can be employed by terrorists as weapons of mass destruction with devastating effects.

The September 11th, 2001 Attack on the World Trade Towers and Pentagon

On September 11th, 2001, airliners were hijacked by terrorists and flown into key U.S. targets, including both towers of the World Trade Towers (Fig. 10.1) and the Pentagon. A similar attack on a fourth target, possibly the White House, was apparently thwarted when passengers overpowered the hijackers and the plane crashed in Pennsylvania. Although it is difficult to place an exact number on the fatalities from these attacks, it appears that as many as 3000 civilian fatalities resulted. In addition, fatalities of emergency services personnel responding to the World Trade Towers incident included 343 firefighters and over 100 law enforcement personnel.

As millions watched the collapse of the World Trade Towers on television in horror, a new phase dawned in America's struggle with terrorism. The unprecedented attacks of September 11th, 2001 served as a major wake-up call and a major challenge to American security. The U.S. government responded with military action abroad and aggressive domestic homeland security provisions. In addition to the obvious death, injury, and property damage, the September 11th attacks also produced widespread fear and panic with far-reaching psychological and economic impacts.

Post-911 Anthrax Attacks

During the fall of 2001 a number of powder-filled envelopes containing anthrax spores were mailed to news media and politicians. As a result, 5 people died and 17 others were sickened. A number of the victims were U.S. Postal Service workers. The resulting contamination required the closing of postal facilities and Capitol Hill office buildings so that they could undergo expensive decontamination procedures. Large numbers of potentially exposed people were tested for anthrax and given precautionary antibiotics.

Coming in the wake of the September 11th attacks, the anthrax letters caused fear and panic throughout the nation. Many Americans were afraid to open their mail, and emergency response organizations were deluged with thousands of anthrax-related calls that turned out to be hoaxes or false alarms. As this book was being written, the FBI continues to investigate, but no arrests have been made in the anthrax attacks.

Future Terrorist Attacks: Anytown, U.S.A

Intelligence sources indicate that, in addition to direct attacks on civilian populations such as those described above, terrorists may broaden their attacks to include other types of targets. Future targets could include:

- Water supplies, which may be intentionally contaminated with toxic materials
- Food supplies, including agricultural and livestock resources, which may be contaminated or genetically altered
- Infrastructure targets, such as bridges; dams; key industrial facilities; and oil, gas, or electric distribution systems

We must realize that future attacks will not necessarily be limited to any particular region or type of occupancy. Terrorists may select "soft targets" or targets of opportunity, simply because they are available and not well defended (Fig. 10.2). Terrorism is psychological warfare, with a major goal of causing fear, panic, and disruption of normal activities. Because of the high level of concern that already exists and the extensive media coverage, attacks need not be large, spectacular, or directed at major targets to achieve this goal. Attacks that produce civilian casualties in small towns or average workplaces may produce widespread fear, panic, and disruption because the attacks involve settings in which most people tend to feel safe.

Figure 10.2 Terrorists may select "soft targets" or targets of opportunity, such as large hazmat containers in populated areas, simply because they are available and not well defended. Note the concrete barriers added to "harden" this potential target.

SPECIAL CONSIDERATIONS FOR TERRORISM EVENTS

WMD incidents differ significantly from other types of incidents involving hazardous materials that we may be called upon to respond to. For this reason, we have to adopt a different mind-set for WMD incident response.

Unique Aspects of WMD Events

To begin with, the WMD incident scene may have the combined characteristics of a disaster area, a hazmat incident scene, and a crime scene. We must treat the scene as a hazmat scene until proven otherwise. Responders must address evidence preservation issues or else important evidence can be needlessly lost. The incident will require a highly coordinated multiagency response involving huge logistical requirements, which will challenge the organizational capabilities of the Incident Management System (see Chapter 4). Mass casualty incidents will require that large numbers of victims be rapidly triaged, treated, and transported to hospitals. Triage for such an incident will require the use of a highly systematic approach, such as the START system described in Chapter 4. Specialized resource requirements, for example, the need for large numbers of nerve agent antidote kits, may be difficult or impossible to achieve. Operational boundaries may be large and poorly or incorrectly defined early in the response, and establishing and maintaining effective site control may be very difficult (see Chapter 11).

Special Hazards to Responders

Responding to a WMD incident may pose exceptional hazards to responders. At a minimum, these hazards can include the hazards of the agents used in the initial attack, because many of them are persistently hazardous. Scene hazards may also include unstable structures or other hazards created by the event. Responders may also be psychologically damaged by the event (see Chapter 16).

Unfortunately, terrorists may directly target responders (Fig. 10.3). This was clearly evident in a string of four bombings that occurred in Georgia and Alabama in 1996 and 1997. The attacks killed a police officer and 2 civilians and injured 130 people, including 23 emergency response personnel. The bomber methodically targeted responding fire and law enforcement personnel in all four cases. Secondary explosive devices were placed and timed specifically to target responders in two of the cases in Georgia, and 13 emergency services personnel were injured as a result. One key point to remember is that if a secondary device is detonated or discovered there may also be other devices on scene as well.

After the attacks of September 11th, 2001, intelligence information indicated that terrorists might attempt to steal emergency response vehicles and covert them into bombs. The information indicated that the terrorists might perpetrate an attack and then drive a vehicular bomb to the scene along with other responding vehicles. The bomb would then be detonated on-scene to devastate the response operation and responding agencies.

Figure 10.3   Terrorists may directly target emergency response personnel.

Responders have been warned to be wary of objects such as flashlights that might be lying around on-scene. This is reportedly due to several instances in the midwestern United States in which pipe bombs constructed with hand lights were left at the scenes of arson fires.

BASIC CONSIDERATIONS FOR WMD AGENTS

The types of WMD agents or materials available for the terrorist's arsenal are the same basic ones that have been used previously in military settings; however, as used by terrorists, they are more likely to be improvised in origin and unconventional techniques may to be used to deploy them. We can categorize the agents as biological, nuclear (or radiological), incendiary, chemical, or explosive in nature. The mnemonic device "B-NICE" is often used for this categorization.

A few basic distinctions apply when we consider WMDs. Incendiaries, explosives, and nuclear bombs do immediate harm through a sudden release of energy. Biological and chemical agents do harm through exposure to the material involved. The same thing applies for radiological dispersal devices, as described below.

It is important for us to be aware of basic considerations such as the likely physical state, general appearance, and container types for the B-NICE agents or devices. For biological and chemical agents, additional considerations apply, including degree of lethality or effectiveness, likely dispersion devices and methods, routes of exposure, persistency in the environment, and whether exposure effects are immediate or delayed. The same considerations will apply when radioactive materials are used in a radiological dispersal device.

Lethality and Effectiveness

Different agents have different levels of inherent ability to do harm. For example, some chemical agents are lethal in amazingly small doses, whereas other agents require a high level of exposure to produce toxic effects. For some biological agents, the victim must inhale or ingest very large numbers of the disease-causing organism, and for other agents a small number of organisms can cause disease. Likewise, for radiological materials, different radioactive isotopes emit different types and levels of radiation. The lethality and effectiveness of an agent will determine the concentration required, and thereby how much agent will be needed to stage an attack.

Routes of Exposure

The routes of exposure of an agent will determine how it will be deployed. Some agents use the respiratory system as a very effective route of entry and may be used to contaminate the atmosphere breathed by intended victims. Other agents are effective through skin contact and absorption and may be used to directly contaminate the skin of victims or to contaminate surfaces that the victims will contact. Agents that are effective through ingestion may be used to sabotage food and water supplies. Some agents are effective through multiple routes of entry. The routes of exposure of an agent have a direct bearing on dispersion techniques and other factors related to use of the agent by terrorists.

Dispersion

For biological and chemical agents to be used effectively as weapons, they must be effectively dispersed. In military applications, chemical agents may be dispersed through point source dissemination with items such as artillery shells, bombs, bomblets or cluster bombs, missiles, rockets, grenades, or mines. In some cases, chemical agents have been released directly from containers such as gas cylinders for point source dissemination. Chemical agents may also be dispersed through line source dissemination, such as from spray tanks attached to aircraft flying along a line oriented perpendicular to the wind direction and located upwind from the intended target. The same general options can be used to disperse biological agents as well. The anthrax attacks in the fall of 2001 relied on natural air currents to disperse the anthrax spores contained in the letters.

For agents targeting the victim's respiratory system, the dispersion method should produce an aerosol in the 1- to 5-micron particle size range. Particles in this size range tend to remain airborne and will be deposited in the lungs most effectively. Particles larger than 5 microns tend to settle out in the victim's upper airway instead of the lungs, and particles smaller than 1 micron tend to remain in the exhaled air. Larger particle sizes are more desirable for agents that target the victim's skin or for contamination of an area. For military purposes, we generally think of agents being dispersed in an outdoor setting, in which case weather conditions such as temperature, relative humidity, and wind speed and direction have a major impact on dispersion.

Terrorists may employ methods similar to those used in a military setting to disperse chemical and biological agents, although their equipment and techniques may be improvised or crude. Improvised explosive devices can be used for point source dissemination of agents, although the explosion may have the effect of incinerating the agent in some cases. Crop dusting equipment could be used for very effective line source dissemination of agents. Spray equipment attached to vehicles or boats, such as those used for mosquito control in some communities, could also be used.

Terrorists may select an indoor target area to make it easier to achieve an effective concentration of agent. In such cases, indoor heating, ventilation, and air-conditioning systems may be used to disperse the agent. Gas cylinders or suitcase-size aerosol generators might be used in such an attack. In reality, any type of spray equipment, such as a hand-pump garden sprayer or even a dry-agent fire extinguisher, could be used.

Persistency

Persistency refers to the length of time after deployment that an agent will continue to be effective in the environment. For chemical agents, persistency is directly related to the volatility of the agent. Chemical agents are generally considered persistent if they remain in a liquid state and able to cause harm for longer than 12 hours after deployment. Chemical agents that evaporate in minutes to hours are considered nonpersistent. Agents that remain in the liquid state for several hours but evaporate in less than 12 hours are considered semipersistent. We can apply the same concept to biological agents. Some biological agents, for example, anthrax spores, can remain viable for years in the environment, whereas some disease organisms die very quickly without a host. Radiological materials remain dangerous for many generations.

Agents or materials that are persistent in the environment pose an ongoing threat to people contacting them. Substances with long-term persistency disrupt normal activities in the contaminated area and may require expensive cleanup operations.

Immediate Versus Delayed Effects

As a general rule, the adverse affects of most chemical agents become apparent immediately after exposure, whereas the adverse effects of biological agents are typically delayed for a number of days. There are exceptions to this general rule.

For example, a lag of several hours typically precedes the onset of signs and symptoms following exposure to chemical mustard agents, as described below. Delayed onset of effects can delay recognition that an attack has occurred, which may have the effect of delaying response to the incident and life-saving medical treatment for victims.

TYPES OF WMD AGENTS AND THEIR HAZARDS

In this section of the chapter, we provide an overview of biological, nuclear or radiological, incendiary, chemical, and explosive agents or devices, using the B-NICE mnemonic. Keep in mind that the information provided here is somewhat generalized and limited because of the scope of this text. Other sources are available that deal solely, and therefore in greater detail, with terrorism and WMDs. Some of those sources are referenced here as appropriate.

Biological Agents

The human race has a long history of using biological agents to achieve military objectives. In the fifteenth and sixteenth centuries, Native Americans were given gifts and trade items that were infected with smallpox. In the early eighteenth century Russian troops catapulted the bodies of plague victims over the walls of a Swedish city under siege. During World War II, Japan used prisoners of war in extensive experimentation with biological agents that resulted in the deaths of as many as 3000 prisoners. Japan is believed to have used a variety of biological agents on Chinese soldiers and civilians, resulting in tens of thousands of deaths, during the war. In more recent times, terrorist organizations have turned their attention to biological agents. For example, Aum Shinrikyo is reported to have attempted to disperse biological agents on ten separate occasions in Tokyo in 1995. Fortunately, no infections resulted.

We can define biological agents as microscopic organisms that are capable of causing illness and/or death in human populations. For our purposes, we will expand the definition to also include toxins of biological origin. One thing all biological agents have in common is that we cannot see them or smell them and detection equipment currently available is of limited usefulness (see Chapter 9).

We can categorize biological agents as bacteria, viruses, rickettsia, or toxins (Table 10.1). The information discussed here is generalized, and only a few of the many potential biological agents will be used as examples for each category. Specific information on a number of biological agents is shown in Table 10.1. References are available that deal specifically with biological agents in greater detail. Examples include the U.S. Army's "blue book" or *Medical Management of Biological Casualties* (U.S. Army, 1996) and Jane's *Chem-Bio Handbook* (Sidell, Patrick, and Dashiell, 2000).

TABLE 10.1 Information on Selected Biological Agents

Agent Type	Disease	Likely Method of Dissemination	Transmissible Human to Human	Infective Dose	Incubation Period	Duration of Illness	Lethality	Persistence	Hazard Class
Bacterial	Anthrax	Spores in aerosol	No (except cutaneous)	8–10,000 spores (inhalation)	1–5 days	3–5 days (usually fatal)	High	Very stable—spores remain viable for years in soil	6.2
Bacterial	Cholera	1. Sabotage (food and water) 2. Aerosol	Rare	$>10^6$ organisms	12 h–6 days	$\geq$1 week	Low with treatment, high without	Unstable in aerosols and fresh water; stable in salt water	6.2
Bacterial	Pneumonic Plague	Aerosol	High	<100 organisms	1–3 days	1–6 days (usually fatal)	High unless treated within 12–24 h	For up to 1 yr in soil; 270 days in bodies	6.2
Bacterial	Tularemia	Aerosol	No	1–50 organisms	1–10 days	$\geq$2 weeks	Moderate if untreated	For months in moist soil or other media	6.2
Rickettsial	Q fever	1. aerosol 2. sabotage (food supply)	Rare	10 organisms (aerosol)	14–26 days	Weeks	Very low	For months on wood and sand	6.2

(continued overleaf)

TABLE 10.1 *(continued)*

Agent Type	Disease	Likely Method of Dissemination	Transmissible Human to Human	Infective Dose	Incubation Period	Duration of Illness	Lethality	Persistence	Hazard Class
Viral	Smallpox	Aerosol	High	Assumed low	10–12 days	4 weeks	High to moderate	Very stable	6.2
	Ebola	1. Direct contact (endemic) 2. Aerosol (BW)	Moderate	1–10 plaque forming units for primates	4–15 days	Death between 7 and 16 days	High to moderate No vaccine	Relatively unstable	6.2
	Venezuelan equine encephalitis	1. Aerosol 2. Infected vectors	Low	Assumed very low	1–6 days	Days to weeks	Low	Relatively unstable	6.2
Biotoxin	Botulinum toxin	1. Aerosol 2. Sabotage (food and water)	No	0.001 µg/kg is LD$_{50}$	Variable (hours to days)	Death in 24–72 h; lasts months if not lethal	High without respiratory support	For weeks in nonmoving water and food	6.1
	T-2 mycotoxins	1. Aerosol 2. Sabotage	No	Moderate	2–4 hours	Days to months	Moderate No vaccine	For years at room temperature	6.1
	Ricin	1. Aerosol 2. Sabotage (food and water)	No	3–5 µg/kg is LD$_{50}$	Hours to days	Days—death within 10–20 days for ingestion	High No vaccine	Stable	6.1
	Staphylococcal enterotoxin B (SEB)	1. Aerosol 2. Sabotage (food supply)	No	Clinical illness from picogram range	1–6 hours	Hours	<1% No vaccine	Resistant to freezing	6.1

Adapted from Medical Management of Biological Casualties (U.S. Army, 1996)

Not all microscopic organisms can be used as weapons. To be effective as a WMD, a biological agent must be capable of producing an event that involves:

(1) Large-scale loss of life
(2) Disruption of normal activities
(3) Panic, and
(4) Overwhelming of health care resources

To produce such an event, a biological agent must be highly lethal, cheap and easy to produce in large quantities, stable in aerosol form, and readily dispersed as an aerosol in the 1- to 5-micron particle range. From the terrorist standpoint, the ideal biological agent would also be communicable from person to person with no treatment or vaccine. Note that an agent need not meet all these criteria to be an effective tool for terrorists. Microorganisms may be genetically "tweaked" to produce strains that are more effective as biological agents. Some very effective biological agents are relatively inexpensive and easy to produce. For this reason, biological agents in general have been referred to as the "poor man's atom bomb."

Biological agents differ from other types of agents in the B-NICE assemblage because the effects of the biologicals are delayed, typically for several days. In contrast, the effects of the other types of agents are usually immediately apparent. In a hypothetical scenario, if a biological agent were released at a large event such as a political rally or sporting event, thousands of people might be infected but none of them would begin to show signs or symptoms until well after the event. By that time, people may have returned to their homes all over the country or possibly to other parts of the world. Initial signs and symptoms would probably be mistaken for common illnesses, causing the victims to self-report to doctor's offices or emergency rooms at widespread locations. Because recognition, response, and treatment would all be delayed, recognizing and responding to a biological attack would be a major public health challenge. To meet this challenge, computer surveillance systems are being brought on line to detect unusual "spikes" of reported illnesses that might be related to a biological attack.

Biological agents vary significantly in appearance. Most Americans seem to think of them as existing in dry powder form, probably because of publicity surrounding the anthrax letter attacks in the fall of 2001. Biological agents may also be encountered in liquid form and are actually cheaper and easier to produce as liquids. Before being dispersed, biological agents typically appear either as liquids or finely divided solids, both of which may vary significantly in coloration.

Bacterial Agents
Bacteria are single-celled organisms capable of independent growth, which means that they can reproduce and grow outside the body of a host organism. Bacterial agents reproduce through cell division or the formation of spores and can therefore be cultured. Although microscopic, bacteria are fairly large as microbes

go and can be seen with a regular light microscope. Examples of diseases caused by bacteria that are considered effective biological agents include anthrax, bubonic and pneumonic plague, brucellosis, glanders, tularemia, and cholera. In this section, we provide information on anthrax as an example.

Anthrax is a disease caused by the bacterium *Bacillus anthracis*, which is endemic in some populations of hoofed animals. *Bacillus anthracis* reproduces through the formation of resistant spores. The spores can remain dormant for years and then become active and infect a host. Anthrax is a zoonotic disease, meaning that it can be transmitted from animals to people. Anthrax occurs in human populations in three forms: cutaneous or skin anthrax, intestinal anthrax, and inhalation anthrax. Only the cutaneous form is communicable from person to person.

Cutaneous anthrax occurs periodically in humans and is usually caused by direct contact with contaminated hides, tissues, or wool from contaminated animals. Symptoms begin from three to five days after exposure with the development of painful ulcerated sores on the skin. A characteristic black scab typically forms over the sore. Sores on the skin may develop into systemic infections, which may be fatal if untreated. The disease is 25% fatal if left untreated. Cutaneous anthrax is readily treatable with common antibiotics. The fatality rate is very low with treatment, even if treatment is not begun until after the onset of signs and symptoms.

Intestinal anthrax is an acute inflammation of the intestinal tract that develops after the ingestion of spores. It is often caused by eating contaminated meat. Signs and symptoms include nausea, loss of appetite, vomiting, and fever followed by abdominal pain, bloody vomiting, and severe diarrhea. Intestinal anthrax is almost 100% fatal if untreated. It can be treated with common antibiotics but is still 26–60% fatal with treatment unless treatment is begun before the onset of symptoms.

Inhalation anthrax is a severe lung infection caused by inhalation of the spores. It begins with a sudden onset of flulike symptoms from one to five days after exposure. Over the following three to five days the patient experiences severe respiratory distress characterized by shortness of breath, a gasping cough, rapid heart rate, sweating, chills, fever, exhaustion, cyanosis, and terminal shock leading to death. In many cases, patients will experience an initial phase of flulike symptoms followed by a symptom-free period, after which the patient will become extremely ill and then die. Inhalation anthrax is almost universally fatal without treatment and is 80–90% fatal even with treatment if treatment is begun after the onset of symptoms. However, prophylaxis, or precautionary treatment after exposure and before the onset of symptoms, with antibiotics is very effective in preventing inhalation anthrax. Fortunately, it is not communicable from one person to another. A vaccine is available to prevent anthrax.

Inhalation anthrax is the form of the disease most likely to be favored by terrorists. They can infect a target population by dispersing spores, which are then inhaled by the victims. An infectious dose, typically 8000 to 10,000 spores, is smaller than a speck of dust. Anthrax can be weaponized by combining spores from especially infectious and lethal strains of *Bacillus anthracis* with some sort of dispersing medium, such as clay. The resulting agent is then milled to produce individual particles in the 1- to 5-micron particle range.

The potential devastation of weaponized anthrax agent is amazing. This was well illustrated by an event that occurred in the Soviet Union in 1979. Weaponized anthrax was apparently released from a secret biological warfare research facility located at the city of Sverdlovsk. Information is sketchy, but apparently a relatively small amount of agent was released—perhaps as little as a gram—and resulted in a number of casualties—estimates range from 120 to over 400 deaths—among residents of Sverdlovsk. The World Health Organization estimates that the release of 110 pounds of anthrax spores along a 1.25-mile line upwind of a city of 500,000 would produce 125,000 anthrax cases resulting in 95,000 deaths.

Viral Agents

Viruses are very small microorganisms that are too small to be seen with a regular light microscope. They are incapable of independent growth and must co-opt a host cell's reproductive mechanism to replicate. Examples of diseases caused by viruses that are considered effective biological agents include smallpox, Venezuelan equine encephalitis, and viral hemorrhagic fevers (Ebola, yellow fever, dengue fever, and hanta). In this section, we provide information on smallpox as an example.

Smallpox is caused by the variola virus and was a serious public health problem during much of the twentieth century. The disease was systematically eradicated by 1980 through aggressive public health practices such as quarantining exposed people and vaccinations. Two laboratory stockpiles of the virus are known to exist, one at the CDC in Atlanta, Georgia and one at Novizbersk, Russia; however, it is suspected that clandestine stockpiles may exist and be available to terrorists.

After a person is infected with the variola virus, a symptom-free incubation period of 10 to 12 days follows. After that, the onset of acute symptoms begins with malaise, fever, headache, backache, and vomiting. Two to three days later an eruption of the mucous membranes begins, which is accompanied by a rash on the face and hands. Over the following week, the rash spreads over the trunk and lesions form that develop into pustules (Fig. 10.4). The pustules scab over and leave scars on healing. Large areas of skin may slough off, and survivors are often severely scarred for life. The mortality rate is 30–40%, with death occurring because of internal lesions and sloughing of tissues within the digestive tract. Treatment for the disease primarily consists of supportive care and measures to prevent secondary infections.

The variola virus is considered an agent likely to be favored by terrorists because it is easily produced in large quantities, very stable, easily disseminated as a respirable aerosol, highly infectious, and communicable from person to person. A vaccine for smallpox exists, but few American citizens are currently vaccinated. Because of heightened concern about a terrorist smallpox attack after the attacks

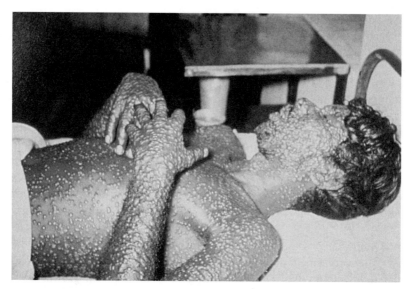

Figure 10.4 Smallpox is a disease likely to be favored for biological terrorism. (Source: Centers for Disease Control and Prevention, www.bt.cdc.gov/agents/smallpox/smallpox images).

of September 11th, 2001, the U.S. Department of Health and Human Services developed a National Smallpox Vaccination Program. The program guidelines called for vaccinations to proceed in three phases, with the first two phases consisting of voluntary smallpox response team members. The three phases were designated as follows.

- Phase 1 included cadres of health care providers in hospitals who would staff wards designated for smallpox patients in the event of an outbreak.
- Phase 2 included first responders and emergency medical personnel.
- Phase 3 is to be a vaccination of the general population that may be called for at a future time.

The initial goal of the program was to vaccinate 450,000 people in the first two phases. Controversy erupted over possible side effects, hazards, and contraindications of the vaccination. As a result, the number of personnel volunteering to be vaccinated to date has been much lower than projected when the program began. Information from the CDC for making an informed decision regarding the vaccination is available at www.bt.cdc.gov.

Rickettsial Agents

Rickettsia are organisms that are similar to bacteria; however, like viruses they are very small and incapable of independent growth. A type of rickettsia causes Rocky Mountain spotted tick fever, a disease spread by ticks. Rickettsial diseases that are considered effective biological agents include Q fever and typhus. In this section, we provide information on Q fever as an example.

Q fever is caused by the rickettsia *Coxiella burnetii* and is endemic in some hoofed animal populations. It is zoonotic, or capable of being transmitted from animals to humans, but is not contagious from one person to another. After a person is infected with Q fever, an incubation period of 10 to 20 days is followed by the development of severe flulike symptoms that last from two days to two weeks. Although incapacitating, Q fever is typically not fatal.

Coxiella burnetii is easy to produce in large quantities and is one of the most robust biological agents. It is so highly infectious that inhalation of a single organism is believed to be sufficient to cause Q fever. These factors make it attractive to terrorists even though it is not inherently lethal. It would most likely be dispersed as an aerosol or possibly used to contaminate food supplies.

Biological Toxins

Some highly toxic chemicals are produced by living organisms such as bacteria, plants, and animals. Examples of biological toxins that have been used by terrorists include ricin, botulinum toxin, and tricothecene mycotoxins. Toxins may be used to directly target people or to sabotage food and water supplies. In this section we provide specific information on ricin as an example.

Ricin is a product of the castor bean plant. It is made from the mash remaining when the castor beans are processed to produce castor oil. Ricin is easy to produce, stable, and readily available in large quantities. It is a cytotoxin, meaning that it kills cells on contact. It is extremely toxic in very low concentrations, and it is reported that as little as a milligram can kill a human being.

Inhalation, ingestion, and injection are potential routes of entry for ricin. It is highly lethal through injection but is much less effective through inhalation and ingestion. In the case of inhalation exposure, symptoms would begin in 1 to 12 hours with weakness, fever, cough, and pulmonary edema. Severe respiratory distress would follow, with death occurring from respiratory failure in 36 to 72 hours. The effects would occur more quickly in a case of ingestion or injection. There is no known antidote or vaccine, and treatment consists of supportive therapy for the patient.

In one well-known case, Bulgarian intelligence operatives used a device disguised as an umbrella to inject a pellet of ricin into the body of Georgi Markov, a Bulgarian political dissident living in London. Markov died one day later of multiple organ failure.

Nuclear or Radiological Devices

Terrorists may attempt to incorporate radioactive materials into their arsenals in a couple of ways. Radioactive materials can be used in nuclear devices to create an atomic or nuclear explosion. Radioactive materials can also simply be scattered by a conventional explosive attached to a radiological source. In this section, we discuss both types of devices.

Nuclear Devices

One of the most devastating scenarios imaginable would be the detonation of a conventional nuclear device in a highly populated area by terrorists (Fig. 10.5). Such an event could produce a level of death and destruction equivalent to that resulting from the atomic bombs detonated at Hiroshima and Nagasaki in Japan during World War II. Such a device produces tremendous blast effect, thermal effect, lethal amounts of radiation, and radioactive fallout. For example, at Hiroshima the death toll within one-half mile of ground zero was 100 percent, blast damage extended to one and one-half miles, and a firestorm caused by the thermal effect burned over four and one-half square miles of the city.

Fortunately, all indications are that a Hiroshima-type device is not likely to be used by terrorists. Conventional nuclear devices are large and expensive, and they require weapons-grade nuclear material and significant infrastructure for production. There is quite a low probability that such a device could be developed or deployed by terrorists, because they must exist as a "shadow network" with the ability to move quickly and frequently to avoid capture. However, the involvement of a "rogue state" or state-supported terrorism might make such a device available to terrorists, so the possibility cannot be completely ruled out. It is possible that an improvised low-yield "mini nuke" could be developed and used by terrorists. For now, government officials consider the likelihood of terrorists initiating an atomic explosion to be low.

Figure 10.5 Detonation of a conventional nuclear device. (Source: Office of Domestic Preparedness).

One type of nuclear device that could be cause for concern is the Special Atomic Demolition Munition (SADM), which is a small tactical nuclear device developed by the former Soviet Union (Fig. 10.6, A and B). SADMs are sometimes referred to as "backpack nukes" or "suitcase nukes". The Soviets intended for the device to be deployed by two parachutists against targets in harbors or other strategic locations. A number of these devices have reportedly not been accounted for in the Soviet arsenal, but at present there is no actual evidence that any have been stolen.

One device comparable to a SADM was the Davy Crockett artillery round developed and tested by the United States in the 1950s. This device is believed to produce about the same effect as a suitcase nuke or a terrorist-constructed mini nuke. In tests of the Davy Crockett device in the 1950s, it produced limited blast and thermal effects but significant amounts of radiation. It is estimated that people within 330 feet of ground zero of the Davy Crockett warhead would experience only second-degree thermal burns but would receive lethal doses of radiation. For people located within 1150 feet (0.2 miles) of ground zero, 95% would die from radiation within days or weeks. Fallout from the blast could result in enough contamination to require exclusion of the public in scattered areas for up to 50 miles downwind.

Most people associate a dramatic mushroom cloud with a nuclear explosion (Fig. 10.5), but not all nuclear devices will produce that effect. Suitcase nukes may produce only a small fireball from the initial blast followed by a "mushroom

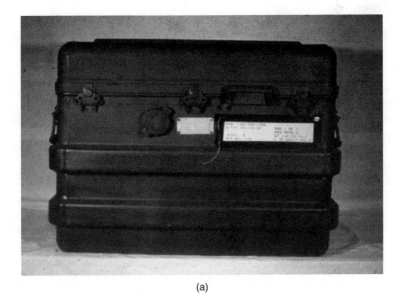

(a)

Figure 10.6 The Special Atomic Demolition Munition (SADM) developed by the Soviet Union has also been referred to as a "suitcase nuke" (A) or "backpack nuke" (B). (Source: Office of Domestic Preparedness).

(b)

Figure 10.6 (*continued*)

cloud" that is actually a vertical column of light brown to white smoke. It could easily be mistaken for a large truck bomb. Keep in mind that most conventional fires and explosions, for instance those involving hydrocarbon products, typically produce dark brown to black smoke that tends to spread out instead of rising in a column. Also, a nuclear device may produce an electromagnetic pulse that knocks out electronics.

Radiological Dispersal Devices

A radiological dispersal device (RDD) or "dirty bomb" is an improvised device in which a conventional explosive device is used to scatter radioactive material and contaminate a target area (Fig. 10.7). The dirty bomb does not involve a nuclear explosion and does not require weapons-grade material. Any type of radiological material could be used, including radioactive sources from hospitals, research labs, and industrial facilities or radioactive waste materials. The size, distribution, and degree of radioactive contamination resulting would vary with factors such as the amount and type of radiological material used, the size and type of explosive charge, and weather conditions at the time of the detonation (Fig. 10.8). The U.S. government currently considers the detonation of dirty bombs probable because such a device would be cheap and easy to produce and deploy.

Figure 10.7 A radiological dispersal device (RDD) or "dirty bomb" is an improvised device in which a conventional explosive device is used to scatter radioactive material and contaminate a target area. (Source: Office of Domestic Preparedness).

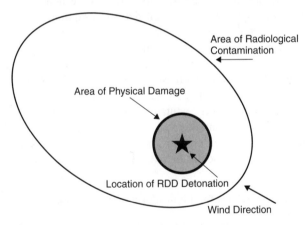

Figure 10.8 In the event of the detonation of an RDD, the immediate harm will probably result mainly from the blast effect associated with the explosive charge. The size, distribution, and degree of radioactive contamination resulting will vary with factors such as the amount and type of radiological material used, the size and type of explosive charge, and weather conditions at the time of the detonation.

A dirty bomb could take many forms ranging from crude to sophisticated and could be difficult to recognize before detonation. In some cases, containers and labels associated with storage and transportation of radioactive materials may be recognized (Fig. 10.7). In other cases, the explosive charge, either improvised or conventional, may be more readily recognized than the radiological material.

Once a dirty bomb is detonated, it would be easy for first responders to initially mistake it for a conventional bombing event.

In the event of the detonation of an RDD, the immediate harm would probably result mainly from the blast effect associated with the explosive charge. Opinions currently vary as to the likelihood that lethal doses of radiation would result. The U.S. Department of Justice's Office of Domestic Preparedness considers it unlikely that a mass casualty incident would result from the radiation dispersed by a dirty bomb. The scattering of the radiological material would tend to dilute the radiation it gives off. However, the explosion alone might produce a mass casualty incident if a large explosive charge is used.

One of the main effects of radiation from a dirty bomb would be psychological. If enough radiation is present to merely be detected with radiation meters, panic of the public and disruption of normal activities will result because the public tends to misunderstand and fear radiation in general. In the event of an RDD event, regional sites equipped for survey and decontamination of personnel may need to be set up. Expensive decontamination operations may be required in contaminated areas.

Considerations for Responding to Nuclear or Radiological Incidents

In responding to any nuclear or radiological incident, it is important to distinguish between safe and unsafe areas. This can only be done with certainty by trained personnel equipped with radiation detection equipment, as described in Chapter 8. EPA has established dose limits for personnel exposed to radiation during emergency procedures, as shown in Table 10.2.

Remember that the three ways of limiting exposure to ionizing radiation are to limit time of exposure, increase distance from the source, and take a position behind available shielding (Chapter 8). Also, it is important to avoid inhaling, ingesting, or becoming contaminated with radioactive material. This is done through avoiding contaminated areas or by using appropriate protective equipment and decontamination protocols when entry into such areas cannot be avoided. Responding to nuclear or radiological incidents requires specialized

TABLE 10.2   EPA Guidelines for Limiting Exposure to Ionizing Radiation During Emergency Procedures

Dose Limit	Emergency Activity Performed	Condition
5000 mrem	All activities	All activities during emergency
10,000 mrem	Protecting major property	Where lower dose not practicable
25,000 mrem	Lifesaving or protection of large populations	Where lower dose not practicable
More than 25,000 mrem	Lifesaving or protection of large populations	Only on a volunteer basis to persons fully aware of the risks involved.

training. One source for such training is the U.S. Justice Department's Office of Domestic Preparedness (www.ojp.usdoj.gov/odp).

Incendiary Devices

Incendiaries are materials that burn readily, such as flammable gases, liquids, and solids. Because these are readily and widely available, they are favored weapons of the terrorist. They can be used to target people and property with devastating effect. By igniting ordinary combustibles, incendiary materials can produce destruction far in excess of what the material alone would be capable of. This was demonstrated in February of 2003 at Daegu, South Korea, when a milk container of flammable liquid was ignited within a subway car. The flammable liquid ignited the car's seats and floor tiles, which were highly flammable. The resulting fire spread through two subway trains and killed 189 people.

From the terrorist's viewpoint, incendiary devices can be simple and inexpensive yet highly flexible, effective, and efficient in inflicting death and property damage. To use incendiary weapons requires only three things: a material to act as a fuel, a triggering mechanism, and a delivery system (Fig. 10.9).

Fuel for an incendiary device can range from a soft drink bottle full of acetone, to a passenger airplane full of jet fuel, to a 600,000-gallon tank of liquefied

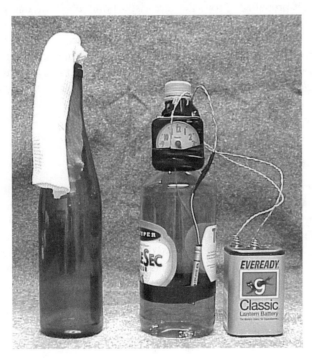

Figure 10.9 Incendiary devices require three things: a material to act as a fuel, a triggering mechanism, and a delivery system.

petroleum gas. To predict what materials may be used by terrorists, look at the materials that are readily available to them. Commonly available fuels include flammable gases, flammable and combustible liquids, and flammable solids. Oxidizers may be used to intensify the combustion of the fuel. Some combinations of materials are hypergolic and will ignite on contact of the components without an external source of ignition. Sophisticated devices of military origin using napalm, thermite, or white phosphorus may be available to terrorists or may be improvised by them.

Triggers for incendiary devices may be chemical, electronic, or mechanical in nature. For simple mechanical ignition, matches may be used to directly ignite a device or match heads may be rigged in a variety of ways to produce ignition mechanically. Electrical devices can be rigged to produce ignition and can be controlled from a remote location or by using a timer for delay. Chemical ignition can be provided by any chemical reaction that produces sufficient heat—for example, an organic liquid mixed with a strong oxidizer, which also produces extra oxygen. An old arsonist's trick is to place the organic liquid inside a latex condom and place the condom with the oxidizer for a delayed ignition. The process is delayed for the length of time it takes for the organic liquid to degrade the latex in the condom. When the condom fails, the two materials intermingle and ignition occurs.

With regard to delivery options, incendiary devices may be planted, hand-thrown, or self-propelled. From the terrorist's point of view, one of the most attractive options may be to select materials that are already at the target location, so that delivery is not required. In other instances, terrorists may carry small quantities of incendiaries to release and ignite at the target or conceal with a delayed ignition device. After the September 11th, 2001 attacks intelligence sources indicated that terrorists might try to stage an attack with hijacked gasoline tankers that could be driven to the target location. More sophisticated incendiary devices may be rocket propelled.

Chemical Agents

The toxic properties of hazardous chemicals were used to achieve military objectives in a number of instances during the twentieth century. The Germans used chemicals such as chlorine, phosgene, and mustard agent with devastating effects during the First World War. Chemical agents were used more recently by the Iraqi regime of Saddam Hussein against the Iranians during the Iran-Iraq war and against Kurdish separatists in northern Iraq. The sarin attack in the Tokyo subway in 1995 made the world painfully aware that chemical weapons could also be employed by terrorists.

In this section, we examine five classes of chemical agents: nerve, blister, blood, choking, and irritant agents. Hazard classification information for chemical agents is shown in Table 10.3. The properties of selected chemical agents are summarized in Table 10.4. In this section we provide a general overview of chemical agents, using a few as examples. More detailed information on a wider

TABLE 10.3   Chemical Agent Hazard Classification Information

Agent Type	Agent Name	Symbol	DOT Class	NFPA 704	DOT ERG
Nerve agent	Tabun	GA	6.1	421	153
	Sarin	GB	6.1	411	153
	Soman	GD	6.1	411	153
	V Agent	VX	6.1	411	153
Blister agent	Sulfur mustard	H, HD	6.1	411	153
	Nitrogen mustard	HN	6.1	411	153
	Lewisite	L	6.1	411	153
	Phosgene oxime	CX	6.1	411	153
Blood agent	Hydrogen cyanide	AC	6.1	442	117
	Cyanogen chloride	CK	2.3	402	125
Choking agent	Chlorine	CL	2.3	300	124
	Phosgene	CG	2.3	401	125
Irritant agent	Tear gas	CS	6.1		159
	Mace	CN	6.1		153
	Pepper spray	OC	6.1		159

array of chemical agents is available through sources such as the U.S. Army's
"green book" or *Medical Management of Chemical Casualties* (U.S. Army, 1995)
and Jane's *Chem-Bio Handbook* (Sidell, Patrick, and Dashiell, 2000). Response
to incidents involving chemical agents requires specialized training. One source
for that training is the U.S. Justice Department's Office of Domestic Preparedness
(www.ojp.usdoj.gov/odp).

Nerve Agents

Nerve agents are similar to organophosphate insecticides, such as Malathion,
Diazinon, and parathion, in terms of the way in which they affect an organism
exposed to them. They restrict the action of enzymes that are critical for the
proper functioning of the nervous system. For a nerve impulse to cross from
the nervous system to a muscle, a neurotransmitter called acetylcholine must be
released at the neuromuscular junction. After this has occurred, it is critical for an
enzyme called acetylcholinesterase to be secreted to deactivate the acetylcholine,
thus deactivating the neuromuscular impulse. The nerve agents inhibit the action
of the acetylcholinesterase, and thereby cause the nervous system to go haywire.
This is why convulsions and spasmodic motions are classic signs of nerve agent
exposure. Other common signs include excessive:

- Salivation, sweating, and secretions from mucous membranes
- Lacrimation (or excessive formation of tears)
- Urination
- Defecation
- Gastric distress

TABLE 10.4 Properties of Selected Chemical Agents from Various Sources

Type	Symbol/Common Name	Volatility Persistency	CAS Number	PEL ppm	IDLH ppm	LCt$_{50}$ ppm	Hazard	Symptoms	Physical Characteristics
Nerve	Tabun/GA	Semipersistent	77-81-6	0.000015	0.03	60	Respiratory, skin, eyes	Pinpointing of the pupils; dimness of vision; runny nose/salivation; Tightness of chest; difficulty breathing; twitching or paralysis; tachycardia; vomiting; loss of consciousness; convulsions	Colorless to lightly colored liquid at normal temperature; G-agents slightly less volatile than water; V-agents about as volatile as motor oil.
	Sarin/GB	Nonpersistent	107-44-8	0.000017	0.03	12	Respiratory, skin, eyes		
	Soman/GD	Semipersistent	94-64-0	0.000004	0.008	9	Respiratory, skin, eyes		
	VX	Persistent	50782-69-9	0.0000009	0.0018	3	Respiratory, skin, eyes		
Blister	Sulfur mustard H/HD	Persistent	505-60-2	0.0005	Unknown	231	Respiratory, skin, eyes	Reddening of skin; blisters; eye pain and reddening; eye damage; Coughing; airway irritation and damage; eye effects may appear in a few hours. Respiratory effects and blisters in 4–24 hours; Can be lethal in large doses.	Oily light yellow to brown liquids with a strong odor of garlic, onion, or mustard; fishy odor for HN series.
	HN—1 Nitrogen mustard	Persistent	538-07-08	0.0004	Unknown	216	Respiratory, skin, eyes		
	HN—2	Persistent	51-75-2	Unknown	Unknown	470	Respiratory, skin, eyes		
	HN—3	Persistent	555-77-1	Unknown	Unknown	179	Respiratory, skin, eyes		

Category	Agent	Persistence	CAS Number				Route	Symptoms	Physical Description
	Phosgene oxime/CX	Persistent	35274-08-9	Unknown	Unknown	687	Respiratory, skin, eyes	Immediate burning; weal-like skin lesions; eye and airway irritation and damage	A solid below 95°F, but vapor can result.
	Lewisite/L	Persistent	541-25-3	0.00035	0.0004	165	Respiratory, skin, eyes	Immediate pain or irritation of skin; other symptoms similar to the H agents.	Oily colorless liquid with the odor of geraniums; more volatile than H
Blood	Hydrogen cyanide/AC	Nonpersistent	74-99-8	10	45	3600	Respiratory	Cherry red skin or lips; rapid breathing; dizziness, nausea, vomiting; Headache; convulsions; death.	Rapidly evaporating liquids; gas or liquefied compressed gas
	Cyanogen chloride/CK	Nonpersistent	506-77-4	0.2	Unknown	4375	Respiratory		
Choking	Phosgene/CG	Nonpersistent	75-44-5	0.1	2	791	Respiratory	Coughing, choking	Odor of freshly mown hay
	Chlorine/Cl	Nonpersistent	7782-50-5	1.0	30	6551	Respiratory	Coughing, choking	Odor of swimming pool

- Emesis (or vomiting)
- Miosis (or pinpoint pupils)

To help remember these signs, use the mnemonic device "SLUDGEM."

Military nerve agents include GA or tabun, GB or sarin, GD or Soman, and VX or V-agent. All have differing specific characteristic as shown in Table 10.4. As a group, they will typically appear as an odorless liquid resembling water or light oil. They may be dispersed by a variety of techniques, as discussed above in this chapter. Nerve agents as a group are quite lethal in very small doses. It has been reported that one fluid ounce of sarin concocted by using information available on the Internet can represent 222,000 fatal doses. Remember that common organophosphate insecticides are similar to, although not as potent as, the nerve agents and may be selected for targets of opportunity by terrorists.

Once a victim has received an effective dose of a nerve agent and is exhibiting strong convulsions and SLUDGEM signs, he or she will most probably die unless quickly given an antidote for the nerve agent. The standard military antidote kit is the Mark I Auto injector, which contains 2 mg of atropine and 600 mg of 2-PAM chloride.

Vesicants

Vesicants are commonly referred to as blister agents because they cause dramatic blistering and irritation to the skin and eyes (Fig. 10.10). Vesicants that might

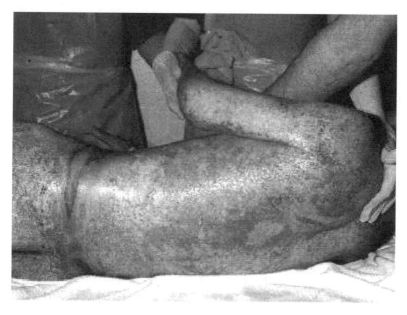

Figure 10.10 Vesicants are commonly referred to as blister agents because they cause dramatic irritation and blistering to the skin. (Source: U.S. Department of Justice and Federal Emergency Management Agency, 1997).

be used by terrorists include sulfur mustard agents, nitrogen mustard agents, lewisite, and phosgene oxime (Table 10.4). The mustard agents, which are commonly referred to simply as "mustard," have previously been used with brutal effectiveness in military operations and are discussed here as an example of the vesicant class.

Mustard agents are heavy, oily liquids that are very persistent in the environment. They are clear liquids in pure form, but may range from yellow to brown or black with impurities, and may have an odor of mustard, onion, or garlic. Routes of exposure include skin contact and inhalation. Characteristic effects include severe irritation to the skin and eyes leading to the formation of blisters and irritation of the upper airway. The gastrointestinal system and central nervous system may also be injured.

Exposure to a few drops of mustard is sufficient to cause severe injury. Mustard is intended to disable and disfigure rather than to kill but may be fatal if large doses are received. Mustard agent begins damage to the cells of the body immediately after exposure, but the victim experiences no signs or symptoms for hours. There is no antidote for the effects of mustard, only supportive therapy. The only means of preventing the damage is immediate decontamination after exposure; however, because of the delayed signs and symptoms, the need for decontamination may not be apparent until too late.

Lewisite and phosgene oxime produce effects similar to those of the mustard agents. However, there is no significant lag between exposure and the onset of signs and symptoms after exposure to lewisite or phosgene oxime.

Blood Agents

Blood agents, or cyanides, act as chemical asphyxiants by interfering with the body's ability to use oxygen at the cellular level. Cyanide agents suitable for chemical warfare include hydrogen cyanide, which is also known as hydrocyanic acid, and cyanogen chloride (Table 10.4). Both are commonly used industrial chemicals.

Hydrogen cyanide is a liquid with a very high vapor pressure (740 mmHg), so it evaporates very quickly. It is highly flammable in addition to being highly toxic. Vapors are lighter than air and reportedly have an odor of bitter almonds, although many people cannot smell the odor. Cyanogen chloride is a highly toxic, nonflammable gas or liquefied compressed gas that is heavier than air and has a pungent odor.

The cyanides kill quickly when inhaled in high concentrations. Skin absorption is an effective route of entry when liquid or high concentrations of gas or vapor contact the skin. The cyanides are not very effective for deployment as chemical agents in an outdoor setting because of the difficulty in maintaining a high enough concentration to be effective. This is especially true of hydrogen cyanide because the vapors are lighter than air. However, the cyanides may be utilized effectively if terrorists release them in an enclosed area. Members of Aum Shinrikyo apparently tried to stage such an attack in the Tokyo subway by mixing a cyanide salt with a strong acid a few weeks after the sarin attack.

The attack failed because the two components failed to mix. Chemical experts have estimated that the amount of hydrogen cyanide produced would have been enough to kill 10,000 to 20,000 people if the attack had succeeded.

Signs and symptoms of cyanide exposure include nausea, weakness, dizziness, and anxiety. In some cases, eye irritation similar to that produced by irritating agents may occur. The victim's skin may have a reddish coloration. Exposure to high concentrations can cause loss of consciousness in a matter of seconds, followed by convulsions and respiratory failure. Treatment for cyanide exposure includes rapid evacuation from the affected area and providing respiratory support with high volumes of oxygen. Cyanide antidote kits containing amyl nitrate, sodium nitrite, and sodium thiosulfate are available.

Choking Agents

Choking agents are gases such as chlorine and phosgene that attack the respiratory system. They are sometimes referred to as pulmonary agents or lung agents. Both chlorine and phosgene were used by Germany as military choking agents in World War I. They are now commonly used as industrial chemicals and may be viewed as agents of opportunity by terrorists. This is especially true of chlorine, because it is commonly stored and used in large quantities near highly populated areas at locations such as municipal water treatment plants.

Choking agents are transported and stored as liquefied compressed gases in various high-pressure containers such as cylinders, ton containers, and pressure railcars. Once released from the container, the liquefied gas will boil to quickly release large amounts of gas, which may form a dense yellowish to greenish cloud. The gas is significantly heavier than air, will tend to accumulate and concentrate in low-lying areas, and can remain at deadly concentrations for significant distances downwind.

The choking agents are acid gases with characteristic pungent odors. Chlorine is commonly used for swimming pool water treatment, and many people are familiar with the odor from that application. Phosgene has the odor of freshly mown hay, but only in nearly dangerous concentrations.

The choking agents form an acidic residue in contact with moist tissues. This has the immediate effect of painful irritation to the eyes and irritation to the respiratory system that results in painful coughing and choking. These symptoms will abate soon after exposure is terminated; however, when inhaled in high concentrations the acid formed within the lungs does severe damage to the lung tissues. As a result, pulmonary edema, or fluid on the lungs, will begin from 2 to 24 hours after exposure. This can result in the victim essentially "drowning" because of pulmonary edema. Skin contact with high concentrations may produce skin irritation, especially if the skin is moist at the time of contact.

Victims may initially have no significant respiratory distress but later become short of breath and begin to cough up fluid. Exposure to high concentrations of choking agents is fatal in many cases and causes permanent lung damage in other cases. Treatment includes termination of exposure, enforced rest and observation for signs of respiratory distress, and supportive treatment including respiratory support with oxygen.

Irritating Agents

Irritating agents are commonly used as "riot control" agents by military and law enforcement organizations. Examples include several military agents such as CS, CR, PS, and CN or Mace. These are collectively known as "tear gas." Capsaicin, or pepper spray, is also commonly used. These agents cause extreme burning of mucous membranes, which is best characterized by lacrimation, or the prolific formation of tears. In high concentrations they can cause irritation to the upper airway and skin and may cause nausea and vomiting. High concentrations may cause blistering of the skin, especially if the skin is moist at the time of contact. Treatment includes immediately removing the patient from exposure.

The irritating agents are generally considered nonlethal. Life-threatening reactions can occur if the victim is exposed to a very high concentration of the agent, such as in a confined area, or has an extreme hyperactive airway reaction. In an incident in Chicago, Illinois, a security guard released pepper spray to break up a fight in a crowded nightclub. As a result, 21 people died and dozens were injured—not from the effects of the pepper spray, but from the resulting panic and stampede that it triggered. Terrorist may use irritating agents to produce the same effects.

Explosives

Explosives are materials that are capable of undergoing violent decomposition with the resulting release of heat, shock, and noise. Explosive devices can be used to target people and/or property through the resulting thermal effect, blast effect, and/or fragmentation effect. They can also be used to disperse biological, chemical, or radiological material (Fig. 10.7). Explosives have long been used widely in both military and civilian applications. Roughly 70 percent of terrorist attacks involve the use of explosives. In some cases, they have been used to directly target first responders.

Explosives used by terrorists may be of commercial, military, or improvised origins (Fig. 10.11). Because explosives are commonly used in mining, quarrying, and construction-related activities, it is not difficult for terrorists to obtain them illegally. Terrorists may also obtain explosives of military origin, either domestic or foreign. Terrorists can also improvise explosives readily with commonly available raw materials. The U.S. Army *Improvised Munitions Handbook* (U.S. Army, 1969) is one source of information on improvised explosives that has been widely disseminated. Other, more sophisticated resources are now readily available on the Internet. In short, the easy availability and effectiveness of explosives ensures their continued utilization by terrorists.

In some cases, explosive devices may be easily recognized by the average responder because of the familiar features of explosive agents, blasting caps or detonators, boosters, and timers. In other cases explosives may not be identifiable as such by the average responder. References are available to assist us in learning to recognize explosive devices. One such reference is the *Explosives Identification Guide* by Mike Pickett (Pickett, 1999).

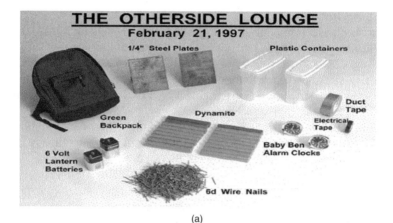

(a)

(b)

Figure 10.11 Explosive devices used by terrorists may include (a) bombs built with commercial explosives, such as the dynamite used in the bombing of the Otherside Lounge in Atlanta, Georgia, and (b) antipersonnel devices of military or improvised origins. [Source of Fig. 10.11(a): U.S. Department of Justice and Federal Emergency Management Agency, 1998].

Explosive devices may range from small pipe bombs, to satchel or "backpack" devices, to car and truck bombs. Specialized devices such as shaped explosive charges may be used to breach containers to create a hazardous materials release. In many cases, the presence of an explosive device will not be obvious and must be inferred based on factors such as:

- Suspicious packages, such as packages with oily stains, chemical odors, protruding wires, etc.

- Abandoned packages, containers, or vehicles in a suspect location
- Devices attached to containers such as gas cylinders, flammable liquid containers, bulk storage tanks, pipelines, or other hazmat containers
- Packages mailed with excess postage and no return address, or an obviously bogus return address
- Trip wires or booby traps
- Written or verbal threats in conjunction with an incident

Explosive devices may be triggered in a variety of ways, such as by trip wires, moving objects, radio control, timers, motion detectors, barometric pressure triggers, sound sensors, or magnetic devices. Keep in mind that if you are a responder, they may be rigged specifically to target you (Fig. 10.3).

When responding to an explosives incident, keep the basic protective factors of time, distance, and shielding in mind. Clear the area of civilians and responders and move to a safe distance as quickly as possible. Specific recommendations for standoff distances are provided in Table 10.5. If possible, assume a position that places shielding between you and the suspected device or location. Always call for expert assistance as soon as you realize that an explosive device may be involved in an incident.

Your organization's standard operating procedures should provide guidelines for responding to known or suspected bomb incidents. Guides 112 and 114 of the DOT *Emergency Response Guidebook* (ERG) provide general guidance on responding to incidents involving explosives. For example, Guides 112 and 114 recommend not using radio transmitters, which includes cell phones, within 330 feet of electronic detonators. Specialized training is required for responding to incidents involving explosives and is available through the U.S. Department of Justice's Office of Domestic Preparedness (www.ojp.usdoj.gov/odp).

BASIC GUIDELINES FOR RESPONDING TO TERRORISM/WMD INCIDENTS

Because WMD incidents differ from ordinary hazmat incidents, our procedures for responding to them must be different, beginning with preemergency planning (see Chapter 3). One important aspect of preparation will be to identify any likely targets and agents of opportunity within your jurisdiction or area of concern. Modify existing emergency response plans or create new plans to include provisions for dealing with incidents involving terrorism and WMDs. Specialized training in WMD response will be required for response roles beyond that of first responder at awareness level.

Using the Federal Job Aid and DOT ERG

At a minimum, every emergency responder should be trained to use the Emergency Response to Terrorism Job Aid (Fig. 10.12). The Job Aid is produced

TABLE 10.5   Safe Standoff Distances for Improvised Explosive Devices

	Threat Description	Explosives Mass[1] (TNT Equivalent)	Building Evacuation Distance[2]	Outdoor Evacuation Distance[3]
High explosives (TNT equivalent)	Pipe bomb	5 lb / 2.3 kg	70 ft / 21 m	850 ft / 259 m
	Suicide belt	10 lb / 4.5 kg	90 ft / 27 m	1080 ft / 330 m
	Suicide vest	20 lb / 9 kg	110 ft / 34 m	1360 ft / 415 m
	Briefcase/suitcase bomb	50 lb / 23 kg	150 ft / 46 m	1850 ft / 564 m
	Compact sedan	500 lb / 227 kg	320 ft / 98 m	1500 ft / 457 m
	Sedan	1000 lb / 454 kg	400 ft / 122 m	1750 ft / 534 m
	Passenger/cargo van	4000 lb / 1814 kg	640 ft / 195 m	2750 ft / 838 m
	Small moving van/delivery truck	10,000 lb / 4536 kg	860 ft / 263 m	3750 ft / 1143 m
	Moving van/water truck	30,000 lb / 13,608 kg	1240 ft / 375 m	6500 ft / 1982 m
	Semitrailer	60,000 lb / 27,216 kg	1570 ft / 475 m	7000 ft / 2134 m

	Threat Description	LPG Mass/Volume[1]	Fireball Diameter[4]	Safe Distance[5]
Liquefied petroleum gas (LPG–butane or propane)	Small LPG tank	20 lb/5 gal / 9 kg/19 l	40 ft / 12 m	160 ft / 48 m
	Large LPG TANK	100 lb/25 gal / 45 kg/95 l	69 ft / 21 m	276 ft / 84 m
	Commercial/residential LPG tank	2000 lb/500 gal / 907 kg/1893 l	184 ft / 56 m	736 ft / 224 m
	Small LPG truck	8000 lb/2000 gal / 3630 kg/7570 l	292 ft / 89 m	1168 ft / 356 m
	Semitanker LPG	40,000 lb/10,000 gal / 18,144 kg/37,850 l	499 ft / 152 m	1996 ft / 608 m

[1]Based on the maximum amount of material that could reasonably fit into a container or vehicle. Variations possible.
[2]Governed by the ability of an unreinforced building to withstand severe damage or collapse.
[3]Governed by the greater of fragment throw distance or glass breakage/falling glass hazard distance. These distances can be reduced for personnel wearing ballistic protection. Note that the pipe bomb, suicide belt/vest, and briefcase/suitcase bomb are assumed to have a fragmentation characteristic that requires greater standoff distances than an equal amount of explosives in a vehicle.
[4]Assuming efficient mixing of the flammable gas with ambient air.
[5]Determined by U.S. firefighting practices wherein safe distances are approximately 4 times the flame height. Note that an LPG tank filled with high explosives would require a significantly greater standoff distance than if it were filled with LPG.

Source: National Ground Intelligence Center (NGIC)

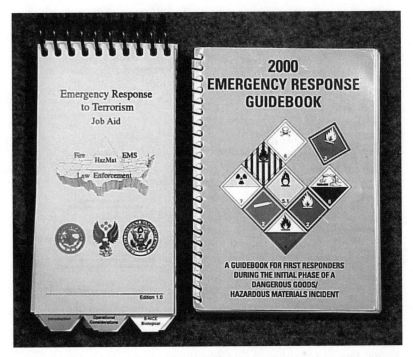

Figure 10.12 Every emergency responder should be trained to use the Emergency Response to Terrorism Job Aid, which is cross-indexed to specific hazard and action guides in the DOT *Emergency Response Guidebook* (ERG).

and distributed jointly by the Federal Emergency Management Agency (FEMA) and the U.S. Department of Justice (DOJ). It is designed specifically to aid fire, hazmat, EMS, and law enforcement personnel responding during the first hour of a terrorist incident. The Job Aid is cross-indexed to specific hazard and action guides in the DOT *Emergency Response Guidebook* (ERG), as shown in Table 10.6, so the ERG will also be an important response resource for first response to a potential terrorist event. For example, Guide 153, as shown in Figure 10.13, is applicable for first response to incidents involving suspected nerve, blister, and some irritant agents until better information is available. In this section, we provide a general description of the Job Aid and how it can be used; however, this is not intended as a substitute for actual training in using the Job Aid.

The Job Aid is a compact and durable document. All pages are plastic so that it can survive rough use in wet environments. It consists of five major color-coded sections.

The Gray Pages: Introduction
The first section of the Job Aid consists of gray pages and contains introductory material. The gray section explains the layout of the entire Job Aid, provides

TABLE 10.6 Relating WMD Incident Types to DOT ERG Guides

Hazard	ERG Guide Number
Biological	158
Nuclear or radiological	163 and 164
Incendiary	
• Flammable gas	118
• Flammable liquid	127
• Flammable solids	134
Chemical	
• Nerve	153
• Blister	153
• Blood	117, 119, 125
• Choking	124 and 125
• Irritant	153 and 159
Explosive	112 and 114

instructions on how to use it, and includes a list of basic assumptions on which the Job Aid was developed and according to which it is to be used.

The Yellow Pages: Operational Considerations

The second section of the Job Aid consists of yellow pages that cover operational considerations for responding to potential terrorism incidents. This section highlights specific strategic and tactical issues that should be assessed. Specific topics covered in the yellow section include:

- Assessing security for response and initial approach
- Command considerations
- On-scene size up
- Incident site management, safety, and security
- Tactical considerations
- Mass decontamination
- Evidence preservation

Assessing security for response and initial approach varies according to whether single or multiple indicators of a possible terrorist event (as described below) are present. For mass decontamination, different guidelines are provided for symptomatic patients, asymptomatic patients who are contaminated or have been exposed, and remote site operations (i.e., hospital emergency rooms).

The White Pages: Incident Specific Actions

The third section of the Job Aid is color-coded white and contains information on incident-specific actions that should be taken based on the type of terrorist incident or agent suspected. Information in the white pages is organized with the B-NICE mnemonic. Specific considerations and guidelines are provided for responding to incidents that may involve biological, nuclear or radiological, incendiary, chemical, and explosive agents or devices. For each type of incident, the white pages direct us to a specific hazard and action guide in the DOT ERG, as shown in Table 10.6.

GUIDE 153	Substances – Toxic and/or Corrosive (Combustible)	ERG 2000

POTENTIAL HAZARDS

HEALTH
- TOXIC; inhalation, ingestion, or skin contact with material may cause severe injury or death.
- Contact with molten substances may cause severe burns to skin and eyes.
- Avoid any skin contact.
- Effects of contact or inhalation may be delayed.
- Fire may produce irritating, corrosive and/or toxic gases.
- Runoff from fire control or dilution water may be corrosive and/or toxic and cause pollution.

FIRE OR EXPLOSION
- Combustible material: may burn but does not ignite readily.
- When heated, vapors may form explosive mixtures with air: indoors, outdoors, and sewers explosion hazards.
- Those substances designated with a "P" may polymerize explosively when heated or involved in a fire.
- Contact with metals may evolve flammable hydrogen gas.
- Containers may explode when heated.
- Runoff may pollute waterways.
- Substance may be transported in a molten form.

PUBLIC SAFETY

CALL Emergency Response Telephone Number on Shipping Paper first. If Shipping Paper not available or no answer, refer to appropriate telephone number listed on the inside back cover.
- Isolate spill or leak area immediately for at least 25 to 50 meters (80 to 160 feet) in all directions.
- Keep unauthorized personnel away.
- Stay upwind.
- Keep out of low areas.
- Ventilate enclosed areas.

PROTECTIVE CLOTHING
- Wear positive pressure self-contained breathing apparatus (SCBA).
- Wear chemical protective clothing which is specifically recommended by the manufacturer. It may provide little or no thermal protection.
- Structural firefighters' protective clothing provides limited protection in fire situations ONLY; It is not effective in spill situations.

EVACUATION
Spill
- See the Table of Initial Isolation and Protective Action Distances for highlighted substances. For non-highlighted substances, increase, in the downwind direction, as necessary, the isolation distance shown under "PUBLIC SAFETY".
Fire
- If tank, rail car or tank truck is involved in a fire, ISOLATE for 800 meters (1/2 mile) in all directions; also, consider initial evacuation for 800 meters (1/2 mile) in all directions.

Source: 2000 Emergency Response Guidebook

Figure 10.13 Guide 153 from the DOT *Emergency Response Guidebook* is applicable for first response to incidents involving suspected nerve, blister, and some irritant agents until better information is available.

ERG 2000 **Substances – Toxic and/or Corrosive (Combustible)** **GUIDE 153**

EMERGENCY RESPONSE

FIRE

Small Fires

- Dry chemical, CO_2 or water spray.

Large Fires

- Dry chemical, CO_2, alcohol-resistant foam or water spray.
- Move containers from fire area if you can do it without risk.
- Dike fire control water for later disposal; do not scatter the material.

Fire involving Tank or Car/Trailer Loads

- Fight fire from maximum distance or use unmanned hose holders or monitor nozzles.
- Do not get water inside containers.
- Cool containers with flooding quantities of water until well after fire is out.
- Withdraw immediately in case of rising sound from venting safety devices or discoloration of tank.
- ALWAYS stay away from tanks engulfed in fire.

SPILL OR LEAK

- ELIMINATE all ignition sources (no smoking, flares, sparks or flames in immediate area).
- Do not touch damaged containers or spilled material unless wearing. appropriate protective clothing.
- Stop leak if you can do it without risk.
- Prevent entry into waterways, sewers, basements or confined areas.
- Absorb or cover with dry earth, sand or other non-combustible material and transfer to containers.
- DO NOT GET WATER INSIDE CONTAINERS.

FIRST AID

- Move victim to fresh air.
- Call 911 or emergency medical service.
- Apply artificial respiration if victim is not breathing.
- **Do not use mouth-to-mouth method if victim ingested or inhaled the substance; induce artificial respiration with the aid of a pocket mask equipped with a one-way valve or other proper respiratory medical device.**
- Administer oxygen if breathing is difficult.
- Remove and isolate contaminated clothing and shoes.
- Incase of contact with substance, immediately flush skin or eyes with running water for at least 20 minutes.
- For minor skin contact, avoid spreading material on unaffected skin.
- Keep victim warm and quiet.
- Effects of exposure (inhalation, ingestion or skin contact) to substance may be delayed.
- Ensure that medical personnel are aware of the material(s) involved, and take precautions to protect themselves.

Figure 10.13 *(continued)*

The Blue Pages: Agency-Related Actions

The fourth section of the Job Aid covers agency-related actions and is color-coded blue. The blue pages provide an overview of considerations and issues that should be assessed by fire, EMS, law enforcement, hazmat, and assisting agencies immediately involved in a potential terrorist incident.

The Brown Pages: Glossary of Terms

The fifth section of the Job Aid contains a glossary of terms and is light brown in color. Many terms of significance to terrorism response are defined and explained in the brown pages. Throughout the Job Aid, a small drawing of an open book is used to identify terms that are included in the glossary.

Recognizing Potential Terrorist Incidents

The Job Aid instructs us to be on the lookout for indicators of a potential terrorist event when responding to emergencies. A possible terrorist event is indicated if the answer to any of the following questions is "Yes":

- Is the response to a target hazard or target event?
- Has there been a threat?
- Are there multiple (non-trauma related) victims?
- Are responders victims?
- Are hazardous substances involved?
- Has there been an explosion?
- Has there been a secondary explosion or attack?

Potential target locations or events could include any that are politically or socially significant or controversial, such as:

- Government offices or installations
- Military facilities
- Political rallies or appearances
- Clinics that provide abortions
- Homes or businesses of controversial figures

It is a good idea to be aware of significant dates as well. For example, the bombing of the Murrah Building in Oklahoma City occurred on the second anniversary of the storming of the Branch Davidian compound at Waco, Texas by federal agents.

Target hazards may include targets of opportunity or convenience, such as chemical storage tanks in populated areas (Fig. 10.2). Consider terrorism a possibility at any mass casualty or other unusual event, even if it is an innocuous setting such as a sporting event or shopping mall, because such locations may be chosen as "soft targets" and for the psychological effect. Also, monitor the current threat level as determined by the Homeland Security Advisory System (Table 10.7) operated by the Department of Homeland Security (www.dhs.gov/dhspublic) and adjust your actions accordingly.

TABLE 10.7   Homeland Security Advisory System

Color	Condition	Meaning
Green	Low	Low risk of terrorist attacks
Blue	Guarded	General risk of terrorist attacks
Yellow	Elevated	Significant risk of terrorist attacks
Orange	High	High risk of terrorist attacks
Red	Severe	Severe risk of terrorist attacks

Security Assessment for Response and Initial Approach to an Incident

The Job Aid recommends that you respond with a heightened level of awareness if there is a single indicator of a potential terrorist incident associated with the event to which you are responding. If multiple indicators are present, the Job Aid recommends that you take the following additional actions:

- Realize that you may be on the scene of a terrorist incident.
- Initiate response operations with extreme caution.
- Be alert for actions against responders.
- Evaluate and implement personal protective measures.
- Consider the need for maximum respiratory protection.
- Make immediate contact with law enforcement for coordination.
- Approach cautiously, from uphill and upwind if possible.
- Consider law enforcement escort.
- Avoid choke points (i.e., congested areas).
- Designate rally points (i.e., regrouping areas).
- Identify safe staging locations for incoming units.
- Establish outer and inner perimeters.

Additional specific guidelines and considerations are provided throughout the Job Aid. First responders from all agencies are instructed to:

- Isolate and secure the scene, deny entry, and establish control zones.
- Establish command.
- Evaluate scene safety and security.
- Stage incoming units.

Subsequent actions will vary according to the response discipline of the first responder (i.e., fire, EMS, or law enforcement) and are also included in the Job Aid.

On-Scene Safety Issues

Early recognition is the most important aspect of responder safety during a WMD event. Look for indications of WMDs on initial arrival at the scene of any incident. These were described above in this chapter when discussing the B-NICE agents and devices and will vary with the type of attack. For example, signs and symptoms of chemical agent exposure (depending on the agent involved) could include:

- Difficulty in breathing
- Redness, burning, and/or itching of the skin and/or eyes
- Irritation of the nose or throat
- Runny nose, excessive tears, and salivation
- Pinpoint pupils
- Pain in the eyes
- Headache
- Vomiting
- Convulsions

Be especially suspicious if multiple patients display the same symptoms. Remember that it is permissible to use common sense and intuition. If things "just don't seem right," they probably aren't.

If you find yourself responding to a terrorist incident, remember to "get your mind right" and keep it that way throughout the incident. During response to such an event, a lot of adrenaline will be flowing. It may be easy to be caught up in the event and become part of the problem. Don't substitute emotion for intellect in decision-making, and don't rush into the situation.

Remember that you must protect yourself in order to help others. Although exposure of victims to the hazards of the incident scene may be unavoidable, exposure of emergency services personnel occurs because of inadequate assessment and failure to comprehend the nature and magnitude of the incident.

A mnemonic device that may be helpful in keeping your priorities straight is SAFE. It is intended to help us remember the following points:

- Safety comes first.
- Assess the situation before taking any actions.
- Focus your efforts on the hazard and avoid exposure to it.
- Evaluate the situation and make the proper notifications.

Safety guidelines or precautions for response activities at the scene of a terrorist attack are included throughout the Job Aid. Examples of some of the precautions follow.

- Initiate on-scene size-up and hazard/risk assessment.

- Ensure that staging areas for incoming vehicles are safe.
- Assess emergency egress routes and position vehicles to facilitate rapid evacuation.
- Designate rally points for reassembly after evacuation in the event that evacuation is required.
- Ensure the use of personal protective measures and shielding.
- Ensure accountability of personnel.
- Designate an incident safety officer.
- Assess command post security.
- Assess decontamination requirements.
- Consider the need for specialized resources such as explosive ordnance disposal, emergency management agency, public health department, environmental agencies, etc.
- Determine life safety threats to yourself, other responders, victims, and the public.
- Consider the potential for a secondary (or tertiary, etc.) attack (e.g., using chemical dispersal devices, secondary explosive devices, or booby traps).
- Reassess initial isolation/standoff distances during the response.
- Dedicate emergency medical services needed for responders.
- Coordinate with law enforcement to provide security and control of perimeters.
- Implement self-protection measures.
- Commit only essential personnel and minimize exposure to the hazard.
- Assess carefully before initiating rescues.

SUMMARY

Terrorist attacks involving WMDs may occur at any time. We can view such events as hazmat incidents that are intentionally perpetrated by terrorists for the purpose of creating mass casualty incidents. These events present unique challenges and hazards for emergency responders. In this chapter, we examined terrorism, WMDs, and some basic considerations for responding to those types of events. In several of the remaining chapters, we will provide additional information related to WMD response. Remember that although hazmat training provides a good foundation for preparing to meet the challenge of terrorism and WMDs, additional specific training is required to meet the WMD challenge.

11

SITE CONTROL

A chemical spill occurred when a delivery hose ruptured while a 10,000-gallon railcar of chlorosulfonic acid was being off-loaded at an industrial facility in Birmingham, Alabama. Approximately 200 gallons of the acid spilled onto the ground before the release was terminated by closing the railcar off-loading valve. The spilled acid generated a dense acidic cloud that threatened plant personnel and off-site population. Injuries occurred because of chemical exposure and an automobile accident resulting from the release. Company employees made the required notifications and initiated the emergency response sequence. Site control was established by creating an isolation perimeter and evacuating areas of potential exposure. Evacuation of 250 off-site residents was required. Access to the scene was restricted to personnel directly involved in the response operation, and control zones were established on site. Facility emergency response personnel began confining and neutralizing the spilled chlorosulfonic acid with sand, soda ash, and lime. Although access to the scene was adequately controlled around the site isolation perimeter, a helicopter operated by a local television news team threatened site security. In attempting to film the incident, the helicopter made a direct approach to the scene. This presented the potential for exposure of media personnel onboard the helicopter. Also, turbulence from the helicopter's propeller could have pushed contaminants from the acidic cloud into the designated cold zone or support zone at the scene, thereby contaminating response personnel and equipment located in that zone. Personnel responsible for coordinating the response operation contacted Federal Aviation Administration officials at the

Emergency Responder Training Manual for the Hazardous Materials Technician, Second Edition, edited by Kenneth W. Oldfield
ISBN 0-471-21387-X   Copyright © 2005 John Wiley & Sons, Inc.

local airport to restrict airspace over the incident. The helicopter was ordered to remain a safe distance from the area of release, and control of the incident scene was thereby restored.

By their nature, hazardous materials incidents such as the chlorosulfonic acid release in Birmingham tend to be out of control. Initial responders to an incident may find a great deal of disorder and confusion in and around the area affected by a release. For example, site personnel, some of whom require medical treatment, may be in the process of fleeing areas immediately affected by the release. Injured personnel may require rescue from hazardous areas. If the incident involves an ongoing, large-quantity release, the migrating hazardous material may represent a significant threat to personnel at the facility, the public off-site, and the environment.

Emergency situations can be worsened by uncontrolled actions taken in response to the incident. For example, facility personnel, in attempting impromptu rescues of injured coworkers, may become additional casualties. Also, curious members of the public who are attracted to the event may wander into hazardous areas or get in the way of responders. As the incident described above demonstrates, members of the press may be an even greater problem than the public with regard to this. Uncontrolled actions of responders may serve to worsen the incident, for example, by tracking contaminants out of hazardous areas and into uncontaminated areas.

To minimize harm resulting from an emergency incident, responders must establish control of the incident scene as soon as possible and maintain that control throughout the response operation. Properly utilized site control procedures will serve to isolate people from hazards related to the incident and allow for an orderly, efficient, and safe response operation.

In recent times, attacks by terrorists using weapons of mass destruction (WMDs) have emerged as a new and very real type of threat that responders must be prepared to deal with (see Chapter 10). Establishing and maintaining site control for WMD incidents and other mass casualty incidents can be extremely challenging. It may involve a number of factors or considerations that are not present in responding to a simple accident involving hazmats. For this reason, site control for WMD incidents and other types of mass casualty incidents are discussed separately within this chapter.

OBJECTIVES OF SITE CONTROL

The very nature of the hazardous materials incident makes site control both difficult and critically important to establish and maintain. For site control measures to be effective, they must successfully address a variety of objectives.

To begin with, site control measures must be designed to minimize chaos and to provide direction and efficiency to the response operation. Site control procedures should allow for accountability so that the location and status of all personnel and equipment on-site are known at all times during the response operation. Successful site control requires preemergency planning (see Chapter 3) and strict adherence to an incident management system that coordinates and controls the efforts of all personnel and agencies on-site (see Chapter 4).

Site control measures must effectively isolate the incident scene to prevent harm to response personnel, employees, and members of the public such as well-intentioned volunteers, media crews, and curious bystanders. Only personnel directly involved in the response should be present at the scene of the incident. The presence of unnecessary people and equipment will only add to the confusion of the incident and increase the likelihood of needless injury. Response personnel not directly involved in activities in the hazard area must be kept at a safe distance to minimize the effects of unexpected events, such as explosions or sudden large-quantity releases that may enlarge the hazard area.

Site control activities must also prevent or minimize contamination of response personnel and tracking of hazardous materials. The uncontrolled movement of personnel and equipment through contaminated areas will result in the tracking of contaminants into clean areas, thus endangering the health of others and increasing the property damage and mitigation costs. Sound site control procedures prevent or minimize contact between contaminated and uncontaminated personnel and equipment. As seen in the incident described at the beginning of this chapter, this aspect of site control may be complicated by unexpected problems such as helicopters and changing wind direction.

CONSIDERATIONS FOR ESTABLISHING SITE CONTROL

The importance of establishing site control procedures cannot be overemphasized. Within the context of an emergency response operation all boundaries and security measures must be strictly followed without regard to an individual's status or position. Persons not involved in the emergency response must be kept outside the site perimeter. This includes onlookers, media representatives, and plant workers not involved in the response operation. Law enforcement or plant security personnel are generally better trained and equipped for crowd control and site security than hazmat team members. Law enforcement officers are also recognized off-site as having the authority to control access to an incident scene.

It is an unfortunate reality that in addition to the initial victims of a hazardous materials incident, others may subsequently become victims through their own actions. The latter group may include hazmat team members and civilians alike. For example, would-be rescuers, acting on impulse rather than in accordance with standard operating procedures (SOPs), may become additional casualties of an incident. Although rescue of injured personnel from hazardous areas should be the first priority considered in initiating a response operation, rescues should only be undertaken after it has been determined that they can be performed without undue risk to responders.

The appropriate mechanism for controlling the activities of response personnel onsite is the Incident Management System (IMS), as described in Chapter 4. It is critical that all activities be coordinated and controlled through this system to prevent response personnel or other involved parties from acting independently and endangering the health and safety of themselves and others. Utilization of a preestablished IMS in managing the incident must promote adherence to SOPs and a highly organized response.

Communication is the key to maintaining control during a response operation. Few things will throw an incident into chaos faster than a breakdown in communication. Ideally, supervisory personnel should monitor all site activities visually. However, this is not always possible. Communication must be maintained between on-site response personnel throughout the incident. Additionally, communication must be available between the command post and off-site agencies and resources. Communication is discussed in greater detail in Chapter 4.

ISOLATION PROCEDURES

The best way to protect people and property from the potential harm of a hazardous material release is to separate them from the released material. This is the purpose of isolation techniques, which may range from using flagging tape to mark off a small area around a leaking drum to complete evacuation of an entire industrial facility and surrounding residential area. Isolation allows response personnel to plan and conduct their activities without having to conduct unnecessary rescue operations.

Access Control

The first responder's initial actions should be to control access to the hazard area and to establish an isolation perimeter. The area located within the perimeter can then be secured in preparation for hazard mitigation operations. Securing occupied areas may involve evacuation or protection in place, as described in the next section. Isolation of the incident scene allows responders an unimpeded field of operations and provides protection to people in case the incident suddenly worsens.

The isolation perimeter can be considered the outer boundary of the site or incident scene (see Figs. 11.1, 11.2, and 11.3). The distance from the point of release to the incident perimeter may be measured in feet or in miles, depending on the specifics of the incident. Two opposing considerations will affect the decision on how much ground to include within the initial perimeter.

On the one hand, it is desirable to make the isolation area as large as possible. It is easier to reduce the isolation area than to expand it as the response operation progresses. As crowds and traffic increase around a site, expansion may involve much more than simply moving barriers. Also, the unpredictable nature of hazmat incidents makes a large buffer zone desirable.

On the other hand, an isolation perimeter requires valuable manpower to patrol. Inadequate perimeter control that allows unauthorized entry to the site may result in a breakdown of site control. A smaller perimeter that is more secure may therefore be preferable to a larger perimeter that is not well controlled.

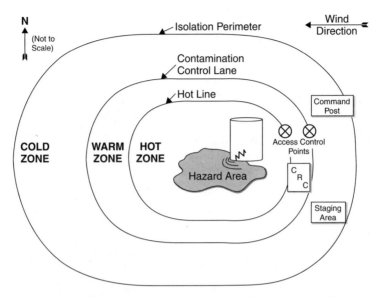

Figure 11.1 Site control and zoning layout for a hazmat incident.

In establishing the isolation area, it may be possible to utilize available geographic and physical barriers such as walls, fences, and bodies of water that will reduce the need for perimeter patrol (see Figs. 11.2 and 11.3). It is important to gain control of access points such as doors, gates, bridges, and intersections as soon as possible. If physical barriers are not available, use barrier tape, rope, sawhorses, traffic cones. or whatever is available to mark the perimeter (see Fig. 11.4).

In planning and initiating site control procedures, response personnel should be aware that ensuring unimpeded access by response personnel or equipment from off-site is just as important as preventing unauthorized entry. Provisions must be made to control traffic and keep vital roads, intersections, and access ways open. Coordination of activities with law enforcement agencies during preemergency planning should address this.

The concept of isolation also applies to incidents inside buildings or structures. Response personnel should initially gain control of entry points (such as doors) to the building and deny unnecessary access; then work can begin to secure the building as necessary. As in out-of-doors incidents, trying to isolate a larger area than can be adequately controlled may lead to a breakdown of site control.

Evacuation and Protection in Place

For spills or releases of solids or liquids with low evaporation rates, outlining and controlling the area of hazard may be a relatively simple process. However, for releases of gases or highly volatile liquids, the contaminant may travel in gas or vapor form over a much greater distance (see Chapter 6). For this reason,

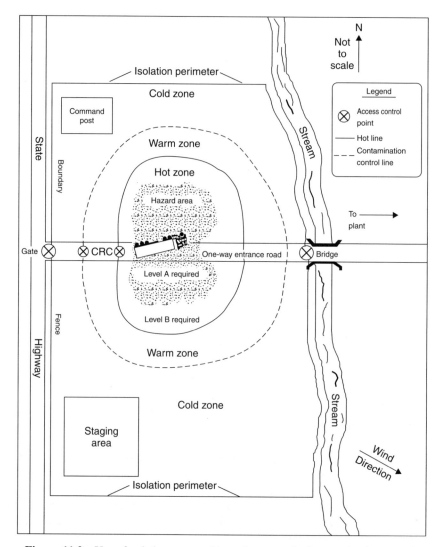

Figure 11.2 Use of existing geographic and man-made features in site control.

the potential hazard area may include inhabited areas within a fixed facility or adjacent to facility property or the location of a transportation accident. In this case, a decision must be made about how to protect the inhabitants. If the individuals are outdoors and the likelihood that the air concentrations will reach a dangerous level is high, complete evacuation of the area is certainly appropriate. Personnel responsible for site control must perform evacuation in a rapid and orderly fashion.

Evacuation can be a difficult and time-consuming activity. Preparations must be made to move people out of the affected area and to a safe location. If the

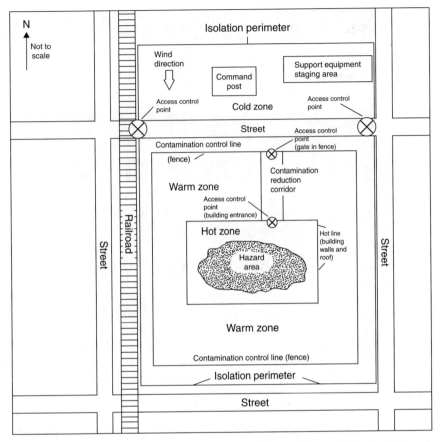

Figure 11.3 Site control for an incident involving a release within a building.

evacuation area extends beyond a fixed facility and into nearby residential or commercial areas, evacuation must be coordinated through local law enforcement agencies and provisions must be made for temporary shelter, food, and security. Evacuation plans should be carefully considered during emergency response planning.

Under certain circumstances, it may not be practical or even safe to evacuate all people from a structure within the evacuation area. In the workplace setting, industrial processes cannot always be quickly shut down and abandoned without causing serious consequences in other areas of the plant. In the community, institutions such as hospitals, detention facilities, and nursing homes present serious problems for evacuation. In some instances, a vapor cloud may be migrating too rapidly to allow complete evacuation of a building. Once a vapor cloud has surrounded a building, safe evacuation may not be possible.

In these cases, it may be appropriate to use protection in place, or shelter in place procedures. These procedures involve sealing off a structure so as to

Figure 11.4 Barrier type can be used to designate areas for various purposes on scene.

minimize the movement of air contaminants into it. Simply closing doors and windows and turning off heating, air conditioning, or ventilation systems may reduce the level of contamination within a building tremendously. However, the type of occupancy threatened must be considered. For example, buildings constructed since the early 1970s tend to be of airtight construction for energy efficiency. Older buildings are not as airtight and don't function as well for shelter in place. Protection in place procedures within an industrial facility may require placement of appropriate protective equipment in vital process control areas to allow time for proper shutdown.

The feasibility of using shelter in place procedures must be assessed during preemergency planning. The decision to use this approach during an incident should be made only after careful consideration of a number of factors related to the potential hazardous materials involved, the population threatened, and the role of weather conditions.

ZONING

Once an isolation perimeter has been established, the area within it can be subdivided into control zones with distinct lines of demarcation (as shown in

Figs. 11.1, 11.2, and 11.3). Zoning is a very useful concept for establishing and maintaining site control for a response operation. Site zones should be plotted on site maps based on information gathered during size-up of an incident and used in planning and conducting response operations. Zones are established based on factors such as:

- Type and degree of hazard (such as contaminant identity and concentration, presence of flammable vapors, etc.) within a given area
- Type and level of personal protective equipment required for safe entry into a given area
- Type of response or response-related activities carried out in a given area

Although various names are used to describe them, three zones are almost universally recognized at hazardous materials incidents—the hot zone, the warm zone, and the cold zone. The incident scene may also be geographically divided in accordance with divisions established by the Incident Management System, as described in Chapter 4.

Figure 11.1 shows a typical zoning layout with related features that are discussed below.

Hot Zone

The hot zone (or exclusion zone) contains the actual hazard area (see Fig. 11.1). This includes the location of the release and any areas to which hazardous substances have migrated, or are likely to migrate, in hazardous concentrations. The hot zone is the area where primary response operations are carried out to mitigate the incident. This is the most hazardous location on-site, and entry requires the use of appropriate protective gear and close adherence to emergency response SOPs. Special areas may be delineated within the hot zone, if different levels of protective gear are required for different tasks and/or work areas within the zone (see Fig. 11.2).

The hot line marks the outer boundary of the exclusion zone (see Figs. 11.1, 11.2, and 11.3). At any given location, the hot line must be far enough from the point of release to prevent exposure of personnel outside the hot zone to unsafe concentrations of hazardous materials released within the hot zone. Location of the hot line should be determined based on hazard and risk assessment (see Chapter 6) of the current site conditions. Any predictable events that may expand the boundaries during response activities should also be considered. Considerations for establishing the hotline include:

- Identity and characteristics of the released material
- Location of the point of release
- Status of the release (i.e., is the leak continuing or has it stopped?)
- Total quantity of material expected to be released as leakage progresses (e.g., based on size of container, location of breach on container)

- Current location of released material and expected migration pattern as the incident progresses
- Presence and extent of airborne hazards, including:
  - Oxygen-deficient or -enriched atmospheres
  - Combustible or flammable atmospheres
  - Toxic gases, vapors, or particulates at harmful concentrations
  - Ionizing radiation
  - Biological agents

No atmospheres that are immediately dangerous to life and health (IDLH) should be present beyond the hot line, with the possible exception of the decontamination area within the warm zone. Obviously, direct-reading hazard detection equipment can be invaluable in establishing the hot line location (see Chapter 9).

Geographic and man-made features must be considered in setting the hot line. Site topography may control the route of migration that a released product will follow. Also, existing structures or natural features may be utilized in confinement operations to limit contaminant movement (see Chapter 14). If so, these locations would need to be included within the hot zone. If fire fighting with water may be involved during incident mitigation, the hot zone must include provisions for containing contaminated runoff. In some instances, it may be possible to adopt existing fence lines, walls, or other active barriers as the hotline to promote adherence to site control procedures (see Fig. 11.3).

The hot zone must include enough room for mitigation activities to take place and must be large enough to provide protection for on-scene personnel outside the zone in the event of an explosion, fire, or unexpected release during response activities. However, the distance that responders must travel from the access control point to the hazard area must also be considered.

The potential effects of changing meteorological conditions such as wind, rain, or high temperatures on contaminant behavior and migration are important considerations for establishing the hot zone. Changes in conditions such as temperature and wind speed and direction may directly affect the generation and dispersion of airborne contaminants (see Chapter 6). A windsock should be used for accurate determination of wind direction during an incident. If no windsock is available, a small length of barrier tape tied to a vehicle antenna or some other fixed object will indicate wind direction.

Warm Zone

The warm zone, or contamination reduction zone (CRZ), is located beyond the hotline and serves as a buffer zone between the exclusion zone and the support zone or uncontaminated area of the site (see Figs. 11.1, 11.2, and 11.3). The warm zone provides an extra margin of safety from the primary hazards of the incident for support and command post personnel located in the cold zone.

Decontamination activities are performed within the contamination reduction corridor (CRC), which is a subpart of the warm zone (see Figs. 11.1, 11.2, and

11.3). Equipment needed to support the primary response operation (such as spare air cylinders, tools, sorbents, fire-fighting equipment, emergency medical supplies) may be staged within the warm zone. As with the exclusion zone, traffic to and from the warm zone must be controlled to prevent the spread of contamination. The contamination control line marks the outer boundary of the warm zone. Contaminated materials should never be transported beyond this line.

The CRC should be upwind (and also uphill, if possible) from the hot zone, as shown in Figures 11.1, 11.2, and 11.3. This is the area where decontamination operations are performed on personnel, tools, and equipment exiting the exclusion zone before entry into "clean" areas of the site (see Chapter 13). The CRC must be able to be deployed quickly as response operations begin, and moved quickly if necessitated by changes in site conditions during response operations. For heavy equipment such as loaders or backhoes used in the hot zone, a separate decontamination corridor should be designated and equipped. Personnel performing decontamination operations in the CRC must use the prescribed protective gear, which depends on the type and degree of hazard present.

Ideally, if the site is properly zoned and site control is maintained, there should be no contaminants present in hazardous concentrations outside the hot zone, except for contaminants that are transported into the CRC by personnel undergoing decontamination. There should certainly be no IDLH atmospheres within the warm zone, with the possible exception of the decontamination area. However, this may not always hold true. If an incident suddenly worsens or wind direction changes abruptly, parts of the warm zone may become dangerously contaminated, necessitating enlargement of the hot zone to incorporate these areas.

Cold Zone

The area of an incident scene located beyond the contamination control line is the cold zone or support zone (see Figs. 11.1, 11.2, and 11.3). Command functions and supporting operations are carried out within the cold zone. The cold zone should remain free of contamination, so that no chemical protective equipment is required for response personnel working in this area. The command post and staging areas for equipment and personnel are located in this zone, and emergency medical services, administrative operations, and various other supporting operations are performed here.

The placement of support facilities and functions within the support zone should take these factors into consideration:

- Wind direction—Support activities should always be located upwind of the hazard area.
- Topography—Support activities should be located uphill of the hazard area, if possible, in most cases.
- Space required—Although support functions are interrelated and should be in good communication, crowding will cause confusion.

- Accessibility—Necessary vehicles and personnel must be able to reach the support zone readily.
- Distances—The support zone should be located far enough away to provide a sufficient buffer from hazardous activities but no further than necessary, so that excessive travel will not be required for entry into the hot zone.

Access Control Points

Access control points are the only locations at which it is permissible to cross zone boundaries (see Figs. 11.1, 11.2, and 11.3). This allows inbound and outbound responders to be checked in and out of specific zones. It is recommended that separate control points be used for inbound and outbound personnel. Access control points allow the use of required PPE and decontamination procedures to be verified. Also, the location of responders can be closely monitored in the event that rescue is required.

Other Zoning Considerations

Although the idealized zoning layouts shown in Figures 11.1, 11.2, and 11.3 may be appropriate, with some modification, for most hazardous materials incidents, there will be incidents where the geography or site conditions will require the use of judgment and creativity in establishing zones. For instance, a release of a low-volatility chemical in the loading dock area of a building may not require a warm zone around the entire area (see Fig. 11.5). Rather, the access to the dock from the building area could be controlled and doors and air vents sealed to create an exclusion zone. A contamination reduction corridor could be set up outside the dock. As long as the objectives of each zone are met, an appropriate zoning plan may take any form.

SITE CONTROL FOR INCIDENTS INVOLVING WMDs AND OTHER MASS CASUALTY INCIDENTS

Site control for WMD incidents has many factors in common with site control for accidental hazmat emergencies, but it also differs significantly (Fig. 11.6). In Chapter 10, we examined a number of unique characteristics of WMD incidents. These characteristics make it even more critical, and more challenging, for responders to establish and maintain site control when responding to WMD incidents. Some of the same challenges will also be present in mass casualty incidents resulting from accidents involving hazardous materials.

Special Considerations for WMD/Mass Casualty Incident Site Control

Emergency responders tend to think in terms of trying to keep people from entering the site of a typical hazmat incident. An even greater challenge at WMD/mass

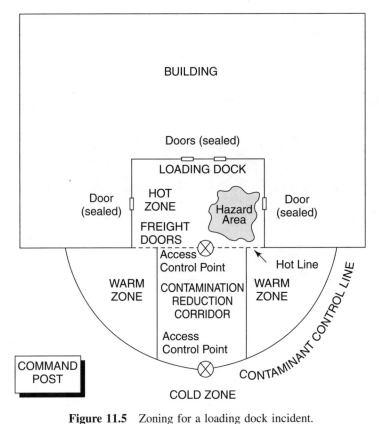

Figure 11.5 Zoning for a loading dock incident.

casualty incident scenes may be to keep exposed victims from leaving the scene without proper decontamination and exposure assessment. If site control fails, chemical, biological, or radiological agents may be widely dispersed by panicked victims fleeing the scene.

A terrorist attack may leave a large geographic area contaminated, and the extent of the contamination may not be immediately apparent. For example, if an explosive charge is used to disperse a chemical agent, it may be easy for responders to treat it strictly as a bombing without being aware of the chemical agent that the bomb dispersed. This can make it easy to underestimate the size of the area affected by the event for site control purposes. Also, the entire area affected by a terrorist attack must be considered a crime scene and issues such as evidence preservation must be addressed.

Incidents involving terrorism and WMDs may pose exceptional hazards to emergency responders. This is true not only because of the high potential hazards of some of the weapons used but also because the terrorists may directly target first responders. Potential hazards to responders can complicate site control. This

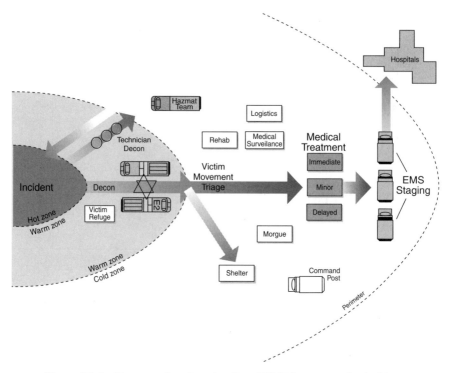

Figure 11.6   Site control and zoning for a WMD/mass casualty incident.

is especially true when secondary devices are considered; for example, terrorists may place secondary devices at probable command post locations.

Incorporating Mass Decontamination, Triage, Treatment, and Transport into Site Control

One of the main challenges of any mass casualty incident is providing triage, treatment, and transportation to hospitals for large numbers of people. This is even more challenging if large numbers of people must first be decontaminated, and still more challenging if they must be rescued from contaminated areas before being decontaminated. The site control and zoning arrangement for WMD/mass casualty incidents must allow for the organization of all these activities at the incident scene (Fig. 11.6).

As discussed in Chapter 13, a mass decontamination corridor will need to be set up in the warm zone to provide rapid decontamination for the large number of victims of the incident. A separate decontamination corridor will need to be set up to provide more rigorous technical decontamination for hazmat team members involved in hot zone entries. The mass decon corridor will require water sprayed in a fog pattern, either provided from fire engines or some other suitable source, making placement of the equipment critical. Access to adequate supplies of water must be ensured. Safe refuge areas may need to be designated for victims to use for shelter from the elements while awaiting mass decon. Shelter may also need to be provided after decontamination for victims who are not immediately transported off-scene.

After mass decon, victims must be triaged and prioritized for transportation with a system such as the START system described in Chapter 4. Site control must ensure an efficient transition of patients through triage, treatment, and transportation (Fig. 11.6) with no bottlenecks or traffic jams. Level II staging, as described in Chapter 4, should be utilized for EMS transport vehicles, with unobstructed scene access and egress maintained for all vehicles. A morgue will probably also need to be established on-scene.

Site control must also accommodate the special resource needs of responders for this type of event. Responders will require adequate types and amounts of protective gear, which may include large amounts of breathing air for SCBAs. Specialized air monitoring equipment may be required for WMD incidents. Chemical agent antidote kits may be needed in large numbers. Decontaminated patients will require gowns or improvised garments for privacy, and in some cases blankets or other means to prevent hypothermia. Site control, in conjunction with IMS, must ensure that these items are properly staged on scene for deployment as needed.

The federal Emergency Response to Terrorism Job Aid provides specific guidelines pertaining to site control for terrorism incidents. The Job Aid is described in Chapter 10.

SUMMARY

Site control procedures range from very simple to very complex based on the severity and complexity of the incident involved. In many cases, isolation procedures may be the most effective action that can be taken to prevent or limit casualties resulting from an incident. The use of effective site control procedures serves not only to prevent the tracking of contaminants and cross-contamination but also to provide for accountability of personnel and equipment and to promote an orderly, well-organized response operation. Site control is also a critical element of WMD/mass casualty incident response. The concept of site control is therefore a very important tool for use in response to any emergency incident involving hazardous materials.

12

PERSONAL PROTECTIVE EQUIPMENT

INTRODUCTION

Hazardous materials response by definition involves working with and exposure to materials that can harm responders. The hazard is the reason hazardous materials technicians are needed. However, a successful response is one in which the responders are not injured or harmed. In other words, although the hazards are present, the responders are adequately protected while they work around the hazards.

Traditional health and safety practice and OSHA regulatory principles recognize a hierarchy for the types of control techniques used to protect workers. The broad categories of protections, listed in descending order of preference, are:

- Engineering controls
- Administrative controls
- Personal protective equipment

The most desirable type of control is the engineering control. Examples of this category of control technique would include redesigning the work process to eliminate the use of hazardous material or using a ventilation system to capture the hazardous vapors before they reach the environment. Engineering controls represent the most effective means of reducing exposure to a hazard in a fixed work environment. However, engineering controls are usually not available or feasible for most chemical releases, especially transportation incidents.

Emergency Responder Training Manual for the Hazardous Materials Technician, Second Edition, edited by Kenneth W. Oldfield
ISBN 0-471-21387-X   Copyright © 2005 John Wiley & Sons, Inc.

Administrative controls include such actions as rotating more workers through an area to reduce the amount of time that any one worker is exposed to the hazard of a particular task. Using good housekeeping to prevent the accumulation of a hazard and using preventive maintenance to keep equipment from breaking down and causing a release are also examples of administrative practices that can reduce exposure to workers. Some administrative control practices such as rotating entry teams to limit exposure can be used at hazardous material responses, but they are not likely to limit the risk of exposure sufficiently by themselves.

Consequently, hazardous materials responders must rely on the least preferable type of hazard control for protection. Why is personal protective equipment (PPE) the least desirable control? In part, because PPE offers no protection until the hazard is at or on the responder. The other controls generally act at the source or along the path (the environment) across which the hazard travels, allowing any hazard that passes the control to dissipate before reaching the responder.

Perhaps the more important shortcoming of PPE is that it requires diligent and consistent use by the wearer while it adds stress and burden. Working with PPE is generally difficult and burdensome compared to working without it. Responders must be aware of activities or conditions that may cause the PPE to fail and guard against them. Also, if the PPE does fail, there is no other protection to prevent exposure.

Although PPE is not the most desirable protection, it is often the only way to protect the responder. Getting the best protection possible from PPE requires the user to understand its capabilities and limitations. The PPE must be properly selected based on the hazards known or suspected to be present at the scene. Responders must be trained to use the equipment properly to prevent failure that would result in exposure.

This chapter discusses the selection and use of PPE in a hazmat response. Because different types of PPE must come together to provide complete protection, the chapter outlines a seven-step process for PPE selection and use. The steps are as follows:

(1) Assess all hazards.
(2) Select respiratory protection.
(3) Select skin protection.
(4) Select personal safety equipment.
(5) Put it all together into levels of protection.
(6) Use the PPE properly.
(7) Store and maintain the PPE.

As with other topics in this book, a separate book could be written on the details related to PPE. This chapter offers an overview that should provide the responder with a solid introduction to the safe and proper use of PPE. The regulatory requirement outlined in HAZWOPER is discussed first, and then the seven-step process follows.

HAZWOPER REGULATORY REQUIREMENTS

Chapters 2 and 3 discuss the regulatory and planning requirements set forth in OSHA's Hazardous Waste Operations and Emergency Response (HAZWOPER) standard. Although only paragraph Q of the standard is said to apply to emergency response operations, paragraph q(10) states that the use of chemical protective clothing during emergency response must meet the requirements of paragraphs g(3) through g(5) of the HAZWOPER standard, which outline the use of PPE. Paragraph g(3)(iv) further incorporates by reference Subpart I of Title 29 of the Code of Federal Regulations, which includes the following standards:

- 1910.132—General Requirements
- 1910.133—Eye and Face Protection
- 1910.134—Respiratory Protection
- 1910.135—Head Protection
- 1910.136—Occupational Foot Protection
- 1910.137—Electrical Protective Devices
- 1910.138—Hand Protection

These standards are discussed further in the appropriate sections of this chapter.

Written PPE Program

The most prominent requirement found in paragraph g(5) of HAZWOPER is that of a written PPE program. OSHA requires that the PPE program address each of the following topics:

- Selection of PPE
- Use and limitations of equipment
- Work mission duration
- Maintenance
- Storage
- Decontamination and/or disposal after use
- Training
- Proper fitting of PPE
- Donning and doffing procedures
- Inspection procedures
- Limitations during temperature extremes
- Program evaluation

The written PPE program must adequately address all the above topics and requirements related to respirators, chemical protective clothing, and other protective equipment as appropriate for the types of protective gear available to response personnel. Specific considerations that the PPE program should address are discussed in the relevant sections of this chapter. The use of respirators is particularly regulated under OSHA's Respiratory Protection Standard (29 CFR 1910.134), which is discussed in the section for Step 2.

Obviously, the written program must be in writing and complete. It may be a stand-alone document or may be included in the employer's emergency response plan as described in Chapter 3. However the program is produced, the required topics listed above should be clearly identified.

Training

One consideration appropriate to the use of all types of PPE is training. 29 CFR 1910.132 requires that the employer ensure that the employee knows at least the following:

- When PPE is necessary
- What PPE is necessary
- How to properly don, doff, adjust, and wear the PPE
- The limitations of the PPE
- The proper care, maintenance, useful life, and disposal of the PPE

The employee must show that he/she understands the training and can use the PPE properly before being allowed to use it in the workplace. The training can take whatever format the employer chooses but should include hands-on practice with the equipment and the opportunity to ask questions of the instructor.

The general PPE standard requires the employer to retrain employees whenever it is apparent that they do not understand how to use the PPE properly. The standard also requires retraining when:

- Changes in the workplace render previous training obsolete.
- Changes in the types of PPE used render previous training obsolete.
- Inadequacies in the employee's knowledge and use of the PPE indicate the need for retraining.

The employer must certify the training through a written certificate that shows the name of the employee, the date, and the subject of the training. In most cases, this training can be documented as being a part of the hazmat training conducted under paragraph q(6) of HAZWOPER (Awareness, Operations, or Technician Level). PPE training does not need to be conducted separately if it is included such hazmat training.

STEP 1—ASSESS ALL HAZARDS

To determine the type and degree of protection a responder needs, the hazards he will face must be known. This is obvious but not always easy to achieve. For the purpose of selecting the appropriate PPE to protect responders, two aspects of the hazard must be known—type and degree.

Type of Hazard

A fundamental principle of hazardous material response is that the released material's identity must be known as soon as possible. Nowhere is this principle more critical than in the protection of the responders who must enter the hazard area as part of the response, because the effectiveness of PPE varies with different chemicals.

Chapter 5 outlines the procedures and clues that can be used to discover and confirm the identity of a released material during a response. In some incidents, the identity will be relatively easy to determine, such as when a properly placarded railroad tanker is involved in a release (Fig 12.1). In other situations, responders may have to make a reconnaissance entry to determine the identity of the released product.

At the minimum, responders should have some idea as to the physical state of the material. The amount and parts of the body that must be protected often depends on whether the release involves an airborne vapor or a splashing liquid. Knowing the physical state helps predict the way the chemical may move and how responders may be exposed.

Figure 12.1 Properly labeled containers often make hazard identification easy.

It is not enough to simply know the material's physical state. Proper selection of the appropriate PPE, particularly the right chemical protective clothing material, requires the responder to know the specific identity of the chemical. As discussed in Step 3, a given chemical protective clothing material will hold out some chemicals well but will let other chemicals permeate through. Also, in the instances where air-purifying respirators are being considered for respiratory protection, the chemical to be stopped must be matched to the proper filtering medium as discussed in Step 2. Therefore, it is critical to know the specific identity of the chemical to choose the proper equipment.

Degree of Hazard

When responders know with reasonable confidence the identity of the material, they will still need to estimate the airborne concentration and the likelihood for the responder to be splashed by liquid product. In other words, they need an estimate of the amount of that chemical that will be present. Skin protection needed for a response involving a possibility of minor splashing as containers are opened would be very different from a response involving the same chemical spewing under pressure out of a ruptured piping system (Fig. 12.2).

Estimating the concentration of chemicals in the air is also critically important for determining whether a given type of respirator will provide sufficient protection to responders. The assigned protection factor of a type of respirator tells how much the respirator can be expected to reduce the exposure concentration. To determine whether the reduction is enough for the responder to breathe

Figure 12.2 Situations involving splashing or pressurized liquids involve a higher degree of hazard.

safely, both the actual concentration and the exposure limit of the specific chemical must be known. The assigned protection factor is discussed in detail in Step 2.

Anticipate the Environmental Conditions and Entry Activities

In addition to assessing the type and degree of chemical hazard, the responder should consider the environmental conditions and the actions to be performed. Environmental conditions such as temperature extremes may affect decisions about the use of PPE. For example, responding to a release on a very hot day will limit the time for which a responder in higher levels of protection can safely work. Responders must consider the impact of stressful environmental conditions on their ability to use PPE safely.

The anticipated work activities during each entry should also be considered. The response actions will likely determine the degree to which responders will come into contact with chemicals. Also, the nature of the activities may dictate that heavier, more durable protective clothing be used to withstand the potential physical abuse. On the other hand, entries involving only reconnaissance with little expected contact with materials may allow for a lower level of protective equipment.

Entering Unidentified Hazard Areas

The paradox many responders face is when the chemical identity and exposure concentration cannot be determined without an entry into the hazard area. If the identity of the chemical is not known, how can the responder know that the chemical protective suit he is wearing into the hazard area will hold the chemical out? How can he know whether the respirator is providing sufficiently clean air to breathe? Incident commanders, safety officers, and entry teams must make every effort to gather as much information as possible from a safe location before resorting to entry into an unidentified hazard area.

However, if it is decided that an entry is necessary, a few precautions can reduce the likelihood of harm from the chemical (Fig. 12.3). Entry teams should:

- Only enter wearing positive-pressure self-contained breathing apparatus as required under 1910.120(q)(3)(iv)
- Wear totally encapsulating chemical protective suits
- Avoid unnecessary contact with any unidentified or unrecognized material
- Enter and remain upwind and uphill of the released material
- Use air monitoring equipment capable of measuring at least oxygen and combustible gases and evacuate if a potentially flammable atmosphere is detected
- Maintain constant communication with the appropriate leader according to ICS protocols

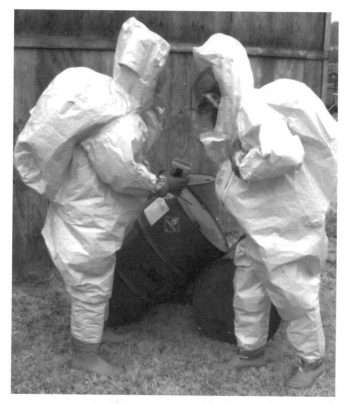

Figure 12.3 Entries into unidentified hazards require the highest levels of protection.

These guidelines should offer sufficient protection for an initial reconnaissance entry, which should be kept brief. Information that is gathered is then used to refine the PPE selection for further response activities.

STEP 2—SELECT THE APPROPRIATE RESPIRATORY PROTECTION

Inhalation is an extremely efficient way for toxic chemicals to enter the body. Therefore, the respiratory system is arguably the most important route of entry to be protected. Emergency responders entering the hot zone during an emergency typically do so with a self-contained breathing apparatus, the highest level of respiratory protection. However, other types of respirator are available and can serve a valuable role in certain response functions. This section discusses all aspects of respiratory protection that may apply to emergency response.

Regulations and Standards

Both legally enforceable regulations and recommended standards guide the use of respirators. This section summarizes the most significant of these, including

OSHA's Respiratory Protection Standard and National Fire Protection Association (NFPA) standards. Other standards may apply to the reader's employer or jurisdiction and should be consulted for their impact on respirator usage.

OSHA's Respiratory Protection Standard
The coverage of responders by either federal OSHA or state occupational safety and health regulators is discussed in Chapter 2. It is assumed that, in either case, a response team will be covered by OSHA's Respiratory Protection Standard (29 CFR 1910.134) or a similar regulation. This standard outlines the basic requirements for the safe use of respirators in any workplace, including hazardous materials response. A full discussion of the standard is beyond the scope of this book. Some key requirements are discussed below, and others are discussed in the appropriate sections of this chapter.

Written Program. The standard requires the employer to develop a written program outlining the procedures for the proper use of respirators. The program is to be based on the employer's specific procedures and equipment. It can be included in the written PPE program described previously or as a separate program and must include the following components:

- Selection procedures
- Medical evaluation
- Fit testing procedures
- Procedures for proper use in routine and emergency situations
- Procedures for cleaning, inspecting, and maintaining respirators
- Procedures to ensure adequate air quality, quantity, and flow for atmosphere-supplying respirators
- Training employees in the hazards to which they are exposed
- Training employees in the proper use of respirators
- Program evaluation

OSHA provides a sample program in the *Small Entity Compliance Guide* that can be downloaded from the OSHA website or obtained from any OSHA office. Commercial vendors also offer software that can be used to develop a program. In either case, the employer must be certain to adapt the generic program to make it applicable to the equipment and procedures of its response team or facility.

Respirator Selection. The standard requires that respirators be selected based on the respiratory hazards present in the workplace. Evaluation of the respiratory hazards includes a reasonable estimate of the responder exposure levels and an identification of the contaminant's physical state. Any time the incident commander cannot identify or reasonably estimate the responder's exposure, he must consider the atmosphere immediately dangerous to life and health (IDLH), requiring the highest level of respiratory protection, as described in Step 1.

Employers must select respirators that are certified by NIOSH. The respirators must be used in compliance with the conditions of their certification. Employers must be certain that respirators are not modified or used in any way that would void the NIOSH certification.

Medical Evaluation. The respirator standard requires employers to ensure that any employee who wears a respirator is physically able to do so without being harmed by the burden that the respirators place on him. Hazardous materials technicians covered under HAZWOPER must be included in a full medical surveillance program in which their fitness has already been determined. Employers should check that their medical surveillance program addresses the requirements of 29 CFR 1910.134(e).

For employees not covered by HAZWOPER, OSHA requires that employer to arrange for a medical evaluation of the employee by a physician or other licensed health care professional (PLHCP). The evaluation is to be performed before the employee is fit tested or required to use the respirator during a response. The purpose is simply to evaluate the employee's ability to use the respirator.

The PLHCP is to use a questionnaire to determine which responders need to receive a physical exam and which responders have no indications of a potential for problems in using respirators. The content questionnaire to be used is dictated by OSHA and must be provided to the responders confidentially in a way they can understand. Based on responses to the first part of the questionnaire, the PLHCP either determines that the responder is fit for using a respirator or recommends follow-up evaluation. The PLHCP provides the employer with a written statement indicating one of the following conclusions:

- The responder is able to wear a respirator.
- The responder is able to wear a respirator with limitations.
- The responder is not physically able to wear a respirator.

The specific test results and findings of the exam are considered privileged and confidential and are not provided to the employer.

There is no requirement for annual or periodic reevaluation of the worker. Reevaluation is called for under any of the following conditions: An employee reports signs or symptoms related to the ability to wear a respirator; the PLHCP, administrator, or supervisor determines it is necessary; information from the respiratory protection program indicates a need for reevaluation; or a change in workplace conditions substantially increases the physiological burden placed on the employee.

Facepiece Seal Protection. OSHA does not allow the use of respirators by responders with facial hair or other conditions that interfere with the face-to-facepiece seal. Where the responder requires corrective eyeglasses, they must be worn in a way that does not interfere with the facepiece seal. This may require a special lens holder available from the facepiece manufacturer that mounts inside the facepiece without temple bars that protrude out of the facepiece. OSHA allows the use of contact lenses while wearing a respirator if they can be worn without problem.

Procedures for IDLH Atmospheres. The respiratory protection standard requires that the employer post at least one responder outside the IDLH atmosphere if responders enter such an area. This is consistent with the HAZWOPER requirement that emergency responders' work in hazardous areas be performed with the buddy system and with properly equipped backup personnel standing by [1910.120(q)(3)(v & vi)].

Training. The standard requires effective training for each responder who is required to wear a respirator. According to the standard, each responder is to be able to demonstrate knowledge of at least the following:

(1) Why the respirator is necessary and how improper fit, usage, or maintenance can compromise the protective effect of the respirator
(2) What the limitations and capabilities of the respirator are
(3) How to use the respirator effectively in emergency situations, including situations in which respirators malfunction
(4) How to inspect, put on and remove, use, and check the seals of the respirator
(5) What the procedures are for maintenance and storage of the respirator
(6) How to recognize medical signs and symptoms that may limit or prevent the effective use of respirators
(7) The general requirements of the section

The training is to be provided before requiring the responder to use a respirator and in a manner that is understandable to him. Retraining is to be conducted annually and whenever there is evidence that it is needed to ensure the safe use of the respirator by the responder. The training can easily be included in HAZWOPER training.

NFPA Standards

NFPA standards are consensus standards that establish a level of competence for firefighters and fire departments. Three standards have sections that specifically pertain to respiratory protection: *NFPA 1404 Standard for Fire Service Respiratory Protection Training 2002 Edition, NFPA 1500 Standard on Fire Department Occupational Safety and Health Program*, and *NFPA 1852 Standard on Selection, Care, and Maintenance of Open-Circuit Self-Contained Breathing Apparatus (SCBA) 2002 Edition.*

NFPA 1404. This standard outlines the important elements of a program to train firefighters in the use of respiratory protection. In addition to describing the requirements for provision and care of respirators used in firefighter training, the standard also lays out the topics that are to be included in the training. These include:

- Recruit training
- Retraining and certification
- Ability to act properly in simulated emergencies
- Recognizing hazards
- Specialized training on SCBAs, SARs, and APRs (discussed below in this chapter)
- Maintenance and testing of equipment
- Inspection

NFPA 1500. This standard covers comprehensive safety and health programs, which includes the use of respiratory protection. Although general in nature, the standard provides some specific guidance affecting SCBA usage. Detailed discussion of the standard is beyond the scope of this book, but some of the key points are discussed below.

NFPA does not allow firefighters to compromise the integrity of the SCBA while operating in a hazardous atmosphere. This disallows the practice during rescue operations of "buddy breathing," the practice of a rescuer and victim sharing a single SCBA by alternately breathing air through the facepiece. Buddy breathing allows a brief exposure of the firefighter to the hazardous atmosphere that may result in his being overcome.

The standard specifically prohibits the use of an SCBA by firefighters with beards or facial hair at the point at which the facepiece is designed to seal with the face. Head coverings or other equipment that passes through the facepiece seal area are also prohibited. These prohibitions apply regardless of what fit test measurements can be obtained. Fit testing is discussed below in this chapter.

NFPA 1852. This standard describes the program elements that are necessary to keep SCBAs, the mainstay of firefighter respiratory protection (Fig. 12.4), ready for use. The emphasis is on the selection, care, and maintenance of the equipment.

Respirator Classification

Selecting a respirator for a particular hazard begins by understanding the options available from which to choose. The following sections describe different classification systems for respirators and their components. Although the classifications are described separately for clarity, they are combined to make up the different types and levels of respirators available to responders. This is addressed in the section on protection factors below in this step.

Figure 12.4   The SCBA is the mainstay of firefighter respiratory protection.

Facepiece Coverage

The first way that respirators are classified is by facepiece design. The facepiece is the barrier controlling the access of air to the entrances to the respiratory system, the nose and mouth. The amount of face covered by the facepiece can be as little as the escape respirator with a mouthpiece and nose clip to as much as the hood respirator that completely encloses the head.

Loose-Fitting Facepieces. Loose-fitting facepieces such as hoods and helmets enclose the user's head. A continuous flow of fresh air is blown across the user's face to dilute or diffuse any contaminants that migrate into the breathing area. Hoods and helmets most frequently are used to protect against particulate contaminants, because these migrate more slowly than gases or vapors. Because these facepieces do not seal tightly to the face, they are often more comfortable for the user than tight-fitting facepieces. However, because there is no tight seal, some contaminants may move into the breathing area, resulting in some exposure to the user. Therefore, they are not frequently used in hazmat response actions.

Tight-Fitting Facepieces. Tight-fitting facepieces form a tight seal with the user's face. Because of this seal, all air entering the facepiece can be controlled. To provide full protection, the seal must be complete, with no points of leakage. These facepieces include filtering facepieces, dust masks, and elastomeric facepieces. The filtering facepiece or dust mask is essentially a filtering medium that is held against the face by elastic bands. The elastomeric facepiece is a more substantial barrier that has an inlet through which clean air enters. Fit testing determines the quality of the seal. Tight-fitting facepieces come in either half-mask or full-facepiece designs (Fig. 12.5).

HALF-MASK RESPIRATORS form a seal around the user's nose and mouth only. Half-mask respirators are less expensive than full-facepiece respirators and allow the user to wear eyeglasses. But because they seal against the curved surface of the nose, they generally provide a lower-quality fit and should not be used in higher-concentration atmospheres. They have limited use in hazmat response.

FULL-FACEPIECE RESPIRATORS cover the entire face, including the eyes. The elastomeric facepiece has a clear visor for visibility. The smoother surface of the forehead and cheeks allow a better fit and a higher degree of protection than the surfaces to which a half-mask facepiece seals. The full-facepiece respirator also protects the eyes from splashing and exposure. Users who wear eyeglasses must use spectacle kits to mount lenses inside the facepiece, because eyeglass temple bars would break the seal of the facepiece to the face.

Method of Protection
The way in which a respirator delivers breathable air, or the method of protection, is the primary way that respirators are distinguished.

Air-Purifying Respirators. Air-purifying respirators (APRs) remove contaminants from the air by means of a filtering medium. They use filtering or adsorption media contained in either cartridges (smaller capacity) or canisters (larger capacity) that are attached to the facepiece inlet. Cartridges typically attach directly to

Figure 12.5 Half-mask and full-face respirators.

the facepiece inlet and are smaller and lighter than canisters. Canisters may attach to the chin of the facepiece or may be mounted on the user's back or chest and attach to the facepiece inlet by a breathing tube. Canisters contain more medium, but they are heavier and more uncomfortable to use.

The medium must be selected based on chemical and physical properties of the contaminant. Air contaminants take the form of either particulates or gases and vapors. Mechanisms for filtering particulates differ from the sorbing methods used for removing gases and vapors.

PARTICULATE RESPIRATORS filter dusts, fibers, fumes, and mists from the air by means of interception or electrostatic capture. Interception involves trapping a particle onto the surface of the filter as air passes through it. In electrostatic capture, the particle has a charge that is attracted to the opposite charge in the filter fibers.

In 42 CFR Part 84, NIOSH classifies particulate respirators according to two criteria, the impact of oil mist in the air on the effectiveness of the filter medium and filter collection efficiency. Oil particles, such as mists of lubricants or cutting fluids, can coat filter fibers as the air is breathed through the filter. This coating can reduce the electrostatic attraction of the particles to the filter, thus affecting some filters' efficiency. Other filters, which rely less on electrostatic attraction and more on interception, are less affected by the presence of oil mist. The three categories of resistance to filter efficiency degradation by oil are labeled N (not resistant to oil), R (resistant to oil), and P (oilproof). N-series filters should not be used when oil mist may be present. R- or P-series filters can be used if oil mist is present, although R-series filters should not be used for more than one work shift.

NIOSH recognizes three levels of filter collection efficiency: 95%, 99%, and 99.97% (rounded up to 100% for labeling). Respirator manufacturers label their particulate cartridges according to both criteria. An N95 cartridge will collect 95% of the particles out of air that has no oil mist present. A P100 cartridge (Fig. 12.6) will collect 99.97% of particles out of the workplace air regardless of the presence of oil mist and is, therefore, the most protective.

GAS AND VAPOR RESPIRATORS remove gases and vapors from the air by sorption or catalysis. These methods involve interaction between the gas and vapor molecules and the granular solid sorbent materials. Sorption involves the attraction of molecules onto the surface of the granular, solid sorbent (adsorption) or into the molecular spaces of the sorbent (absorption). Catalysis uses a catalyst to initiate a reaction between the contaminant and another chemical. For example, hopcalite initiates the reaction between carbon monoxide and oxygen to form the less toxic carbon dioxide.

Atmosphere-Supplying Respirators. Atmosphere-supplying respirators provide a complete atmosphere from outside the confined space. They do not purify or otherwise use the air inside the space, so they can be used in situations where APRs cannot. The air must be of at least Grade D quality as specified by the

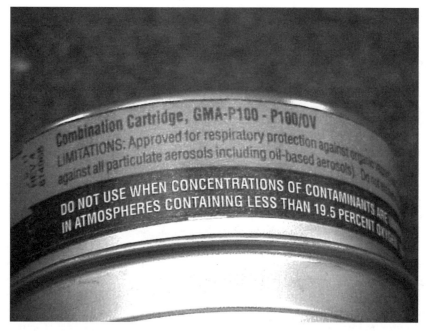

Figure 12.6 The label will identify the type of particulate cartridge.

TABLE 12.1 Common Air Quality Specifications

	NFPA 1500	CGA Grade D	CGA Grade E
Oxygen (%)	19.5–23.5	19.5–23.5	20–22
Carbon dioxide (ppm)	1000	1000	1000
Carbon monoxide (ppm)	10	10	10
Total volatile hydrocarbons (ppm)	–	–	25
Oil mist + particulate (mg/m^3)	5	5	5
Pronounced odor	None	None	None

Compressed Gas Association Commodity Specification for Air, G-7.1–1989. Table 12.1 shows common air quality specifications. Figure 12.7 shows the two categories of atmosphere-supplying respirators—the supplied-air respirator (SAR) and the self-contained breathing apparatus (SCBA).

SUPPLIED-AIR RESPIRATORS (also called air line respirators) provide breathable air to the responder directly from a source outside the hazard area, such as a compressor with a purification unit or from large air cylinders. The air is delivered to the user through a hose or air line by way of a regulator that controls the air flow. SARs must meet NIOSH criteria for the minimum and maximum air flow rates into the facepiece for all hose lengths. This system provides a practically limitless work period if used with a compressor.

Figure 12.7   SCBA and SAR.

SARs can be used for entry into an IDLH atmosphere only if they are equipped with a self-contained auxiliary air supply for escape, according to the OSHA respirator standard. These auxiliary air cylinders are often referred to by the estimated breathing time they provide, typically 5-, 10-, or 15-minute escape bottles. It is important to remember that individual breathing and air consumption rates vary widely, so these times should not be taken literally. If the primary breathing air supply fails, entrants must exit the space immediately.

SELF-CONTAINED BREATHING APPARATUS (SCBAS) provide breathing air to the responder from a compressed gas cylinder worn in a harness on the responder's back. A regulator is mounted either on the facepiece or on a harness strap with a breathing tube connecting to the facepiece. The regulator controls the flow of air into the facepiece as described below in this chapter. An audible and/or vibrating alarm warns the user when the amount of air pressure (representing the amount of breathable air) drops below 25% of the working pressure.

Certain safety requirements apply to the use of SCBAs. NIOSH (1987) guidelines for the use or respiratory protection the following requirements for all certified SCBAs:

- Pressure gauges visible to the wearer that indicate the quantity of gas remaining in the cylinder

- Alarms that signal when only 20–25% of service time remains
- Bypass valves that allow the user to get breathing air in the event of regulator failure
- Breathing air fittings that are incompatible with other gas fittings

SCBAs are classified as "closed circuit" or "open circuit" according to how they handle the air exhaled by the user.

Closed-circuit SCBAs (or "rebreathers") capture and reuse the breathing gas exhaled by the user (Fig. 12.8). Exhaled air contains some oxygen (15–17%) plus carbon dioxide added by the user. Closed-circuit SCBAs recirculate the exhaled air through a "scrubber" to remove the carbon dioxide and then add oxygen from a compressed gas cylinder to render the air breathable. Closed-circuit SCBAs can be used for extended entries into the hazard area, sometimes as long as four hours, because almost all of the breathing air volume is conserved.

Open-circuit SCBAs direct all of the exhaled air out of the facepiece through an exhalation valve. Open-circuit SCBA cylinders are constructed of steel or aluminum wrapped in fiberglass, with working pressures of 2200, 3000, or 4500 pounds per square inch (psi). Because all of the air is exhausted to the environment, they have rated service lives of between 30 and 60 minutes.

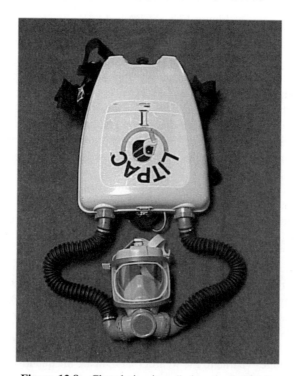

Figure 12.8   Closed-circuit or "rebreather" SCBA.

ESCAPE SCBAS are designed only for escape from atmospheres that suddenly become hazardous. These units are compact and can be donned quickly. Escape-only SCBA units (ESCBAs) have a very limited air supply, usually 5–15 minutes. Many ESCBAs have a clear plastic hood with an elastic collar that the user dons during an emergency, and breathing air is fed into the hood in a continuous flow. These units are not suitable for entry into a hazardous atmosphere, but they may find use in hazmat response scenes where an IDLH atmosphere may develop quickly and migrate out of the hot zone.

HYBRID SAR-SCBA UNITS are essentially SCBAs with a port on the regulator for attaching an air line (Fig. 12.9). This allows the user to enter a hazard area while breathing from the SCBA cylinder and then connect to an air line for the work operation. When used in this way, no more than 20% of the air supply should be used during entry.

QUICK-FILL® SYSTEMS sold by Mine Safety Appliance Company (MSA) use a special quick-connect high-pressure hose to refill SCBA cylinders while the user continues to wear and operate the SCBA. The high-pressure hose coming from a cylinder bank or other breathing air source attaches to a special adapter installed on the high-pressure side of the regulator. The cylinder is recharged without requiring the wearer to remove the unit or stop using the SCBA (Fig. 12.10). According to MSA, NIOSH has extended certification to SCBAs equipped with the Quick-Fill® system.

A transfill hose allows a rescuer equipped with the Quick-Fill® system to share air from his cylinder with another similarly equipped user who is trapped

Figure 12.9 This regulator has an attachment for an air line as well as a Quick-Fill adapter.

Figure 12.10   SCBAs with a Quick-Fill adapter can be refilled while in use.

or otherwise unable to get to a source of clean air. In this case, the donor and recipient cylinders equalize pressure between them. This is potentially very dangerous if the rescuer shares so much of his breathing air that what remains is inadequate for him to safely exit the hazard area. A transfill operation should be performed only when a rapid exit of the rescuer can be ensured.

Mode of Operation
Based on the pressure generated within the facepiece during use, respirators are classified as operating in either a negative-pressure or a positive-pressure mode of operation.

Negative-Pressure Respirators. With a negative-pressure respirator, the user inhales and creates a negative pressure or vacuum inside the facepiece. This draws air into the facepiece through the inlet. If the respirator is an APR, the air comes from the environment through the purifying element. If an atmosphere-supplying respirator is equipped with a regulator operating in the demand mode, the vacuum opens a valve that allows air in from the cylinder or air line. Once the air is replaced, the pressure equalizes and air flow into the facepiece stops.

Any leak or break in the facepiece seal offers an easier path for air to enter. A negative-pressure respirator with a poorly fitting facepiece will allow a potentially significant exposure to the user. Negative-pressure respirators have a lower protection factor (discussed below in this chapter) than positive-pressure respirators.

Positive-Pressure Respirators. Positive pressure respirators maintain a slight positive pressure within the facepiece at all times. If there is a leak in the facepiece seal, air should blow out of the facepiece through the leak and prevent contaminated outside air from entering. Respirators that operate in the positive-pressure mode provide more protection than equivalent respirators operating in the negative-pressure mode. Entry into IDLH conditions requires the use of either SCBAs or SARs that operate in the positive-pressure mode.

Powered air-purifying respirators (or PAPRs) are APRs that use a fan or pump to draw air through the filtering medium and into the facepiece. This creates a positive pressure within the facepiece that forces air out of any leaks in the seal. Because the user does not have to inhale the air through the filtering medium, the PAPR is less stressful to use than negative-pressure APRs.

Two positive pressure designs are currently used:

- Continuous-flow respirators continuously push air into the facepiece under pressure.
- Pressure-demand respirators are designed to maintain a slight positive pressure within the facepiece. Inhalation reduces that positive pressure and opens the regulator to allow replacement air to enter. These units use less air because the breathing air flows only during inhalation.

One concern regarding the use of positive-pressure respirators is that heavy breathing while performing strenuous tasks may cause the entrants to temporarily "overbreathe" the positive pressure within the facepiece. If there is a poor facepiece seal, this could lead to brief exposures to toxic contaminated air. Even when using positive-pressure respirators, entrants must use a properly-fitted facepiece.

Selection and Use Considerations for Respirators for Hazmat Response

Each type of respiratory protective equipment has certain advantages and limitations that must be considered when making a respirator selection. Responders must select respirators that provide adequate protection against all of the airborne hazards that have been identified under the environmental and task conditions of the response (as described below in this chapter). The following discussion considers the major advantages and disadvantages of each type of respirator.

Selection Considerations for APRs

Air-purifying respirators are an option for protecting emergency responders. Because responders may be working in high concentrations of chemicals, care must be used in selecting and using APRs.

Advantages of APRs. The advantages of APRs include their light weight, ease of use and maintenance, low price, and minimal restriction of the user's movement (Fig. 12.11). These advantages make it desirable to use APRs whenever they provide adequate protection to responders.

Disadvantages of APRs. Despite their advantages, there are a number of limitations to using APRs that restrict their use in hazmat response situations. The limitations include the following:

- APRs cannot be used in oxygen-deficient atmospheres (atmospheres containing less than 19.5% oxygen). Therefore, they cannot be used with the totally encapsulating suits described in Step 3.
- All potential atmospheric hazards in the atmosphere must be identified before entry to select the proper purifying medium and ensure that APRs will offer adequate protection (see Table 12.2 for the color codes for APR cartridges and canisters).
- APRs cannot be used if the identified contaminants are highly toxic in small concentrations (such as hydrogen cyanide).
- APRs cannot be used in exposures that are or may become higher than the contaminant's IDLH level.

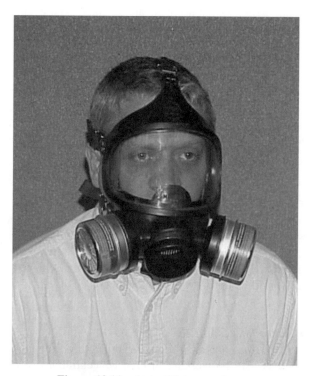

Figure 12.11   Air-purifying respirator.

TABLE 12.2   Color Codes for Respirator Cartridges and Canisters

Atmospheric Contaminants to be Protected Against	Colors Assigned*
Acid gases	White
Hydrocyanic acid gas	White with 1/2-inch green stripe completely around the canister near the bottom
Chlorine gas	White with 1/2-inch yellow stripe completely around the canister near the bottom
Organic vapors	Black
Ammonia gas	Green
Acid gases and ammonia gas	Green with 1/2-inch white stripe completely around the canister near the bottom
Carbon monoxide	Blue
Acid gases and organic vapors	Yellow
Hydrocyanic acid gas and chloropicrin vapor	Yellow with 1/2-inch blue stripe completely around the canister near the bottom
Acid gases, organic vapors, and ammonia gases	Brown
Radioactive materials, excepting tritium and noble gases	Purple (magenta)
Particulates (dusts, fumes, mists, fogs, or smokes) in combination with any of the above gases or vapors	Canister color for contaminant, as designated above, with 1/2-inch gray stripe completely around the canister near the top
All of the above atmospheric contaminants	Red with 1/2-inch gray stripe completely around the canister near the top

*Gray shall not be assigned as a main color for a canister designed to remove acids or vapors. Orange shall be used as a complete body or as a stripe color to represent gases not included in this table. The user will need to refer to the canister label to determine the degree of protection the canister will afford.

Source: 29 CFR 1910.134

- A leak in the facepiece seal may allow a significant amount of contaminant to enter, because most APRs work in the negative-pressure mode.
- The service life of the cartridges or canisters (discussed below) must be calculated so they can be changed before breakthrough occurs.
- Breathing through negative-pressure APRs is difficult; PAPRs can be used if needed.
- Environmental conditions, such as high temperature or humidity, can reduce the effectiveness of the purifying medium.

Other Considerations in Using APRs. Methods for removing gases and vapors depend on properties that vary with different chemicals. Cartridges can be designed to remove individual chemicals, such as mercury or ammonia, or groups of chemicals, such as organic vapors or acid gases. However, none is universally effective against all contaminants. To select the proper cartridge, the user must know the identity of the contaminant.

A gas and vapor respirator effectively removes contaminant from the air until its medium becomes saturated. At that point, contaminant passes through the cartridge and enters the facepiece to expose the user. The time needed for contaminant to saturate the medium, called the *service life of the cartridge*, depends on factors such as:

- Concentration of chemical in the air
- Amount of sorbent material in cartridge
- Breathing rate of the user
- Humidity
- Temperature

The OSHA standard requires the employer to calculate a schedule for changing cartridges based on the service life of the cartridge. Respirator manufacturers can provide computer software to aid in the calculation of such a schedule based on their cartridge and the conditions during the response.

Good warning properties allow the APR user to smell, taste, or experience irritation in the event of breakthrough or leakage at concentrations of the contaminant below appropriate exposure limits. Generally APRs should be used only for contaminants having good warning properties. Users should leave the hazard area immediately at the first sensation of a warning property and evaluate the cause. It may be cartridge saturation or a poorly fitting facepiece.

The Role of APRs in Hazmat Response. The role of APRs in the area of emergency response tends to be highly restricted. Given the large number of restrictions on the use of APRs, the process of determining that APRs will provide adequate protection in a specific situation is necessarily complicated and time consuming. The time-critical nature of emergency response will frequently not allow for this process. Also, hazard levels may drastically increase unexpectedly, for example, because of container failure during entries. In some instances, entry into hazardous atmospheres may be required before information such as identity and concentration of contaminants can be gathered. For such entries, only positive-pressure SCBA is allowable under OSHA regulations. Because of these restrictions, APRs are not typically used for primary response operations.

APRs may provide adequate protection for personnel involved in secondary response operations, such as decontamination, which are carried out in areas of reduced hazard on site. However, from a strictly legal standpoint, it should be determined that all restrictions on the use of APRs are met by conditions in the work area before APRs are assigned.

Figure 12.12   Supplied-air (air line) respirator.

Selection Considerations for Supplied Air Respirators

Supplied air respirators (Fig. 12.12) supply a complete breathable atmosphere rather cleaning the existing air and therefore are often a better choice for responders. The following should be considered before choosing SARs.

Advantages of SARs. SARs offer several advantages for protection of responders during a hazmat response. They provide a much longer work time by using large cylinder banks or compressors as breathing air sources. They are less cumbersome and have a lower profile, allowing the user to fit through tighter spaces. They are also lighter to wear and place less physical demands on the responder.

Disadvantages of SARs. Problems with the use of SARs in emergency response primarily involve the air line. The air flow resistance inside the air line limits the length of the hose to 300 feet. The air line also forces the responder to exit the hazard area by the same path he entered. The air line can become tangled, cut by physical hazards, or contaminated and permeated by chemical hazards if used in the hazard area. SAR systems are expensive to purchase and require extensive maintenance. They can be used in IDLH environments only if they are equipped with an escape air supply, such as a 5-minute escape air cylinder. Careful consideration must be given to the increased stress levels placed on entry personnel by extended duration of entry made possible by SARs.

The Role of SARs in Hazmat Response. SARs offer some advantages over SCBAs for response activities in which APRs cannot be used. These advantages include:

- Extended air supply
- Reduced apparatus weight
- Easier movement through small passages or entryways

However, in most cases air line-related problems (such as the potential for entanglement or contamination of the air hose) outweigh these advantages for entry into the hot zone. Like the APR, their use is likely to be limited to the more controlled conditions of the warm zone and the decontamination area.

Selection Considerations for SCBAs

The self-contained breathing apparatus provides the highest and broadest protection for emergency activities. The following considerations may help determine whether SCBA use is appropriate for an entry into the hazard area.

Advantages of SCBAs. The primary advantage of the SCBA is that it provides breathable air from a source that is carried and controlled by the responder (Fig. 12.13). This provides the benefit of atmosphere-supplying respiratory protection without the air line restrictions. SCBAs offer the highest level of respiratory protection available to responders entering the hot zone.

Disadvantages of SCBAs. Disadvantages of SCBAs are primarily related to the air cylinder, which is both heavy and cumbersome. It limits the responder's entry time based on the capacity of the air cylinder. SCBAs are expensive and require detailed maintenance procedures.

The Role of SCBA in Emergency Response. Because of the high protection factor and lack of air line-related mobility restrictions, positive-pressure SCBA tends to be the mainstay of respiratory protection for emergency response personnel. Because of this, OSHA regulations contained in 29 CFR 1910.120 require that positive-pressure SCBA be used during emergency response operations in hazardous areas until such time as the incident commander determines, based on air monitoring results, that some less protective type of respiratory protective equipment is appropriate.

Respiratory Protection for WMD Incidents

NIOSH, the U.S. Army Soldier Biological and Chemical Command (SBCCOM), and the National Institute for Standards and Technology (NIST) are developing

Figure 12.13   Self-contained breathing apparatus.

CBRN Agent Approved

See Instructions for Required Component
Part Numbers, Accessories, and Additional
Cautions and Limitations of Use

Figure 12.14   Label for respirators certified for use against chemical and biological agents.

standards for certifying respirators as protective against chemical, biological, radiological, and nuclear (CBRN) agents. The work is being done through NIOSH's National Personal Protective Technology Laboratory (NPPTL) and is part of its mission to test and certify PPE.

The CBRN certification program involves an extensive battery of tests based on performance specifications. Respirators are submitted by manufacturers for testing by NPPTL, and those that pass the test are certified as *CBRN-Agent Approved* (Fig. 12.14). The standards are established for each different respirator type. At the time of this writing, standards were developed or in process for:

- Powered air-purifying respirators
- Escape respirators (APR and self-contained escape respirators)

- Full-facepiece APRs
- SCBAs

The SCBA standard requires a minimum service life of 6 hours against distilled sulfur mustard (HD) and sarin (GB) agents without breakthrough. (See Chapter 10 for a discussion of these agents.) Full-facepiece APR must meet the P100 level of particulate protection and provide at least 8 hours of service life without breakthrough. These service life values are based on exposure to fixed challenge concentrations under laboratory conditions, so they should not be used as absolute values for field conditions. (See the discussion of cartridge service life calculations above.) However, the tests demonstrate that the approved respirators will meet a standard of performance. This will give the responder increased confidence in the use of the equipment for responding to a WMD incident.

Respirator Fit Testing

A properly selected respirator will not provide adequate protection if a leaking facepiece allows contaminated air to enter. A tight-fitting facepiece must form a complete, unbroken seal with the face to prevent the outside air from bypassing its protective barrier. Unfortunately, even relatively large leaks can go undetected by the user. To be certain they are getting the best protection, responders must use an approved protocol to test the quality of the facepiece fit.

OSHA Regulation

In the respiratory protection standard (29 CFR 1910.134), OSHA includes specific requirements for respirator fit testing by employers. These are outlined in Paragraph (f) and Appendix A of the standard, which describes:

- The kinds of fit tests allowed
- The procedures for conducting them
- How the results of the fit tests must be used

Both NFPA and OSHA require that employers must fit test responders before the initial use of a respirator and at least annually thereafter. In this way, the responder is assured of a proper fit from the beginning, and the annual retest should catch changes in the responder that affect the quality of the fit. If the responder or respirator program administrator detects a condition (such as significant weight change or facial scarring) that may change the fit between fit tests, a retest should be performed earlier.

Only the fit test protocols approved by OSHA and described in Appendix A of the standard can be used. The descriptions of the protocols are detailed, and commercially available fit test kits meet these requirements. The discussion of these methods is based on the requirements in the standard.

Fit testing of tight-fitting atmosphere-supplying respirators must be performed in the negative-pressure mode, regardless of the normal mode of operation (negative or positive pressure). This may require the use of an adapter that allows the use of APR cartridges with the SCBA or SAR facepiece. Respirator manufacturers can provide such adapters for their equipment.

Respirator Fit Checks

Two simple fit checks must be performed each time a responder dons a respirator (Fig. 12.15). These fit checks do not qualify as the respirator fit tests that are described below; however, they do help the responder ensure that the mask is properly positioned and that there are no major leaks. OSHA describes these tests in Appendix B-1 of the standard and requires that respirator users perform them each time they don a respirator.

The positive-pressure check is performed by sealing the exhalation valve with the palm of the hand and exhaling into the facepiece. The responder should feel the facepiece expand away from the face and remain that way for several seconds. If there is a leak, air will leak out and the facepiece will collapse back to its original position. The responder may be able to feel the air flow on the skin and thus locate the leak.

In the negative-pressure check, the responder lightly covers the cartridges, canister inlet, or other respirator inlet with the palm of the hand, inhales slightly, and holds his breath. The facepiece should collapse against the face and hold the position for several seconds. A leak will allow air to enter, and the facepiece will relax away from the face.

In addition to these tests, users can employ isoamyl acetate (or banana oil) ampoules to check the respirator fit. This is a quick check and not the qualitative fit test that is described below. The responder must be wearing an APR with organic vapor cartridges. He breaks a small ampoule of the sweet-smelling vapor

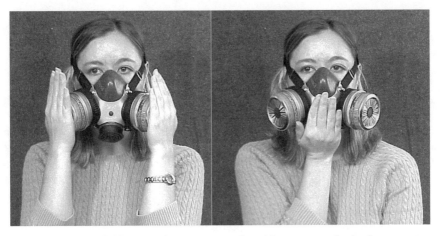

Figure 12.15 Negative-pressure and positive-pressure fit checks.

and waves it slowly around the facepiece. If the vapor is detected, the facepiece seal has a leak and it should be adjusted; however, a saturated cartridge may allow the vapor to pass through and falsely indicate a leak. This check should be performed with fresh cartridges.

Approved Fit Test Protocols

The fit of any tight-fitting facepiece to the responder's face must be tested according to either a qualitative or quantitative fit test protocol. Loose-fitting facepiece respirators do not require fit testing. The specific procedures of each protocol are spelled out in Appendix A of the standard. Each protocol includes a challenge agent and a set of exercises that is to be performed by the user during the test to challenge the fit.

Qualitative methods can be used to fit test all positive-pressure respirators (PAPRs and atmosphere supplying) and any negative-pressure APR that must achieve a fit factor of 100 or less. The required fit factor of 100 or less derives from multiplying a protection factor of 10 by a safety factor of 10 to account for the difference between fit testing results and the actual protection achieved during work. In other words, a negative-pressure APR can be qualitatively fit tested if it will be used only in atmospheres where the concentration of hazardous contaminant is no more than 10 times its exposure limit. Quantitative methods must be used when a negative-pressure respirator will be used in atmospheres that contain a hazardous contaminant at a level greater than 10 times the exposure limit.

Qualitative Methods. Qualitative methods involve the exposure of a person wearing a respirator to a challenge agent that can be tasted or smelled or will cause a response from the user (Fig. 12.16). If the user does not detect or respond to the agent, an acceptable fit has been achieved. This is referred to as a "pass/fail" test.

The approved qualitative methods use one of these challenge agents:

- Saccharin mist
- Isoamyl acetate (banana oil)
- Bitrex
- Irritant smoke

The saccharin mist and banana oil tests rely on the user's detection of sweet-tasting or -smelling agents. The Bitrex and irritant smoke agents evoke an involuntary response by the user.

All of these methods include a sensitivity test as part of the protocol. The user enters an enclosure or dons a hood (except for irritant smoke tests) without wearing the respirator, and the agent is introduced into the space until he detects it. The sensitivity test determines that the user can detect the agent and at about what level. Next the user dons the appropriate respirator and enters the enclosure or dons the hood again. A 100-times stronger concentration of the agent is introduced into the space, and the user performs a prescribed series of exercises to challenge the fit. If the user detects or responds to the agent, the fit is considered

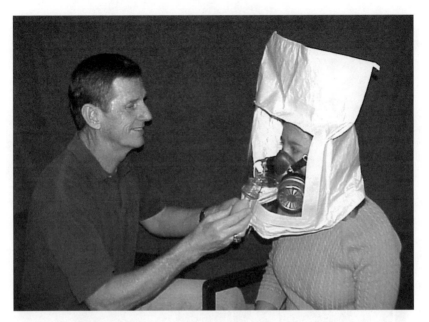

Figure 12.16 Qualitative respirator facepiece fit test.

inadequate and the respirator is adjusted or exchanged. If the user does not detect the agent, the fit is considered adequate.

Quantitative Methods. Quantitative methods measure the quality of the fit and provide a numerical *fit factor*. The fit factor is a direct assessment of the fit of a particular facepiece model and size to the individual user. The tests can be performed on the user's personal mask or on a representative mask of the same brand, model, and size (Fig. 12.17).

According to OSHA regulations, when quantitative methods are used, half-mask respirators must achieve a measured fit factor of at least 100 and full-facepiece respirators must achieve a fit factor of at least 500.

OSHA has approved three quantitative methods:

- Generated aerosol
- Condensation nuclei counter
- Controlled negative pressure

The required procedures are described in Appendix A and are based on the instructions from the test equipment manufacturers.

THE GENERATED AEROSOL METHOD entails the generation of a known concentration of an aerosol such as corn oil mist inside a test chamber. The user enters the chamber wearing a facepiece fitted with a port that allows sampling of the

Figure 12.17 Quantitative respirator facepiece fit test.

air inside the mask. A real-time aerosol monitor measures the concentrations of the aerosol inside the enclosure and inside the facepiece. The fit factor is the ratio of the aerosol concentration of the enclosure to the concentration inside the facepiece.

THE CONDENSATION NUCLEI COUNTER technology often is referred to by the brand name of the tester equipment, PORTACOUNT® by TSI. It uses ambient dust as the challenge agent and alternately samples air from outside and inside a facepiece equipped with P100 particulate filters and a sampling probe. The sampled air passes through an alcohol-rich atmosphere and then through a condenser, where alcohol vapors condense around the dust particles. The air stream then passes through a laser beam, where the dust/-alcohol nucleus diffracts the beam and creates a flash of light. A photo-counter detects the flash, and each particle is counted. Here again, the ratio of the concentration of aerosol outside the facepiece to the concentration inside the facepiece determines the fit factor.

THE CONTROLLED NEGATIVE PRESSURE technology is sold as the Fit Tester 3000® by Occupational Health Dynamics. The tester uses normal air as its challenge agent. Special adapters replace the cartridges to block the respirator inlets. During the fit test the user draws a normal breath and holds it. Tubing attached to the adapters then allows the tester to draw a target negative pressure inside the facepiece and try to maintain it during the 8-second test. If leakage in the seal allows a volume of air to enter the facepiece, the tester draws an equivalent volume of air out to maintain the target negative pressure. The additional pump flow rate needed to maintain the pressure represents the leakage flow rate, that which is compared to the breathing flow rate to establish an equivalent fit factor.

Protection Factors

The protection factor (PF) describes the overall performance of the respirator in terms of a ratio of the concentration of an agent outside the facepiece to the concentration inside the facepiece. The factor accounts for the quality of the fit as well as the penetration through purifying elements and other factors.

The assigned protection factor (APF) is defined as a measure of the minimum anticipated workplace level of respiratory protection provided by a properly functioning respirator or class of respirators to properly fitted and trained users. APFs are typically estimates, not absolute values based on extensive research. They are useful for selecting a type of respirator for a given hazard as long as some safety margin is included.

For instance, half-mask air-purifying respirators have an APF of 10. A user could expect that if he uses a half-mask APR with an organic vapor cartridge in an atmosphere containing 40 ppm of dimethylaniline, he should not actually breathe more than about 4 ppm ($40 \div 10$) of dimethylaniline. Although this would be below the OSHA PEL of 5 ppm, the user may not actually experience that exact level of protection and so should consider a full full-facepiece respirator with an APF of 50 to provide for extra safety.

Maximum-use concentration (MUC) is the highest concentration of a particular chemical in which a class of respirator can be used. The APF of a respirator is multiplied by the exposure limit for the chemical. In the example of dimethylaniline above, the MUC for full-facepiece respirators would be 250 ppm (50×5 ppm).

At this writing, OSHA has proposed a rulemaking to add APF values to the Respiratory Protection Standard. Table 12.3 shows the proposed APFs for some types of respirators. The final values may be different from those proposed.

STEP 3—SELECT CHEMICAL PROTECTIVE CLOTHING

Chemicals may cause harm directly on contact with the skin or may pass through the skin to harm other parts of the body. Some chemicals very readily absorb through the skin. These chemicals have a "skin" notation with their airborne exposure limits (e.g., PEL) to remind health and safety professionals to protect the skin as well as the inhalation route of entry diligently.

Chemical protective clothing (CPC) provides a barrier between the responder's body and chemicals. If properly selected and used, CPC effectively prevents the chemicals from making contact with the skin. Responders must consider many factors when selecting the appropriate type and amount of protective clothing.

TABLE 12.3 Assigned Protection Factors

Type of Respirator[1,2]	Half-Mask	Full Facepiece
Air-purifying respirator	10^3	50
Powered air-purifying respirator (PAPR)	50	1000
Supplied-air respirator (SAR) or air line respirator:		
• Demand mode	10	50
• Continuous-flow mode	50	1000
• Pressure-demand or other positive-pressure mode	50	1000
Self-contained breathing apparatus (SCBA):		
• Demand mode	10	50
• Pressure-demand or other positive-pressure mode (e.g., open/closed circuit)		10,000[4] (maximum)

[1] Employers may select respirators assigned for use in higher workplace concentrations of a hazardous substance for use at lower concentrations of that substance or when required respirator use is independent of concentration.

[2] The assigned protection factors in Table 12.3 only apply when the employer implements a continuing, effective respirator program as specified by OSHA's Respiratory Protection Standard at 29 CFR 1910.134, including training, fit testing, maintenance and use requirements.

[3] This APF category includes quarter masks, filtering facepieces, and half-masks.

[4] Although positive-pressure SCBAs appear to provide the highest level of respiratory protection, a SWPF study of SCBA users concluded that all users may not achieve protection factors of 10,000 at high work rates. When employers can estimate hazardous concentrations for planning purposes, they must use a maximum assigned protection factor no higher than 10,000.

CPC must be selected properly if it is to provide any protection to the responder. In fact, improper CPC can actually be more harmful to the user than no protection at all, if it allows a chemical to reach the skin and then holds it there. The type and material of clothing determine how well the responder is protected from a particular chemical.

The following factors must be considered when selecting protective clothing:

- The identity of the chemical contaminants and their physical state (solid, liquid, or gas)
- The clothing's resistance to chemical attack and physical damage
- The duration of the entry and the exposure time of the clothing
- Potential for direct exposure by splashing or spraying
- The degree of stress (particularly heat stress) placed on the responder by the protective clothing
- The clothing's effect on the mobility of the responder

This section presents an introduction to the basics of CPC selection. The selection should be made only by trained and knowledgeable individuals, given the complexities involved and the potential consequences of improper selection. The incident commander and the safety officer should confirm the appropriateness of the CPC material and clothing type before an entry is allowed.

Chemical Attacks on CPC

To provide adequate protection to the wearer, the material from which the CPC is constructed must resist attacks from the chemicals in the space. Chemicals move through or damage the CPC material by means of three mechanisms (Fig. 12.18). Understanding these mechanisms is helpful when selecting and using any type of CPC.

Permeation is the movement of chemicals through the CPC material on a molecular level. Think of permeation as the chemical "soaking" through the material in that there is no permanent change in the material and an opening is not necessary. Individual molecules of contaminant pass between the molecules of the CPC material. This occurs when the physical properties of the chemical and the material allow the chemical from outside the garment to dissolve into the material and then "off gas" on the inside of the garment without producing any visible effect on the material.

Degradation occurs when there is a chemical interaction between the chemical and the material that causes the material to change form or lose physical integrity. Degradation typically does produce a visible or otherwise detectable change in the CPC material.

Penetration refers to the bulk movement of chemical through an opening or flaw in the material. The term "leaking" comes to mind with penetration, which may occur through imperfect seams, small holes, zippers, or any other opening. There is no chemical or physical process other than simple leakage involved with penetration.

Permeation and degradation involve a direct interaction between the chemical and the CPC material that is based on their physical and chemical properties.

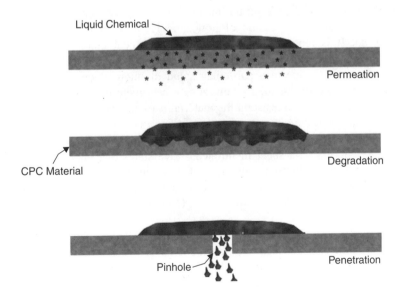

Figure 12.18   Chemical attacks on CPC.

Because these properties vary with the different chemicals and materials, the severity of these effects also varies as different chemicals contact different materials. The appropriate material must be purposefully selected for each chemical that is known or expected to be present.

Measurement of Chemical Attack

Research and development for chemical protective clothing is performed by manufacturers, academic institutions, and government agencies. A large number of CPC materials are tested against a vast number of chemicals. These tests follow standard practices laid out by such organizations as the American Society for Testing and Materials (ASTM) and the National Fire Protection Association (NFPA) to ensure consistency and accuracy. The data generated by these tests are used by manufacturers to develop selection charts that are discussed in a later section of this chapter.

In the typical test procedure for permeation, a swatch of material is mounted between two glass hemispheres. The test chemical (or challenge agent) is circulated in either liquid or vapor form through one hemisphere. Air or another carrier medium is circulated on the other side of the material and returned to a detection device that will measure the chemical. The test begins with the first exposure of the chemical to the material and ends when the chemical is detected on the other side.

The results of the chemical resistance studies are expressed as *breakthrough time* and *permeation rate*. Breakthrough time is the time that it takes for the chemical to pass through the test material and be detected on the other side. It is reported in seconds, minutes, or even hours. Generally the longer the breakthrough time, the more resistant the material is to the chemical.

Permeation rate is a measure of how quickly the chemical moves through the material. It is reported as the amount of chemical that moves through a surface area of material per unit of time. For example, one vendor reports that ethyl acetate will permeate a particular brand of butyl glove material (17-mil thickness) at a rate of 3.4 micrograms/square centimeter/minute. The lower the permeation rate, the more resistant the material is to the chemical. Be sure that all of the measurement units match when comparing the rates. Also, do not assume that all brands of gloves made of the same material will perform in the same way. The same vendor's data sheet mentioned above states that ethyl acetate (same chemical) will move through another, thicker (25 mil) brand of butyl glove at a rate of 500 micrograms/square centimeter/minute (much faster permeation rate.) Selection should be made of a specific CPC product (not just a material type) whenever possible.

The tests are performed in laboratories under controlled conditions. The testing can be assumed to have been done at about 70°F unless otherwise stated. The results also should indicate the thickness of the material that was tested. The significance of these factors in selection is discussed below in the section on using the selection information.

CPC Material Selection Information

Many sources of information are available to responders for selecting CPC. Once the chemical has been identified, CPC can be selected by using information from the sources described below.

Qualitative and Quantitative Data

Selection information typically is provided in either qualitative or quantitative forms. *Qualitative* recommendations typically rate the material's resistance to chemicals with subjective terms such as "excellent," "good," "fair," "poor," or "not recommended." Such guides are usually tables that show the effectiveness of a material against a number of representative chemicals or chemical families (Table 12.4). The responder would choose the material with the highest or most favorable rating for the chemical of interest.

Specific criteria for assigning each rating may or may not be provided with the listing. Responders may be somewhat uncomfortable making selection decisions based on these ratings if the criteria for each rating are not provided. The ratings are not arbitrary but are based on an assessment of permeation or degradation data, even if they are not provided.

Quantitative selection guides give the actual breakthrough time and, usually, the permeation rate data for each chemical-material combination. The information is given in a table with columns for each CPC material and rows showing data for each of a number of representative chemicals (Table 12.5).

Some guides now include information on degradation, often in the form of qualitative rating. Manufacturers assign this rating based principally on the amount of weight gained when the material is immersed in the chemical. When

TABLE 12.4   Qualitative Data on Chemical Resistance of CPC

Chemical	PVC	Viton	Butyl
Acetaldehyde	I	P	E
Acetone	P	P	E
Benzyl chloride	P	E	F
Butyl acetate	P	P	G
Calcium hypochlorite	E	E	E
Diisocyanate	I	I	I
Ethylene dichloride	P	G	F
Hydrofluoric acid	E	G	G
Methylene chloride	P	G	F
Propyl alcohol	E	E	E
Xylene	P	E	P

E—No effect
G—Minor effect
F—Moderate effect
P—Severe effect
I—Insufficient data available

TABLE 12.5   Typical Vender-Provided Quantitative Data on Chemical Resistance of CPC

Chemical Permeation Resistance Chart	VAUTEX 23 mils thick (19.1 oz/yd^2)		BETEX 19 mils thick (15.9 oz/yd^2)	
Chemical	Breakthrough Time (min)	Permeation Rate (μg/cm^2/min)	Breakthrough Time (min)	Permeation Rate (μg/cm^2/min)
Acetone	15	33.4	135	10.8
Ammonia gas	240+	NA	240+	NA
Carbon disulfide	240+	NA	<5	5.8
Chlorine gas	240+	NA	240+	NA
Dimethyl formamide	30	7.7	240+	NA
Ethyl acetate	15	57.2	30	5.7
Gasoline	240+	NA	25	287
Isobutylamine	45	16.0	50	6.8
Nitric acid	240+	NA	240+	NA

provided, this rating is often an addition to quantitative information from permeation tests.

Sources of Selection Information

Manufacturers. As might be expected, CPC manufacturers and vendors provide most of the selection information available to the CPC user. They perform most of the research and development work on new clothing materials. The data generated during the research and in subsequent testing are used to determine its applicability to different workplace settings. The manufacturers provide the selection data in print form or in electronic database form. Printed selection information is provided as a table relating one or more products to a list of chemicals, which may represent a group of similar chemicals. The recommendations may be either qualitative or quantitative in nature. The tables usually are offered free of charge, often as a part of marketing information.

Computerized CPC selection guides are available from most manufacturers. The guides allow the user to search a database for recommendations of the manufacturer's products best suited to protecting against a particular chemical. The computer program usually includes some additional information about the chemical, such as chemical and physical properties and exposure limits. Most manufacturers provide the programs free on request, although some charge a fee for the software. Many manufacturers make their selection guides available on their websites as well.

Independent Sources. Some selection guides have been produced by independent sources that are not associated with CPC manufacturers. Most notable of these is the *Quick Selection Guide to Chemical Protective Clothing, Fourth Edition* (Forsberg, 2002). This guide is a compilation and evaluation of CPC test data from sources all over the world.

The *Emergency Action Guides* published by the Association of American Railroads include CPC recommendations. The guides are emergency response guides, and the CPC recommendations are just one piece of the information they provide. Other published chemical references may include CPC selection data as well.

The material safety data sheet (MSDS) may also recommend CPC material. Not all MSDSs provide specific recommendations for CPC. For instance, many simply recommend "rubber gloves," using the term generically to mean "chemical resistant" without reference to a specific material. Responders should check an MSDS if it is available for the released chemical. However, given the potential inadequacies and inaccuracies of MSDSs, responders should confirm the CPC recommendations with other sources whenever possible.

Government Resources. NIOSH has made available the *Recommendations for Chemical Protective Clothing: A Companion to the NIOSH Pocket Guide to Chemical Hazards.* This electronic database was developed by a contractor to provide recommendations for CPC materials for each of the approximately 600 chemicals in the *Guide.* The database can be accessed from NIOSH's website and is available as part of the *NIOSH Pocket Guide to Chemical Hazards* CD-ROM.

CAMEO® is a system of software applications used widely to plan for and respond to chemical emergencies (see Chapter 3). It is distributed by the EPA free of charge to hazmat responders from its website. The chemical information database includes CPC recommendations and can be a useful reference resource.

Using CPC Material Selection Information

Responders should bear in mind some factors affecting how the data are generated when selecting and using CPC. Some of these limitations are discussed below, followed by practical guidance for using selection data.

Limitations of the Information
Small Amount of Test Material. Permeation tests use a small swatch of material without seams or other inconsistencies. However, the clothing made from the material uses larger swaths of material, often with sewn seams, zippers, and other potential breaks. Despite manufacturers' extensive quality control procedures, some variation among garments does occur. Some garments have integrated boots, gloves, and visors that may be made of a different material than the suit itself. Responders must be certain that these components are included in the manufacturer's selection information.

Seams, fasteners, zippers, and accessories are possible opportunities for leakage into the garment. Seams are sewn and taped to prevent penetration, and flaps cover zippers to prevent penetration by splashed liquids.

Single-Chemical Tests. Permeation tests are generally performed one chemical against one material, although some common mixtures such as gasoline may be tested. There are tens of thousands of chemical compounds in use, so the number of potential mixtures would be impossible to test. Reactions between chemicals could significantly change the permeation of one or both of the chemicals and cause a more rapid breakthrough. Because without direct testing of mixtures it is impossible to know how much the permeation of chemicals in a mixture would be affected, responders should apply safety factors to change schedules or other decisions about CPC used against mixtures.

Laboratory Conditions. Laboratory tests of the material are conducted under controlled environmental conditions. However, the conditions of the workplace can be extreme and may change during the use of CPC. Very high temperatures may decrease the breakthrough time because of an increased permeation rate. In addition, the flexing and folding of the material can cause weakening of the material that may allow chemical permeation to occur more quickly than during static permeation testing.

Manufacturer-Supplied Data. Much of the selection data available for use come from CPC manufacturers and vendors that want to sell their product. Although this may seem like a conflict of interest, potential liability provides a strong motivation for the manufacturer to provide accurate selection information. Making false claims about a product could result in injury to a responder; therefore, they follow standard, approved protocols when conducting the permeation tests. Nonetheless, responders should take vendors' marketing claims about their products with caution and rely only on technical data or recommendations for selection.

Use of Breakthrough Times

Responders may be tempted to treat breakthrough times as safe time limits for using the selected CPC. Considering the limitations described above, this can be dangerous. Breakthrough times are provided only for comparing one material with another to select the most effective protection. The breakthrough times should be considered "best-case" times because actual response conditions are likely to be worse than the test conditions.

When selecting CPC material, users should remember these basic principles:

- No material provides universal protection against all chemicals. The identity of the chemical must be known and the proper material selected.
- No material protects forever. It is only a question of time before a chemical will break through the material.
- The responder must select the material that provides the longest protection against the most chemicals present.

Types of Protective Clothing

Chemical protective garments can provide protection to any or all parts of the body. The degree and nature of the hazards in a space and the work to be done determine the degree of protection needed for the entrant. As a rule, the more of the body to be protected, the greater the burden placed on the wearer. A basic guideline for selecting the type of clothing to be used is that chemical protection must be adequate but not overly burdensome to the entrant.

Chemical Protective Garments

Chemical protective suits are designed to protect the body from exposure to damaging chemicals. The suits may completely enclose the wearer or simply shield the body from splash. The properties of the chemical and the type of work to be done will determine which type of suit is required.

Totally Encapsulating Chemical Protective Suits. Totally encapsulating chemical protective (TECP) suits completely enclose the wearer and seal out gases and vapors as well as liquid splashes and particulates (Fig. 12.19). They provide complete protection to all of the body and skin. They are selected when:

- Chemicals are highly toxic or damaging to the skin.
- Chemicals are readily absorbed through the skin into the rest of the body.
- Work conditions involve a high degree of splashing by a skin-damaging chemical.

They are designed to provide a gas-tight barrier between the wearer and the environment. They have seams that are sewn, glued, and taped to prevent vapor

Figure 12.19 TECP suits completely enclose the responder.

intrusion. The zippers on such garments are made gas tight by pressing material against material rather than simply interlocking metal teeth. The suits have a clear face shield for vision outside the suit.

The TECP suit may have attached boots for chemical and physical protection or simply enclose the foot in a bootee that provides chemical protection and slips into a substantial outer boot for physical protection. Likewise, substantial chemical-resistant gloves may be attached to the suit or the suit may have an inner glove and a solid, smooth cuff ring over which an outer glove is stretched and sealed.

Being a vapor-tight barrier, the TECP suit requires the use of atmosphere-supplying respirators, either SCBA or SAR. If an SAR is used, a pass-through port must be installed in the suit to connect the air line to the SAR unit inside. Air exhaled from the atmosphere-supplying respirator facepiece into the suit creates a positive internal suit pressure that is controlled through pressure-relief valves in the suit body. By having a slight positive internal pressure, the suit provides additional protection in that any leakage will be from the inside out.

TECP suits may be designed as either reusable or limited-use (disposable). Reusable suits may be substantially more expensive than disposable suits initially; however, the savings gained by reusing the suit may actually make them more economical for longer responses requiring repeated entries. Before a suit is reused, it must be properly decontaminated as described in Chapter 13.

Nonencapsulating Suits. Nonencapsulating chemical protective (NECP) suits, also known as splash suits, cover most of the body with gaps around the neck, or hood if so equipped (Fig. 12.20). They provide good protection from moderate liquid splash and particulate contamination but allow gases and vapors to enter the suit. Splash protection can be enhanced by sealing the cuffs to the boots and gloves, but this does not convert a splash suit into a gas-tight TECP suit.

NECP suit material must still be properly selected, even though the suits are not gas tight. This prevents degradation and slow permeation and provides better protection to the responder. These garments should be worn when protection from liquid contact is needed but some vapor contact with the skin should cause no harm.

Disposable Overgarments. Lightweight, disposable garments may be used to protect the more expensive reusable garments. The inexpensive disposable garment takes most of the physical abuse and minimizes the splash, while the reusable garment provides the primary chemical protection. The disposable garment is discarded in the decontamination process, and the reusable garment is decontaminated.

This increases the chemical protection and extends the useful life of the reusable garment. Because the overgarment is lightweight, it does not add significantly to the heat stress of the wearer, who is already in a vapor-tight environment. A similar approach is used when outer gloves and boots are used to take the brunt of the contamination and physical abuse of the parts of the body most likely to be contaminated, the hands and feet.

Figure 12.20   The NECP suit has some openings and is not vapor tight.

Partial Body Coverings. In addition to suits for protecting the whole body, there are protective clothing items including aprons, sleeves, and leggings to protect portions of the body. These provide localized splash and contact protection where minimal splashing is expected. Proper selection of these articles is important to reduce permeation and prevent degradation by chemical contact.

Boots and Gloves. The hands and feet are the parts of the body most likely to be contaminated while also being subject to the most physical action and abuse. Obviously, it is very important to choose the proper material to provide chemical resistance. But responders must also choose the construction that will provide adequate durability without being too cumbersome and awkward.

In some situations where multiple chemicals have been identified, it may be necessary to layer gloves of two different materials to get adequate protection. Each material provides protection from some chemicals, and together the range of chemicals blocked is increased. This works best where the tasks to be performed do not require fine finger dexterity.

Responders need to change the gloves on a schedule that will prevent permeation and exposure. The required frequency is determined by the chemical(s) present, the physical demands of the work, and the likelihood of direct contact with the chemical. A greater likelihood of contamination of the hands may make it necessary to change gloves even more frequently than other CPC, such as the suit.

A discussion of hand and foot protection for nonchemical hazards follows.

Protective Clothing for Nonchemical Hazards

Hazmat responses are by definition primarily concerned with hazardous materials. However, other, nonchemical hazards may be present that require special protective clothing. Examples of such hazards include fire, temperature extremes, water, and radiation. The protective clothing for such hazards may have specifications beyond the scope of this discussion, but basic descriptions are provided below.

Firefighter's Protective Clothing. Firefighters wear a number of different types of protective clothing. Basic turnout or bunker gear is composed of an insulating inner liner and a fire-retardant outer shell. It is intended to provide limited protection from heat, hot water, embers, and debris. It does not provide protection from chemical gases or vapors. Exposure to hazardous materials will likely result in chemicals permeating into the fabric, making decontamination of firefighter's turnout gear difficult at best. It is not designed or intended to provide chemical protection.

Chemical protective clothing material itself is very flammable. Flash covers made from a flame-retardant material with an aluminized exterior (Fig. 12.21) may be worn over CPC garments when entry must be made into a potentially

Figure 12.21 Aluminized flash covers can provide some protection from a flash fire.

flammable environment. They are intended to provide brief protection to the responder in the event of a flash fire to allow for a rapid escape. They are not intended to allow responders to enter an active fire environment.

Temperature Extremes. Protective clothing may be needed to protect responders entering extremely hot or cold environments. Such clothing is used for extreme conditions outside the normal variation of outdoors environments. Cooling garments may hold frozen cooling packets close to the body or circulate chilled water or air through tubing over the body. New high-tech garments use a cooling material that is not as cold as ice. These garments are intended for brief use only.

Layers of insulating clothing that will wick moisture away from the skin can protect responders in extremely cold environments. The multiple layers provide insulation by creating protected space where air that is warmed by the body stays near the body, slowing the heat loss. Outer shells that block the wind and keep moisture out are also helpful. Prolonged contact with cold metal surfaces will result in significant heat loss by conduction, so clothing barriers that prevent such contact should be worn.

Radiation. Radiation and radioactive materials are discussed in Chapter 8. Radiation is the spontaneous emission of energy from a source. The energy is given off as either a particle (alpha or beta) or as a wave (gamma or X ray). Protective clothing worn for radiation work is designed to prevent contamination of the worker by radioactive alpha or beta particles. Clothing provides no protection from gamma radiation. If radiation is known or suspected to be involved in a response, no one should approach the hazard area until the radiation hazard is evaluated by the appropriate radiation expert, who can then recommend protective clothing, if appropriate.

STEP 4—CHOOSE PERSONAL SAFETY EQUIPMENT

The respiratory and skin protective equipment described above incidentally protects against some other hazards as well. For instance, a full-face respirator covers the eyes and therefore protects against debris and flying objects that may injure the face and eyes. However, responders may face nonchemical hazards that require additional safety equipment. Step 3 in the use of PPE is to identify any remaining hazards not addressed by the respirators and CPC that have been selected. The following discussion is a brief summary of personal safety equipment and key features or specifications to be considered in their selection and use.

Head Protection

Neither respiratory protection nor CPC will protect the head from injury when a wrench is dropped from above. Head injuries may occur at a hazmat response from accident types such as:

- Falling or flying objects

- Being struck by moving equipment
- Bumping the head in low or tight spaces

Head protection must be added to prevent injury.

Regulations and Standards

OSHA regulates the use of helmets or "hard hats" in its *Head Protection Standard* (29 CFR 1910.135), which refers to ANSI's *American National Standard for Industrial Head Protection* (Z89.1–2003). The standard for head protection for firefighters is part of NFPA 1971: *Standard on Protective Ensemble for Structural Fire Fighting*. The standard specifications for effective head protection are given in these standards.

OSHA Standard. In the head protection standard, OSHA has two requirements:

- Employers must ensure that each affected employee wears a protective helmet when working in areas where there is a potential for injury to the head from falling objects.
- Employers must ensure that a protective helmet designed to reduce electrical shock hazard is worn by each affected employee when near exposed electrical conductors that could contact the head.

Hazmat response potentially involves injury from falling objects, although the electrical shock hazard may be less frequent. OSHA requires that employers use helmets that meet the criteria of ANSI Z89.1–1986.

ANSI Standard. ANSI Z89.1–2003 represents the consensus as to the required performance characteristics of protective helmets. It describes the types and classes of helmets and their required components as well as what accessories are allowed. The standard describes two classifications of helmets. First, with regard to the type of impact:

- Type I helmets are designed to reduce the force of impact resulting from a blow to the top of the head only.
- Type II helmets reduce the force of impact that may be received off center or to the top of the head.

The standard also classifies helmets according to their protection from electrical contact.

- Class C (conductive) helmets are not intended to provide protection against contact with electrical conductors.
- Class G (general) helmets reduce the danger of contact exposure to low-voltage conductors.
- Class E (electrical) helmets reduce the danger of exposure to high-voltage conductors.

In addition to classifying the helmets, the standard describes specific test procedures for assessing the performance of helmets with regard to such things as:

- Flammability
- Force transmission to the head and impact energy attenuation
- Apex and off-center penetration
- Chin strap retention
- Electrical insulation

Manufacturers whose helmets meet the specifications set out in the standard will label them according to the type and class with which they comply, such as Type I, Class C or Type II, Class E.

NFPA Requirements. NFPA 1971 provides the same type of performance standard for structural firefighter ensembles, including the helmet. Given the demanding environments in which firefighters work, the testing procedures and performance requirements for firefighter helmets are more rigorous than industrial helmets. Firefighter helmets meeting these rigorous requirements are labeled as NFPA 1971 compliant.

Types of Head Protection
Helmets. The most common form of head protection is the helmet, also known as a "hard hat" (Fig. 12.22). Helmets provide protection by absorbing the impact

Figure 12.22   Helmet or "hard hat".

of flying objects and preventing them from penetrating the head. The basic construction of the helmet includes:

- Outer shell to prevent penetration
- Adjustable headband to hold its position
- Suspension system to absorb impact

The outer shell typically is made of rigid materials such as high-density polyethylene, polycarbonate, or acrylonitrile butadiene styrene (ABS) plastic. These materials typically do not conduct electricity but must be tested and labeled according to their classification. The straps of the suspension system distribute the impact force over a larger area of the head and absorb some of the energy and must maintain a minimum clearance specified by the manufacturer between the shell and the head during normal use. The headband must be adjustable in size and designed to keep the helmet in the proper position on the wearer's head. Sizing must be consistent with normal hat sizes. The headband may be covered with terry cloth or other material to absorb sweat.

Bump Caps and Hoods. Bump caps do not meet the ANSI specifications for protection from impact or penetration by flying objects or electrical hazards. They provide protection for the head from bumps when working in tight, low spaces. Hoods or soft caps made of special fabrics may be used to protect the head and neck from airborne hazards such as sparks, dust and debris, splashes, or temperature extremes. This headwear may be worn by itself or as an accessory to other primary head protection, such as the hardhat liner worn in cold environments.

Special Considerations for Head Protection
The helmet headband can be adjusted in size to provide a secure fit. The force distribution function of the helmet only works when the helmet and its suspension system are positioned properly on the head. Respirator facepieces can prevent the helmet from seating securely on the head. Adjustable chin straps may be necessary to hold the helmet on the top of the head.

Some wearers choose to put their helmets on backwards to avoid the problem of the brim. If this is done, the suspension system should be reversed so that the head is properly cradled. Responders should consult the manufacturer to see whether this affects the ANSI rating of the helmet.

Inspect head protection before each use to look for signs of wear, aging, or damage. Discard or repair helmets with cracking, discoloration, or texture changes of the shell or torn or broken suspension system components. Replace any helmet that sustains a substantial blow or impact, because imperceptible damage may have been done.

There is no required replacement time for head protection. Rely on inspection to indicate the need for replacement. Hardhats worn regularly in harsh or extreme environments or in direct sunlight may need to be replaced more frequently because of the damage that may be done to the shell.

Eye and Face Protection

The face and eyes are among the most sensitive exposed areas of the body. The skin can tolerate contact with dust, dirt, and debris without significant problem; however, even such small foreign bodies in the eye can cause damage or can create safety hazards by affecting the responder's vision. Add to that the increased vulnerability to chemical exposure, and one can see that it is important to provide protection of these areas.

Standards

OSHA and ANSI have written standards for eye and face protection. OSHA standards describe required employee protection, and reference ANSI standards for the performance of protective equipment.

OSHA Standard. OSHA's *Eye and Face Protection Standard* (29 CFR 1910.133) regulates the protection of the face and eyes. The standard requires the employer to ensure that each affected employee uses face or eye protection when exposed to:

- Flying particles
- Molten metal
- Liquid chemicals
- Acids or caustic liquids
- Chemical gases and vapors
- Potentially injurious light radiation

Any of these might be present or involved in a hazardous materials release. Employers must ensure that responders are protected from them by providing effective eye protection.

In most cases, hazmat responders will be wearing full-face respirators where hazards are present. Because the eyes are already protected by the mask, no additional eye protection is needed. However, some operations, such as welding, may require special forms of protection. In these cases, the OSHA standard requires that the eye protection meet the specifications of the ANSI standard.

ANSI Standard. ANSI Z87.1–2003, *American National Standard Practice for Occupational and Educational Eye and Face Protection*, sets the specifications for face and eye protection. The standard provides a helpful description of protective devices, which they classify as:

- Spectacles
- Goggles
- Faceshields
- Welding helmets
- Full-face and loose-fitting respirator faceshields

For each classification, the standard provides testing procedures and the required standard of performance in each of these categories:

- Impact resistance
- Optical (vision) requirements
- Flammability resistance
- Corrosion resistance
- Cleanability

A detailed discussion of the tests is not necessary and is beyond the scope of this book. However, one change made in the most recent revision of the standard is worth noting. There are now two levels of impact resistance—basic impact and high impact. Basic-impact lenses will be marked "Z87," and high-impact lenses will be marked "Z87+" on the frame and/or the lenses. Responders selecting eye and face protection that is labeled as compliant with ANSI Z87.1–2003 can have confidence that the equipment will provide good protection against the hazards.

Types of Face and Eye Protection

Safety Glasses. Spectacles (or safety glasses) are made up of lenses joined by a bridge over the nose and supported by temple bars. The lens may be flat with separate sideshields or curved as a single piece around the side of the eye (Fig. 12.23).

Figure 12.23 Safety glasses.

New materials and stylish designs make spectacles much more comfortable than the old, stiff safety glasses of the past. This is evidenced by the fact that many employees are choosing to wear tinted safety glasses as sunglasses. Therefore, all employees should always wear their eye protection in hazardous areas.

Spectacles do not fit tightly against the face, so they offer limited splash protection and no protection from gases and vapors. They are intended primarily for impact protection. ANSI-compliant industrial safety glasses have lenses designed to withstand much greater impact force than common sunglasses or even recreational eye protection. Employers must ensure that only approved eye protection is worn by employees in eye hazard areas.

Special coatings are now being applied to lenses that enhance resistance to scratching and fogging as well as absorbing almost all of the damaging ultraviolet (UV) light. These coatings enhance the function and useful life of the lenses. Many spectacles have replaceable lenses to reduce cost.

Many spectacles have adjustable components for achieving the best size and fit. These adjustments allow the glasses to sit comfortably in the proper position on the wearer's face. Improved comfort and attractive styling have contributed greatly to workers' acceptance and use of proper eye protection.

Goggles. Goggles are protective eyewear that fit against the face. Because they fit tightly against the face, goggles provide excellent protection against splash, dust, and debris as well as frontal impact. They provide limited protection from gases and vapors if they are unventilated. Ventilated goggles allow air flow behind the lens to reduce fogging. Goggles can be worn over prescription glasses to protect them from damage.

Faceshields. Faceshields are a transparent vertical barrier surrounding the wearer's face. The shield may rest on its own suspension system or attach to the brim of a helmet. Shields sit out from the face to accommodate other equipment such as safety glasses being worn underneath. Chin and neck protection can be added to the shield as well.

Faceshields protect the face against major splashing and flying objects. Faceshields are not primary protectors, but they should always be worn with eye protection such as spectacles or goggles. For additional protection when liquids are being poured or when flying objects may injure the face, faceshields provide wide coverage with minimal discomfort to the wearer.

Welding Shields. Welding shields are specialized faceshields that protect welders from impact and from direct radiant energy from certain welding operations. According to OSHA's welding standard (29 CFR 1910.252), the helmet must be positioned to protect the head, neck, and ears from direct radiant energy. They may be suspended from a hardhat, supported by their own suspension system, or hand held. They must have filter plates (shaded appropriately) and protective cover plates.

Hearing Protection

The hazards of noise and the requirements of OSHA's Hearing Conservation Standard are discussed in Chapter 8. When the noise level during response actions exceeds the allowable dose and engineering controls cannot adequately reduce the levels, hearing protective devices may be required (Fig. 12.24).

Types of Hearing Protection

Earplugs are made of sound-absorbing foams or other moldable materials that insert into the ear canal and fit tightly against the walls. The wearer rolls the foam earplug into a small cone, inserts it deep into the ear canal, and lets it expand to fill the canal. Flange-type plugs are mounted on firm stems that allow easy insertion into the ear canal. If a good fit and insertion are achieved, the plug blocks the passage of the sound waves into the ear.

Semi-inserts or hearing bands are soft pluglike inserts on either end of a tension headband. They block off the external opening of the ear canals and prevent the sound from entering. Because they do not form as tight a seal as earplugs, these devices do not reduce the noise levels as much as earplugs. However, they are very easy to put on and take off and therefore are useful protection for intermittent exposures to lower noise levels.

Earmuffs are padded cup-shaped enclosures lined with sound-absorbing material that completely cover the outer ear. They fit tightly against the head of the wearer to prevent sound waves from entering the ear. Any breaks in the seal, such as from eyeglass temple bars, reduce the effectiveness of the hearing protection. Some hardhats are designed with attachment points for arms that hold the earmuffs in position against the head without compromising the positioning of the hardhat.

Some particularly noisy environments may require the use of a combination of earplugs and earmuffs. This combination enhances the protection but does

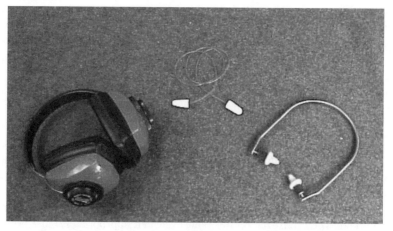

Figure 12.24   Types of hearing protection—earmuffs, plugs, and inserts.

not double the noise reduction as one might think. The effectiveness of such a combination of protective devices should be evaluated by a trained occupational noise professional to determine whether the noise reduction will be adequate.

Noise Reduction Rating

EPA developed the noise reduction rating (NRR) system to report the noise-reducing capability of hearing protectors. Hearing protection manufacturers use this system to determine the level of noise reduction in decibels that their products provide and mark the NRR value on the packaging of the devices.

Responders can use the NRR to estimate the noise exposure they can expect while wearing the protectors. Appendix B of OSHA's *Hearing Conservation Standard* outlines the procedures for using the NRR to determine the actual noise exposure, taking into account the appropriate correction factors when A-weighted or C-weighted measurements are used.

Hand Protection

The hands are vulnerable to many different hazards. The OSHA Hand Protection Standard (29 CFR 1910.138) requires that the employer select and enforce the use of appropriate hand protection when hands might be exposed to any of these hazards:

- Chemical exposure
- Severe cuts or lacerations
- Severe abrasions
- Punctures
- Thermal burns
- Cold exposure

Any or all of these hazards may be a part of responding to a hazardous materials release. This brief standard simply requires that the employer base the selection on:

- Performance characteristics of the hand protection
- Tasks to be performed
- Conditions present
- Duration of use
- Hazards identified

Types of Protective Gloves

A number of glove types are available to provide protection from the hazards. Although they are discussed separately, some glove designs provide overlapping or multiple protections.

Chemical Protective Gloves. Responders must select gloves made of a material that protects the hands from contamination by hazardous chemicals according to the guidelines for selecting CPC material in Step 3. Because the hands are not only more likely to contact chemicals but also to do so while doing work that may cause physical damage, durable gloves should be chosen (Fig. 12.25). Other characteristics are also discussed above in the section on boots and gloves in Step 3.

Electrical Insulating Gloves. Some gloves act as a barrier to prevent the flow of electricity from an electrical conductor through the hand. OSHA's *Electrical Protective Equipment Standard* (29 CFR 1910.137) provides detailed requirements for the design, testing, and use of gloves and other electrical protective equipment.

Figure 12.25 Heavy contamination and rough work make glove selection important.

Gloves are classified according to the minimum test voltage they will block, with Class 0 withstanding 5000 volts AC and Class 4 up to 40,000 volts AC. A leather protector glove must be worn over insulating gloves used on voltages higher than 250 volts AC except where high finger dexterity is required. The protector glove receives the brunt of physical damage, allowing the rubber insulating gloves to remain intact and function as the primary electrical barrier.

Cut-Resistant Gloves. Responders may have to work in situations where sharp metal edges or other hazards may cut the hand. Gloves made of strong, cut-resistant materials can protect the hands. Whereas some gloves are constructed out of stainless steel mesh, many gloves are made out of strong fabrics such as Dupont's Kevlar® that are woven with stainless steel threads. Cut-resistant gloves protect against minor cuts or lacerations, severe abrasions, and punctures. Effective engineering controls such as machine guarding are the best protection against major cuts or amputations.

Cloth and Leather Gloves. Simple leather gloves provide excellent protection against abrasion and physical exertion. Leather is useful in keeping hands warm in cold environments. Cotton and polyester cloth gloves can protect against abrasion, exertion, and other minor hazards.

Rescue Gloves. Rescue and rope-handling gloves are thin and tight fitting with reinforced padding across the palms. They protect the wearer from rope burns and abrasions while providing the high dexterity needed for rescue operations.

Temperature-Resistant Gloves. Gloves can insulate the hand from contact with extremely hot or cold objects. Gloves made of materials like Kevlar® and Zetex®, a new fabric made of silica, provide good heat resistance. Thick insulation aids in both heat and cold resistance. Gloves for use in cold environments should also prevent water contact, which causes rapid heat loss.

Foot Protection

Besides the potential for chemical exposure mentioned in Step 3, the feet can be injured by other hazards at a hazardous materials response, such as falling objects, compression by heavy objects, punctures, and others. Properly selected safety shoes can prevent or reduce many of these injuries.

OSHA Foot Protection Standard

OSHA's *Foot Protection Standard* (29 CFR 1910.136) simply requires that foot protection be worn by employees where foot injuries may result from rolling or falling objects, objects piercing the sole, or electrical hazards. It requires that all footwear comply with ANSI Z41-1991. The current revision of the ANSI standard was issued in 1999.

ANSI Standard

ANSI Z41-1999, *American National Standard for Personal Protection—Protective Footwear,* provides the performance requirements for industrial foot protection, including requirements for testing and classifying all footwear according to the minimum compression and impact protection at the toes. These classifications describe the minimum force the shoes will withstand without causing damage to the foot and are for impact protection:

- I/75—75 foot-pounds
- I/50—50 foot-pounds
- I/30—30 foot-pounds

For compression, the classifications and values are:

- C/75—2500 pounds
- C/50—1750 pounds
- C/30—1000 pounds

The other sections of the standard address the following as they apply to shoes being tested:

- Metatarsal protection
- Conductive and static dissipative footwear to prevent static electricity buildup and discharge
- Electrical hazard protection
- Puncture resistance

ANSI-compliant footwear is labeled according to a specific format to communicate the relevant ratings. An example of the label might be:

ANSI Z41 PT 99

M I/75 C/75

Mt/75 EH

PR

This label indicates that this male (M) shoe was tested according to the protective (PT) section of ANSI Z41-1999. It received the highest ratings for impact (I/75), compression (C/75), and metatarsal (Mt/75) support. It also passed tests for electrical hazards (EH) and puncture resistance (PR) as dictated in the standard. Such a label is required to be stamped or sewn on the inside of the shoe, and it provides useful information to the wearer.

Features of Protective Footwear

Steel-toe shoes and boots have a steel cup in the front of the shoe that protects the toes of the wearer from injury by impact or compression by heavy objects.

Some shoes also provide protection for the metatarsal area of the foot as well. Puncture-resistant footwear has a steel plate in the insole that prevents sharp objects from penetrating the sole and injuring the foot.

Electrical protective footwear is designed to prevent the flow of electricity through the shoe in the event of contact with an electrical conductor. The sole and heel of the shoe must be made of nonconductive material. The electrical hazard protection of the shoes deteriorates in wet environments and when there is excessive wear on the sole and heel.

Conductive and static dissipative shoes are designed to prevent the buildup of static electricity in the body. The soles of the shoes conduct any static electricity to the ground, where it dissipates. This footwear may be needed in flammable atmospheres where static discharge could ignite a fire and in work around electronics and other equipment that is sensitive to a static electricity discharge.

Rubber boots keep feet dry in wet and muddy environments. Some footwear includes design features to increase traction or reduce slipping on slippery floors. The shoes may be marked as oil resistant if they do not absorb oils and become slippery. If rubber boots are selected for protection against chemical hazards, they must be properly selected according to the CPC selection principles in Step 3.

PASS Alarms

The Personal Alert Safety System (PASS) device is designed to serve as a locator device if a responder is trapped or loses consciousness during an entry. The device includes a motion sensor that automatically detects if the responder is motionless for an extended period. The device may also be activated manually by a responder who is trapped. It then emits a very loud audible alarm and flashes bright light to lead rescuers to the responder's location.

STEP 5—PUT IT ALL TOGETHER INTO LEVELS OF PROTECTION

PPE has so far been described in the separate categories of respiratory protection, chemical protective clothing, and personal safety equipment. However, PPE is worn together in a personal protective ensemble. Systematic approaches to the classification of PPE ensembles include (1) EPA levels of protection and (2) NFPA certification standards for chemical protective garments. Both of these systems offer levels of protection that can be selected according to the type and degree of protection needed.

The concept of levels of protection is a helpful way to balance the need for adequate protection from the hazards against the increased stress placed on the responder by the PPE. In other words, although the highest degree of respiratory protection and chemical protective clothing will ensure that responders are not harmed by the chemical, all of that PPE places added weight, insulation, and mobility problems on the responder. In some situations, the potential for harm such as heat injury imposed by the PPE may be greater than the harm likely to occur from exposure to the chemical. In this case, more is not necessarily better.

The best approach to PPE selection is to provide adequate protection without placing too much stress or burden on the responder. Each of the two systems provides a quick and standardized way to select the appropriate degree of protection.

EPA Levels of Protection

EPA established its levels of protection system as a means to protect its personnel working at hazardous waste sites. With the promulgation of its HAZWOPER standard, OSHA recognized the system as appropriate for the protection of all workers covered by the standard. The levels are as follows.

Level A Protection

Level A protection provides the highest possible degree of both respiratory protection and skin and eye protection (Fig. 12.26). The typical ensemble includes the following items:

- Totally encapsulating chemical protective (TECP) suit
- Atmosphere-supplying respirator (SCBA or air line)
- Inner chemical-resistant gloves
- Chemical-resistant safety boots
- Hardhat
- Two-way radio communications
- Inner coveralls or thermal underwear (optional)
- Disposable gloves and boot covers (optional)

The TECP suit can be considered the definitive protective item of the Level A ensemble.

Level A protection is required for entries into atmospheres containing high concentrations of unidentified contaminants. Level A protection is also required for entries into atmospheres containing identified airborne contaminants that are known to pose a high degree of hazard to the skin. Such contaminants may be present as gases, vapors, or particulates and may attack the skin directly on contact or pass through the skin to attack other organs. Level A protection may also be required if liquids hazardous to the skin are likely to be encountered in an operation involving major splashing or high-pressure spraying (e.g., valve repair).

Level B Protection

Level B protection provides the same degree of respiratory protection as Level A but provides a lesser degree of skin protection (Fig. 12.27). The Level B ensemble includes:

- Nonencapsulating chemical protective suit (splash suit)
- Atmosphere-supplying respirator (SCBA or air line)
- Inner chemical-resistant gloves

Figure 12.26   Level A ensemble.

- Outer chemical-resistant gloves
- Chemical-resistant safety boots
- Hardhat
- Two-way radio communication

Level B is used whenever the conditions for the use of APRs are not met and threats to the skin are non-IDLH and in the form of liquid splashes or particulate contaminants.

Figure 12.27   Level B ensemble.

Level C Protection

Level C Protection involves the same degree of skin protection as level B but a lesser degree of respiratory protection (Fig. 12.28). The Level C ensemble includes:

- Chemical-resistant splash suit
- Air-purifying respirator
- Chemical-resistant gloves (inner and outer)
- Chemical-resistant safety boots
- Hardhat
- Two-way radio communication
- Hearing protection (if needed)

Level C should only be used in work areas where it can be determined that all criteria for the use of APRs are met. For this reason the role of Level C in emergency response tends to be highly restricted (see the section of this chapter on APRs).

Level D Protection

Level D personal protective equipment provides protection only against "normal" workplace safety hazards (Fig. 12.29). The Level D ensemble includes:

- Cotton coverall

Figure 12.28   Level C ensemble.

Figure 12.29   Level D ensemble.

- Hard hat
- Steel-toed boots
- Safety glasses
- Work gloves

Level D should be used only in work areas such as the support zone, in which both respiratory hazards and skin hazards are absent.

An important aspect of PPE is protection from "normal" workplace safety hazards, as well as the chemical-related hazards typical of the emergency operations area. All levels of protection should incorporate the basic safety equipment (as represented by the Level D ensemble) as needed to ensure responder safety from all site hazards.

Modified Levels of Protection

In considering levels of protection, it should be noted that the four levels as presented here are highly generalized and should be fine-tuned to provide the specific degree of protection required for a specific task. Any number of modified levels of protection may be used in emergency response.

NFPA Levels

NFPA has developed three performance-based standards that require that chemical protective suits provide a specified minimum level of protection to be certified for use. This process involves a comprehensive approach requiring whole-ensemble evaluation, as opposed to merely specifying the configuration that makes up a level. The certifications represented in the three standards are for CPC ensembles and assume that firefighters generally and hazardous materials responders specifically will primarily use the highest level of respiratory protection, namely the SCBA. Therefore, they resemble EPA Levels A and B in their configuration but go further to ensure a consistent level of performance against most classes of chemicals.

NFPA 1991: Vapor-Protective Suits for Hazardous Chemical Emergencies

For certification under 1991, vapor-protective suits must be able to pass a pressure test conducted in accordance with ASTM F 1052, *Practice for Pressure Testing of Gas-Tight Totally Encapsulated Chemical Protective Suits*. Also, suits must be able to pass a "shower test" for penetration resistance as specified by NFPA.

The primary material of which a suit is constructed (i.e., the barrier fabric), as well as visors, boots, and gloves, must be able to resist breakthrough for a least one hour when tested for permeation by each of 21 chemicals specified in ASTM Standard Guide F 1001, which are considered to be representative of the major chemical classes encountered during hazmat incidents. The test battery for certification under NFPA 1991 includes the following chemicals:

- Acetone

- Acetonitrile
- Anhydrous ammonia
- 1,3-Butadiene
- Carbon disulfide
- Chlorine
- Dichloromethane
- Diethyl amine
- Dimethyl formamide
- Ethyl acetate
- Ethylene oxide
- Hexane
- Hydrogen chloride
- Methanol
- Methyl chloride
- Nitrobenzene
- Sodium hydroxide
- Sulfuric acid
- Tetrachloroethylene
- Tetrahydrofuran
- Toluene

The suit must also offer adequate permeation resistance for any additional chemicals for which the suit is certified. Any ensemble components worn outside the primary suit material must also meet the minimum requirements for chemical resistance.

All suit components must meet minimum performance-based requirements for certification under 1991. These components include visors, valves, seams, closures, gloves, and boots. The suits must also meet minimum requirements for durability and flammability resistance.

NFPA 1992: Liquid Splash-Protective Suit for Hazardous Chemical Emergencies

NFPA 1992 is, like NFPA 1991, a whole-ensemble, performance-based standard. However, the major focus of 1992 is on protection in chemical splash environments. Thus no requirements related to gas-tight integrity are applicable under 1992. However, overall suit water penetration resistance must be demonstrated through "shower testing."

NFPA 1992 also requires that splash-protective ensembles adequately resist penetration by chemicals. The NFPA battery of challenge chemicals used for this standard was developed by deleting from the ASTM F 1001 chemical assemblage those chemicals known to be skin toxins or known or suspected of causing cancer. The test battery for certification under NFPA 1992 includes the following chemicals:

- Acetone
- Acetonitrile
- Ethyl acetate
- Hexane
- Sodium hydroxide
- Sulfuric acid
- Tetrahydrofuran

All ensemble components, including barrier fabric, seams, closures, gloves, boots, visors, and respirator components (if worn outside the chemical protective garments), must resist penetration by these chemicals and any additional chemicals for which the suit is certified for at least one hour. All suit components must also meet minimum requirements for durability and flammability resistance for certification under NFPA 1992.

NFPA briefly maintained *NFPA 1993: Support Function Protective Garments for Hazardous Chemical Operations*. Ensembles certified under NFPA 1993 were intended only for use in supporting functions, as opposed to use in an actual "hot zone" situation. Examples of such support functions are decontamination or cleanup operations, in which all activities are carried out in controlled environments where conditions are known. However, this standard no longer exists, and if these functions are performed in areas where an actual or potential chemical threat exists, NFPA 1991 or NFPA 1992 garments would be appropriate.

NFPA 1994: Protective Ensembles for Chemical/Biological Terrorism Incidents

In 2001, just before the terrorist attacks of September 11 and the anthrax incidents that soon followed, NFPA issued *NFPA 1994: Protective Ensembles for Chemical/Biological Terrorism Incidents*. This standard recognizes the unique hazards and circumstances of responding to these incidents that are maliciously intended to be lethal to innocent civilians and responders. It is also a whole-ensemble performance standard that ensures adequate protection for different levels of exposure. Three classes of ensembles are distinguished not only on the level of threat and exposure but also on the condition of victims at the scene. Table 12.6 describes the anticipated conditions for using each class of ensemble.

Class 1 Ensemble. Class 1 ensembles are intended for the worst-case situations in which there is an immediate threat and agents are unidentified or have not been measured. These ensembles must be gas tight, allowing no more than 0.02% intrusion of a gas, and constructed of materials that provide the highest permeation resistance to chemical and biological agents and high-threat industrial chemicals. They are to include any necessary protective gloves, visors, footwear,

TABLE 12.6 Hazard Criteria of NFPA Classes for Ensembles for Chemical and Biological Incidents

Class	Inhalation Threat	Skin Contact by Liquid	Condition of Victims
1	Unknown or unverified	Expected and not permitted	Unconscious, not ambulatory, but showing no symptoms
2	IDLH	Probable	Symptomatic but not ambulatory
3	STEL	Possible	Symptomatic and ambulatory

and other components necessary to provide whole-body, gas-tight protection. All components must show high resistance to abrasion, tearing, puncture, and other physical damage.

Class 2 Ensemble. Class 2 ensembles are intended for situations where the initial release has subsided and conditions are identified and better quantified. These ensembles provide limited vapor protection, allowing no more that 2.0% vapor intrusion and permeation resistance to the same agents as Class 1 ensembles but tested at lower concentrations. They must pass a shower test by allowing no liquid intrusion when sprayed from several different directions. They also must show resistance to physical damage, but the criteria are slightly lower than those for Class 1 ensembles.

Class 3 Ensemble. Class 3 ensembles are intended for use in low-exposure conditions after the release has passed or on the periphery of the response scene. They are not expected to provide vapor protection, but their material must still resist chemical permeation. They must resist physical damage, but the criteria are lower than Class 2 ensembles.

3/30 Rule

The "3/30 Rule" is a common reference to guidelines that were issued by the U.S. Army Soldier and Biological Chemical Command (SBCCOM) Domestic Preparedness Chemical Team in its document titled *Guidelines for Incident Commander's Use of Firefighter Protective Ensemble (FFPE) with Self-Contained Breathing Apparatus (SCBA) for Rescue Operations During a Terrorist Chemical Agent Incident.* This report offers guidelines to incident commanders as to the length of time that variations of the typical FFPE can be used in response to a chemical terrorist incident.

SBCCOM conducted tests on volunteer career firefighters performing rescue activities in simulated chemical agent atmospheres to determine how long first responders could be exposed without experiencing harmful effects. The tests were performed on four "quick fix" configurations of FFPE:

- Turnout
- Self-taped turnout
- Buddy-taped turnout
- Tyvek suit under turnout

The report provides maximum reconnaissance times for each configuration in each of four chemical agent atmospheres (GB, GD, VX, and HD). The 3/30 reference derives from the times referenced for actions in unknown environments. These are:

- Self-taped turnout gear with SCBA provides sufficient protection in an unknown nerve agent environment for a *3-minute reconnaissance* to search for living victims (or a 2-minute reconnaissance if HD is suspected).
- Standard turnout gear with SCBA provides a first responder with sufficient protection from nerve agent vapor hazards inside interior or downwind areas of the hot zone to allow *30 minutes of rescue time* for known live victims.

These guidelines are based on the highest possible exposures of nerve agents in closed environments, so they would be protective for all response conditions. They assume that the responders would not enter the environment until at least 10 minutes after the initial release and that a living victim indicates rescue would be prudent. Because nerve agent attacks on unprotected persons are fatal within seconds, the discovery of living victims indicates that the exposure to protected responders would be low and there is hope to be able to help other victims. The 3-minute reconnaissance applies to environments where there is no living victim to serve as an exposure indicator. The report emphasizes that these are only general guidelines and should only be used until more scene-specific information becomes available.

STEP 6—USE THE PPE PROPERLY

Once the PPE has been selected and assembled, the responder must then use it properly. The most important rule in using PPE is to follow the manufacturer's guidelines. The equipment has been thoroughly tested under a variety of use conditions by the manufacturer, and recommendations for its use are based on these tests. Any limitations or instructions given by the manufacturer should be taken seriously if the maximum protection is to be obtained.

In addition to the manufacturer's guidelines, other factors should be considered. Factors limiting work mission duration and in-use monitoring are two examples. These and others are discussed below.

Factors Limiting Safe Work Mission Duration

Work mission duration must be estimated before work in PPE actually begins. Factors limiting the length of time that one can safely remain in a contaminated area are discussed below.

Air Supply Consumption

In situations in which the SCBA is used, short duration of air supply can be a major problem and, in some instances, a major threat to the safety of responders. Air supply consumption with an SCBA unit may be significantly increased (thus reducing time on task) by factors such as strenuous work rate, lack of fitness of the user, and/or large body size of the user. Shallow, rapid, irregular breathing patterns, or hyperventilation, can also lead to rapid air consumption. These conditions may result from heat stress, anxiety, lack of acclimatization, or lack of familiarization with the SCBA.

Permeation and Penetration of Protective Clothing or Equipment

Work mission duration cannot safely exceed the length of time during which items of CPC can be expected to provide adequate protection. Thus penetration and permeation of CPC (as described previously) are of major concern. Penetration may occur because of leakage of fasteners or valves on PPE, particularly under extreme temperature conditions. Permeation may occur because of improper selection of material, or prolonged use of equipment in a given atmosphere.

Ambient Temperature Extremes

Ambient temperature extremes can affect responder safety and safe work duration in a number of ways. For example, the effectiveness of PPE may be reduced as hot or cold temperatures affect:

- Valve operation on suits and/or respirators
- Durability and flexibility of CPC materials
- Integrity of fasteners on suits
- Concentration of airborne contaminants
- Breakthrough time and permeation rates of chemicals

In many instances, heat stress is the most immediate hazard to the wearer of PPE and the greatest limitation on work mission duration. Coolant supply will directly affect mission duration in instances in which cooling devices are required to prevent heat stress. Specific methods of dealing with heat stress are discussed in Chapter 8.

Personal Factors Affecting Respirator Use

A number of personal use factors may diminish the effectiveness of respirators. For example, any facial hair or long hair that comes between the respirator facial seal gasket and the wearer's skin will prevent a good respirator fit. Beards are not allowable for response team members required to use respirators. Facial features, such as scars, hollow cheeks, deep skin creases, missing teeth, etc., may also prevent a good respirator fit. Chewing gum and tobacco should be prohibited during respirator use.

Likewise, the temple pieces on conventional eyeglasses interfere with respirator fit. However, spectacle kits are available that can be used to mount the corrective lenses within the facepiece. Contact lenses can be worn with a full-facepiece respirator under OSHA regulations (29 CFR 1910.134), provided the wearer can demonstrate the ability to do so without problems.

Donning PPE

In donning an ensemble of PPE, an established routine should be followed. All equipment should be inspected as a part of the donning procedure. Donning and doffing of PPE should always be done with the aid of an assistant. Field check (i.e., positive- and negative-pressure tests) for respirator fit should always be performed as part of the donning procedure. After donning, all ensemble components should be checked for proper fit, proper functioning, and relative comfort before entering a hazardous area.

Inspection of PPE

PPE should be fully inspected before each use (Fig. 12.30). PPE inspection checklists, such as the following, may be used.

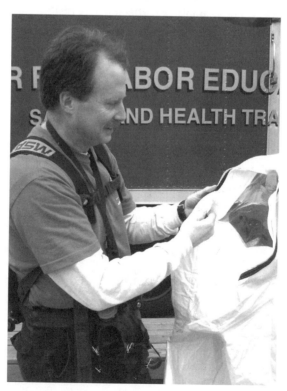

Figure 12.30   PPE must be fully inspected before use.

Inspecting CPC

General Inspection Procedure (applicable to all items of CPC)

- Determine that the clothing material is correct for the specified task at hand.
- Visually inspect for imperfect seams, nonuniform coatings, tears, and malfunctioning closures.
- Hold up to a light and check for pinholes.
- Flex the product and observe for cracks and observe for other signs of shelf deterioration.
- If the product has been used previously, inspect inside and out for signs of chemical attack, such as discoloration, swelling, and stiffness.

Inspecting Fully Encapsulating Suits

- Check the operation of pressure relief valves.
- Inspect the fitting of wrists, ankles, and neck.
- Check faceshield, if so equipped, for cracks, crazing and/or fogginess.
- TECP suits require periodic pressure testing or whole-suit in-use testing (as described in appendix A of 29 CFR 1910.120).

Inspecting Gloves

- Before use, check for pinholes. Blow into glove, then roll gauntlet toward fingers and hold under water. No air should escape.

Inspecting Respirators

General Procedures (applicable to all types of respirators)

- Check material condition of harness, facial seal, and breathing tube (if so equipped) for pliability, signs of deterioration, discoloration, and damage.
- Check faceshields and lenses for cracks, crazing, and fogginess.
- Check inhalation and exhaust valves for proper operation.

Inspecting Air-Purifying Respirators

- Inspect APRs:
  - Before each use to be sure they have been adequately cleaned
  - After each use, during cleaning
  - At least monthly if in storage for emergency use
- Examine cartridges or canisters to ensure that:
  - They are the proper type for the intended use.
  - The expiration date has not passed.
  - They have not been opened or used previously.

Inspecting SCBAs

- Inspect SCBAs:
  - Before and after each use
  - At least monthly when in storage
  - Every time they are cleaned
- Check air supply.
- Check all connections for tightness.
- Check for proper setting and operation of regulators and valves (according to manufacturers' recommendations).
- Check operation of alarms.

Inspecting Supplied-Air Respirators

- Inspect SARs:
  - Daily when in use
  - At least monthly when in storage
  - Every time they are cleaned
- Inspect air line before each use, checking for cracks, kinks, cuts, frays, and weak areas.
- Check for proper setting and operation of regulators and valves (according to manufacturers' recommendations).
- Check escape air supply (if applicable).
- Check all connections for tightness.

In-Use Monitoring of PPE

While working in PPE, responders should constantly monitor equipment performance. If indications of possible in-use equipment failure are noted, they should exit the contaminated area immediately and investigate.

Degradation of CPC Material
Degradation of ensemble components during use may be indicated by discoloration, swelling, stiffening, and softening of materials. Likewise, any tears, punctures, or splits at seams or zippers should be noted, as should any unusual residues on items of PPE.

Symptoms of Exposure
Perception of odors; irritation of skin, eyes, and/or respiratory tract; and general discomfort may be indications of equipment failure. Also, symptoms commonly associated with chemical exposure and oxygen deficiency may be the end result of equipment failure. These symptoms include rapid pulse, nausea, chest pain, difficulty in breathing, or undue fatigue. Restrictions of movement, vision, or communication may also result from equipment failure.

Heat Stress and PPE

Heat stress and other physiological factors directly affect the ability of personnel to operate safely and effectively while wearing PPE. The body has several mechanisms for the control of temperature in the body, as discussed in Chapter 8. A heat-induced illness (heat strain) can result any time these mechanisms are compromised or overloaded.

PPE Compounds Heat Stress. The use of PPE, especially impermeable clothing, affects the body's ability to cool itself in several ways. First, the blood carries heat from deep inside the body to the skin. There, the natural principle of convection causes the heat to be released to the lower-temperature air outside the body. Impermeable clothing, especially totally encapsulating ensembles, prevents the heated air from leaving the skin surface, thus preventing this natural heat transfer.

Also, the evaporated sweat from the body quickly saturates the air near the body, thus preventing this vital cooling mechanism. The body continues to produce sweat, which drips from the body wasted, and dehydration and loss of important electrolytes can result.

The physical stress of wearing the PPE including chemical protective clothing and respiratory protection can add weight and reduce mobility, thereby causing the responder to work harder. This increased work results in increased metabolic heat and increased demand for oxygen as energy is produced. The heat produced adds to the load on the cooling system of the body. Because the increased demand for oxygen usually occurs at a time when much of the body's blood is out at skin level and not available for the heart to pump, the heart must pump more frequently to move enough blood through the lungs to receive oxygen. Thus a higher pulse rate usually accompanies heat stress.

Monitoring for Heat Stress. Given these examples of how use of PPE can promote heat stress, the need for special attention to potential heat conditions is obvious. Generally speaking, the use of ambient temperature, humidity and wind speed as indicators of heat conditions has limited use at best, because the PPE essentially creates its own environment inside. Therefore, heat stress monitoring and control activities should focus on the individuals using the equipment and not environmental parameters. For instance, medical monitoring using such parameters as heart rate, oral temperature, and body weight loss as described in Chapter 8 should be instituted whenever the work area temperature exceeds 70°F.

Heat Injury Prevention with PPE. Some encapsulating suit manufacturers have begun designing temperature control capabilities into their suits. Such designs include tubing for distribution of cooled air or water throughout the suit. The air distribution systems typically require the use of an air line and large quantities of respirable air. Water cooling systems require ice storage or refrigeration units and a pump, which add bulk and weight to the suit. These suits present challenges for emergency response work, but they may be needed for extreme temperatures.

Another cooling option is the use of cooling garments or vests containing ice packs. These garments provide cooling to the wearer and do not require air lines or water cooling systems. However, they are fairly heavy and require an on-site source of ice or the ability to freeze coolant gel packs. Also, once the cooling medium has lost its cooling effect (e.g., the ice has melted), the cooling garment becomes an insulator that holds heat close to the body and compounds the heat stress problem. Therefore, the cooling medium must be regularly checked and replaced during the use of the garment.

Doffing PPE

Like donning, doffing of PPE should be done according to an established routine. Furthermore, doffing routines should be well integrated with decontamination and disposal procedures for used PPE (Fig. 12.31).

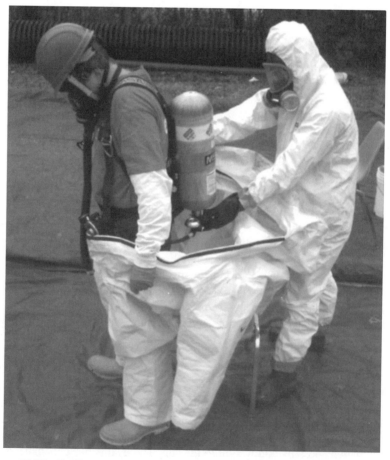

Figure 12.31   Doffing procedures for PPE must be integrated with decontamination.

STEP 7—STORE AND MAINTAIN PPE PROPERLY

Some PPE is disposable and not intended for multiple uses. It must be stored properly until it is needed so that it is not damaged or degraded. Other PPE, such as SCBAs, will be reused and must be stored and maintained properly to be ready for the next use.

Storing PPE

Storage is an important aspect of PPE use for emergency responders, because hazmat equipment is typically not used on a regular basis. Improper storage may lead to damage due to contact with dust, moisture, sunlight, damaging chemicals, extreme temperatures, and physical abrasion.

The following considerations should be observed in storing CPC:

- Potentially contaminated clothing should be stored in an area separate from street clothing.
- Potentially contaminated clothing should be stored in a well-ventilated area, with good air flow around each item, if possible.
- Different types and materials of clothing and gloves should be stored separately to prevent issuing the wrong material by mistake.
- Protective clothing should be folded or hung in accordance with manufacturers' recommendations.

The following considerations should be observed in storing respirators:

- SCBAs, supplied-air respirators, and air-purifying respirators should be dismantled, washed, and disinfected after each use.
- SCBAs should be stored in storage containers supplied by the manufacturer.
- Air-purifying respirators should be stored individually in their original cartons or carrying cases.
- All respirator facepieces should be sealed inside a plastic bag for storage.

Reuse of CPC

Items of CPC must be completely decontaminated before reuse. Otherwise, the items cannot be considered safe to use. If matrix permeation is possible, the article of CPC should be hung in a warm, well-ventilated place to allow the item to release the permeated contaminant.

In some instances, contaminants may permeate the CPC material and be difficult or impossible to remove. Such contaminants may continue to diffuse through the CPC material toward the inner surface during storage, posing the threat of direct skin contact the next time it is worn.

Extreme care should be taken to ensure that permeation and degradation have not rendered CPC unsafe for reuse. It should also be noted that permeation and degradation may occur without any visible indications. If complete decontamination cannot be confirmed, any CPC that is questionable should be taken out of service.

Maintenance of PPE

Effective maintenance is vital to the proper functioning of PPE. Thus the employer's PPE program should include specific maintenance schedules and procedures in accordance with manufacturers' recommendations for all reusable items of PPE.

Maintenance can generally be divided into three levels as follows:

- Level 1: User or wearer maintenance, requiring a few common tools or no tools at all
- Level 2: Shop maintenance, which can be performed by the employer's maintenance shop
- Level 3: Specialized maintenance that can be performed only by the factory or an authorized repair person

In some instances, it may be advisable for an employer to send selected employees through a manufacturer's training course in order to establish an in-house PPE maintenance program. Otherwise, the equipment must be serviced by the manufacturer or its authorized representatives. Repairs or maintenance performed by untrained individuals can damage or alter the PPE in ways that can pose a danger to responders who rely on the equipment in highly dangerous situations.

SUMMARY

Although not the most desirable choice, in most instances PPE will constitute the responder's only line of defense against the hazards associated with a hazardous materials incident. This chapter provided a seven-step process for selecting and using PPE during a hazmat response. The PPE to be selected includes respirators, chemical protective clothing, and personal safety equipment such as head, eye, and face protection. The chapter also looked at issues related to the proper inspection, storage, and maintenance of the equipment. When properly selected and used according to standard procedures, PPE should provide adequate protection for responders as they go about their response to a dangerous situation.

13

DECONTAMINATION

INTRODUCTION

An in-house hazmat team responded to a methyl ethyl ketone release of approximately 250 gallons that occurred at a paint blending plant. The responders wore butyl rubber boots, which are appropriate for use in working around the chemical. Although it was necessary for team members to walk through pooled liquid to approach the point of release, they did not decontaminate their boots after the incident. When the boots were examined the next day, the soles had been destroyed.

Decontamination is an important part of any hazardous materials incident response. It is not difficult to learn or to do, as long as plans and training are completed well in advance of the need for it. In this chapter we examine the rationale for effective technical decontamination and suggest criteria for determining the most efficient way to accomplish decontamination during an incident. We also discuss principles and important differences in techniques used to decontaminate large numbers of people (mass decontamination) and look at recommendations for removing chemical or biological agents from those involved in a WMD incident as described in Chapter 10.

TECHNICAL DECONTAMINATION: DEFINITION AND JUSTIFICATION

"Decontamination" is defined as the removal or neutralization of contaminants that have accumulated on tools, clothing, personnel, and vehicles that have been

Emergency Responder Training Manual for the Hazardous Materials Technician, Second Edition,
edited by Kenneth W. Oldfield
ISBN 0-471-21387-X   Copyright © 2005 John Wiley & Sons, Inc.

inside the hot zone where hazardous materials are present. Because the hot zone boundary is set at a safe distance from the hazard (see Chapter 11, Site Control), people or equipment crossing this boundary would potentially carry contaminants from a "dirty" zone to a "clean" one. The goals of the decontamination process are:

- Remove contaminants from these carriers
- Confine the contaminants to the decon area
- Contain and dispose of contaminated materials in a safe and legal manner

The overall response goal in any incident is to protect people, both responders and members of the community, from exposure to hazards. Responders must ensure that chemical hazards are not spread outside the contaminated zone and that they themselves do not carry any chemicals away with them, putting themselves and their families at risk from chemical exposure. An effective decontamination plan, efficiently carried out, will eliminate the spread of chemical hazards.

METHODS OF DECONTAMINATION

There are many ways to remove chemicals from people and equipment, and each has a use in certain situations. We can group all these methods into two basic types: physical decontamination methods and chemical decontamination methods.

Physical Decontamination

Physical decontamination includes all methods that manually separate a chemical from the surface to which it adheres. These are basically scrubbing techniques that force the chemical to loose its grip on the surface. Physical methods are effective for removing the major portions of thick, gooey materials such as mud and sludge and will entirely remove many contaminants.

Advantages
Physical decontamination has far more advantages than disadvantages, because it is usually harmless in itself and can be used without causing any collateral harm. The advantages of physical decontamination are:

- Immediate removal of contaminants that, if left in contact with chemical resistant fabrics, will be more likely to permeate (diffuse through) the fabric
- Reduction of reliance on chemical solutions that may be harmful to responders, decontamination station personnel, and suit fabrics
- Ready availability of water streams and scrubbing tools
- High effectiveness against many kinds of chemicals

Disadvantage

The disadvantage of physical decontamination is that it may not completely remove all residues of some chemicals, especially those that are oily. If too aggressive, physical methods may damage softer materials. Physical removal should, in these cases, be followed by washing with a chemical solution.

Equipment

Equipment used in physical decontamination includes, but is not limited to, the following five types.

Water Streams. It is almost impossible to decontaminate equipment and personnel thoroughly without a water stream of some sort. This can be delivered by a garden hose, fire hose, shower, or, less effectively, a portable pressure-pump sprayer. The garden hose should be provided with a spray nozzle to increase the pressure of delivery, preferably a nozzle with a trigger grip that can be activated only when needed. If a fire hose is used, pressure and volume should be stepped down through smaller hoses by means of adaptors.

Specially designed showers made of PVC pipe and connectors are available that can be easily snapped together. Quick-release connectors at strategic points make showers easy to break down and transport on a hazmat vehicle. Some shower walls collapse upon themselves for ease of assembly and disassembly. Some shower designs offer water spray patterns that are quite effective for personnel decontamination. The best showers are designed so that the spray hits the person from several different directions at two or more levels and are able to maintain enough pressure to accomplish active physical removal. Some sort of easily accessible water flow control valve is advisable, so that the shower only runs when needed.

Scrapers and Scrubbers. Many types of scrapers and scrubbers (Fig. 13.1) can be used in conjunction with a water spray to aid in removal of thick contaminants. For heavy mud or sludge, a smooth-edged scraper such as those sold in tack shops for scraping sweat from horses is appropriate. No scraper should be used on protective clothing that might degrade the surface of the fabric. A brush is handy for scrubbing; those with long handles allow the decon personnel to avoid contact with the chemical.

The palms of the gloves and the soles of the outer boots require special attention when scrubbing; as they are the generally the most contaminated (Figs. 13.2 and 13.3). Care should be taken to brush all areas of the suit, including the arms and between the legs.

Dry Brushes. If the contaminant is in dry form, it can be removed by dry brushing. If the material is water reactive, dry brush off the major portion and rinse the remainder with copious amounts of water to dissipate the heat from any exothermic reaction.

Figure 13.1 Physical decontamination makes use of brushes to remove contamination.

Figure 13.2 It is important to scrub the palms of the gloves because they may be heavily contaminated.

Steam. Steam can be used on contaminated equipment that can tolerate high temperatures. Steam from a pressure jet adds high temperature into the balance between surface adhesion and separation. High temperature may soften the contaminant as its temperature approaches its boiling point and encourage it to

Figure 13.3 Soles of boots are another area that is usually heavily contaminated.

evaporate off the surface. Steam cannot be used on people, and is likely to be harmful to protective fabrics, but is appropriate for use on some tools, parts of vehicles, and heavy equipment. Use of steam may increase the concentration of the contaminant in air, thus increasing the inhalation hazard to decon personnel. This must be considered when selecting protective equipment.

Freezing. Freezing or otherwise lowering the temperature will sometimes enhance the removal of certain highly viscous contaminants from contaminated equipment. When the temperature approaches the freezing point of the chemical, the latter will become solid rather than sticky and may let go of the surface. Care should be taken in using this method, as certain chemical-resistant fabrics that decon personnel are wearing may become brittle, crack, and be irreparably damaged.

Chemical Solutions

For situations where physical decontamination and water alone will not remove contaminants, chemical solutions, such as detergents, may be appropriate. If decontaminating personal protective equipment (PPE), consult the manufacturer's recommendations. Some decon solutions may actually degrade some types of PPE, so care must be taken in their selection.

Advantages

Chemical solutions can be formulated that will change one chemical into another that is less harmful or change its physical state. In some cases, the use of solutions makes contaminant removal much more efficient than physical decontamination methods and water alone. Most chemical solutions used for decontamination are relatively inexpensive and are readily available. Some are sold at any grocery store or home store, whereas other specialty solutions have to be ordered from a distributor. Planning ahead is important. Know what types of chemicals are in your jurisdiction and have the proper decon solutions on hand.

Disadvantages

The major disadvantage to the use of chemical solutions is the possible harm they pose to responders and decontamination personnel. Only two types of solutions are discussed here: detergent solutions and neutralizing solutions.

Detergent Solutions. Detergent solutions work by chemical means to remove a contaminant from a surface. They are most effective against oil-based chemicals, which are the most difficult to remove by purely physical means. Detergents work by reducing surface tensions that form between oil and water or between oil and any other surface. These tensions hold the two together, sometimes quite strongly.

Detergents, including soaps, emulsifiers, and surfactants, are made up of molecules that have two very different ends. One end is a long carbon chain, represented by the straight lines in Figure 13.4; the other end, represented by the balls, can be made up of many different types of atoms. The arrangement or geometry of the atoms in detergent molecules is such that one end of the molecule has a positive electrical charge and the other side has a negative charge. If this is the case, the molecule is called a polar molecule, meaning that it has electrical poles or that its positive and negative charges are permanently separated. In the

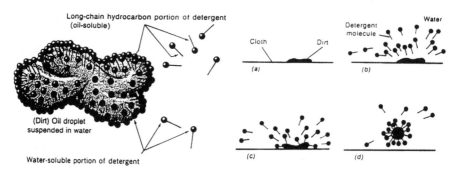

Figure 13.4   Detergent removes oily and greasy contaminants from surfaces.

case of detergent molecules, the carbon chain end (which looks like the sucker stick in the drawing) is attracted to oils and the other end (the round part) is attracted to water. Many, many molecules of detergent surround the droplets of oil, separating them and floating them right off the surface of whatever they are adhered to, whether it be greasy dishes or a moon suit. Flushing with water will then wash away the droplets of oil with their surrounding detergent molecules.

Neutralizing Solutions. Neutralizing solutions can be used to alter a contaminant chemically and render it harmless. They are most useful in ensuring the total cleanliness of sample bottles when laboratory analysis of samples is desired or for decontamination of sampling equipment to prevent cross-contamination of these devices. If further information is desired regarding other solutions, a chemist should be consulted. It should also be remembered that each solution of this sort must be tailored to the specific chemical contaminant one wishes to alter.

Levels of Decontamination Required

How much decontamination is enough? It is often difficult to be certain of the level of decontamination required to remove all chemicals adequately. Many contaminants are not visible, and those that are may give little or no indication of subsurface contamination, such as permeation or penetration of clothing.

The only way to be sure that a suit or tool is clean after it has been decontaminated might be to conduct a "swipe test." A swipe test is performed by wiping a decontaminated item with a bit of material known to be clean and then sending the material to a lab to be analyzed for contaminants. No information is immediately available, but when the information does come back, it will indicate whether the procedures outlined in the decon plan have been effective or need to be altered for future use.

Two factors must be considered in writing a decontamination plan: the type and the amount of contamination.

Type of Contamination

Responders should consider the chemical itself, its health hazards, and the likelihood that it has gotten on clothing and equipment surfaces. To estimate the amount that may have gotten on surfaces, they should learn the physical state of the contaminant and the environmental substances, like mud or water, with which it has mixed.

The type of contamination also determines the degree of hazard it presents; as has already been discussed, this information can be gained only after the chemical has been identified. Extremely hazardous chemicals require extremely vigorous decontamination methods and a higher level of protection for decon personnel than might otherwise be needed. More stations may be required than typically recommended, wash/rinse steps repeated, and more distance left between the decontamination stations to protect decon personnel at adjacent stations.

Solids. Solids are most likely to get onto the surfaces of suits or respirators if they are powdery and suspended in air. Particulates released in a fire or in a plume from a spill may also cover almost every surface. Solids in a surface release, unless carried by wind or dissolved in mud or water, will primarily contaminate boot soles, gloves if they are touched, and any tools or wheels that encounter them.

Liquids. Liquids may splash anywhere, especially on the clothing of responders who approach the point of the release. Obviously, avoidance of splashing or walking through puddles will limit contamination. Liquids that splash onto the suit will run down into creases in the suit and possibly into boots. Wearing the proper size of chemical protective suit will decrease folds and creases in the protective clothing, making decontamination easier and more effective. The pants leg of the chemical protective suit should be pulled over boots and never tucked into boots. This will eliminate the possibility of contaminants draining down the leg into the boot. Chemical-resistant tape, or in some cases duct tape, can be used to hold the pants leg of the suit securely over the boot around the ankle or a glove to the arm sleeve around the wrist.

Gases. Unless dissolved in rain or humid air, or on the naturally moist surfaces of skin, gases will result in little surface contamination. In cold weather, condensation on any surfaces that are cooler than ambient temperature may occur, creating an avenue for gases to contaminate surfaces.

Amount of Contamination

The amount of contamination will depend partly on the physical state of the chemical and partly on the task performed by the responder or his equipment. It has already been mentioned that liquid or sludge materials are most likely to cause heavy contamination.

Responders who walk into puddles, pick up contaminated objects, or perform patching or plugging operations on spewing liquids must be assumed to be heavily contaminated. The use of disposable protective outer gloves and boots should be considered in these situations, as well as for any entry that must be made into an area of contaminated mud or oil. In some situations where gross contamination of protective clothing has occurred, it may be more efficient and cost-effective to discard the entire suit after the initial decon and suit removal than to potentially expose others who will try to decon the suit further.

SETTING UP A CONTAMINATION REDUCTION CORRIDOR (CRC)

Location

The location of the contamination reduction corridor (CRC) or decontamination area in the warm zone will depend on certain parameters of the incident location that are discussed in Chapter 11, Site Control. The decision about the location of the CRC should be made as rapidly as possible, because it is an axiom of emergency response that no entry into a potentially contaminated area should be made until the decon area is set up and ready.

Incident Parameters

The environmental conditions in which the incident occurs are important considerations in placing the CRC. Notable factors are the nature and degree of the hazard presented by the chemicals, the wind conditions present, and the topography of the incident scene.

The Nature and Degree of the Hazard. If the chemical is a solid or nonvolatile liquid, the hazard zone may be relatively small and the decon area close to the release. A highly volatile liquid, vapor, or other gas will create a larger hazard zone, as does the threat of fire or explosion. The shorter the distance responders must walk to be decontaminated, the longer their time on task can be and the sooner decon can begin. Once the chemical has been identified, the degree of hazard can be assessed further.

Wind. Wind direction is the most critical weather parameter for determining the location of the decon area. The decon area should be located upwind of the hazard area to protect personnel from contaminants carried out of the hazard area by the wind. Because wind direction is subject to change, the decon area should be located in the place most likely to remain safe if the wind changes. This can be very difficult if the response is in an urban area with tall buildings. Wind direction can change 180° without notice. The decon plan should address this for those operating in urban areas.

Topography. The topography of the land on which the incident has occurred should also be considered. Because decontamination always includes water that must be contained, the area should be set up on level ground so that this task is made easier. Water runs downhill, and contaminated water cannot be allowed to flow into clean areas or onto soil or other permeable surfaces.

Access to Water and Equipment

Access to water and other necessary equipment is vital. If a mobile tanker is the water source, it must be able to be moved into hose range of the area and conserving water is of utmost importance. Turn shower valves and hoses off when not in use. This will also limit the amount of contaminated wash/rinse water that will have to be tested and disposed of if hazardous. If water is provided by a fixed hydrant, hose length will limit access. Other equipment that may be needed in close proximity to the decon area includes the air tank refill system, emergency medical equipment, and salvage drums.

Personnel Decontamination Line

The line where people and their suits are cleaned should be set up separately from the area where tools, equipment, and vehicles are cleaned. The area should be clearly marked, and access control points should be set up to restrict personnel to those wearing the appropriate protective clothing. Vehicle access should be completely separate from personnel access, for obvious safety reasons. Personnel entry should be separate from exit to avoid cross-contamination.

The decontamination line should consist of a number of distinct stations set up in such a way as to reduce contamination as the responder moves down the line. Stations should be clearly separated from each other by enough distance to eliminate any splashing from one station to another. No flow of liquid from one station to another should be permitted.

Stations in the Line

The number of stations required for adequate decontamination varies, depending on the nature and degree of hazard presented by the chemical(s) involved in the incident. Hazmat teams may also find themselves limited by the number of people available to staff the stations; however, it is wiser to find ways to gain additional personnel than to limit the number of stations.

Recommendations for maximum and minimum decontamination station layouts and steps at each station can be found in the *Occupational Safety and Health Guidance Manual for Hazardous Waste Site Activities*, released in 1985 by the EPA, USCC, NIOSH, and OSHA. These recommendations range from a maximum 19-station decon layout for decontamination of Level A entry teams to the minimum of a 7-station layout for responders leaving the contaminated zone. Seven stations will suffice for most incidents in which responders have not been heavily contaminated or contaminated with a highly toxic substance (Fig. 13.5).

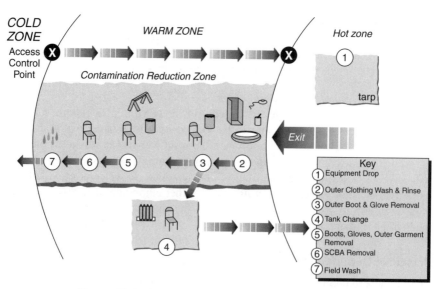

Figure 13.5 Seven-station decontamination line layout.

TABLE 13.1   Suggested Decontamination Equipment at Each Station (For a Seven-Station, Level A Protective Clothing Line)

Station	Equipment
1	Plastic ground sheet
2	Container of detergent solution
	Source of running water
	Containment basin
3	Containers (lined) for gloves, boots, tape
	Bench or stool
4	Full air tanks
	Outer boots
	Outer gloves
	Tape
	Optional: resting bench, drinking water and cups, medical monitoring equipment
5	Containers (lined) for boots and gloves
	Bench or stool
	Container or rack for suits
6	Plastic ground sheet
7	Soap
	Running water
	Container for runoff
	Towels

The overall goal at each station is to remove contamination from the garment or other item in such a way that the contaminant will not touch the worker as he doffs the garment. Specific objectives at each station in a seven-station layout to decontaminate Level A protective clothing are listed below. They can be used with appropriate modifications for a seven-station line for responders wearing Level B protection. Table 13.1 lists suggested equipment for each station.

Station 1: Equipment Drop

This station consists of a tarp or sheet of plastic just inside the hot zone, where tools or monitoring equipment can be left by a team as they exit. The equipment can be picked up and used by the next entry team, and it and the tarp can be decontaminated or disposed of at the termination of the incident. Equipment requirements at this station can be as minimal as a plastic sheet or tarp.

Station 2: Outer Clothing Wash and Rinse

This is really two stations in one, because wash water will probably include detergent and rinse water will not. Scrubbing or scraping can accompany both processes. The objective is the removal of contaminants from all outer garments, including suit, outer gloves, and outer boots (Fig. 13.6; see also Fig. 13.10 below). The equipment requirements at this station include

Figure 13.6   Gross decontamination is done at Station 2.

- A container of detergent solution (e.g., 5-gallon bucket, garden sprayer, drum)
- A source of copious amounts of water for rinsing
- A containment basin (e.g., kiddy pool, shower enclosure, lined pit, leak-protected drum pallet)
- Long-handled brushes (handle made of plastic, not wood)

Station 3: Outer Boot and Glove Removal

Outer boots and gloves are removed here and are placed in a plastic-lined container for disposal or later decon and/or drying. Responders' boots can be removed easily if they are able to sit down as shown in Figure 13.7. Equipment needed at Station 3 includes

- Containers (one for gloves, one for boots, one for used tape)
- Plastic liners (garbage bags)
- A bench or stool

Figure 13.7 Decontamination personnel remove boots without help from the responder.

Station 4: Tank Change

This station is located off to the side of the decon line and is a clean area. Station 4 is visited only by responders who are going back on standby to return to the hot zone (Fig. 13.8). They may return after replacing their SCBA cylinder with a full one, donning clean outer gloves and boots, and being retaped. This station is placed after Station 3 so that it will remain clean, because suits are unzipped here and responders are taken off of SCBA air and may be exposed to any contaminants present in the area. Equipment needed at this station includes

- Full air cylinders
- Clean outer boots
- Clean outer gloves
- Tape
- Optional: bench for resting, drinking water and disposable cups so that responders can hydrate, trash can, medical monitoring equipment

Station 5: Outer Garment and Inner Glove Removal

The outer garment, in this case the Level A suit, and inner gloves are removed and deposited in separate containers lined with plastic. Assistants should help the responder with the doffing so that he does not touch the outer surfaces of the garment (Fig. 13.9). Decon personnel should peel the suit away from the responder

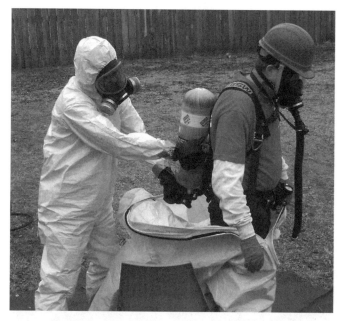

Figure 13.8 Tank changes should be accomplished without allowing contamination of any surface inside the suit.

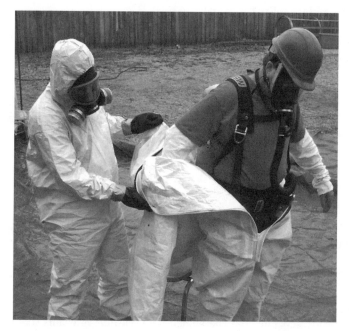

Figure 13.9 The responder is warned not to help in the removal of his suit and not to wipe his face until he has washed his hands.

as one would peel a banana. A bench, stool, or chair should be provided so the responder can sit down once the suit has been lowered past his midriff. Once the responder is sitting, the suit can be pulled easily from the legs. Equipment needed at this station includes

- Containers
- Plastic liners
- Bench or stool
- Container or hanging rack for nondisposable suits

Station 6: SCBA Removal

The SCBA backpack and facepiece are removed by an assistant, and the responder is reminded not to touch his face until he has washed his hands. The unit is placed on plastic, to be cleaned later. The only equipment needed here is a plastic sheet on the ground.

Station 7: Field Wash

Soap and water are provided here for the responder to wash his hands and face, and clean towels should be supplied. Ideally, a private shower would be available; lacking this, responders should take a full shower as soon as possible. Equipment at this station includes

- Soap and running water (e.g., garden sprayer, hose)
- A container to catch runoff water

Decontamination of Tools, Equipment, and Vehicles

The decontamination of tools and equipment can be made much easier by protecting them from contamination. Some air-monitoring devices, in fact, are practically impossible to decontaminate if allowed to be splashed with liquid chemical. All sensitive equipment should be protected by clear plastic; a large bag will fit such units as the photoionization detector (PID), the flame ionization detector, and the Geiger counter. A rubber band or long twist tie can be used to close the bag while allowing the probe to extend through the opening of the bag. Care should be taken to collect only air samples; liquids will damage not only the outside of the probe but also the sensor.

The ground rules for avoiding boot contamination by not walking into pooling product or contaminated mud should be extended to vehicles. Even drivers of fire trucks and ambulances should take care to avoid contaminated ground.

Equipment decontamination should be handled at the end of the incident, although gross decontamination, such as removal of contaminated mud, should be done as soon as possible to reduce cross-contamination of personnel. It should be noted that certain tools, for example those with wooden handles, may have soaked up contaminants that cannot be removed. These should be disposed of as hazardous waste.

Metal or plastic tools can be scrubbed with cleaners that are too strong for skin or protective fabrics; the only restrictions are those of incompatibility and limits on the exposure of people using them. Plastics are degraded by some organic solvents; corrosive acids and bases may damage certain metals. If a chemical solution is used, it should be evaluated by the same criteria used in assessing the hazards of any chemicals in the incident and protection of decon personnel must be appropriate.

MANAGEMENT OF THE DECONTAMINATION AREA

In considering criteria for good management of the decon area, it is appropriate to recall that the overall goal of the process is to eliminate transfer of contaminants from contaminated equipment and clothing to any other surfaces that cannot be contained and disposed. These other surfaces include all areas outside the decon line, all materials that will later leave the decon area, and all personnel who attend or walk through the decon line.

Orderly Cleaning and Doffing

Let us first consider how exposure to the responder can be prevented as he removes contaminated garments. This goal can be accomplished through orderly cleaning and doffing of the items the responder is wearing. (We assume that he is not carrying anything, having left all tools and equipment behind at Station 1).

Touching Contaminated Surfaces
The responder never touches anything inside the decon area. Doffing assistants unzip, untape, and remove all his protective clothing for him. At Station 3, the assistant removes the outer boots and gloves without touching the suit. Station 5 assistants remove all other garments, and Station 6 operators remove the SCBA. At none of these stations does the responder help, but rather he avoids touching any of his equipment. He touches nothing until he has at least washed his hands and face at Station 7, at which time he exits the decon corridor.

At no time does a doffing assistant touch a part of the clothing not necessary in the removal appropriate to his station. Nor does he touch the skin or inner clothing of the responder.

Physical Safety
The decon stations are set up with the responder's safety in mind. All tripping hazards are eliminated, and no station requires him to hold someone or something to keep his balance. A seat is available wherever outer or inner boots must be removed.

Personnel Movement
No decon personnel should move between stations; they stay at the station to which they have been assigned. Contamination should decrease throughout the

line; assistants can easily spread the chemical by walking down the line. It is easy to understand how an operator at one station would wish to assist others when he is not busy, but this must be prevented.

Protection of Decontamination Personnel

People who operate the decon stations must be protected from exposure to the chemical being removed and, if potentially harmful chemical solutions are used on vehicles or equipment, from these solutions as well. They must therefore be trained in and equipped with the appropriate PPE. As a general rule, decon assistants wear protective clothing one level below that of the responders they are cleaning, but ideally selection will be based on the type and amount of hazard the chemical presents. Chapter 12 provides a good discussion of hazard-based PPE selection.

Is the amount of the chemical on the surface being decontaminated large enough to be splashed onto the assistant as it is washed or rinsed? If so, splash protection should be worn by the assistant.

Can the chemical be expected to volatilize under decon conditions? If so, respiratory protection must be worn.

Is the chemical one that is known to allow secondary contamination of assistants? Chemicals that fall into this category include corrosives, phenols, pesticides, PCBs, asbestos, hydrogen cyanide gases and salts, and hydrofluoric acid solutions. Protection from these chemicals, and others judged to be potential causes of secondary contamination, must be afforded to decon workers.

Containment of Liquids

All contaminated materials, including all wash and rinse fluids, must be contained for proper disposal. This will necessitate pumping or otherwise removing wash and rinse water from confinement pools as they fill and providing a large container to hold it for disposal. Because all wash and rinse waters generated in the decontamination process are contaminated with the incident chemicals, they must be contained until they are disposed of as hazardous waste, treated to render them nonhazardous, or analyzed and determined to be safe for release into the environment. Close attention to where these liquids are flowing during decontamination will suggest capture methods.

Several containment devices can be purchased, and others can be fairly easily built (see Figs. 13.6 and 13.10). Children's swimming pools work well and are inexpensive and readily available. The inflatable type is convenient if the hazmat brigade carries its equipment on a truck or van, as this pool takes up little space when deflated and can easily be inflated with an SCBA cylinder. The rigid type may last longer, because it is not as likely to be punctured, but is harder to handle and store.

Portable shower stalls are available for purchase through safety products distributors (Fig. 13.10). Some hazmat units have made their own from PVC pipe covered with sewn and seam-sealed plastic.

Figure 13.10 A portable shower stall prevents splashing of the ground or nearby personnel.

The entire contamination reduction corridor (CRC) should be lined with plastic, the edges of which can be turned up and supported to direct overflow into a catchment basin if it occurs.

EMERGENCY MEDICINE AND DECONTAMINATION

Decontamination and emergency medical services (EMS) may have to be performed on anyone who becomes ill or is injured in a hazardous area. When someone in the hot zone needs medical help, he or she cannot be treated like an ordinary patient. There is a good possibility that this person could be contaminated with enough hazardous material that then could be spread to rescuers and first aid providers.

Protect the Responder

No person, including EMS personnel, should enter the hot zone for any reason unless the proper protective gear is worn, and a person unfamiliar with the gear should not attempt to wear it. It must be emphasized that EMS responders who have not been trained to wear respirators and protective clothing should stay out of the hot zone and decontamination area. Sophisticated protective gear should only be used by those with proper knowledge and experience. If decontamination is carried out before delivery of the victim to the EMS personnel at the perimeter, no special gear may be needed.

Primary goals for emergency personnel in a situation involving hazardous materials include:

- Termination of exposure to the patient
- Removal of the patient from danger
- Patient treatment—while not jeopardizing the safety of rescue personnel

Decontamination and medical personnel should do everything possible not to come in contact with any potentially hazardous substances to avoid exposure. Also, every effort should be made to see that the patient is decontaminated before transport and delivery to the hospital emergency room to avoid secondary contamination of emergency health care workers and other patients in the hospital.

If EMS responders undertake primary assessment of a victim who is still in the hot or warm zone, they should only do so if they are properly trained and wearing appropriate PPE. They must give priority to the victim's ABCs: airway, breathing, and circulation. Other personnel may, at the same time, begin to decontaminate the patient so that the victim's and responders' protective gear may be downgraded as contamination is removed. A call to the local or regional poison control center may provide information useful in determining whether medical personnel are at risk of secondary contamination from the patient; this information is available only if the name of the chemical is known.

Stop Contamination and Prevent Secondary Contamination

During initial patient stabilization and decontamination, all clothing suspected of being contaminated should be removed or cut away. Any clothing and other items removed such as watches or jewelry should be placed in plastic bags and tagged. Figure 4.9 in Chapter 4 shows an example of a triage tag used for this purpose. In this example, the "Contaminated Evidence" tab along the side of the tag would be removed and placed into the bag with the clothing removed from the victim so that it can later be identified and traced back to the victim. The "personal property" receipt can be placed in or attached to the bag containing items such as watches, wallets, or jewelry for the same purpose. Any clothing that cannot be removed from the victim should be wrapped with plastic or whatever is available in order to contain the contaminants and prevent their migration to others.

Patient decontamination should be an orderly process. It should begin at the head, with particular attention to the eyes, and move down the body. Because intact skin is, in most cases, more resistant to contaminant permeation than eyes or damaged skin, eyes and open wounds must be thoroughly flushed. Wounds should be covered with appropriate dressing after washing.

Decontamination should be performed with the least aggressive methods. Mechanical or chemical irritation to the skin must be limited to prevent increased permeability. Contaminated areas should be washed under a gentle spray of warm (never hot) water, wiping with a soft sponge and using a mild soap such as dishwashing liquid. Care should be taken that contaminants are not introduced into open wounds. The degree of decontamination should be based on the nature of the contaminant, the form of contaminant, and the patient's condition. It is important that runoff water be contained, if possible, because it may be contaminated and will require treatment before disposal.

Patient Treatment

Treatment of the chemically exposed patient will include primary and secondary surveys; the primary survey can be accomplished simultaneously with decontamination, and secondary surveys should be completed as conditions allow. Unless required by life-threatening conditions, invasive procedures such as intravenous medication or intubations should be performed only in fully decontaminated areas. These procedures may create a direct route of introduction of the hazardous substances into the patient's body.

In the case of a victim of accident or illness in a contaminated zone, it may not be known immediately whether the patient is suffering from chemical exposure. Whatever the results of later assessment, the names of all potentially hazardous chemicals to which the patient may have been exposed must be determined as quickly as possible. Because exposure of EMS personnel must be prevented and decontamination procedures started, references must be consulted immediately. This is possible only if the chemicals known or suspected to be present are identified. It may also be necessary to continue to monitor the patient for latent effects of chemical exposure. Symptoms of exposure and the references where they can be found are discussed in Chapter 7.

Patient Transport

If it becomes necessary to transport a contaminated patient by ambulance, special care should be exercised to prevent contamination of the ambulance and subsequent patients. Exposed surfaces that the contaminated patient is likely to come into contact with should be covered with plastic sheeting. The patient should be as clean as possible before transport, and further contact with contaminants should be avoided. Responders should make every attempt to prevent the spread of contamination and, at the very least, should remove patient clothing and take other means to prevent further contamination of the patient and cross-contamination of themselves.

In an ambulance during transport, personnel must use appropriate respiratory protection if contaminants are present. If weather conditions permit, opening windows in the patient's and driver's compartments will provide maximum fresh air ventilation. The receiving hospital should be contacted as soon as possible and given any information that has been gained about the chemicals involved.

Transportation by helicopter presents special, more serious hazards; therefore, most air ambulance services will not transport patients that have been contaminated. Exposure of the flight crew may interfere with their ability to fly safely. It is even possible that downdraft from the helicopter could spread contamination on the site. This means of transport should not be used without careful consideration of all factors.

Medical Management Guidelines

The Agency for Toxic Substances and Disease Registry (ATSDR) has developed Medical Management Guidelines (MMGs) for acute chemical exposures to aid emergency department physicians and other emergency health care professionals who manage acute exposures resulting from chemical incidents. The MMGs are intended to aid health care professionals involved in emergency response to effectively decontaminate patients, protect themselves and others from contamination, communicate with other involved personnel, efficiently transport patients to a medical facility, and provide competent medical evaluation and treatment to exposed persons. These guidelines can be found at ATSDR's website, http://www.atsdr.cdc.gov.

POSTINCIDENT MANAGEMENT

When the incident is over, decon assistants must go through an orderly process of cleaning and doffing their own protective clothing. Starting with the most highly contaminated station, the operator bags up any disposed or cleaned materials, labels them, and cleans all reusable equipment and containers. He then goes through the remainder of the line, being decontaminated as he goes. The other operators do the same, moving from most to least contaminated. Toward the clean end of the decontamination line heavy contamination has been removed and secondary contamination of decon personnel should be minimal. Consequently, these decon personnel may not require a full-immersion decon but may be able to simply remove their contaminated PPE and proceed through a field wash at the final station.

The definitions to be used in determining whether wash and rinse water and trash must be disposed of as hazardous wastes are the same definitions used in designating any hazardous waste. It is not up to the responder to make these decisions.

Contaminated water and solutions, disposable clothing, tape, and any other items removed in the decon process should be contained, closed, and clearly labeled for proper testing and disposal.

The last job for the decontamination team is to restore readiness for the next incident. This requires thorough cleaning of all equipment, assessment of breakage, maintenance and oiling of mechanical equipment, and replacement of any damaged or consumed materials. All equipment and materials are made ready and stored in their proper location. The decon team members can then meet and evaluate their part in the incident to determine how they can help to make the next response even more efficient.

MASS DECONTAMINATION

Thus far in this chapter, we have been concentrating on the technical decontamination process and relevant considerations of a typical hazmat incident. This process works well on a controlled site with a finite number of personnel that will need to be decontaminated. The question is, "How well would it work during a mass casualty incident (MCI)?" Probably not well at all. Technical decontamination alone, as described above in this chapter, would consume most if not all available resources (personnel and equipment) and still not be adequate. Technical decontamination procedures require approximately 15 minutes per person to complete. Using a conservative number of a hundred victims that need to be decontaminated, you are looking at 25 hours to complete with one technical decon line. Mass decontamination, on the other hand, should take somewhere in the range of 3 minutes per person. Time is the biggest consideration. Removing as much contaminant from those exposed as quickly as possible will reduce the effects of the exposure to the patient and reduce the extent of secondary contamination to others.

As described in Chapter 11, most emergency responders tend to think in terms of trying to keep people from entering the site of a typical hazmat incident, whereas the challenge of a mass casualty incident is to keep exposed victims from leaving the scene without proper decontamination and exposure assessment. Because of this, we need to be able to decontaminate quickly and effectively a large number of victims that have been potentially exposed to harmful or life-threatening substances.

Mass Decontamination Corridor (MDC)

For decontamination during mass casualty incidents, a mass decontamination corridor will need to be set up in the warm zone to provide rapid decontamination for the large number of victims of the incident (see Fig. 11.6 in Chapter 11). In January, 2000, the U.S. Army Soldier and Biological Chemical Command released a guidance document entitled *Guidelines for Mass Casualty Decontamination During a Terrorist Chemical Agent Incident* that can be found on the web at http://hld.sbccom.army.mil. In this guidance document they addressed the issue of how to effectively decontaminate large numbers of victims and identified five general principles to guide emergency responder policies, procedures, and actions after a chemical agent incident. These are

- Expect a 5-to-1 ratio of unaffected to affected victims.
- Decontaminate victims as soon as possible.
- Disrobing is decontaminating; from head to toe, removing more clothes is better.

- Water flushing generally is the best mass decontamination method.
- After a known exposure to a liquid chemical agent, emergency responders should be decontaminated as soon as possible to avoid serious effects.

Methods of Mass Decontamination

The mass decontamination corridor will require water sprayed in a low-pressure (30 psi is recommended by the Emergency Response to Terrorism Job Aid) overlapping fog pattern, provided from either fire apparatus (Fig. 13.11) or some other suitable source. This makes placement of the equipment critical. It was found in the guidelines mentioned above that simply removing clothes (at least down to undergarments) and flushing or showering with water alone provides enough shear force and dilution to physically remove chemical agents from skin. Other sources report that just by disrobing victims can remove up to 80% of the contaminant. For suspected biological or radiological incidents, wetting down potential victims before they disrobe is recommended. This prevents the particles from becoming air borne while clothes are being removed and thus keeps the person from breathing in the radiological or biological contaminant. However, this is not recommended for chemical agent incidents, where the water can actually spread the chemical agent through the clothes to the skin or to others in the decon process.

Flushing or showering with water alone was found to be the best, most effective method of mass decontamination. Soap or mild detergents resulted in a marginal improvement because of its polarity as described above in this chapter. Liquid soap was found to be quicker to use than solids and reduced the need for scrubbing, which can further damage the skin. A 0.5% bleach (sodium hypochlorite) and water solution also is known to remove, hydrolyze, and neutralize most

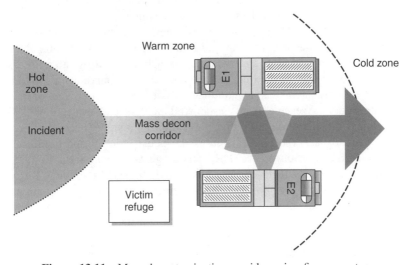

Figure 13.11 Mass decontamination corridor using fire apparatus.

chemical agents. Such a solution is a commercially available laundry bleach such as Clorox that has been diluted with 9 parts water to 1 part bleach. (Consult http://www.bt.cdc.gov for specific recommendations.) *Jane's Chem-Bio Handbook* also mentions the use of a granular form of bleach. See that document for more information.

There are a few disadvantages to using soaps and bleach solutions that need to be considered. First and foremost is availability, having an adequate supply on hand. Second, extra time would be taken using these, delaying other victims from reaching the decon corridor. Third, using soaps may actually increase the effects caused by blister agents by hydrating the skin. Fourth is the availability of equipment to apply bleach solution to victims. Finally, the contact time that is necessary for a 0.5% bleach solution to work is longer than the expected shower duration.

Recommended Equipment for use in Mass Decontamination Operations

Large volumes of water are needed to provide a low-pressure water spray to form a large-capacity shower. Ladder pipes, monitor nozzles, and fog nozzles attached to fire apparatus can be strategically positioned (approximately 16–20 feet apart) to create a corridor for large numbers of victims to pass through. (see Figs. 11.6 and 13.11.) Once the corridor is formed, it is suggested that water be sprayed from every direction feasible to ensure maximum effectiveness. Multiple corridors can be established for ambulatory and nonambulatory or male and female victims depending on availability of equipment and personnel. Salvage covers draped over ladders and ropes suspended between apparatus can form privacy barriers if time allows. Remember that the most imperative principle of mass casualty decontamination is the expedient and effective use of water in the removal of the contaminants from the victims.

The Interim Guidelines for PPE and Decontamination for Public Safety First Responder Personnel to use in Responding to Chemical Terrorist Situations published in December 2001 (http://www.dhs.state.or.us/publichealth/bioterrorism/chemppe.pdf) recommend that decon personnel responding to a chemical terrorist situation wear Level B for all warm zone activities. Most references state that victims whose skin or clothing is contaminated with nerve agents, hydrogen cyanide, or mustard agents can contaminate rescuers by direct contact or through off-gassing vapors. Wearing pressure-demand SCBAs and protecting the skin with chemical-protective clothing and butyl rubber gloves is necessary because vapors are readily absorbed by inhalation and ocular contact and liquid readily absorbs through skin. (Source: ATSDR, Nerve Agents: Tabun, Sarin, Soman, and VX, NIOSH Emergency Response Card: Mustard, NIOSH Emergency Response Card: Hydrogen Cyanide.) For specific information and recommendations for responding to incidents in which chemical and biological agents of mass destruction are involved, go to the CDC's website and consult the appropriate NIOSH Emergency Response Card. The web site address is http://www.bt.cdc.gov; buttons on the left of the page will send you to information on biological agents and chemical agents.

In mass casualty incidents where human life is in jeopardy, there probably will not be time to consider strategies to mitigate runoff from mass decontamination operations that may negatively impact the environmental. Emergency responders are protected from most environmental and other liabilities through a "good Samaritan" provision in the Comprehensive Environmental Response, Compensation, and Liability Act (CERCLA), section 107 (d) Rendering Care or Advice. This provision states that "No person shall be liable under the sub chapter for costs or damages as a result of actions taken or omitted in the course of rendering care, assistance, or advice in accordance with the National Contingency Plan (NCP) or at the direction of an on-scene coordinator appointed under such plan, with respect to an incident creating a danger to public health or welfare or the environment as a result of any releases of a hazardous substance or the threat thereof." Once the immediate threat to human life is gone, all possible action to stop or mitigate environmental damage must be demonstrated.

Other Mass Decontamination Resources

In January, 2002, the U.S. Army Soldier and Biological Chemical Command (SBCCOM) published *Guidelines for Cold Weather Mass Decontamination During a Terrorist Chemical Agent Incident.* This document states that "regardless of the ambient temperature, people who have been exposed to a known life-threatening level of chemical contamination should disrobe, undergo decontamination with copious amounts of high-volume, low-pressure water or alternative decontamination method, and be sheltered as soon as possible." In this document they suggest alternative decontamination methods, give recommendations based on outside ambient temperatures, and discuss signs, symptoms, and treatments of cold shock and hypothermia. This document can be found at the SBCCOM website (http://hld.sbccom.army.mil).

In October 2001 the U.S. Department of Justice, National Institute of Justice published a *Guide for the Selection of Chemical and Biological Decontamination Equipment for Emergency First Responders, Volumes I and II.* This guide is accessible over the Internet at http://www.ncjrs.org.

SUMMARY

In typical hazardous material incidents, chemicals released may contaminate the tools and equipment used by emergency responders or get on the outside of their protective clothing. These materials can be prevented from leaving the scene of the release. If responders follow prearranged SOPs for technical decontamination, and use materials and methods appropriate to the chemicals encountered, all contaminants will be collected on disposable materials or in wash and rinse solutions to be handled as waste. Proper technical decontamination protects responders, members of their families, and the people who live in the community.

Mass casualty incidents involve the coordinated mass decontamination of large numbers of people to remove contaminants. Technical decontamination procedures would not typically be the most effective means to decontaminate large numbers of people. Remember these important facts: Victims should be decontaminated as soon as possible to prevent further damage due to effects of exposure; removal of clothing can remove up to 80% of the contaminant from the victim; and flushing skin with large volumes of water is the best mass decontamination method. Emergency responders must practice all of the procedures described in this chapter if decontamination is to be effective in the field.

14

BASIC HAZARDOUS MATERIALS CONTROL

INTRODUCTION

At an industrial park located in Newark, New Jersey, a ruptured seam on a storage tank released approximately 2000 gallons of hydrochloric acid. Vapors generated from the surface of the spilled acid produced an acidic cloud that drifted across nearby tracks of the major rail artery linking Boston, New York, Newark, and Washington, DC. The rail lines were temporarily closed, leaving thousands of commuters stranded. By spraying the surface of the spilled hydrochloric acid with hazardous materials foam, responders were able to suppress vaporization of the acid so that the rail lines could be reopened.

As commuters stranded by the vapor cloud probably realized, bringing the hazardous materials involved in an incident under control is a critical aspect of incident mitigation. Effective control operations can prevent or minimize undesirable outcomes that may result from the release, or potential release, of hazardous materials. For this reason, hazmat control operations should be initiated as soon as appropriate procedures can be selected and determined to be safe through hazard and risk assessment and the use of decision making guidelines such as the DECIDE process.

For the purposes of this textbook, hazardous materials control procedures can be generally divided into two categories: confinement procedures (defensive) and containment procedures (offensive). Confinement procedures are utilized to limit the migration of a hazardous material within the environment to the smallest area

Emergency Responder Training Manual for the Hazardous Materials Technician, Second Edition, edited by Kenneth W. Oldfield
ISBN 0-471-21387-X   Copyright © 2005 John Wiley & Sons, Inc.

possible after a release. We typically think of confinement procedures as being performed in response to spills, which exist when a substance has escaped from its container. In contrast, containment procedures are performed to terminate a release of hazardous materials at the source. We tend to think of containment procedures as being utilized to stop container leaks that exist when a substance is in the process of escaping from its container.

Confinement procedures differ fundamentally from containment procedures. Procedures used in confinement are considered to be defensive in nature, because they can be performed at a location remote from the actual point of release of hazardous materials. Although confinement procedures require that a certain amount of ground be sacrificed to the released substance, responders involved in confinement operations are typically exposed to lower hazard levels than responders involved in containment operations. Procedures used in containment are considered to be offensive in nature, involving work in close proximity to the point of release. Prompt containment operations will minimize the amount of released material that must be contained. However, these operations typically involve a higher potential level of exposure of personnel to the hazards involved and may require special tools, equipment, and protective gear, as well as special training.

This chapter is intended to provide knowledge of basic hazardous materials control procedures as utilized in confinement and other defensive operations. This basic knowledge level is appropriate for emergency response personnel who will function in the role of first responder at the operations level. Advanced hazardous materials control procedures, as utilized by hazardous material technicians in containment operations, are covered in Chapter 15.

THE ROLE OF HAZARD AND RISK ASSESSMENT AND DECISION MAKING IN HAZARDOUS MATERIALS CONTROL

To be safe and effective, hazardous materials control procedures must be selected based on factors such as (1) the properties of the released material (solid vs. liquid, flammability), (2) characteristics of the release event (total tank failure vs. small leak) and subsequent migration of the material (topography, water drainage routes), and (3) site-specific conditions and considerations (weather conditions, soil types). Information gathered during the hazard and risk assessment process is vital. This information should be used in conjunction with decision-making guidelines (such as Benner's DECIDE process) in order to select and implement appropriate control procedures while ensuring the safety of personnel involved in control operations.

Types of Releases

In selecting appropriate confinement procedures, the type of release involved is a critical consideration. For the purposes of this chapter, releases will be classified as land, air, or water releases and appropriate confinement techniques will be presented for each type of release. As discussed in Chapter 6, these releases may occur as follows.

Land Releases

Land releases occur anytime a container is breached and the contents spill out onto the ground. In the case of a liquid product, the released material may then migrate freely along the ground, perhaps reaching the surface water system to create a water release. Land spills may also vaporize, forming air releases. Land releases typically lead to soil contamination and may also produce groundwater contamination, both of which must be addressed in postemergency response cleanup operations and are therefore beyond the scope of this text.

Water Releases

Water releases can occur when a hazardous material is released directly into a stream or other body of surface water or migrates to reach a body of water after a land release. Contaminants in groundwater may also pollute surface water bodies.

Air Releases

Air releases typically occur when a hazardous material escapes its container in a gaseous form or when a liquid vaporizes after escaping the container. The contaminant is then able to migrate through the atmosphere, as discussed in Chapter 6.

Other Considerations

It is important to note the relationship between the different types of releases as shown in Figure 6.5. For example, a single large-scale land release may conceivably result in air and water releases, as well as soil and groundwater contamination, as an incident runs its course. To be fully effective in minimizing harm, responders may be required to engage in different types of confinement operations simultaneously at multiple locations. Thus it is important that responders fully assess all potential outcomes of the release event before selecting confinement procedures intended to intervene in the course of events. Otherwise, available resources may not be used to the best advantage.

As in all other aspects of hazardous materials emergency response, the safety of response personnel should be given the highest priority in selecting and performing control operations.

PROCEDURES AND CONSIDERATIONS FOR BASIC HAZARDOUS MATERIALS CONTROL

As discussed above, the ultimate goal of basic hazardous materials control procedures is to utilize low-risk, defensive operations to prevent or minimize harm resulting from a release event. Ideally, this involves using appropriate procedures to retain or capture the released material at a given location so that it can be collected, treated, or otherwise dealt with.

In some cases, it may be impossible to capture a released product, as when large quantities are involved and resources for the initial response are in short supply. In such cases, responders may be able to divert the flow of product away from the predicted migration pathway and into an area where it will do less harm. As an example, responders in some incidents have created diversion dikes or ditches (as described below) to shunt flowing product away from streams or storm sewer inlets and prevent contamination of a municipal water supply.

In worst-case situations, there may be very little that responders can do to prevent exposure of an area to hazardous materials released in an incident. For example, it may be impossible to prevent a vapor cloud from migrating into an occupied area. In such situations, responders may still be able to utilize defensive tactics to minimize harm resulting from the incident. For example, after establishing control of the potentially affected area, responders may act in advance of the vapor cloud to limit property damage by removing mobile equipment and supplies to a safe area, covering stationary equipment to minimize exposure, and isolating buildings by closing doors and windows and shutting off heating, air conditioning, and ventilation systems.

In situations in which adequate planning and preparation for emergency response have been carried out before an accidental release, responders may be able to quickly confine and collect the released material so that minimal environmental exposure occurs. In other cases, the available resources for initial response may be greatly overwhelmed. In such cases, tough decision making will be required in determining how available resources can be best utilized so as to most effectively limit harm resulting from the incident. Figure 14.1 illustrates how some of the procedures described in this chapter might be applied in the situation shown in Figure 6.5.

Controlling Land Releases

When hazardous materials are spilled onto the ground, the speed and direction of their migration within the environment will be largely dependent on the physical properties of the material involved and the topography, or shape of the Earth's surface, in the vicinity of the release location. Spilled solids may accumulate on the ground immediately adjacent to the point of release, allowing responders to simply collect the materials with shovels and place them in a suitable container. In contrast, spilled liquid products may flow for great distances and be very difficult to confine and place into containers.

Spilled liquid products tend to flow from higher to lower points of elevation under the influence of gravity. In many instances, the liquid product will enter and flow along drainage ways, such as ditches, gutters, hollows, or dry stream beds. In so doing, the material will tend to migrate as a narrow stream of product down the drainage way. In the absence of a well-defined drainage way, spilled product may tend to flow in a braided pattern or as a sheet of product in the downslope direction.

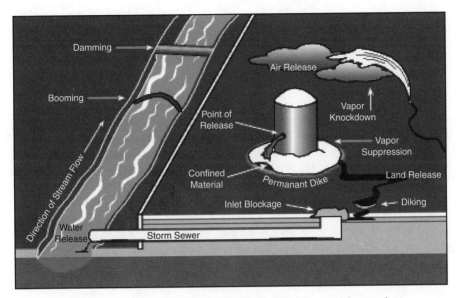

Figure 14.1 Overview of basic hazardous materials control procedures.

Effective control of land releases requires that we predict the pathway that the released material will follow and attempt to intercept and confine the product at some downstream point along that route. This interception point must be far enough in advance of the spill to allow adequate lead time for responders to create confinement structures, such as dikes, before the arrival of the product. This will maximize confinement efficiency and minimize exposure of the responders involved. However, excessive lead time should be avoided, as this sacrifices needless ground to the hazardous material.

Basic Control Procedures

A variety of procedures can be utilized for controlling spilled materials. In many instances, these procedures involve the creation of temporary control structures that are constructed with whatever materials are available or can be rapidly obtained. Ingenuity on the part of response personnel may be required. Equipment and supplies that may be used in performing basic control procedures include the following:

- Shovels
- Dry granular sorbent
- Picks or mattocks
- Salvage pumps
- Sandbags
- Salvage drums
- Bagged materials (such as mulch bags)

- Plastic sheets, tarps, and/or salvage covers
- Sorbent socks, pillows, and/or sheets
- Sorbent booms
- Barrier booms

Diking. Diking involves the construction of relatively impermeable barriers located so as to block the flow of a spilled hazardous material across the Earth's surface. Dikes can be constructed in straight, V-shaped, or circular configurations as shown in Figure 14.2. Dikes must be of the appropriate size and configuration for the characteristics of the spill involved. For example, material flowing along a ditch may be readily confined by constructing a simple dike or dam across the ditch. In contrast, for product migrating in "sheet flow" down a gradual slope or along a broad shallow depression, a longer V-shaped dike may be called for. For small or slow-moving spills on flat terrain, it may be possible to build a circular dike that completely surrounds the point of release. Dikes should be used in series, so that incidental leakage from one structure will be confined by the next. This also provides for backup in case of dike failure.

Diversion. Diversion involves the placement of control structures so as to divert liquid products away from places where we do not want them to go (such as storm sewer inlets) and into locations where we do want them to go (such as

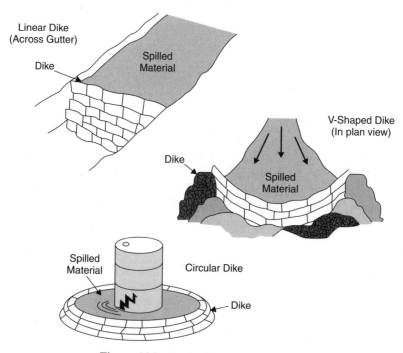

Figure 14.2 Basic dike configurations.

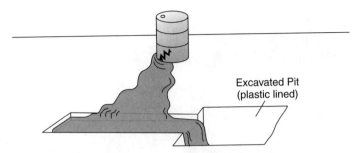

Figure 14.3   Use of a ditch and dike to divert spilled liquid into an excavated pit.

pits excavated as "catch basins" in the ground). For example, ditches or trenches can be created in advance of a spill to divert the spill so as to prevent a significant exposure or to direct the spill into a confinement structure as shown in Figure 14.3. Such a confinement structure may be a diked area or a catch basin. Linear dikes and barrier booms (as described below) may also be employed in diversion operations. Barrier booms can be improvised by using sections of fire hose that are wrapped in plastic (to minimize contamination), placed in the path of the spill, and charged.

Inlet Blockage (Retention). Inlet blockage, or what is sometimes called retention, is used to prevent hazardous materials from entering storm sewer systems or other highly undesirable locations. For example, plastic sheets, tarps, salvage covers, and/or neoprene mats can be used in conjunction with soil, sand, or other suitable materials, as shown in Figure 14.4, to completely block off gutter inlets, drainage grates, and manhole covers that feed into the storm sewer system and thereby the surface water system. Self-sealing products are available that conform to the surface and are suitable for many different applications. They are manufactured in at least two types of materials (polyurethane and PVC) and in many different shapes (examples including conical to act as drain plugs and mats that can be laid over manholes). Inflatable bags that may be placed into drain and sewer pipes and inflated to form a tight seal are commercially available (see Fig. 14.12 below). In some instances, barrier booms or sorbent booms (as described below) may be placed around entry points to confine or divert the product.

Implementing Control Procedures
Simple construction operations are usually required for implementing basic control procedures. Earth moving is frequently required so that, as a minimum, shovels and picks or mattocks will be needed. However, response personnel's ability for manual earth-moving may be quickly overtaxed if anything beyond a small dike or ditch is required, especially if work must be carried out in personal protective equipment. For this reason, it is recommended that earth moving equipment, such as backhoes, loaders, and dump trucks, be located during pre-emergency planning and placed on standby if it is anticipated during size-up that larger control structures may be required.

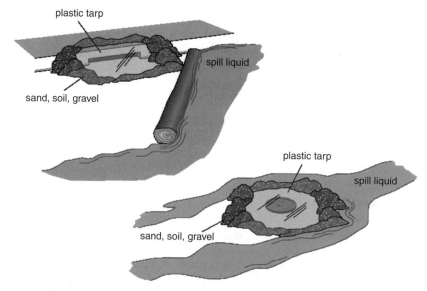

Figure 14.4   Use of inlet blockage for diversion and confinement.

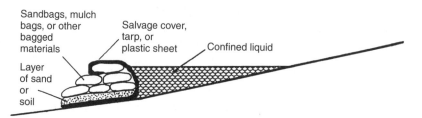

Figure 14.5   Improvised dike using bagged materials, shown in cross-sectional view.

If manual construction techniques are to be used, it is advisable to try to utilize bagged materials, such as sand bags or mulch bags, in creating control structures as shown in Figure 14.5. This will be much faster and easier than moving dirt or sand by the shovel load.

Whenever possible, responders should use available items such as plastic sheets, tarps, or salvage covers to enhance the effectiveness of control structures. For example, plastic sheets can be used to prevent liquid product from soaking into and through soil or other materials used to construct a dike. Likewise, these items can be used to line diversion ditches or catch basins so as to minimize soil contamination and related cleanup costs. In all cases, the potential for reaction between materials used in creating confinement structures and the material to be confined must be assessed.

It should be noted that commercially available products for dike construction (such as foamed concrete or polyurethane) work very well in some instances.

However, use of these products requires equipment that is not commonly available and personnel with special training. Booms (both barrier and sorbent types) and the other commercially available products described below may also be utilized in basic confinement operations on land.

Controlling Water Releases

In selecting appropriate techniques for mitigating releases of hazardous materials into water bodies, important size-up considerations include the degree of water solubility and the specific gravity of the product involved. These properties are discussed fully in Chapter 6. According to its solubility, the hazardous material may tend to intermix with, or remain separate from, the water body involved. Heavy insoluble products, or those with a specific gravity greater than 1, will tend to sink to the bottom of a body of water. In contrast, light insoluble products, or those with a specific gravity of less than 1, will tend to float on the surface of the water. Based on these properties, a number of options may be available to the responder attempting to control the hazardous material. It must be remembered that agitation of a water body, such as through turbulence or wave action, may have the effect of physically mixing an insoluble contaminant with water.

Controlling Heavy Insoluble Materials

When a heavy insoluble material enters a water body such as a stream, it tends to migrate to the stream bottom and then along the bottom under the influence of gravity and water current. Movement will typically be in the downstream direction for flowing streams. This behavior allows several procedures to be used in confining the product involved so that it can be collected and placed in a suitable container for recycling or disposal.

Overflow Dams. Overflow dams can be used in some instances for confining heavy insoluble materials. This procedure involves constructing a dam, as shown in Figure 14.6, across the stream bed at some point downstream from the point of release. If properly located and constructed, the overflow dam should confine the released material at the base of the dam while allowing uncontaminated water

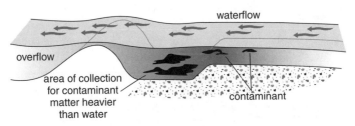

Figure 14.6 Use of an overflow dam and pit in a stream bed to confine a heavy, insoluble material.

near the surface of the stream to flow over the dam and continue downstream. This is highly preferable to attempting to capture the entire stream flow.

Catch Basins. Catch basins are another type of structure that may be effective in the confinement of heavy insoluble materials. This procedure requires the excavation of a pit or catch basin in the floor of the stream in advance of the material. Ideally, the product will accumulate because of the influence of gravity and remain at this location until collected. Usually, if this technique is used, several catch basins in sequence are placed in the steam floor.

Both of these procedures for confining heavy insoluble materials tend to work best with small, slow-moving streams. When streams that are larger or fast flowing are involved, these procedures may be ineffective or extremely difficult to implement.

Controlling Light Insoluble Materials

When light insoluble materials enter a stream, river, or lake they tend to float on the surface of the water. Direction and speed of migration of the material is then influenced by factors such as stream flow and wind. Several procedures may be utilized in confining this type of material.

Booms. Booms can be very efficient devices in some situations for confining light insoluble materials, as shown in Figure 14.7. Simply stated, booms are elongated tubular devices that float on the surface of a water body. These devices are commercially available or can be improvised, for example, by capping and inflating sections of hose. Barrier booms are designed simply to confine a product. In contrast, sorbent booms are filled with a sorbent material designed to soak up the spilled product (typically a petroleum-based product) while remaining impervious to water.

Booms can be connected end to end and deployed so as to surround a "slick" of floating product on a water body such as a lake or large, slow-moving river. Booms can also be deployed across a stream at locations downstream from a point of release to achieve confinement.

One major weakness of booming as a confinement procedure is that wave action and stream turbulence tend to splash the floating material over the booms. This problem may be addressed by placing booms in series. However, if the

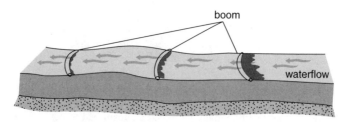

Figure 14.7   Use of booms to confine a light, insoluble material.

water is sufficiently rough, booms may simply not be effective in confining the released material.

Underflow Dams. Underflow dams are quite similar to the dams or dikes discussed above, with the exception that underflow dams incorporate pipes or tubing sections, as shown in Figure 14.8. The pipes are placed so that the downstream ends are elevated (angled) relative to the upstream ends. This configuration causes the floating contaminant to be confined behind the dam while allowing uncontaminated water along the base of the stream to flow through the dam and continue downstream. Underflow dams are frequently used in conjunction with booms.

Filter Fences. Filter fences are simple improvised confinement devices constructed with posts, fencing, and suitable sorbent materials, as shown in Figure 14.9. The fencing acts to hold the sorbent material in place, and the sorbent material acts to confine and soak up the contaminant as the stream flows through the filter fence. Like dikes or dams, filter fences should be placed in series for greater efficiency and to provide backup in case of structural failure.

Controlling Materials That Are Water Soluble

Materials that are soluble in water or have a specific gravity of approximately 1 can be extremely difficult to control when released into a body of water. Because these types of products tend to disperse rapidly throughout a water body, confinement techniques typically involve capturing the entire volume of contaminated water so that it can be treated to remove the contaminant. This will most probably involve expertise and equipment, such as mobile treatment plants, that are not readily available to the responder.

Sealed Booms. Sealed booms may be utilized to limit dispersion of soluble contaminants in some instances. These booms support curtains that are anchored to the bottom. This allows the entire depth of water to be confined within the boomed area.

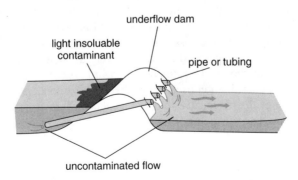

Figure 14.8   Use of underflow dam to confine a light, insoluble product.

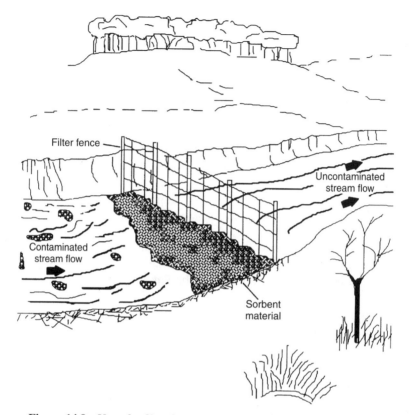

Figure 14.9   Use of a filter fence to confine a light, insoluble material.

Dams. Dams may be constructed for the purpose of confining the entire volume of contaminated water for treatment. However, this is usually an effective approach only for small streams and for a short period of time. Otherwise, the size of the dam required would be beyond the capabilities of the response team.

Chemical Treatment. Chemical treatment processes may be a viable option for dealing with soluble contaminants in water. For example, neutralization may be utilized to adjust a corrosive contaminant's pH. Likewise, flocculants may be used to cause precipitation of a dissolved contaminant out of the water column. The precipitant can then be pumped or dredged from the bottom for disposal. It is important to note that chemical treatment for hazard control requires specialized expertise and close coordination of activities between responders, personnel with expert knowledge of the chemicals involved, officials of environmental agencies having jurisdiction, and cleanup personnel.

Controlling Air Releases

In selecting procedures to control airborne gases and vapors, it is important to consider the vapor density of the material involved relative to that of air. As discussed in Chapter 6, lighter-than-air gases and vapors (i.e., those with a vapor density of less than 1) tend to migrate upward into the atmosphere and disperse. In contrast, heavier-than-air gases and vapors (i.e., those that form a vapor-air mixture that is heavier than air) tend to migrate to the ground and may accumulate in low-lying areas.

In response to air releases of hazardous substances, there may be little that responders can do other than to terminate the release at the source, if possible (see Chapter 15), evacuate the potentially affected area, and allow the released vapors to disperse. However, it must be remembered that heavier-than-air vapors do not tend to disperse readily. Restriction of airspace above the incident may be required, if lighter-than-air vapors are involved. In some instances, specific procedures, such as those discussed below, may be utilized in controlling air releases.

It is important to determine whether or not the substance involved is flammable. If so, it will be necessary to eliminate all potential sources of ignition in areas where a flammable atmosphere may develop. Otherwise, ignition of the vapor-air mixture can occur, with disastrous results.

Vapor Knockdown or Dispersion

In some instances, it may be possible to knock down a vapor cloud by directing fog patterns from fire hoses through the cloud as shown in Figure 14.10. This is most effective with water-soluble materials. For air releases of substances that are not water soluble, the air turbulence created by the fog patterns may enhance the dispersion of vapors, thus reducing the atmospheric hazard resulting from the release. In using this technique, large quantities of contaminated water are produced. The contaminated runoff should be confined to minimize environmental pollution (Fig. 14.10). The Association of American Railroads' *Emergency Action Guides* is a good source to consult to determine whether or not this technique is applicable for a given substance. Fans, blowers, and compressed air have also been used to disperse vapors in some instances.

Vapor Suppression

For incidents in which vaporization of a spilled liquid is producing an air release, vapor suppression at the surface of the liquid pool may be a viable control technique. Vapor suppression techniques use some sort of foam blanket to cover a volatile material and prevent the evolution of vapors from its surface (Fig. 14.11). Firefighters are familiar with fire fighting foams. Hazardous materials foams utilize different agents but require some of the same techniques and equipment for application. Specific training is required for their safe and effective use.

Fire Fighting Foams. Aqueous film-forming foam (AFFF) is one example of fire fighting foam that can also be used as a vapor suppressant on certain kinds of spills. When AFFF is applied to a spill, draining foam bubbles create a continuous membrane of aqueous, viscous solution on the surface of the spilled liquid. This membrane acts as a vapor barrier. AFFF is often used on flammable liquid spills to suppress vapors and prevent ignition. However, if the flammable liquid is a

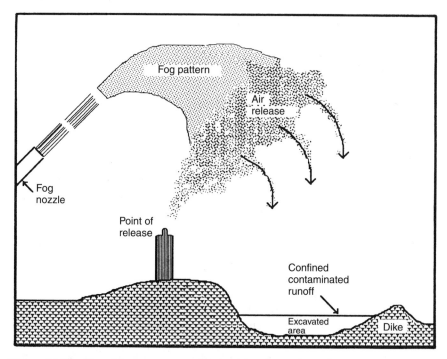

Figure 14.10   Use of fog patterns to control an air release, shown in cross-sectional view.

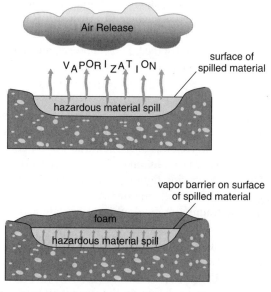

Figure 14.11   Vaporization and vapor suppression of a hazardous material spill, shown in cross-sectional view.

polar solvent or has a polar solvent added, the aqueous film will be dissolved by the polar solvent, rendering the foam ineffective. Such liquids are considered to be foam destructive. This characteristic can cause problems during vapor suppression procedures in response to fuel spills because polar solvents like ethanol and methanol are added to gasoline and some other fuels.

Alcohol-resistant aqueous film-forming foam (AR-AFFF) is effective on both hydrocarbon and polar solvent fuel spills. These foams form a polar solvent-resistant membrane on the surface of the spill and are therefore resistant to the destructive properties of polar solvent fuels.

Hazardous Material Foams. Hazardous material foams are of specialized formulations developed for use against nonflammable chemicals that emit large quantities of vapors. They are not suitable for use in fire fighting. These specially formulated foams are designed to handle hazardous materials other than fuels. One type of agent is used for acidic materials and another for alkaline materials. Hazmat foams are resistant to pH extremes that would rapidly destroy fire fighting foam. They are applied with air-aspirating foam nozzles. In some cases, these new foams convert the spilled material to a less hazardous one.

Considerations for the Use of Foams. Using any foam to mitigate a volatile spill requires expertise in choosing the foaming agent and selecting the application technique and specialized training in using foams for vapor suppression. Training in the use of foams for fire fighting is not sufficient. It is important to remember that foams are mostly water and will add considerable volume to confinement and cleanup liquids.

For effective use of foams in vapor suppression, the following materials must be available at the scene:

- Appropriate application equipment for the foam selected
- Sufficient quantities of foam concentrate to reapply the foam blanket every 30 minutes, or as needed, until the spill stops vaporizing or is cleaned up
- Enough water to maintain the critical ratio of foam concentrate to water

It is also important to have a plan that lists foam suppliers or alternate sources so that the supply of foam can be restocked quickly if depleted during an incident.

COLLECTION TECHNIQUES

After spills have been confined, the hazardous substances involved must be collected for recycling, treatment, or off-site disposal. In the case of spilled liquids this is typically done by using pumps or vacuum trucks to vacuum up the product or sorbent materials to soak up the product. In the public sector, these types of operations are typically performed by cleanup crews who are called in after the emergency is stabilized. However, in the private industrial sector, response personnel may also be involved in cleanup operations.

Pumping and Vacuum Collection

Pumping is the method of choice for collecting large-quantity spills after confinement. Liquid materials can be pumped from land, from the bottoms of water bodies, or from the water surface.

A variety of pumps that may be used in collecting liquids are available. General duty pumps, which are commonly powered by electric motors or gasoline engines, may be suitable for collecting some hazardous materials. Other materials may require specialized equipment, such as manually or pneumatically powered pumps designed for transferring flammable products. In all instances, the potential for reactivity between the product being transferred and the material from which the pump, hoses or tubing, and fittings are constructed must be considered.

Also, when transferring flammables, appropriate grounding and bonding procedures, as described in Chapter 15, must be utilized for equipment and containers used. Figure 14.12 illustrates the use of a pneumatic double-diaphragm salvage pump.

Figure 14.12   Use of an inflatable bag to block the inlet and a pneumatic double diaphragm salvage pump to collect material.

Usually, a residue of product will remain after pump collection. This residue can be removed by using sorbent materials or other options, as described below.

Sorbent Collection

When a spill is small, you may be able to use specific materials to soak up the product, as shown in Figure 14.13. Strictly speaking, these types of materials are classified as either absorbents or adsorbents, depending on the specific process by which they collect and retain the spilled material. For the purposes of this text, the generic term "sorbent" will be used in reference to both types of materials.

A variety of commonly available and commercially available products can be utilized as sorbent materials. Sorbents that will best handle the materials expected to be involved in incidents on the site should be chosen during preemergency planning. The physical and chemical interactions between the sorbent and the spilled material are important considerations. If sorbents are improperly selected,

Figure 14.13   Use of sorbent sheets to remove the residue remaining after pump collection of material.

undesirable chemical reactions can occur, producing toxic by-products and heat, which may result in fire.

Rate of absorbance, convenience in application and collection, and available disposal options are basic criteria for selection. Materials with low pH, low viscosity, and high temperature will absorb more quickly. Bulk materials are hard to carry and clean up, and they cannot be used in water. Sorbents contained in sheets, rolls, sausages, pillows, socks, or booms are more convenient to use.

Available types of sorbent materials can be generally classified as vegetable by-products, mineral products, or synthetics (Table 14.1). Sorbents in granular form such as vermiculite, fly ash, perlite, granular clays, crushed limestone, and activated carbon are used extensively in spill control operations. These materials have the effect of producing a solid or semisolid mass that can be easily collected and may reduce the hazard associated with the product involved, as when crushed limestone is applied to an acidic spill. However, under existing environmental regulations the resulting material may require further treatment before disposal in a hazardous waste landfill.

Solidification

Certain granular sorbent materials react chemically to completely solidify a spilled liquid in a continuous matrix. This may allow the product to be legally placed in a hazardous waste landfill without further treatment. A number of commercially available solidifiers can be selected specifically for use with spills of acids, bases, fuels, oils, solvents, organic materials, and aqueous products. Selection of a solidification material can only be made by a disposal expert, in conjunction with personnel at the facility to which the waste is to be sent.

Gelation

Gelation is a technique in which a gel-forming chemical is introduced into the liquid pool of contaminant or contaminated water. The resulting gel is a semisolid and is therefore easier to remove.

TABLE 14.1   Types of Sorbents

Type	Material
Absorbents	Vermiculite, clay, diatomaceous earth, perlite (volcanic shale rock), crushed limestone, cellulose, corn cobs, polypropylene, polyolefin, sphagnum peat moss
Adsorbents	Activated carbon, activated alumina
Neutralizers	Sodium bicarbonate, limestone, magnesium oxide based (acids), citric acid based (caustics), carbon based (organic solvents and fuels)
Encapsulating/suppressing	Hydrolysate-based hazardous materials foams, fire fighter's foams, acrylic polymer-based gels
Other	Biocides and fungicides as biological agent decontaminants

Chemical compatibility is the most important consideration when choosing gelation. The gelling agent must be compatible with the chemical to which it is applied. The gel formed by this method is a hazardous material and must be disposed of as such.

Other Techniques and Considerations

Dispersants

After a floating, oily substance has been removed from the land or water surface by vacuuming and/or sorption, a dispersant can be applied to the thin slick remaining. Dispersants are emulsifiers that break up an oily contaminant. They are generally applied to insoluble liquids spilled into water, like an oil slick on the surface of a water body, or onto land, as when fuel spills occur on paved surfaces.

Dispersants are applied with standard fire fighting equipment, by using a foam eductor or by pumping from a booster tank into which the dispersant has been directly poured. A fog nozzle is used to apply the stream straight into the spill. Agitation by the hose stream helps the dispersant to mix into the spilled material.

As the oily contaminant is broken up, the continuous slick will be broken into tiny droplets that will then be dispersed into the environment. Manufacturers of dispersants state that the contaminant will then be biodegradable (removed from the environment by bacteria or fungi), and it is true that tiny droplets of chemicals allow better access by microorganisms. Microbes are selective about what they degrade, however, and are unable to metabolize many chemicals. This process should never be used, other than for contaminant residue after collection, without consultation with the appropriate environmental agency, because the contaminant is not removed from the environment, only dispersed.

Dispersant streams can also be directed into manholes or duct openings to emulsify floating layers of volatile fuel oils. The emulsion formed can be recovered by pumping. Chemical dispersants can also be used to remove an oil residue from vegetation, buildings, and vehicles. The emulsified chemical is then rinsed off into the soil. The same environmental considerations discussed above are important.

Always check with your state or local environmental agency before buying and using dispersants. Some states have forbid their use.

Dilution

Dilution is the process of adding large quantities of water to a spilled material that is hazardous because of its concentration. Diluting the material to a less concentrated solution reduces the hazard.

Dilution with water is only effective when the material is water soluble. The hazard is reduced, but the volume of material that must then be confined and cleaned up is enlarged.

In some situations, the possibility exists that dilution can render the material nonhazardous, if its hazard was due only to concentration. However, the appropriate environmental agency must be consulted before using this method to flush the spill down a drain or leaving it unconfined in the environment.

Dilution with water should never be considered under the following circumstances:

- The material has not been identified.
- The material is a strong acid.
- The material is water reactive.

SUMMARY

Basic hazardous materials control operations are performed to confine the hazardous materials released in an incident to the smallest possible area. The considerations and procedures included in this chapter are intended to aid emergency response personnel in selecting and implementing appropriate basic control procedures. Although these are relatively simple procedures, personnel expected to perform them must be provided sufficient equipment and field training to perform them safely and efficiently. These concerns must be addressed during preemergency planning (see Chapter 3). Even though confinement options may be limited in some instances, responders should identify and perform any defensive actions that can be safely performed at locations remote from the point of release to minimize damage resulting from the incident.

15

ADVANCED HAZARDOUS MATERIALS CONTROL

Having addressed considerations and procedures for basic, or confinement-related, hazardous material control operations in Chapter 14, we now turn our attention to advanced, or containment-related, control operations. As noted in the Introduction to Chapter 14, and illustrated by the following scenario, both types of operations are critical to incident mitigation.

As a result of a faulty weld, an oil storage tank located in Jefferson Borough, Pennsylvania, failed. The tank collapsed almost instantaneously, releasing almost one million gallons of heavy oil in a 30-foot wave. The initial surge of oil washed over stationary confinement berms, battered nearby tanks, and twisted pipelines before entering the Monongahela River. This release resulted in one of the better-publicized environmental disasters of previous years. Despite a massive effort to confine and collect the spilled oil, approximately one-half of the almost one million gallons initially spilled was not recovered and is presumed to be lost in the environment. As the released oil migrated down the Monongahela River and entered the Ohio River, thousands of people went without water as suppliers along the rivers were forced to shut their intakes off. However, the incident could have had even more devastating consequences were it not for the actions of response personnel at the scene of the release. The oil-soaked vicinity of the failed tank had a strong odor of gasoline when responders first arrived. This was considered to be very serious because a nearby gasoline storage tank appeared to have been damaged by the release event. After diligent searching, the responders located the source of the gasoline release: a leak in a pipeline damaged by the

Emergency Responder Training Manual for the Hazardous Materials Technician, Second Edition, edited by Kenneth W. Oldfield
ISBN 0-471-21387-X   Copyright © 2005 John Wiley & Sons, Inc.

sudden release of oil. Response personnel were able to stop the gasoline leak by using a golf tee as a plug. The ingenuity of the responders in stopping the leak removed the threat of impending explosion and fire, which would have made the incident on the Monongahela an even greater disaster.

By terminating a release at the source, containment operations offer the advantage of minimizing environmental contamination. Also, if offensive actions are not taken to terminate a sustained, large-quantity release at the source, the confinement capabilities of response personnel may be overwhelmed. Thus containment operations should be undertaken in all instances in which response personnel can safely perform them. However, because responders must work in close proximity to the actual point of release to perform containment operations, they are likely to be exposed to hazards of the incident. Safety of personnel must be given the highest priority in these types of operations.

This chapter is intended to provide knowledge and understanding of advanced hazardous materials control procedures as utilized in containment and related operations. These operations typically require special tools, equipment, protective gear, and expertise to be performed safely and effectively. Only those personnel who have received training to at least the level of hazardous materials technician should attempt to perform the types of procedures described in this chapter.

ASSESSMENT AND DECISION MAKING FOR CONTAINMENT OPERATIONS

Like all aspects of emergency response, containment operations must evolve directly from the hazard and risk assessment process, as described in Chapter 5. This process will provide basic information about the incident that can be used, in conjunction with decision-making guidelines (such as Benner's DECIDE process), to select appropriate containment procedures.

Containment-Related Planning and Decision Making

Because emergency response is time critical, containment procedures must be carried out as efficiently as possible. Thus preplanning is essential. The emergency response plan should incorporate general procedural guidelines for those containment operations that could be required during an emergency. However, specific procedures must be determined based on specific information gathered during size-up of the actual incident. Through preplanning and the use of decision-making guidelines responders should be able to:

- Determine strategic containment objectives
- Determine tactical options for containment
- Choose the best containment procedure and alternate procedures
- Choose equipment and tools needed to perform the procedure selected
- Choose required personal protective equipment
- Perform the chosen containment procedure

- Evaluate progress as the containment procedure is performed
- Change plans or call for additional assistance and equipment as needed to safely terminate the release

Assessment Considerations for Selection of Containment Procedures

Hazardous materials incidents may involve uncontrolled releases from a variety of containers, including drums, vats, tanks, cylinders, pipes, and transport vehicles. To identify procedural options for containing these releases, a number of specific factors related to the release event must be considered. These factors can be generally classified as follows:

- Properties of the hazardous material involved, including:
  - Physical state (solid, liquid, gas)
  - Physical properties (vapor pressure, density, etc.)
  - Toxicity (as related to PPE requirements)
  - Flammability (requiring use of nonsparking tools and intrinsically safe electrical equipment)
  - Corrosivity (which may destroy materials used in patches or plugs and damage tools used)
  - Reactivity (including reactivity with available patch materials)
  - Composition (aqueous, petroleum based, etc.)
- Characteristics of the container, including:
  - Size of container
  - Amount of contents remaining
  - Configuration of container
  - Material of which container is composed
  - Overall integrity of container
- Characteristics of breach and resulting release, including:
  - Size of opening
  - Shape of opening
  - Location of opening on container
  - Volume and rate of release
  - Expected duration of release
  - Actual or potential pressure at point of release

Considerations for Hazards to Personnel

Because containment operations require personnel to work in highly hazardous areas, safety should be the most important consideration in planning containment operations and choosing containment procedures. A careful assessment of the nature and degree of stress to containers damaged in an incident is required to ensure that a sudden violent release will not occur during the course of operations. This is especially important for operations involving work on pressurized containers. Additional safety considerations are also required for situations in which fire or explosion could occur during the operation.

If information gathered during size-up indicates that a containment procedure being considered may place responders at undue risk, other less hazardous offensive options should be considered. Response personnel should never attempt to perform containment operations at the point of release before options for stopping the release remotely have been explored. This can be accomplished in some instances in the industrial setting by simply closing valves at some other location within the facility. However, close coordination with personnel having knowledge of the design of systems within the facility is required for this type of operation.

Containment procedures should never be attempted unless the equipment, protective gear, and training provided to response personnel are adequate to allow the procedure to be carried out in a safe, effective manner. For situations in which containment cannot be achieved without risks to the safety of response personnel, it is permissible to remain in a defensive mode and explore additional options as the incident runs its course.

EQUIPMENT, SUPPLIES, AND TOOLS USED IN CONTAINMENT

A variety of items may be used in containment procedures (Fig. 15.1 and Tables 15.1 through 15.5). Some of these are general-purpose items that are

Figure 15.1 Overview of useful tools, equipment, and supplies for hazmat containment operations.

TABLE 15.1   Plugs, Patches, and Related Items

Items/Description	Function/Use	Procedural Considerations/Comments
Expandable pipe plugs	Placed inside pipe and tightened; internal expansion of plug blocks pipe.	Various sizes available; must be appropriate size for pipe to be plugged. Some types require a special tool for installation. Some are fitted with vent pipes and/or valves for pressure installation.
Pipe repair clamps (e.g., "pipe saver" or "Band-Aid" clamps)	Placed around leaking pipes, fastened, and tightened to clamp off leaks	Clamps must be selected for specific outside diameter of leaking pipe. Gasket used must be compatible with leaking material.
Leak-sealing bandages	Used to stop pipe leaks	Perform same function as pipe clamps but are pneumatically operated.
Boiler plugs	Screwed or driven into holes in containers, pipes, etc.	Various sizes available
Sheet metal screws with gasket	Screwed into small, simple punctures in drums	Can be improvised readily. Gasket materials used must be compatible with material to be contained.
Toggle bolt patches	Toggle is pushed through a small hole in the container wall, and patch is tightened into place to cover the hole.	Can be fitted with wing nut, washer, and neoprene pad, lab stopper, or rubber ball for patching small holes. Avoid overtorquing, as toggle may fail.
T-bolt patches	T is placed through small, elongated tears in container walls and rotated 90°. Patch is then tightened into place to cover the hole.	Can be fitted with wing nut, washer, and neoprene pad, lab stopper, or rubber ball for patching small holes. Avoid overtorquing, as T may be pulled through thin container walls.
Wooden plugs and wedges	Used individually or in combination to plug holes of various sizes and shapes	Use with suitable gasket or sealant material to stop leakage. Soft wood should be used. Wooden plugs may react with some corrosives and ignite.

(*continued overleaf*)

TABLE 15.1 (*continued*)

Items/Description	Function/Use	Procedural Considerations/Comments
Plugs and wedges of neoprene, viton, etc.	Used to plug small leaks of various shapes	Material used must be compatible with chemicals involved. Foam materials used must be closed cell. Can be preshaped to fit expected configuration of hole (e.g., forklift tine).
Drift pins	Driven into small holes in containers	Should be nonsparking for flammables.
Metal sheeting	Used as backing in conjunction with gasket materials, sealant, or adhesive to patch leaks of various sizes and shapes	Must be flexible or preshaped. Can be held in place with toggle bolts, T-bolts, ratchet straps, ropes, etc.
Strap iron "Band-Aid"	Used to patch large and/or irregularly shaped leaks	Can be attached by toggle bolts, T-bolts, ratchet straps, ropes, etc. Can be fitted with relief pipe and valve for application under pressure.
Ratchet straps, banding straps, etc.	Used to hold large patches in place on containers	Can be used with adhesive patches and removed after adhesive cures.
Dome cover clamps	Used to clamp tank truck dome covers into closed position	Can be improvised or fabricated in-house.
Leak-sealing bags (product of Vetter & Company)	Used to stop leaks from holes in tank trucks or rail tankers	Bags are strapped in place over breach and inflated to stop leaks.

readily available, whereas others are highly specialized and available only from specific sources.

Equipment that may be used in containment operations can be roughly divided by function into the categories discussed in the following sections.

Plugs, Patches, and Related Items

Plugs are designed to stop leakage by filling holes in containers through which materials are escaping. Patches stop leaks by covering holes in containers. Clamps, fasteners, straps, and similar items are used to hold patches or plugs in place. Plugs, patches, and related items are described in Table 15.1.

Adhesives, Sealants, and Gaskets

Adhesives, sealants, and gasket materials are frequently used in conjunction with patches and plugs. For example, adhesives can be used to secure patches in place on a container. Some adhesives, such as hazmat epoxies or fuel tank epoxies, may be applied directly to breaches in containers. Sealants and gaskets are often used to enhance the sealing action of patches and plugs. Adhesives, sealants, and gaskets are described in Table 15.2.

Tools

Tools of various types can be used for purposes such as preparing a surface for placement of adhesive or sealant, screwing or driving plugs into place, installing patches, and tightening leaking pipe connections. Tools that may be used during containment operations are listed in Table 15.3.

Other Containment-Related Items

Additional items that do not fall within the categories above may be utilized during containment operations. For example, drip pans may be used to confine leaking product at the point of release during plug/patch procedures. Grounding and bonding equipment is required for safe transfer of a flammable liquid to a sound container (see below in this chapter). Overpack drums may be required for transportation of leaking drums after patching (see below in this chapter). Any number of additional items may be required for safe and effective containment operations. Some of these items are listed in Table 15.4.

Containment Kits

Preassembled kits containing a variety of general-purpose containment-related items are commercially available (Table 15.5). Examples include the Hazardous Materials Response Kits available from Edwards and Cromwell Spill Control of Baton Rouge, Louisiana. It is also possible, although less convenient, to assemble response kits in-house using supplies, equipment, and tools that are locally available. "Specialty kits," such as the Chlorine Institute Emergency Kits used to control leaks from chlorine containers, are designed for use only in highly specific applications.

CONTAINMENT PROCEDURES

A great variety of procedures and techniques may be utilized for containment during hazardous materials incidents. Procedures that are used must be tailored to the specifics of each situation. Thus it is entirely possible that no two containment procedures used in the field will be exactly alike. Ingenuity, resourcefulness, and flexibility are called for on the part of the responder.

TABLE 15.2   Adhesives, Sealants, and Gaskets

Item/Description	Function/Use	Procedural Considerations/Comments
Plastic steel sealant (e.g., Devcon brand) Aluminum putty sealant (e.g., Devcon brand) Plastic carbide sealant (e.g., Devcon brand)	Used as sealant in conjunction with various plug and patch materials	Should be selected based on composition of container material and degree of durability required.
5-Minute epoxy and epoxy gel (e.g., Devcon brand)	Used as adhesive in conjunction with various plug and patch materials	Syringes provide a convenient means of premixing resin and hardener.
Hazmat putties (various types sold under various trade names)	Used to stop leaks of liquid products	Can be held in place by various patches (e.g., fiberglass fabric and epoxy or tape).
Plug-n-Dike Putty	Used to stop leaks, as well as to construct small dikes	Can be held in place by various patches. Available in dry granular or premixed form.
Pig Putty (product of New Pig Corp.)	Used to stop leaks of liquid products	Epoxy, with resin and hardener contained in round stick
Fuel tank epoxy	Used to stop leaks of petroleum-based products	Epoxy, with resin and hardener in small color-coded blocks
Viton sheets, patches, and washers; closed-cell neoprene sheets, patches, and washers.	Used as gaskets to enhance seal of patches, plugs, etc.	Material used must be compatible with chemicals involved. Can be used as whole sheet or cut to the size and configuration needed, depending on the application. Thickness varies with application.
Duct tape	Provides temporary seal for cracks and small leaks	Can be used over putty or foam packing as a temporary patch. Product may dissolve glue holding tape in place.
Aluminum repair tape	Provides temporary seal for cracks and small leaks	Backless type preferred because of difficulty in peeling backing from tape while wearing gloves.
Teflon sealant tape	Used to enhance seal of threads on plugs and pipe connections	

TABLE 15.2 (*continued*)

Item/Description	Function/Use	Procedural Considerations/ Comments
Lead wool	Used as packing to stop leakage along container seams (e.g., drum chimes)	
Lead foil	Used as packing or to wrap wedges or plugs	
Cloth or felt	Used to wrap wedges and plugs for enhanced seal	
Tubeless tire plug/patch kit	Used to stop small leaks in a variety of containers	
Plumber's oakum	Used as filler material and wrapping for plugs and wedges	Resin-impregnated, fibrous substance; swells when wet
Fiberglass fabric	Used in conjunction with epoxy to hold patch/plug materials (e.g., putty) in place	
Assorted O-rings, washers, nuts	Used in conjunction with various plugs and fasteners to enhance seal.	
Bar soap (e.g., Octagon brand)	Used as packing to stop leaks in containers such as fuel tanks	For temporary use only

It is not possible to describe in detail all the specific containment operations that responders may be called on to perform in the field. However, general procedures that may be applicable to various types of releases are incorporated in the information that follows.

General Procedures

In all containment operations, follow general procedures for responder safety and minimization of the spread of contamination. Use basic safe work practices to minimize personal exposure. For example, avoid walking through or standing in puddles of spilled product, if possible. Also, try to minimize direct contamination of protective clothing through splash or spray during procedures such as patch installation or valve repair.

Try to minimize spillage during the course of all containment operations. This will minimize contamination of the work area and thereby reduce the likelihood

TABLE 15.3 Tools That May be Required for Containment Operations

Item/Description	Function/Use	Procedural Considerations/Comments
Acid brushes	Application of adhesives to containers and patches	
Utility shears	Cut and shape gasket materials	
Cotter key extractor	Placement or removal of gaskets, packing, etc.	
Wire brushes	Surface preparation before application of adhesives	Nonsparking recommended for flammable hazards
Emery cloth	Surface preparation before application of adhesives	
Putty knives	Surface preparation or application of sealants and adhesives	Nonsparking recommended for flammable hazards
Hammers, various sizes and types Sledge Mallet Ball-Peen Claw Deadblow Etc.	Used in installation of wedges and plugs, as well as for various general uses	Nonsparking (e.g., brass, rubber) recommended for flammable hazards. Dead blow hammer is loaded to prevent rebound and possible injury.
C-clamps	Used to hold patches in place while adhesive cures	
Tubing clamps (e.g., Entex brand)	Used to pinch leaking tubing closed	Other items, such as vise grip pliers, can sometimes be adapted for this function.
Bung wrench	Tightening of drum bungs to stop leakage; removal of bungs for transfer of product to sound container	For flammable product, use nonsparking wrench, bond, and ground containers.
Expandable pipe plug wrench	Required for installation of some expandable pipe plugs	
Pipe wrench Adjustable wrench (e.g., Crescent brand) Combination wrench set Socket wrench set	Various general applications related to confinement procedures	Nonsparking recommended for flammable hazards

TABLE 15.3 (*continued*)

Item/Description	Function/Use	Procedural Considerations/Comments
Slip-joint pliers Needle-nose pliers Vise-grip pliers Lineman's pliers Wire cutters Punch-chisel set Hacksaw Flat bastard file Round bastard file Screwdriver set Pop-rivet tool	Various general applications related to confinement procedures	Nonsparking recommended for flammable hazards

TABLE 15.4 Other Containment-Related Items

Item/Description	Function/Use	Procedural Considerations/Comments
Drums and other containers	Transfer product from leaking container to sound container	Must be bonded and grounded for transfer of flammables.
Salvage drums	Storage of used sorbents or overpacking of damaged drums for transport	Drums must be DOT certified for materials to be transported.
Drip pans	Confinement of leaking product before and during patching procedure	Must be bonded and grounded for use with flammables.
Sorbent materials and related equipment	Confinement and collection of product at point of release in conjunction with containment procedures	See Chapter 14
Grounding and bonding equipment	Prevent ignition through discharge of static electricity during operations involving flammables	

of contamination of personnel. If possible, change the position of the container involved to place the point of release in the vapor phase. Because a given volume of spilled liquid may produce a much larger volume of vapor (in some cases several hundred times the original volume of liquid), this change in position will significantly reduce the concentration of hazardous vapors in the work area. For

TABLE 15.5   Examples of Specialized Containment Kits and Equipment

Item/Description	Function/Use	Procedural Considerations/Comments
Vetter containment kits (produced by Manfred Vetter GmbH & Company Zülpich, Germany)	Various kits for plugging/patching pipes, transport vehicles, and various large containers	Contains inflatable patches/plugs that can be placed and inflated with compressed air.
Hazardous materials response kits, series A through F (product of Edwards and Cromwell of Baton Rouge, LA)	Used for containing leaks from a variety of sources, including drums, large containers, and pipes, depending on the kit selected	Series A used for leaks of various sizes from containers of various sizes Series C used for pipe leaks Series D used for small containers Series F used for large leaks in large containers
Chlorine A kit	100- and 150-lb chlorine cylinders	Specialty kits, specifically designed for chlorine containment by the Chlorine Institute, Washington, DC
Chlorine B kit	1-ton chlorine containers	
Chlorine C kit	Chlorine railcars	
Chlorine D kit	Chlorine barges	
Cylinder Recovery Vessel ("Cylinder Overpack")	Contains leaks in gas cylinders.	The leaking cylinder is placed in the recovery vessel, which is sealed to contain the leak. Vessel must be designed for the gas it contains.
Midland Emergency Response Kits	Contains leaks from chlorine railcar domes.	Product of Midland Manufacturing

containers that cannot be repositioned, it may be possible to place a suitable container beneath the point of release to confine the escaping material.

In some situations involving pressurized containers or systems, it may be possible to reduce the amount of pressure involved before the onset of operations by manipulating valves or pump pressures. Obviously, these operations will require expert knowledge of the system involved. Container pressures may be lowered significantly by cooling the containers involved, such as through the use of hose streams. Cooling containers may be vital for situations in which container temperatures are elevated, such as when there is a fire nearby.

For all operations involving flammable materials, extra precautions must be followed. Direct-reading air monitoring equipment for detecting flammable atmospheres, as described in Chapter 9, is required. Container grounding and bonding procedures, as described below, should be utilized. All tools used should be of nonsparking design, and any electrical equipment used (such as direct-read atmospheric instruments, radios and flashlights) must have the proper safety ratings. Furthermore, suitable protective equipment should be utilized, with appropriate fire-suppression equipment and personnel on standby in case of flash fire.

In all containment operations, compatibility between the leaking material and the components of plugs, patches, and related items used is a critical consideration. If any evidence of a reaction is noted, cease the procedure immediately and explore other options.

Procedures for Small Containers

For the purpose of this text, small containers will be considered those that can reasonably be repositioned by a two-person entry team. A very commonly encountered container of this type is a 55-gallon drum.

Small Simple Punctures in Small Containers

A variety of procedures may be used for plugging or patching small simple punctures in small containers. Boiler plugs or sheet metal screws fitted with a suitable gasket material may be very effective for this type of leak (Fig. 15.2). Likewise, plugs or wedges of wood or other materials may be driven into the hole (Fig. 15.3). Small toggle bolt patches may be installed if the opening is large enough to allow the toggle to pass through in the closed position (Fig. 15.4).

In some cases a suitable epoxy, such as fuel tank repair epoxy, may be placed over small holes in containers and allowed to set. Pig Putty®, marketed by the New Pig Corporation of Tipton, Pennsylvania, is a hazmat epoxy that comes in a tube consisting of two elements—a resin and a hardener. One element forms the core of the tube and is surrounded by a mantle of the other element. By breaking off a length of the tube and kneading the two elements together, the epoxy is activated and begins to set.

Another option is to pack the opening with putty or closed-cell foam. It may be desirable to precut items such as neoprene foam or neoprene wedges into

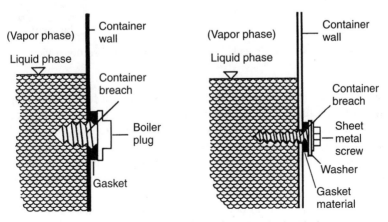

Figure 15.2   Screw plugs, shown in cross-sectional view.

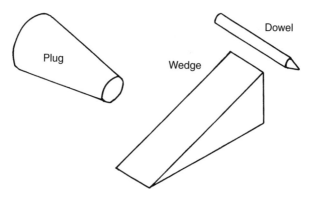

Figure 15.3   Basic shapes of plugs and wedges for driving into breaches in containers.

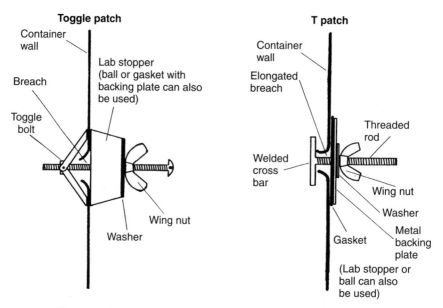

Figure 15.4   Toggle and T patches, shown in cross-sectional view.

shapes anticipated to be required, such as the shape of a forklift tine. Tape may be utilized to hold the packing material temporarily in place. The tape can then be saturated with epoxy, or an epoxy and fiberglass patch (as described below) can be placed over the tape.

After the container is repaired, it may be desirable to place it in an overpack container. If so, the patch or plug installed must not interfere with the overpacking procedure. For this reason, it may be desirable to trim items such as toggle bolts or wooden wedges flush with bolt cutters or a saw. Overpacking procedures are described below in this chapter.

Large or Irregular Holes in Small Containers

In repairing large or irregularly shaped holes in small containers, techniques similar to those described above are applicable with certain modifications. Toggle patches or T patches, fitted with a backing plate and gasket of suitable size and shape, may work for some leaks of this type (Fig. 15.4). It may also be possible to place a number of wooden wedges, of a variety of sizes and shapes, in the breach and drive them tight (Fig. 15.5). For very large leaks, it may also be possible to place a patch consisting of a suitable gasket with a sheet metal backing plate over the breach and then tighten the patch into position with worm drive clamps or load binding straps (Fig. 15.6). PSI Urethanes, Inc. of Austin, Texas manufactures the Barrel Patch Safety Seal, which uses a ratcheting nylon strap to tighten a sticky, flexible patch over the breach in a drum (Fig. 15.6).

In some cases, it may be desirable to utilize a semipermanent patching procedure, such as a fiberglass-epoxy patch (Fig. 15.7). This type of patch may be installed as follows:

(1) After the container has been grounded (if required) and repositioned, as discussed above, pack the breach with closed cell foam, putty, or other suitable material (Fig. 15.7a).

(2) Prepare the surface of the container for the adhesive by scraping, brushing, or sanding to remove loose paint, rust, scale, or dirt. Use nonsparking items for this if the container's contents are flammable. The entire area to which adhesive is to be applied should be clean and dry.

(3) Prepare the required amount of epoxy adhesive by mixing resin and hardener. The fast drying "5-minute" type epoxy works well but requires you to work quickly. Small, stiff-bristled brushes, such as acid brushes, are good for mixing and applying adhesive material.

(4) Apply a thin film of epoxy to the area of the container surrounding the breach (Fig. 15.7b).

(5) Place a single layer of fiberglass cloth (as used in boat and automobile body repair) over the breach and surrounding area (Fig. 15.7c). The epoxy film applied in Step 4 will hold the cloth in place.

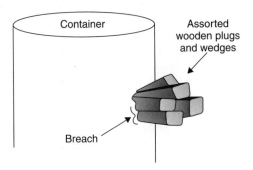

Figure 15.5 Use of multiple wedges and plugs to close a large opening.

Figure 15.6   The PSI Barrel Patch Safety Seal (lower repair item) is designed to quickly and conveniently stop leaks in drum sidewalls. The same repair may be improvised by using a patch of gasket material and a backing plate, with devices such as worm gear clamps or ratchet straps used to clamp it over the breach (upper repair item).

(6) Completely saturate the fiberglass cloth and surrounding areas of the container surface with the remaining epoxy (Fig. 15.7d).

(7) Allow the epoxy to harden before moving the container.

Other Small Container Leaks

You may encounter a variety of other types of leaks in small containers. For example, leaks may occur along the chime of steel drums. The chime is the crimped area where the top or bottom head is joined to the cylindrical side or wall section of the drum. These types of leaks may be patched by peening a suitable material, such as lead wool, into the chime along the area of leakage.

Closure failure may also occur in small containers. For example, the threads of drum bungs may be stripped or otherwise damaged, especially if the threads are corroded. In such situations, the bung hole in the drum can be closed by driving a plug of wood or some other suitable material into place.

In some cases, drums may be in such a deteriorated condition that repair is not feasible. For example, corrosion may produce pinholes in numbers too great to be patched or weaken the container wall to the point that it cannot reasonably be

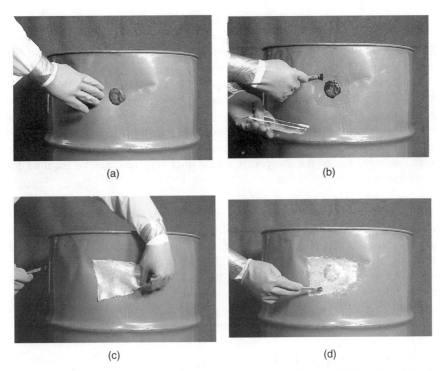

(a)　　　　　　　　　　(b)

(c)　　　　　　　　　　(d)

Figure 15.7 Fiberglass-epoxy patching consists of several steps. (a) The breach in the container is packed with a suitable material, and the area around it is cleaned and dried. (b) A thin film of epoxy is smeared around the breach. (c) A layer of fiberglass cloth is placed over the breach. (d) The fiberglass cloth is completely saturated with epoxy, which is then allowed to harden.

patched. In such cases, transfer of the drum's contents into a sound container is probably the best option, because it requires minimal disturbance of the container. Overpacking may also be an option if the container has enough integrity to survive being moved (see below in this chapter for overpacking procedures). If deteriorated containers are involved, overpacking and any other operations attempted should be performed with extreme care, because total container failure may occur.

Procedures for Large Containers

Procedures for controlling releases from large containers are generally analogous to those described for leaks from small containers. However, larger containers impose certain limitations. For example, unless heavy equipment is available, it is typically not possible to change the position of these containers to place the point of release in the vapor phase. We may have no choice but to terminate a release while a hazardous material is actively flowing through the breach. Transferring the contents to a sound container may be a viable option to perform in conjunction with, or instead of, containment operations for large container leaks. However, this should only be attempted by personnel with appropriate knowledge of the valves, fittings, and transfer procedures for the containers involved.

Transfer is considered a specialized operation. Unless you have special training in the transfer operation required, or have knowledge of the operation gained in the course of your routine work activities, call in personnel with the appropriate experience and expertise to perform the transfer operation. If transport vehicles such as tank trucks or railcars are involved, coordination with the transport companies and/or the shipper of the product involved may be required for product transfer.

Small Simple Punctures in Large Containers

Any of the procedures described for small simple punctures in small containers may also work for stopping leaks from small simple punctures in large containers. However, these procedures may be more difficult to perform for larger containers, and they may involve a higher potential for exposure of personnel if the point of release is located below the liquid-vapor interface of the product in the container. In these instances, as in all containment operations, the general safety procedures discussed above should be followed carefully.

Large Irregular Holes in Large Containers

Releases from large or irregularly shaped holes in large containers can be especially difficult to control, given the high potential flow rates from such openings. However, a variety of options exist for addressing this type of problem.

Large or irregular holes may be patched in some instances by placing a metal plate and gasket over the breach and tightening the plate against the container wall utilizing ratchet-equipped load binding straps or chains or ropes tensioned with load binders or power pullers (Fig. 15.8). A similar alternative to this type of procedure is the use of a "strap iron Band-Aid' patch that is held in place by T bolts placed through the breach (Fig. 15.9). For either of these options to work well, the backing plate used must either be preformed to the shape of the container or flexible enough to assume that shape under tension.

In either of the procedures described above, it may be possible to utilize a suitable adhesive to seal the packing plate to the container. The backing plate can be fitted with a valved outlet for use in transfer of product and for pressure relief during installation (Figs. 15.8 and 15.9). In the latter case, the patch should be installed with the valve in the open position. After installation, once the gasket has sealed, it may be possible to close the valve and completely stop the release.

Commercially available patch items suitable for large releases from large containers can also be used. One such item, designed with railcar, tank truck, and storage tank releases in mind, is an inflatable patch that can be bound to the container wall over the breach with ratchet straps and then inflated to seal off the leak (Fig. 15.10). Manfred Vetter GmbH & Company of Zülpich, Germany

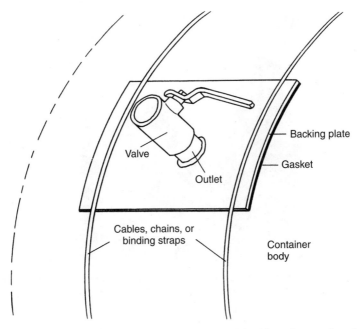

Figure 15.8 Large container patch using a gasket and backing plate equipped with an outlet valve.

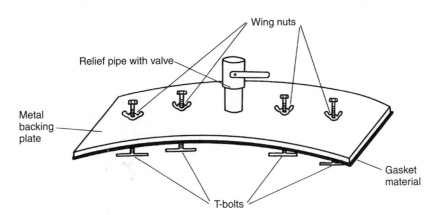

Figure 15.9 Strap iron "Band Aid" suitable for long tears in large containers.

is a leading supplier of his type of item as well as other pneumatic patch and plug items.

Another patch item consists of a pad that is held in place by a vacuum created by a special vacuum pump. The pad is fitted with a valved outlet line for off-loading the container contents. Patch items held to ferric metal containers by magnetic force are also available.

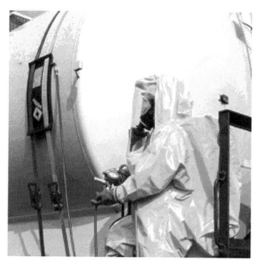

Figure 15.10 Inflatable patches are positioned over a breach in a container and inflated to seal off the leak. (Courtesy of Safety Solutions Int. Inc./Manfred Vetter GmbH & Comp.).

Dealing with Other Large Container Leaks

Transportation accidents may result in leakage from various fittings on containers such as cargo tanks or tank cars. Specialized training and a thorough knowledge of the specific fittings involved may be required to contain such leaks. In some cases, simply closing valves or tightening fasteners may be sufficient to contain a release. In other cases, specialized equipment may be required.

For example, in rollover accidents involving DOT-406 tank trucks, leakage from dome lids on overturned tankers is common. Traditionally, dome lid clamps have been used to stop these leaks simply by putting more pressure on the dome lid to clamp off the leak (Fig. 15.11a). In recent years, a newer device, the Lid-Loc, has been marketed by Safe Transportation Training Specialists of Carmel, IN as an alternative to the conventional dome lid clamp (Fig. 15.11b). The Lid-Loc is sold in sets of three for sealing off dome leaks on DOT-406 and DOT-407 cargo tanks.

For leaks in large containers transporting liquefied compressed gases, specialized procedures and equipment may be required. For example, for containers transporting chlorine, kits provided by the Chlorine Institute (see Figs. 15.15 and 15.16 below) or Midland Manufacturing may be required. These kits are discussed below in this chapter.

Procedures for Plumbing Leaks

A variety of hazardous materials releases may originate from valves, piping, and other components of plumbing systems in an industrial facility. In all cases, it is highly preferable to shut down the system involved to terminate the release and allow for repair or replacement of defective or damaged parts. However, in

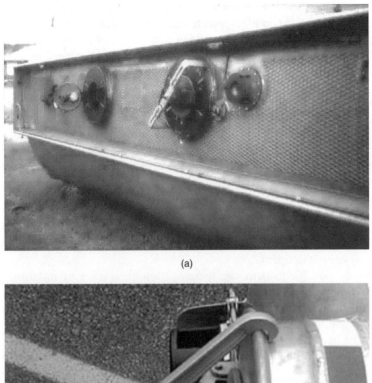

(a)

(b)

Figure 15.11 Tools for stopping leaks from cargo tank truck domes include (a) the dome lid clamp, which has been used traditionally, and (b) the Lid-Loc, which offers a convenient alternative for stopping dome lid leaks. (Lid-Loc photograph courtesy of Safe Transportation Training Specialists.)

some instances, immediate shutdown may not be a feasible option and response personnel will be called on to respond to an ongoing release from a piping system. In responding to such releases, a working knowledge of the system involved is vital. Response personnel should possess this knowledge firsthand, or coordinate closely with personnel who do, for initiation of control procedures.

Leaking Pipe Connections

Loose pipe connections or fittings in plumbing systems can cause significant leaks, especially if the system involved is under pressure. In many instances, these leaks can be stopped simply by tightening connections such as unions or flanges with pipe wrenches or other suitable tools. Persistent leaks at pipe connections may require shutdown for repair or replacement of defective components.

Holes in Piping

Holes in piping may occur because of accidental physical damage to plumbing systems. Holes may also develop through routine wear and tear or corrosion while a system is in operation.

It may be possible to terminate releases from these holes by installing a pipe repair clamp or "pipe-saver" (Fig. 15.12). These clamps must be selected based on the outside diameter of the leaking pipe. They are installed by placing the device around the pipe so that the gasket material covers the leak, latching the bolt(s) into place to fasten the clamp around the pipe, and tightening the bolts so as to compress the clamp and stop the leak. It is worth noting that if a pipe repair clamp is not available, a similar item can be improvised, as shown in Figure 15.12, using suitable gasket material with a backing plate and small worm gear clamps (such as automobile radiator hose clamps).

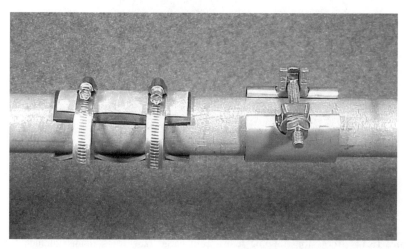

Figure 15.12 Pipe repair clamps (right) are available for patching pipe leaks. An improvised repair may be possible using gasket material, a sheet metal backing plate, and worm gear clamps (left).

Broken Piping

Pipes may be completely broken as a result of accidental damage to plumbing systems. A handy device for temporarily stopping the flow of hazardous materials from broken pipes is the expandable pipe plug (Fig. 15.13).

These plugs must be selected based on the inside diameter of the broken pipe. Pipe plugs are installed by inserting the plug several inches into the pipe through the broken end. The plug is then tightened, shortening the distance along its length and thereby forcing it to swell or expand around its circumference. As a result of this swelling the plug should tighten against the inside of the pipe, thus

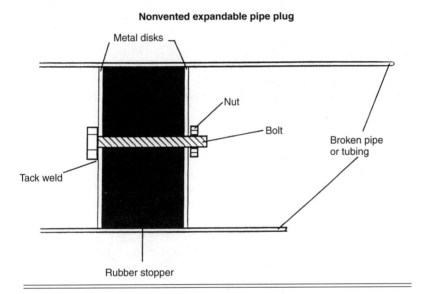

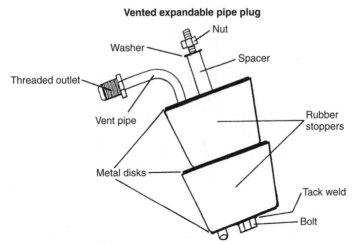

Figure 15.13 Expandable pipe plugs, shown in cross-sectional view.

sealing off the leak. Some plugs of this type can be installed by using a socket wrench with a drive extension, whereas others (such as those used by public utility companies) require a special tool for installation. They should be installed in pairs, with one plug directly behind the other.

Standard pipe plugs are not suitable for applications involving a significant amount of pressure. However, for pressure applications, a vented pipe plug may be used. This plug is fitted with a vent tube, as shown in Figure 15.13. Edwards and Cromwell Spill Control markets a kit containing multiple sizes of this type of plug. Flexible tubing can be fitted to the threaded nipple on the end of the vent pipe so that the escaping liquid product can be diverted into a secondary container, or a gaseous product can be vented at a safe location or neutralized. It is also possible to fit the nipple with a valve. The plug can then be installed with the valve in the open position to vent the pressure during installation. After installation, it may be possible to stop the release by closing the valve. However, in this procedure, as in all operations involving pressure, extra caution is required. It must be assumed that the plug may be blown out at high velocity at any time during the operation. This can result in serious injury to responders.

Leaking Valves

Leakage of hazardous materials from valves on containers or in piping systems may occur because of a variety of causes. Efficient response to releases involving damaged or malfunctioning valves may require specific knowledge of the valves involved and special tools, equipment, and parts. Coordination with knowledgeable personnel (such as plant maintenance personnel) should be initiated during the preemergency planning phase.

Leakage past valve seats (Fig. 15.14) sometimes occurs when valves are in the closed position. In addressing valve leaks, it should initially be determined

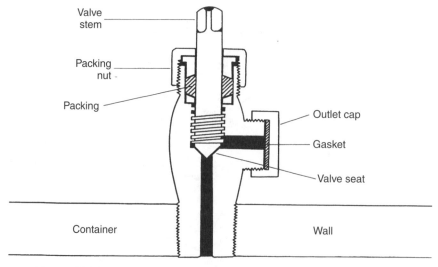

Figure 15.14   Common valve components, shown in cross-sectional view.

that the valve is fully closed, because failure of personnel to close valves fully is a common cause of accidental releases. In some instances, a piece of scale or debris within the system may have lodged in the valve seat so as to prevent full closure. In such a case, slightly opening the valve may dislodge the debris and allow the valve to be fully closed.

Leakage around valve packing glands is also a common cause of release at valve locations (Fig. 15.14). In some cases this may be remedied by simply tightening down on the packing nut to increase the pressure on the packing.

Valve leakage due to problems such as worn or damaged valve seats or stripped valve stem threads will require extensive repair or replacement of the valve involved. This will typically require shutdown of the system involved. However, simple procedures, such as capping outlets on leaky valves, may work as temporary measures.

Procedures for Pressurized Containers

A variety of hazardous materials may be transported, stored, and used in pressurized containers. Pressurized containers come in a variety of sizes and configurations, such as cylinders, rail tank cars, cargo tanks, and stationary storage tanks (see Chapter 5). Design and level of pressurization vary according to the material the cylinder is designed to contain, which may be a compressed gas, a liquefied compressed gas, or a cryogenic liquid.

Pressurized containers normally contain materials with high vapor pressures. Such materials exist as gases at atmospheric pressure. However, within the pressurized containers, these substances may exist as liquids because of the pressure or temperature under which they are stored. Some cylinders contain gases under pressures exceeding 2000 psi. A breach in one of these cylinders can result in a sudden release of this pressure, with disastrous results. Damaged pressure cylinders have been known to act as projectiles, flying through the air for great distances and smashing through solid objects. In addition to the hazardous properties (reactivity, toxicity, flammability, corrosivity, etc.) of the materials stored in pressurized cylinders, the level of pressurization involved can be extremely hazardous to response personnel. Keep in mind also that liquefied compressed gases can produce very cold temperatures if released. This may result in frostbite to responders or damage to personal protective equipment resulting from the extreme cold. Response to incidents involving pressurized containers requires great caution.

Releases from pressurized containers can occur from leaks in container walls or leaks in attachments such as fittings, valves, and pressure relief devices. If liquid material is escaping from a pressurized container, the container should be repositioned, if possible, to place the point of release in the vapor phase so that only gas escapes from the breach. This is required because of the potentially huge liquid-to-gas expansion ratios involved. Other measures, such as cooling a container, may work to reduce pressure, thus reducing the rate of release. Also, it may be possible to move a leaking container into a safer area for repair or to allow the contents to vent.

For the most part, response operations involving pressurized containers will require specialized knowledge and may require special equipment (Figs. 15.15 and 15.16). In some instances minor problems, such as leaks due to loose connections or valves that are not fully closed, may be remedied by using general skills and commonly available hand tools. Other procedures may require special expertise and special equipment, such as the special containment kits and cylinder overpacks, or "caskets," described in Table 15.5.

Specialized containment kits are available for dealing with liquefied compressed gas releases. The Chlorine Institute provides four kits for containing leaks in chlorine containers, including:

Figure 15.15 Chlorine Institute Emergency Kit "A" is designed for containing leaks from 100- and 150-pound chlorine cylinders.

Figure 15.16   Chlorine Institute Emergency Kit "C" is designed for containing leaks from chlorine rail tank cars.

- Emergency Kit "A" for 100- and 150-pound cylinders (Fig. 15.15)
- Emergency Kit "B" for ton containers
- Emergency Kit "C" for tank cars and tank trucks (Fig. 15.16)
- Emergency Kit "D" for barges

Indian Springs Specialty Products, Inc. of Baldwinsville, New York markets an Anhydrous Ammonia Cylinder Emergency Kit that is very similar to the Chlorine Institute "A" kit, only designed for anhydrous ammonia. They market Chlorine Institute "A", "B", and "C" kits for chlorine repair and also provide gasket packages to allow them to be used for containing sulfur dioxide releases. Midland Manufacturing of Skokie, Illinois markets Emergency Response Kits similar to the Chlorine Institute "C" kit designed for capping leaks from rail car domes.

Materials contained in pressurized containers may require special handling and storage procedures. Acetylene, for example, is unstable in the pure form and may become shock sensitive if improperly stored. For response to incidents involving pressurized cylinders, hazmat team members should have a thorough knowledge of the specific materials, containers, and procedures involved. Unless hazmat team members have applicable knowledge gained through their normal work activities, specific training must be provided.

GROUNDING AND BONDING FLAMMABLE LIQUID CONTAINERS

Whenever a liquid product is in motion, such as when they are pumped, flow through pipes or hoses, or simply fall freely through air, it is possible for the liquid to assume a static electrical charge. Even the oscillation of product in a container as it is repositioned to place a leak in the vapor phase can produce a static charge. If an electrostatically charged product contacts an object that is grounded or has a lower electrical potential, static discharge, accompanied by a spark, can occur. If the liquid product involved is flammable, the spark resulting from the static discharge may serve as an ignition source for a flammable vapor-air mixture produced by vaporization of the product, resulting in a flash fire. This is a great concern for the safety of responders involved in operations such as container handling and product transfer when flammables are involved.

To safely handle flammable liquids, static electrical charges must be suppressed, or conducted safely to ground, as soon as they are formed. Static suppression is achieved through the utilization of grounding and bonding equipment and procedures.

Grounding and Bonding Equipment

Various equipment items will be needed for grounding and bonding procedures (Fig. 15.17). These items are available from suppliers such as Stewart R. Browne Manufacturing, Inc. of Atlanta, Georgia and various safety equipment suppliers.

Grounding Clamps and Cables

Several types of clamps may be used for grounding and bonding operations, depending on the types of containers involved. Basic clamp configurations include

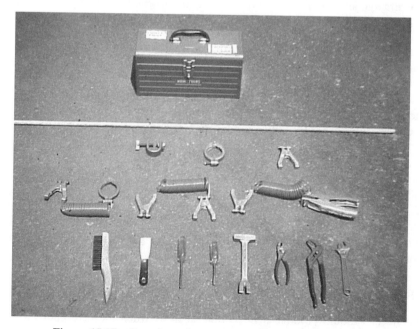

Figure 15.17   Overview of grounding and bonding equipment.

plier clamps (which are spring loaded) and c-clamps and pipe clamps (both of which are screw tightened). Plier clamps are equipped with sharp replaceable points and powerful springs, so that direct metal-to-metal contact can be made through paint, rust, grease, and other foreign material on containers. Grounding clamps must be maintained in good condition, or else replaced. Under no circumstances should cheaper, unsuitable items such as automotive jumper cables be utilized in grounding operations.

Cables of suitable length, quality, and durability should connect grounding clamps. Because good conduction by the cables is vital, they should be inspected regularly. Some emergency responders use only uninsulated cables because they can be readily inspected. Others use only cables with clear or "see-through" insulation. Preuse inspection should incorporate a careful examination of the clamp-to-cable connection. They should be checked periodically for electrical resistance from clamp to clamp using an ohmmeter or a multimeter. Resistance should be less than 1 ohm.

Grounding Electrodes

Grounding electrodes are the final components of grounding/bonding systems that transfer static charges to earth. In a field operation, the grounding electrode will typically be a ground rod. This is a copper-coated metal rod specifically designed to be driven into the ground to conduct electrical charges into the soil. Stationary objects, such as an underground metal water pipe, may also be used in the field setting if good conductance to ground can be ensured. During a response at an industrial facility, responders may be able to utilize permanently installed grounding systems used in routine dispensing of fuels and solvents.

It is critical that the grounding electrode provide a good ground. A metal object in contact with the soil will not necessarily provide a good ground. This is especially true in dry regions, during prolonged droughts, or in soil types that are poor conductors. One device that can take the guesswork out of establishing a good grounding electrode is the megohmmeter or "megger," a testing device that can indicate whether or not charges will actually be conducted to ground through a given grounding electrode.

Other Items Needed for Grounding and Bonding Operations

A variety of tools or other items may be needed for grounding/bonding operations. For example, scrapers or wire brushes may be needed for removal of paint, rust, or other material from container bodies to ensure good metal-to-metal contact with grounding clamps. It is critical that these items be of nonsparking design.

Procedures for Grounding and Bonding

Static suppression for hazardous materials operations involves two distinct procedures, bonding and grounding. These are discussed below.

Bonding Procedures

Bonding refers to the process of equalizing the electrical potential between two containers. This is done by connecting a cable, with suitable clamps, between

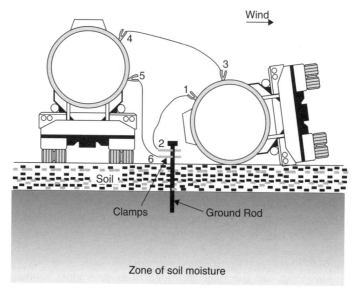

Figure 15.18   Grounding and bonding containers before product transfer is intended to safely conduct static electrical charges to ground (numbers show order of attachment of clamps).

the two containers (Fig. 15.18). It is important to always make the connection at the hazardous location first. In this way, if static discharge occurs as the circuit is completed, the resulting spark will be at the "safe" container. When attaching clamps it is important to ensure direct metal-to-metal contact between the container and clamps. For plier-type clamps, use a back and forth motion to work the points through paint, rust, and dirt on containers. For some installations, such as when pipe clamps are used, surface preparation may be required to expose bare metal before clamp installation.

Grounding Procedures

Grounding is the process of connecting a container or other object to ground so as to suppress any static charges that may develop. This is accomplished by connecting a grounding cable, fitted with suitable clamps, between the container and a suitable grounding electrode or grounding system (Fig. 15.18). Concerns for ensuring metal-to-metal contact at clamp locations, as described for bonding procedures, apply in grounding as well.

When attaching grounding cables, always make the connection to the grounding electrode last, in case a spark occurs as the circuit is completed. For transferring operations, both containers should be grounded. Thus, if one of the lines involved in grounding and bonding the containers is accidentally disconnected during the operation, charges should be safely conducted to ground as long as the other two lines remain in place (Fig. 15.18). Pumps used in product transfer should also be grounded.

When utilizing a ground rod as the grounding electrode, it is important that the rod be driven deeply enough to ensure contact with moist soil for adequate grounding. For field operations, a standard 8-foot ground rod should be driven to the hilt, or as deeply as possible into the ground. The ground rod should then be checked with a megohmmeter, as described above, especially in dry regions or during extended periods of dry weather.

Other Safety Precautions for Transferring Flammables

It is important to use good site control procedures for transfer operations. Place the sound container that will receive the transferred product upwind from the damaged container. Try to locate ground rods, pumps, and areas where people will be working upwind from the damaged container as well.

During transfer operations involving flammable liquids, personnel can minimize the amount of electrostatic charge generated by transferring the product slowly and minimizing splashing, spraying, and swirling of the product during the operations. All pipes or hoses, pumps, and containers used should be clean. Other safety precautions that have been used include the wearing of special clothes designed to suppress electrostatic buildup and the use of inert gas blanketing during transfer operations.

OVERPACKING DAMAGED CONTAINERS

In handling hazardous materials in small containers that have been damaged, it may be desirable to overpack the damaged containers. Basically, this involves placing the damaged container into a slightly oversized open-topped salvage container. For example, 55-gallon drums are commonly overpacked with 85-gallon salvage drums. The space between the inner and outer-drums is packed with a suitable sorbent material, and a tight-fitting lid is clamped to the salvage drum. The salvage container may then be safely handled and transported. For purposes of transportation, note that DOT regulations related to certification of salvage containers for the product shipped, labeling of the containers, and shipping papers must be strictly adhered to. If the product is considered a hazardous waste, EPA requires that a hazardous waste manifest accompany it during transportation, as in shipment to a disposal site.

Safety Considerations for Overpacking

In overpacking operations, as in all aspects of hazmat response, safety should be the foremost consideration. Chemical threats to personnel involved in these operations are posed by the materials involved, which may be toxic, flammable, corrosive, or otherwise dangerous. If feasible, damaged containers should be temporarily repaired before overpacking to minimize spillage and resulting chemical hazards during the procedure. Safety considerations for the use of PPE and safe handling of hazardous materials, as previously covered in this text, must also be addressed.

Personnel are also at risk from physical safety hazards related to the mechanics of the overpacking procedure. A 55-gallon drum of liquid product typically weighs several hundred pounds, so that crushing-type injuries, especially to hands and feet, are definitely possible. The use of safety gear such as hard hats and safety shoes is important for physical protection. Responders must also bear in mind that improper lifting can produce back strain, or similar injuries, when drums are handled manually.

Procedures for Placing Damaged Containers into Overpack Containers

For the purpose of describing the following procedures, it will be assumed that the damaged containers involved are 55-gallon drums. In handling containers of this size, some type of lifting equipment should be used, if at all possible, so that personnel will not have to manually handle the drums.

Procedures Using Heavy Equipment
If feasible, equipment such as an overhead crane, fork truck, backhoe, or front-end loader should be employed for lifting drums and placing them into the salvage container. A device suitable for placement around the container to be lifted, such as a nylon sling, should be attached to the lifting equipment. The damaged container should be lifted and lowered gently into the salvage container. The operation can then be completed by adding the sorbent packing and placing the lid on the salvage container. It is vital that basic safety procedures be followed during these operations. Personnel should never be located beneath a suspended container. Also, general safety rules related to work around heavy equipment should be followed closely.

Procedures for Manual Overpacking
When overpacking must be performed manually, the weight involved can result in back strain or other types of strain injuries. Also, the threat of crushed hands, feet, or digits is very real. Extreme care and the use of appropriate protective gear are called for. Never place parts of your body where they could be injured. For example, placing hands in possible pinch points, such as between the salvage drum and damaged drum, should be avoided during overpacking operations.

In performing manual overpack operations, personnel should lift in teams of two, acting in unison, to reduce individual strain. The use of commercially available drum upender bars (Fig. 15.19), or similar improvised items (such as boards), to provide leverage is highly recommended. "Manhandling" of drums should be avoided in all cases.

The Inverted Overpack Technique. This technique involves placement of the overpack container, in the inverted position, over the damaged drum (Fig. 15.20). The overpack container is then rotated manually into the upright position (Fig. 15.19), packed with sorbent material, and closed.

Figure 15.19 A drum upended bar provides leverage for rotating a drum into the upright position.

Figure 15.20 Overpacking a 55-gallon drum using the inverted overpack technique.

The Roller Technique. This technique involves the use of 1.5- to 2-in.-diameter tubing, such as PVC plastic tubing, to serve as rollers. The damaged drum is placed on its side, with the rollers underneath it (Fig. 15.21). The salvage drum is then positioned so that the damaged drum can be rolled most of the way into it. The salvage drum can then be turned upright with a suitable lever, and the remaining steps of the process can be performed. If rollers are not available, a 2 × 4-in. board or similar item can be utilized to elevate one end of the damaged drum to start it into the salvage drum.

Angle Roll Technique. In using this technique, the damaged drum and the salvage drum are placed end to end at an obtuse angle, as shown in Figure 15.22. The drums can then be simultaneously rolled, so that the damaged drum tends to work its way into the salvage drum. At some point, the inner drum will "foul out" inside the salvage drum and inward progress will stop. At this point, the drums can be repositioned to reverse the original angle and rolled in the opposite direction. Through continuing this process, the damaged drum can be worked most of the way into the salvage drum. The salvage drum can then be lifted into the upright position, with a suitable lever, so that the remaining steps of the process can be performed.

Figure 15.21 Overpacking with the roller technique.

Figure 15.22 Overpacking with the angle roll technique.

SUMMARY

Advanced hazardous materials control operations are performed to contain an uncontrolled release at the source. These offensive operations can be very effective in minimizing the damage resulting from an incident but typically involve a high exposure potential for personnel who must work at the point of release.

Size-up considerations, procedures, equipment, and tools that may be required for containment operations were described in this chapter. However, this information is somewhat generalized and should be supplemented as appropriate for the specific containment operations that may be required. In all instances, field training will be required before the performance of containment operations in emergency situations. These concerns must be addressed during preemergency planning if response personnel are to be adequately trained and equipped.

In planning and performing advanced control operations, the safety of response personnel must be given the highest priority. No operations that place responders at undue risk should be attempted. Responders should bear in mind that, in some situations, containment will not be a safe option and responders will have to rely on defensive tactics alone in mitigating the incident.

16

MENTAL STRESS IN EMERGENCY RESPONSE

INTRODUCTION

Emergency response work has many of the characteristics associated with high-stress occupations. Although firefighters in a large urban setting will have different problems than a hazardous materials technician worker in a remote rural hazardous material facility, both may suffer from job and job-related pressure. Hazardous materials technicians, fire service members, and many other categories of first responders are often placed in situations where there is a potential for a traumatic incident. Every person who is involved in a traumatic incident will react differently psychologically and has the chance of being negatively affected by traumatic stress.

Traumatic stress, also commonly referred to as critical incident stress, is the response produced when a person or group of people is exposed to a traumatic event. Traumatic stress can cause a person to develop physical and psychological changes that affect his or her working, family, and social environment. This chapter describes the basics of traumatic stress and methods of coping with traumatic stress.

STRESS BASICS

Stress is a response from the mind and the body to stressor stimuli. The response can be to a positive and motivating stress called eustress or a negative, harmful stress

Emergency Responder Training Manual for the Hazardous Materials Technician, Second Edition, edited by Kenneth W. Oldfield
ISBN 0-471-21387-X   Copyright © 2005 John Wiley & Sons, Inc.

called distress. A certain amount of eustress keeps people motivated to work well and follow through with tasks. Too much eustress can cause a buildup of physical and emotional stressors, which then can cause distress and adverse health effects.

Sources of stress can be biogenic or psychosocial. Biogenic stimuli cause a biochemical reaction in the human body and can include caffeine, nicotine, and certain medications. Psychosocial stimuli do not directly cause stress but may create conditions that are right for events to become stimuli. Events that can become psychosocial stressors are viewed as challenging or threatening.

Cognitive and emotional interpretations of either the biogenic or psychosocial stressor determine whether an event will result in a stress response for an individual. Cognitive interpretations will lead to a stressful situation if the event is interpreted to be meaningful, potentially challenging, and/or threatening. Emotional interpretations will lead to a stressful situation if we choose to or have learned to interpret the event as challenging, threatening, or adverse.

Cognitive and emotional interpretations will determine whether the body is stressed and will then cause cognitive, physical, emotional, or behavioral responses. Certain target organs act in response to stress and give off signs and symptoms of being under stress. Table 16.1 lists the common responses to a traumatic event.

Traumatic Stress

A traumatic event is an event, or series of events, that causes moderate to severe stress reactions. Traumatic stress is a normal reaction of normal people to an

TABLE 16.1   Common Responses to a Traumatic Event

Cognitive	Physical	Emotional	Behavioral
– Poor concentration	– Nausea	– Shock	– Suspicion
– Confusion	– Lightheadedness	– Numbness	– Irritability
– Disorientation	– Dizziness	– Feeling overwhelmed	– Arguments with friends and loved ones
– Indecisiveness	– Gastrointestinal problems	– Depression	– Withdrawal
– Shortened attention span	– Rapid heart rate	– Feeling lost	– Excessive silence
– Memory loss	– Tremors	– Fear of harm to self and/or loved ones	– Inappropriate humor
– Unwanted memories	– Headaches	– Feeling nothing	– Increased/decreased eating
– Difficulty making decisions	– Grinding of teeth	– Feeling abandoned	– Change in sexual desire or functioning
	– Fatigue	– Uncertainty of feelings	– Increased smoking
	– Poor sleep	– Volatile emotions	– Increased substance use or abuse
	– Pain		
	– Hyperarousal		
	– Jumpiness		

Adapted from www.cdc.gov

overwhelming experience or abnormal circumstance that significantly, and usually suddenly, impacts their life and well-being. Traumatic stress is a reaction that can include the cognitive, physical, emotional, and behavioral signs and symptoms of stress discussed above. The stress is typically brought on as a reaction to a traumatic event that responders are not used to dealing with on a day-to-day basis. The stress reaction can be immediate or delayed, and many people recover from it within a few weeks of the critical event.

Posttraumatic Stress

If a traumatic event has caused someone to experience traumatic stress and this stress is left untreated, then it can develop into posttraumatic stress. Posttraumatic stress reactions include posttraumatic stress disorder, brief psychotic reactions, dissociative disorders, adjustment disorders, acute stress disorder, and borderline personality disorder. The most common type of posttraumatic stress is posttraumatic stress disorder (PTSD).

PTSD is an intense physical and emotional response to thoughts and reminders of the event that last for many weeks or months after the traumatic event. PTSD is often written about in histories of wars as well as large fires such as the Great Fire of London in 1666. The Vietnam War resulted in many soldiers experiencing PTSD and prompted the psychiatric community to officially recognize posttraumatic stress as a psychiatric disorder. Although there are many references to combat cases of PTSD, it is important to note that PTSD can develop out of many noncombat traumatic events. PTSD can be brought on by various events including:

- A serious accident
- The destruction of one's home or community
- Disasters
- Terrorism
- Natural catastrophes
- Rape
- Criminal victimization
- The threat of harm to oneself or others
- Experiencing/witnessing actual physical harm to oneself or others.

For a diagnosis of PTSD, a person must be experiencing symptoms that fall into three categories: reliving, avoidance, and increased arousal. The symptoms of reliving include flashbacks, nightmares, and extreme emotional and physical reactions to reminders of the event. Emotional reactions include feeling guilty, extreme fear of harm, or numbing of emotions, and physical reactions include uncontrollable shaking, chills or heart palpitations, and tension headaches.

Symptoms of avoidance include staying away from activities, places, thoughts, or feelings related to the trauma or feeling detached or estranged from others. The symptoms of increased arousal include being overly alert or easily startled, diffi-culty sleeping, irritability or outbursts of anger, and lack of concentration. Other symptoms that have been linked with PTSD include panic attacks, depression, suicidal thought and feelings, drug abuse, feelings of being estranged and isolated, and not being able to complete daily tasks. If these symptoms are present for a period of time longer than one month, then the person is experiencing PTSD.

Victims of Traumatic Stress

Victims of traumatic stress can be anyone involved in a traumatic event. The typical victims include emergency service personnel, public safety personnel, nurses, physicians, disaster workers, and any "victims" or bystanders involved in the traumatic event. Victims of traumatic stress can be affected by large traumatic events that affect large groups of people, or they can be affected by small events in which there may be a personal connection to either the victims of the incident or the situation of the victims. Traumatic events can affect survivors, rescue workers, and friends and relatives of victims who have been directly involved.

METHODS OF COPING WITH TRAUMATIC EVENT STRESS

Stress resulting from a traumatic event can cause adverse physical and psycho-logical affects if left untreated or unnoticed. Various methods of coping with traumatic stress have been developed and are utilized during large events such as a terrorist incident or plane crash and also during small events, such as a car wreck, the death of a child, or incidents where a response scene is more grue-some than usual. Two of the methods of coping with traumatic stress discussed below are the Critical Incident Stress Management program and the Centers for Disease Control and Prevention advice for assessing mental health.

Critical Incident Stress Management

Critical incident stress management (CISM) is a comprehensive program used to help people recover from traumatic stress, also called critical incident stress. The main goals of CISM are to alleviate traumatic stress in a timely manner so that affected people can continue on with work and life within a relatively short time after a traumatic event and to prevent the development of posttraumatic stress. CISM is an entire system of interventions provided by a CISM team designed to be used before, during, and after a traumatic event. The program of CISM grew out of the idea of the Critical Incident Stress Debriefing process developed by Dr. Jeffrey T. Mitchell in 1983 (Mitchell, 1996).

History of Critical Incident Stress Management
Many events have occurred in history where people were affected by stress from work situations that included both "typical" response activities and "disaster"

response activities. The following is a description of four large-scale events that demonstrate the development of critical incident stress management.

San Diego, California. One of the first accounts of emergency responders experiencing negative effects from critical incident stress occurred as a result of a large plane crash in San Diego, California. On September 25, 1978, Pacific Southwest Airlines Flight 182 crashed into a heavily populated area of San Diego. The crash resulted in a four-alarm fire and the loss of 144 lives. The incident included very little or no stress management as part of the recovery for the responding personnel. Many firefighters and police personnel experienced emotional problems, and several disability retirement claims were filed because of the effects of stress from the incident.

Washington, DC. On January 13, 1982, Air Florida Flight 90 crashed into the Fourteenth Street bridge over the Potomac River, resulting in 76 deaths. Within a few weeks after the traumatic event, emergency personnel were showing symptoms of stress and were asking for help. The CISD process was still relatively new and had not been officially used on a large-scale incident. The CISD process was formally implemented for the first time on a large-scale traumatic event. The emergency personnel who participated in the CISD process reported that the CISD process was very helpful in treating their stress-related problems.

Kenner, Louisiana. Another incident in which emergency responders experienced negative affects from a critical incident occurred in Kenner, Louisiana. On July 9, 1982, Pan American Flight 759 crashed into a populated area not far from the New Orleans International Airport. The crash instantly killed all people on board and eight Kenner residents on the ground, for a death toll of 162 people. The Kenner Fire Department responded to the scene along with fire servicemen, police officers, ambulance personnel, and many other rescue workers from surrounding areas.

Shortly after the plane crash the local parish substance abuse clinic scheduled seminars to help rescue workers, families of the victims, and community residents deal with the traumatic events that occurred in their neighborhood. In this event, the incident commander on the scene recognized the mental trauma involved and had many of his rescue personnel attend an open meeting to discuss the critical incident stress. The incident commander was aware of the problems and lack of counseling after the San Diego crash and wanted counseling available for the rescue personnel if needed.

Cerritos, California. The final event discussed in the development of critical incident stress management occurred on August 31, 1986, when a small aircraft collided with an Aero Mexico jetliner over Cerritos, California. The incident resulted in the death of 67 people on board both airplanes and 15 people on the ground when the airplane crashed into a residential area. Again, with this incident many different types of emergency service personnel responded to the crash and the Los Angeles firefighters worked in the same type of disaster conditions as the San Diego and Kenner firefighters. The Los Angeles County Fire Department had developed a critical incident stress plan and was able to put it into operation shortly after the incident occurred. The critical incident plan was based on the problems associated with the San Diego and Kenner events.

Critical incident stress management more recently has been applied to the Oklahoma City bombing in 1995, the Texas A&M bonfire collapse in 1999, and the Space Shuttle *Columbia* disaster in 2003. Although the events discussed so far describe the use of critical incident stress among emergency personal for large-scale events, it is important to note that critical incident stress can affect response personnel even if the emergency situation is on a smaller, more personal scale.

Elements of Critical Incident Stress Management
CISM is a comprehensive program that includes the following services:

- Preincident education
- Demobilizations
- Crisis Management Briefing (CMB)
- Defusing
- Critical Incident Stress Debriefings (CISD)
- Individual crisis intervention
- Family critical incident stress management
- Organizational consultation
- Follow-up/referral

As mentioned above in this chapter, a traumatic event is an event, or series of events, that causes moderate to severe stress reactions. The following is a list of types of traumatic events after which a group of emergency personnel supervisors might want to consider implementing a critical incident stress management program:

- Line of duty deaths
- Serious line of duty injuries
- Suicides of emergency personnel
- Disasters
- Law enforcement shootings
- Accidental injuries to others caused by one's actions
- Significant events involving children
- Prolonged incidents that end in loss of life
- Events with excessive media coverage
- Life-threatening experiences
- Severe abuse

- Homicides
- Terrorism
- High-publicity crimes of violence
- Knowing the victim of the event

The CISM Team. The CISM team is composed of professional support personnel, such as mental health professionals, clergy, and peer support personnel (typically emergency service personnel) who are trained in CISM techniques. Victims of critical incident stress respond better to peer support personnel if they are not from their own service unit or if they are not familiar with them on a personal level. The CISM team is trained to recognize when someone is not responding well to the CISM process and needs further assistance beyond what they can provide. The peer support personnel usually come from the same types of emergency services as the group receiving the CISD.

CISM Interventions. A CISM team can provide such CISM interventions as preincident education, demobilizations, critical incident stress debriefing, crisis management briefing, defusing, individual crisis intervention, family critical incident stress management, and organizational consultation. The most important task of a CISM team is to assess which intervention is most appropriate for the situation and to ensure that the intervention is done properly by trained personnel.

CRITICAL INCIDENT STRESS DEBRIEFING (CISD) is the most recognized program in CISM. CISD is a structured process that was developed to help people involved in critical incidents mitigate any stress reactions they may experience and to keep them from developing posttraumatic disorders. The CISD process is the most difficult CISM intervention to apply, and it consists of seven phases:

(1) Introduction
(2) Fact
(3) Thought
(4) Reaction
(5) Symptoms
(6) Teaching
(7) Re-entry

A CISD is provided for emergency personnel by a group of people who make up the CISM Team.

DEMOBILIZATION is an intervention designed to help emergency personnel cope with massive traumatic events. Massive traumatic event stress can often be difficult for many different groups of people to cope with. Instead of taking the time necessary to effectively complete a debriefing, the shorter type of intervention called demobilization is often offered for emergency service personnel.

Demobilization is applied immediately after the emergency services personnel are released from their duties at the scene of the event. The demobilization is composed of two segments, (1) a short (10–15 min) information segment on understanding and managing stress followed by (2) a slightly longer segment (20 min) in which the personnel are given time to rest and eat before they return to their normal routine. Demobilization is always followed by a debriefing at a later date, and it gives the emergency personnel a chance to temporarily cope with their stress before returning to their next work shift.

CRISIS MANAGEMENT BRIEFINGS (CMBS) are used if a traumatic event affects nonemergency personnel such as civilians, schools, and businesses. This type of briefing can be done for many different groups in a population that has experienced a traumatic event. Everyone who attends the briefing should leave with a fact sheet that includes signs and symptoms of stress, stress management techniques, and local resource information in dealing with stress in case they need to seek further assistance. Crisis management briefings involve four phases:

- Phase one—Gather a group of people together who have experienced the same event.
- Phase two—Explain the facts of the crisis event.
- Phase three—Explain the most common reactions that may result from the crisis event.
- Phase four—Discuss simple and practical stress management strategies that may be useful to the affected population.

The final step of any critical incident stress intervention is the follow-up/referral step. This critical part of CISM ensures that participants who were identified as needing further help are given the right information. CISM team members are responsible for making sure the participants in need have access to a higher level of care.

Running a CISM Program. CISM teams with the proper training in the various methods and techniques of operating a crisis intervention are available for most fire, police, and community organizations. In assessing the need for a CISM team involvement, a call is placed to a dispatcher who asks questions and gathers information about the incident. The team coordinator then contacts the original caller and discusses whether and what type of intervention is needed based on the information known about the incident. CISM teams must be properly trained, and they are not a replacement for professional help that may be needed by someone who is experiencing traumatic stress.

If your place of employment wants to start a CISM team or work with a locally established CISM team, the most useful step is to contact the International Critical Incident Stress Foundation (International Critical Incident Stress Foundation, Inc., 3290 Pine Orchard Lane, Suite 106, Ellicott City, MD 21042. Phone Number: 410-750-9600. www.icisf.org).

Centers for Disease Control and Prevention Strategy for Coping with Traumatic Stress

The U.S. Centers for Disease Control and Prevention (CDC) Injury Center has established guidelines that can be followed when coping with traumatic stress resulting from massive traumatic events. The CDC recommendations are to screen, support, and track people who may be negatively affected by an exposure to a massive traumatic event. A survey tool for briefly assessing the mental health of a population after a traumatic event has been developed to help health departments and other decision-makers assess the psychological impact of a massive traumatic event (*Survey Tool for Briefly Assessing the Mental Health of a Population Following a Traumatic Event.* www.cdc.gov/masstrauma/response/mhsurvey. htm). The CDC recommends the following for helping people cope with a traumatic event:

- Refer patients to a mental health professional in your area who has experience treating the needs of survivors of traumatic events.
- Provide education to help people identify symptoms of anxiety, depression, and PTSD.
- Offer clinical follow-up when appropriate, including referrals to mental health professionals.

Professional help may be needed if symptoms affect relationships with family and friends or affect work activities. There are a few things that can be done to help cope with traumatic events:

- Understand that your symptoms may be normal, especially right after the trauma.
- Keep to your usual routine.
- Take the time to resolve day-to-day conflicts so they do not add to your stress.
- Do not shy away from situations, people, and places that remind you of the trauma.
- Find ways to relax and be kind to yourself.
- Turn to family, friends, and clergy person for support, and talk about your experiences and feelings with them.
- Participate in leisure and recreational activities.
- Recognize that you cannot control everything.
- Recognize the need for trained help, and call a local mental health center.

Table 16.2 is a list of agencies the CDC recommends as resources for help and information for coping with stress due to a traumatic event.

TABLE 16.2   Resources for Help and Information for Coping with Traumatic Event Stress

Agency Name	Web Site
American Red Cross	www.redcross.org/services/disaster
Anxiety Disorders Association of America	www.adaa.org
National Center for Post-Traumatic Stress Disorder	www.ncptsd.org
National Institute on Mental Health	www.nimh.nih.gov
Posttraumatic Stress Disorder Alliance	www.ptsdalliance.org
Substance Abuse and Mental Health Services Agency	www.samhsa.gov

SUMMARY

A traumatic event can cause survivors and affected people to experience trauma-related stress. The event can be large or small and can affect a group of people or an individual who may have a stressful response. People who suffer from trauma-related stress must be identified and given the opportunity to receive help and advice from properly trained professionals. Everyone reacts to stress differently, and the amount of help needed will be different for every person and every situation. If stress is properly recognized and treated, emergency responders will be able to continue to work in a potentially stressful profession while being able to control the negative physical and psychological effects that traumatic stress can cause.

REFERENCES

Action Training Systems, *Diking, Diverting and Retaining Spills* (video cassette), Action Training Systems, Seattle, WA, 1989.

American Conference of Governmental Industrial Hygienists (ACGIH), *Air Sampling Instruments for Evaluation of Atmospheric Contaminants*, 9th Edition, ACGIH, Cincinnati, OH, 2001.

American Industrial Hygiene Association (AIHA), *Emergency Response Planning Guidelines Series*, AIHA Press, Fairfax, VA, 2002.

American National Standards Institute (ANSI), *American National Standard for Industrial Head Protection, Z89.1-2003*. American National Standards Institute, New York, 1997.

ANSI, *American National Standard for Personal Protection—Protective Footwear, Z41-1999*. National Safety Council, Itasca, IL, 1999.

ANSI, *American National Standard for Respiratory Protection, Z88.2-1992*, American National Standards Institute, New York, 1992.

ANSI, *Practice for Occupational and Educational Eye and Face Protection, Z87.1-2003*. American National Standards Institute, New York, 2003.

B. A. Plog, Niland, J., and Quinlan, P. J., eds., *Fundamentals of Industrial Hygiene*, 4th Edition, National Safety Council, Itasca, IL, 1996.

Benner, L., *Hazardous Materials Emergencies*, Internet Edition, Starline Software Ltd., Oakton, VA, 2000.

Bureau of Explosives, *Emergency Action Guides*, Association of American Railroads, Washington, DC, 1990.

Callan, M., *Street Smart Approach to Haz Mat Response: A Common-Sense Approach to Handling Hazardous Materials Emergencies*, Red Hat Publishing, Annapolis, MD, 2002.

Emergency Responder Training Manual for the Hazardous Materials Technician, Second Edition, edited by Kenneth W. Oldfield
ISBN 0-471-21387-X   Copyright © 2005 John Wiley & Sons, Inc.

Cox, D. B. and Eckmyre, A. A., *Hazardous Materials Management Desk Reference*, McGraw-Hill, New York, 2000.

Environmental Protection Agency, *Hazards Analysis on the Move*, EPA Document Number 550-F-93-004, Office of Solid Waste and Emergency Response, Government Printing Office, Washington, DC, 1993.

Federal Emergency Management Agency (FEMA), *Hazardous Materials Exercise Evaluation Methodology (HM-EEM) and Manual*, FEMA, Washington, DC, 1989.

Forsberg, K. and S. Z. Mansdorf, *Quick Selection Guide to Chemical Protective Clothing*, 4th Edition, John Wiley & Sons, New York, 2002.

GATX, *GATX Tank and Freight Car Manual*, 6th ed., General American Transportation Corporation, Chicago, 1994.

Hawley, C., *Hazmat Air Monitoring and Detection Devices*, Delmar Learning, Clifton Park, NY, 2001.

Henry, M. F., ed., *Hazardous Materials Response Handbook*, National Fire Protection Association, Quincy, MA, 1989.

Hildebrand, M. S. and Noll G. N., *Storage Tank Emergencies: Guidelines and Procedures*, Red Hat Publishing, Annapolis, MD, 1997.

International Fire Service Training Association, *Hazardous Materials for First Responders*, Fire Protection Publications, Stillwater, OK, 1988.

Kelly, R. B., *Industrial Emergency Preparedness*, Van Nostrand Reinhold, New York, 1989.

Maslansky, C. J. and Maslansky, S. P., *Air Monitoring Instrumentation: A Manual for Emergency, Investigatory, and Remedial Responders*, John Wiley & Sons, New York, 1993.

Meyer, E., *Chemistry of Hazardous Materials*, Prentice-Hall, Englewood Cliffs, NJ, 1977.

Mitchell, J. T. and Everly, G. S., *Critical Incident Stress Debriefing: CISD An Operations Manual for the Prevention of Traumatic Stress Among Emergency Service and Disaster Workers*, 2nd Edition, Revised, Chevron Publishing. Ellicott City, MD, 1996.

National Fire Protection Association (NFPA), *NFPA 471 Recommended Practice for Responding to Hazardous Materials Incidents, 2002 Edition*, National Fire Protection Association, Quincy, MA, 2002.

NFPA, *NFPA 472 Standard for Professional Competence of Responders to Hazardous Materials Incidents, 2002 Edition*, National Fire Protection Association, Quincy, MA, 2002.

NFPA, *NFPA 473 Standard for Competencies for EMS Personnel Responding to Hazardous Materials Incidents, 2002 Edition*, National Fire Protection Association, Quincy, MA, 2002.

NFPA, *NFPA 1201 Standard for Developing Fire Protection Services for the Public, 2000 Edition*, National Fire Protection Association, Quincy, MA, 2000.

NFPA, *NFPA 1404, Standard for Fire Service Respiratory Protection Training 2002 Edition*, National Fire Protection Association, Quincy, MA, 2002.

NFPA, *NFPA 1500 Standard on Fire Department Occupational Safety and Health Program, 2002 Edition*, National Fire Protection Association, Quincy, MA, 2002.

NFPA, *NFPA 1521 Standard for Fire Department Safety Officer, 2002 Edition*, National Fire Protection Association, Quincy, MA, 2002.

NFPA, *NFPA 1561 Standard on Emergency Services Incident Management System, 2002 Edition*, National Fire Protection Association, Quincy, MA, 2002.

NFPA, *NFPA 1852 Standard on Selection, Care, and Maintenance of Open-Circuit Self-Contained Breathing Apparatus (SCBA), 2002 Edition*, National Fire Protection Association, Quincy, MA, 2002.

NFPA, *NFPA 1991 Standard on Vapor-Protective Ensembles for Hazardous Materials Emergencies, 2000 Edition*, National Fire Protection Association, Quincy, MA, 2000.

NFPA, *NFPA 1992 Standard on Liquid Splash-Protective Ensembles and Clothing for Hazardous Materials Emergencies, 2000 Edition*, National Fire Protection Association, Quincy, MA, 2000.

NFPA, *NFPA 1994 Protective Ensembles for Chemical/Biological Terrorism Incidents*, National Fire Protection Association, Quincy, MA, 2001.

National Fire Service Incident Management System Consortium, Model Procedures Committee, *Model Procedures Guide for Hazardous Materials Incidents*, Fire Protection Publications, Stillwater, OK, 2000.

National Institute for Occupational Safety and Health (NIOSH), *Elements of Ergonomics Programs, A Primer Based on Workplace Evaluations of Musculoskeletal Disorders* (Publication no. 97-117), NIOSH, Cincinnati, OH, 1997.

NIOSH, *NIOSH Guide to Industrial Respiratory Protection*, DHHS (NIOSH) Publication No. 87-116, Government Printing Office, Washington, DC, 1987.

National Response Team, *NRT-1 Hazardous Materials Emergency Planning Guide, 2001 Update*, Government Printing Office, Washington, DC, 2001.

National Response Team, *NRT-2 Developing A Hazardous Materials Exercise Program—A Handbook For State and Local Officials*, Government Printing Office, Washington, DC, 1990.

National Response Team, *The National Response Team's Integrated Contingency Plan Guidance, Federal Register* Volume 61, Number 109, 61 FR 28642, Government Printing Office, Washington, DC, 1996.

Ness, S. A., *Air Monitoring for Toxic Exposures*, Van Nostrand Reinhold, New York, 1991.

Noll, G. N., Hildebrand, M. S., and Yvorra, J. G., *Hazardous Materials: Managing the Incident*, Fire Protection Publications, Stillwater, OK, 1995.

Noll, G. N., Hildebrand, M. S., and Donahue, M. L., *Gasoline Tank Truck Emergencies: Guidelines and Procedures*, 2nd Edition, Fire Protection Publications, Stillwater, OK, 1996.

Noll, G. N., Hildebrand, M. S., and Donahue, M. L., *Hazardous Materials Emergencies Involving Intermodal Containers: Guidelines and Procedures*, Fire Protection Publications, Stillwater, OK, 1995.

Pickett, M., *Explosives Identification Guide*, Delmar Publishers, Albany, NY, 1999.

Sidell, F., Patrick, W. C., and Dashiell, T. R., *Janes' Chem-Bio Handbook*, Jane's Information Group, Alexandria, VA, 2000.

S. R. DiNardi, ed., *The Occupational Environment: Its Evaluation, Control, & Management*, 1st Edition, American Industrial Hygiene Association Press, Fairfax, VA, 1997.

U.S. Army Medical Research Institute of Chemical Defense, *Medical Management of Chemical Casualties*, U.S. Army, Aberdeen Proving Ground, MD, 1996.

U.S. Army Medical Research Institute of Infectious Diseases, *Medical Management of Biological Casualties*, U.S. Army, Fort Detrick, MD, 1996.

U.S. Army Technical Manual TM 31-210, *Improvised Munitions Handbook*, U.S. Army, Frankford Arsenal, Philadelphia, PA, 1969.

U.S. Department of Justice and Federal Emergency Management Agency, *Emergency Response to Terrorism: Basic Concepts*, U.S. Department of Justice, Office of Justice Programs and U.S. Fire Administration, National Fire Academy, Washington, DC, 1997.

U.S. Department of Justice and Federal Emergency Management Agency, *Emergency Response to Terrorism Self-Study Guide*, U.S. Department of Justice, Office of Justice Programs and U.S. Fire Administration, National Fire Academy, Washington, DC, 1998.

U.S. Office of Domestic Preparedness, *Weapons of Mass Destruction (WMD) Radiological/Nuclear Awareness Course*, Instructor Manual, Version 5, U.S. Department of Justice, Washington, DC.

Veasey, D. A., McCormick, L. C., Hilyer, B. M., Oldfield, K. W., and Hansen, S., *Confined Space Entry and Emergency Response*, McGraw-Hill, New York, 2002.

INDEX

Emergency Responder Training Manual for the Hazardous Materials Technician, Second Edition,
edited by Kenneth W. Oldfield
ISBN 0-471-21387-X   Copyright © 2005 John Wiley & Sons, Inc.

NFPA 704 Label

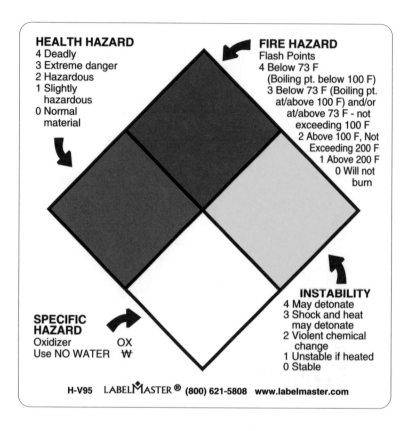

HEALTH HAZARD
4 Deadly
3 Extreme danger
2 Hazardous
1 Slightly
  hazardous
0 Normal
  material

FIRE HAZARD
Flash Points
4 Below 73 F
  (Boiling pt. below 100 F)
3 Below 73 F (Boiling pt.
  at/above 100 F) and/or
  at/above 73 F - not
  exceeding 100 F
2 Above 100 F, Not
  Exceeding 200 F
1 Above 200 F
0 Will not
  burn

INSTABILITY
4 May detonate
3 Shock and heat
  may detonate
2 Violent chemical
  change
1 Unstable if heated
0 Stable

SPECIFIC HAZARD
Oxidizer   OX
Use NO WATER   W̶

H-V95   LABEL✦MASTER ® (800) 621-5808 www.labelmaster.com

Copyright ©2001, National Fire Protection Association, Quincy, MA 02269. This warning system is intended to be interpreted and applied only by properly trained individuals to identify fire, health and stability hazards of chemicals. The user is referred to the recommended classifications of certain chemicals in the NFPA Guide to Hazardous Materials which should be used as a guideline only. Whether the chemicals are classified by NFPA or not, anyone using the 704 system to classify chemicals does so at their own risk. This label is used with NFPA's permission.

Labels supplied courtesy of Labelmaster, Div. of American Labelmark Co., Inc.